Conceptual Electromagnetics

Conceptual Electromagnetics

Branislav M. Notaroš

CRC Press
Taylor & Francis Group
Boca Raton London New York

CRC Press is an imprint of the
Taylor & Francis Group, an **informa** business

CRC Press
Taylor & Francis Group
6000 Broken Sound Parkway NW, Suite 300
Boca Raton, FL 33487-2742

Printed on acid-free paper
Version Date: 20170523

International Standard Book Number-13: 978-1-4987-7066-8 (Hardback)

Visit the Taylor & Francis Web site at
http://www.taylorandfrancis.com

and the CRC Press Web site at
http://www.crcpress.com

To Olivera, Jelena, and Milica

CONTENTS

PREFACE

Electromagnetic theory is a fundamental underpinning of technical education, but, at the same time, one of the most difficult subjects for students to master. In order to help address this difficulty and contribute to overcoming it, here is a textbook on electromagnetic fields and waves for undergraduates completely based on conceptual understanding of electromagnetics, and so entitled, simply, *Conceptual Electromagnetics*. This text provides engineering and physics students and other users with an operational knowledge and firm grasp of electromagnetic fundamentals aimed toward practical engineering applications by combining fundamental theory and a unique and comprehensive collection of as many as *888 conceptual questions and problems in electromagnetics*. Essentially, the book presents and explains the entire undergraduate electromagnetics, and all of its topics and aspects, solely using conceptual questions interwoven with the *theoretical narrative and basic equations*. The goal is that students develop a stronger intuition and a deeper understanding of electromagnetics and find it more attractive and likable.

This book provides abundant opportunities for instructors for innovative lecturing and in-class and homework assignments and testing, including online instruction and distance education, and for students for independent learning. Conceptual questions are also ideal for interactive in-class questions, explorations, and discussions (usually referred to as *active teaching and learning*), for student-to-student interaction and students teaching one another (so-called *peer instruction*), and for team work and exchange of ideas (*collaborative teaching/learning*). Generally, all these pedagogical techniques and approaches have recently gained a lot of attention by educators in science and engineering, and are paving their way as a preferred mode, or a major component, of class delivery and instruction. Multiple studies and classroom experiences across science and engineering have indicated that these novel *learner-centered pedagogies and practices*, and active teaching/learning in particular, are very effective, motivational, and positively evaluated by students. In addition, conceptual questions are perfectly suited for class *assessment*, namely, to assess students' performance and evaluate the effectiveness of instruction, as well as the success of programs and curricula, which is especially important in light of *ABET* and similar accreditation criteria (the key word in these criteria is "assessment").

Conceptual Electromagnetics has a twofold intent. It is a self-contained textbook that can be used either as an *independent resource* or as a *supplement* to any available undergraduate electromagnetics text (e.g., [1]–[18] in the Bibliography). In other words, it is designed either to serve as a principal textbook for a concepts-based electromagnetic fields and waves course (or sequence of courses) or to complement another (currently used or to be adopted) textbook and a variety of teaching styles, as a comprehensive companion adding a very significant conceptual component to the course. In either way, conceptual questions are designed to strongly enforce and enhance both the theoretical concepts and understanding and problem-solving techniques and skills in electromagnetics.

In addition, respective parts of almost all chapters of the book can be effectively incorporated *also in higher-level courses* on antennas, microwave theory and engineering, wave propagation and guidance, advanced electromagnetic theory, computational electromagnetics, electrical machines, signal integrity, etc. (for instance, see [19]–[34]). Furthermore, the book may be used by students outside of any particular course arrangements and by practicing engineers and scientists as well – to review and solidify the knowledge of fundamentals of electromagnetic fields and waves or certain aspects of electromagnetic theory and applications, now with an emphasis on conceptual understanding. Finally, because of its conceptual ("quickly and straight to the point") philosophy and effective multiple-choice format, the book may be useful for distance learning, online courses, and other forms of *nontraditional course delivery*.

Conceptual questions are multiple-choice questions that focus on core concepts of the material, requiring conceptual reasoning and understanding, and no (or very little) calculations. Pedagogically, they are an invaluable resource. They can be given *for homework and on exams*, as well as in *class presentations and discussions* – to be combined with traditional lecturing. In fact, the entire course can be taught exclusively using the theoretical parts and conceptual questions provided in this book. Namely, the course topics can be comprehensively lectured through conceptual questions presented in class with a theoretical introduction (also from the book), which can be combined with a reading assignment beforehand. Homework assignments and tests can be composed either purely of conceptual questions or as a combination of conceptual questions (problems) and traditional computational problems.

For instance, one of the many possibilities of active learning and peer instruction using this material would imply posing a conceptual question to the class, taking a "vote" on it, and then having a discussion of different answers and approaches, ideally with a resolution of disagreements between students within groups of peers (in the spirit of Eric Mazur's *Peer Instruction* in introductory physics). The students and the instructor discuss why some (incorrect) answers appeared attractive and seemed right, and ultimately what is (or should be) the reasoning behind the choosing of the one correct answer. Overall, with this material, electromagnetics classes can be very educational and productive on one side and real fun (for both students and instructors) on the other.

Additionally, this material may align very well with a novel teaching approach called *inverted or flipped classroom*, where students preliminary learn the class content outside the classroom, by reading the theory (from this book or another text or lecture notes) or watching video lectures, which then frees more face-to-face time in the classroom for active and problem-based learning – using conceptual questions and problems.

Of course, the whole process can be tied to the available classroom (and other) technology, which can make it both very efficient and appealing to students. It can as well be made a part of an existing virtual (electronic) *learning management system*. Moreover, with so many (888) conceptual questions available in the book, there are more than enough of them for in-class presentations and discussions, for homework assignments, for tests and assessment, and for additional independent study and practice, respectively.

Conceptual Electromagnetics is designed primarily (but by no means exclusively) for *junior-level undergraduate students* in electrical and computer engineering, physics, and similar departments, for either two-semester course sequences or one-semester courses, and/or equivalent quarter arrangements. However, it can also be used earlier and later in the curriculum. It covers all important theoretical concepts, methodological procedures, and solution approaches in electromagnetic fields and waves for undergraduates – organized in 12 chapters on electrostatic fields; steady electric currents; magnetostatic fields; time-varying electromagnetic fields; uniform plane electromagnetic waves; transmission lines; waveguides and cavity resonators; and antennas and wireless communication systems. It largely follows the organization of *Electromagnetics* by Branislav M. Notaroš, published in 2010 by *Pearson Education*.

On the other hand, the book allows a lot of *flexibility and many different options in coverage* of the material, including the transmission-lines-early and transmission-lines-first approaches. Namely, Chapter 10 (Circuit Analysis of Transmission Lines) discusses only pure circuit-theory concepts, so that it can be taken at any time, along with Sections 6.9 and 6.10, which introduce phasors and complex representatives of time-harmonic voltages and currents.

Moreover, *Conceptual Electromagnetics* may be very effectively combined with MATLAB® computer exercises, tutorials, and projects provided in *MATLAB®-Based Electromagnetics* by Branislav M. Notaroš (*Pearson Education*, 2013).

Each section within each chapter of the book starts with theoretical material for the topic, which is then followed by conceptual questions, strongly coupled to the theory. Many components and derivatives of the theory, along with abundant applications, are introduced through conceptual questions themselves. Most importantly, the book is *absolutely self-sufficient*: students and other readers will be able to answer all questions based only on the theoretical material and equations in the book.

Conceptual questions in the book are also aimed at helping students actively integrate conceptual knowledge into the *problem-solving process*. Many of the questions require the student to perform conceptual, strategic, and qualitative analyses of problem situations, which are then of great and immediate help with quantitative analyses and calculations in standard computational problems. Without a conceptual knowledge structure to which they can be tied, equations are meaningless and quickly forgotten during and after the course. The conceptual questions in conjunction with the computational problems will help students to develop problem-solving strategies based on conceptual analysis.

Furthermore, in *assessments of student learning* using conceptual questions, the performance of students and the effectiveness of instruction are usually evaluated as the "gain" between the course "pretest" and "posttest" scores. Selected conceptual questions from the large collection provided in the book can readily be used by instructors as partial and final assessment instruments for individual topics at different points in the course and for the entire class. For the purposes of ABET (or similar) *accreditation compliances*, conceptual questions can, obviously, be easily implemented to precisely and directly assess students' understanding and mastery of individual principal *course concepts* (electromagnetic field and wave concepts). They can then be directly converted (mapped) to quantitative assessment of individual course objectives (that every student passing the course should meet, at a prescribed level). Course objectives are mapped to *Student Outcomes* (e.g., *outcomes a through k* in ABET terminology) for the entire program (e.g., electrical engineering program), and finally to the *Program Educational Objectives*. It is also possible to directly map conceptual questions to some of the student outcomes. Based on quantitatively assessed students' performance on individual course concepts, the instructor can modify instruction, delivery mode, assignments, and tests, and even the course content. With this, conceptual questions become the main part of the assessment feedback mechanism (described, for instance, by the ABET *Continuous Improvement* criterion, which arguably is the most challenging one to achieve, document, and comply with).

The multiple offered answers for each question are designed to emphasize true understanding of the material as well as severe misconceptions. Some of the offered incorrect answers for each question are designed to serve as *"distractors"* of different types, which are meant to identify common misconceptions and errors by learners and to, simply, draw a student with no or poor understanding of the concept away from the correct choice (that might otherwise be selected as a random guess).

Not less important, in terms of *grading of tests, exams, and homework*, it is so much easier and faster to grade multiple-choice questions than traditional computational problems. Moreover, since the questions in this book are conceptual, choosing a wrong answer does not mean a simple error in computation but a misunderstanding of a concept or a major conceptual error regarding an equation or a solution procedure. Consequently, tests with conceptual questions can indeed be graded and the knowledge assessed on an on/off (correct/incorrect) basis considering only the provided answer choice, and not the full work. In addition to being efficient, such grading is the most fair and objective, and it eliminates the need for any discussions and interpretations of the student's work on the test.

Conceptual questions in this book follow the intent and form of the questions on the *Electromagnetics Concept Inventory (EMCI)*. The EMCI (author Branislav Notaroš) is an assessment tool designed to measure students' understanding of fundamental concepts in electromagnetics. This work was done as part of the NSF Foundation Coalition project. The EMCI is motivated by the Force Concept Inventory (FCI), created by Hestenes and Halloun, and its impact on physics education.

Conceptual questions of this scope and intent are *completely new in the electromagnetics area*, and in practically all electrical and computer engineering areas. Also, this is one of the most complete and ambitious attempts to use them in science and engineering education overall. So far, conceptual electromagnetics has been extremely well received by students. I hope that this book will help that conceptual questions and problems in electromagnetics become a widely adopted and used pedagogical tool and practice in electromagnetics education – in instruction, learning, and assessment – and that more and more students start liking and appreciating this fascinating discipline with endless impacts.

Please send comments, suggestions, questions, and/or corrections to notaros@colostate.edu.

Branislav M. Notaroš
Fort Collins, Colorado

About the Author

Branislav M. Notaroš is Professor in the Department of Electrical and Computer Engineering and University Distinguished Teaching Scholar at Colorado State University, where he also is Director of Electromagnetics Laboratory. He received a Ph.D. in electrical engineering from the University of Belgrade, Yugoslavia, in 1995. His research publications in computational and applied electromagnetics include more than 180 journal and conference papers. He is the author of textbooks *Electromagnetics* (2010) and *MATLAB®-Based Electromagnetics* (2013), both with Pearson Prentice Hall, as well as *Conceptual Electromagnetics* (2017), with CRC Press. Prof. Notaroš served as General Chair of *FEM2012*, Colorado, USA, and as Guest Editor of the Special Issue on Finite Elements for Microwave Engineering, in *Electromagnetics*, 2014. He is Editor of *Electromagnetics, Wireless, Radar, and Microwaves Series* with CRC Press. He was the recipient of the 1999 Institution of Electrical Engineers (IEE) Marconi Premium, 2005 Institute of Electrical and Electronics Engineers (IEEE) MTT-S Microwave Prize, 2005 UMass Dartmouth Scholar of the Year Award, 2012 Colorado State University System Board of Governors Excellence in Undergraduate Teaching Award, 2012 IEEE Region 5 Outstanding Engineering Educator Award, 2014 Carnegie Foundation and CASE USPOY Colorado Professor of the Year Award, 2015 American Society for Engineering Education (ASEE) ECE Distinguished Educator Award, 2015 IEEE Undergraduate Teaching Award, and many other research and teaching awards. He is a Fellow of IEEE. For more information, see www.engr.colostate.edu/~notaros.

Prior Publication of Parts of *Conceptual Electromagnetics* Material

Many of the conceptual questions in *Conceptual Electromagnetics*, including the associated figures, are adapted, by permission granted by Pearson Education, from *Electromagnetics* by Branislav M. Notaroš (Pearson Prentice Hall, 2010; [1] in the Bibliography), where they appear as an e-supplement (B. M. Notaroš, *Conceptual Questions in Electromagnetics*, 158 pages) on the companion website of the book.

Many parts of theoretical narratives starting the sections of *Conceptual Electromagnetics*, including the associated figures, are adapted, by permission from Pearson Education, from *MATLAB®-Based Electromagnetics* by Branislav M. Notaroš (Pearson Prentice Hall, 2013; [2]).

A number of figures in *Conceptual Electromagnetics* are adapted, by permission from Pearson Education, from *Electromagnetics* (B. M. Notaroš, Pearson, 2010; [1]).

1 ELECTROSTATIC FIELD IN FREE SPACE

Introduction

Electrostatics is the branch of electromagnetics that deals with phenomena associated with static electricity, which are essentially the consequence of a simple experimental fact – that charges exert forces on one another. These forces are called electric forces, and the special state in space due to one charge in which the other charge is situated and which causes the force on it is called the electric field. Any charge distribution in space with any time variation is a source of the electric field. The electric field due to time-invariant charges at rest (charges that do not change in time and do not move) is called the static electric field or electrostatic field. This is the simplest form of the general electromagnetic field, and its physics and mathematics represent the foundation of the entire electromagnetic theory. On the other hand, a clear understanding of electrostatics is essential for many practical applications that involve static electric fields, charges, and forces in electrical and electronic devices and systems.

1.1 Coulomb's Law

Coulomb's law states that the electric force $\mathbf{F}_{e12}$ on a point charge Q_2 due to a point charge Q_1 in a vacuum or air (free space) is given by[1] (Figure 1.1)

$$\mathbf{F}_{e12} = \frac{1}{4\pi\varepsilon_0} \frac{Q_1 Q_2}{R^2} \hat{\mathbf{R}}_{12} \quad \text{(Coulomb's law)} . \tag{1.1}$$

With $\mathbf{R}_{12}$ denoting the position vector of Q_2 relative to Q_1, $R = |\mathbf{R}_{12}|$ is the distance between the two charges, $\hat{\mathbf{R}}_{12} = \mathbf{R}_{12}/R$ is the unit vector[2] of the vector $\mathbf{R}_{12}$, and ε_0 is the permittivity of free space,

$$\varepsilon_0 = 8.8542 \text{ pF/m} \quad \text{(permittivity of free space)} . \tag{1.2}$$

[1]In this text, vectors are represented by boldface, regular (non-italic) symbols, e.g., **F** and **R**, whereas the magnitudes of vectors, as well as scalar quantities, are denoted by italic, non-bold symbols, e.g., F, R, and Q. Of course, the boldface (**F**) notation for vectors, which is common in typewritten work in general, corresponds to the usual notation with vectors designated by placing a right-handed arrow over the symbol, as $\vec{F}$, in handwritten work.

[2]All unit vectors in this text will be represented using the "hat" notation, so the unit vector in the x-direction (in the rectangular coordinate system), for example, is given as $\hat{\mathbf{x}}$ (note that some of the alternative widely used notations for unit vectors would represent this vector as $\mathbf{a}_x$, $\mathbf{i}_x$, and $\mathbf{u}_x$, respectively).

By point charges we mean charged bodies of arbitrary shapes whose dimensions are much smaller than the distance between them. The SI (International System of Units) unit for charge is the coulomb (abbreviated C), while the unit for force (**F**) is the newton (N).

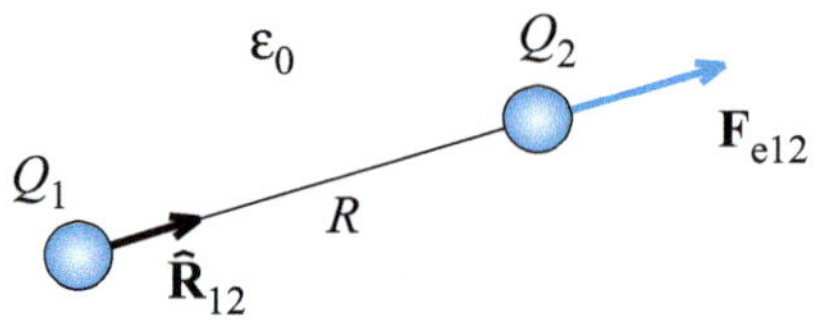

Figure 1.1 Notation in Coulomb's law, given by Eq. (1.1).

If we have more than two point charges, we can use the principle of superposition to determine the resultant force on a particular charge – by adding up vectorially the partial forces exerted on it by each of the remaining charges individually.

CONCEPTUAL QUESTION 1.1 **Two point charges at square vertices.** Two small charged bodies are placed at two vertices of a square in free space (Figure 1.2). The electric force between the charges is stronger for[3]

(A) case (a).

(B) case (b).

(C) The forces are equal in magnitude for the two cases.

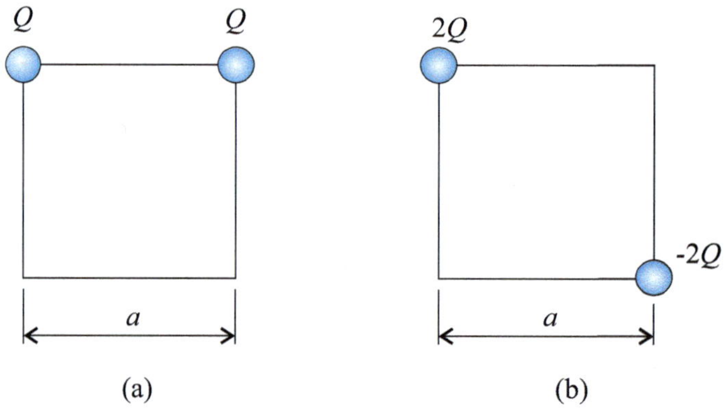

Figure 1.2 Two point charges at vertices of a square; for Conceptual Question 1.1.

[3] *For every conceptual question in this text, exactly one answer is correct.*

CONCEPTUAL QUESTION 1.2 **Three point charges at triangle vertices.** Three point charges of unequal magnitudes and polarities are placed at vertices of an equilateral triangle (Figure 1.3). The electric force $\mathbf{F}_\mathrm{e}$ on the lower right charge is

(A) as in Figure 1.3(a).

(B) as in Figure 1.3(b).

(C) as in Figure 1.3(c).

(D) as in Figure 1.3(d).

(E) zero.

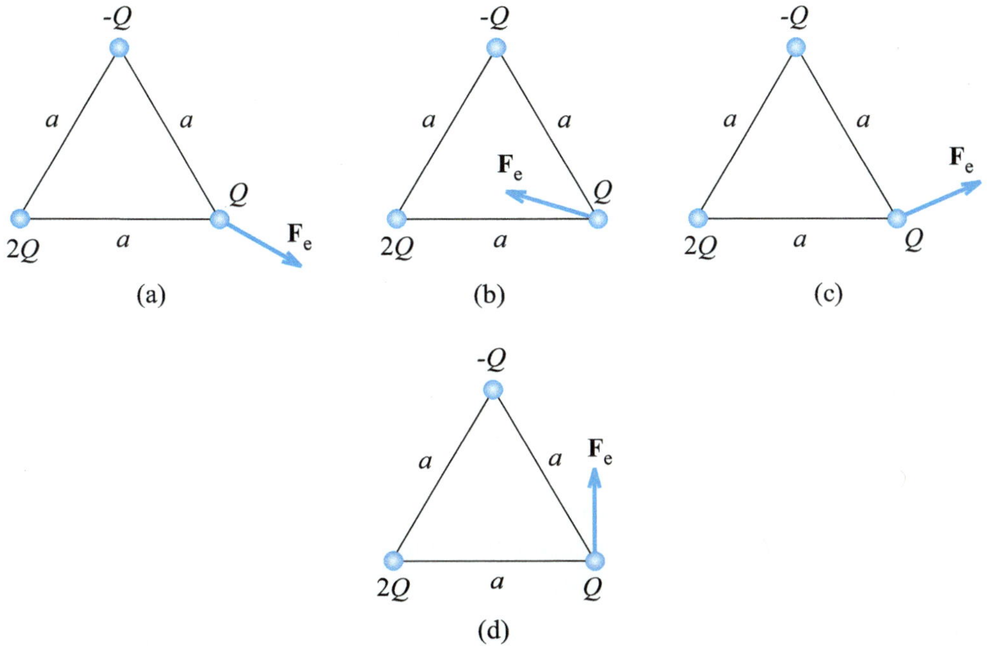

Figure 1.3 Three unequal point charges at vertices of an equilateral triangle; for Conceptual Question 1.2.

1.2 Electric Field Intensity Vector Due to Given Charge Distributions

The electric field is a special physical state existing in a space around charged objects. Its fundamental property is that there is a force (Coulomb force) acting on any stationary charge placed in the space. To quantitatively describe this field, we introduce a vector quantity called the electric field intensity vector, **E**. By definition, it is equal to the electric force $\mathbf{F}_e$ on a small probe (test) point charge Q_p placed in the electric field, divided by Q_p, that is,

$$\mathbf{E} = \frac{\mathbf{F}_e}{Q_p} \quad (Q_p \to 0) \quad \text{(definition of the electric field intensity vector; unit: V/m)}. \tag{1.3}$$

The unit for the electric field intensity is volt per meter (V/m). From the definition in Eq. (1.3) and Coulomb's law, Eq. (1.1), we obtain the expression for the electric field intensity vector of a point charge Q at a distance R from the charge (Figure 1.4)

$$\mathbf{E} = \frac{1}{4\pi\varepsilon_0}\frac{Q}{R^2}\hat{\mathbf{R}} \quad \text{(electric field due to a point charge)}, \tag{1.4}$$

where $\hat{\mathbf{R}}$ is the unit vector along R directed from the center of the charge (source point) toward the point at which the field is (to be) determined (field or observation point).

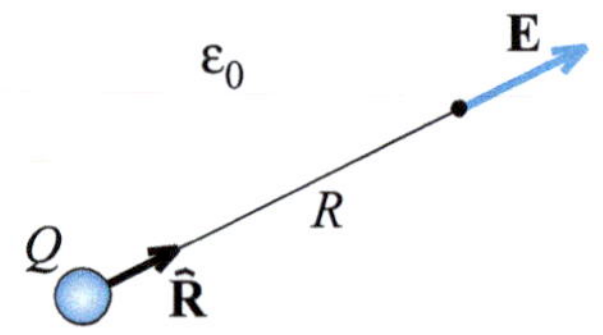

Figure 1.4 Electric field intensity vector due to a point charge in free space.

In the general case, charge can be distributed throughout a volume, on a surface, or along a line. Each of these three characteristic continuous charge distributions is described by a suitable charge density function. The volume charge density (in a volume v) is defined as [Figure 1.5(a)]

$$\rho = \frac{dQ}{dv} \quad \text{(volume charge density; unit: C/m}^3\text{)}, \tag{1.5}$$

the surface charge density (on a surface S) is given by [Figure 1.5(b)]

$$\rho_s = \frac{dQ}{dS} \quad \text{(surface charge density; unit: C/m}^2\text{)}, \tag{1.6}$$

and the line charge density (along a line l) is [Figure 1.5(c)]

$$Q' = \frac{dQ}{dl} \quad \text{(line charge density; unit: C/m)}. \tag{1.7}$$

Note that the symbol ρ_v is sometimes used instead of ρ, σ instead of ρ_s, and ρ_l instead of Q'. In addition, by Q' ($Q' = \text{const}$) we also represent the so-called charge per unit length (p.u.l.) of a long uniformly charged structure (e.g., thin or thick cylinder), defined as the charge on one meter (unit of length) of the structure divided by 1 m,

$$Q' = Q_{\text{p.u.l.}} = \frac{Q_{\text{along } l}}{l} = \frac{Q_{\text{for 1 m length}}}{1\,\text{m}} \quad \text{(charge per unit length, in C/m)}, \quad (1.8)$$

and hence Q' numerically equals the charge on each meter of the structure.

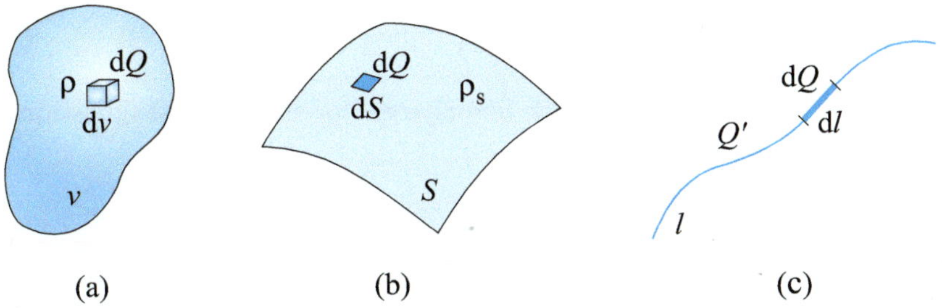

Figure 1.5 Three characteristic continuous charge distributions and charge elements: (a) volume charge, (b) surface charge, and (c) line charge.

By virtue of the superposition principle, the electric field intensity vector due to each of the charge distributions ρ, ρ_{s}, and Q' can be regarded as the vector summation of the field intensities contributed by the numerous equivalent point charges making up the charge distribution. Thus, by replacing Q in Eq. (1.4) with charge element $\mathrm{d}Q = Q'\,\mathrm{d}l$ in Figure 1.5(c) and integrating, we get

$$\mathbf{E} = \frac{1}{4\pi\varepsilon_0}\int_l \frac{Q'\,\mathrm{d}l}{R^2}\,\hat{\mathbf{R}} \quad \text{(electric field due to line charge)}, \quad (1.9)$$

and similar expressions for the electric field due to surface and volume charges.

CONCEPTUAL QUESTION 1.3 Electric field due to a uniform charge along a semicircle. A uniform line charge of density Q' (Q' = const) has the form of a semicircle of radius a, lying in the xy-plane, in free space, as shown in Figure 1.6. If $Q' > 0$, the electric field intensity vector $\mathbf{E}$ due to this charge at the point on the z-axis defined by the coordinate $z = a$ can be represented as

(A) $\mathbf{E} = E_x\,\hat{\mathbf{x}}$, where $E_x \neq 0$.

(B) $\mathbf{E} = E_z\,\hat{\mathbf{z}}$, where $E_z \neq 0$.

(C) $\mathbf{E} = E_x\,\hat{\mathbf{x}} + E_y\,\hat{\mathbf{y}}$, where $E_x > 0$ and $E_y > 0$.

(D) $\mathbf{E} = E_x\,\hat{\mathbf{x}} + E_y\,\hat{\mathbf{y}}$, where $E_x > 0$ and $E_y < 0$.

(E) $\mathbf{E} = E_x\,\hat{\mathbf{x}} + E_z\,\hat{\mathbf{z}}$, where $E_x > 0$ and $E_z > 0$.

(F) $\mathbf{E} = E_x\,\hat{\mathbf{x}} + E_z\,\hat{\mathbf{z}}$, where $E_x < 0$ and $E_z > 0$.

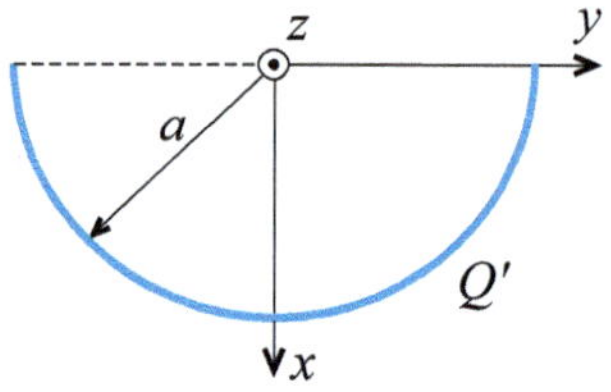

Figure 1.6 Uniform semicircular line charge; for Conceptual Question 1.3.

CONCEPTUAL QUESTION 1.4 **Half-positive, half-negative finite line charge.** A line charge of finite length in free space has a density Q' ($Q' > 0$) along one half and $-Q'$ along the other, as depicted in Figure 1.7. The associated electric field intensity vector $\mathbf{E}$ at a point M equally distant from the line ends is

(A) as in Figure 1.7(a).

(B) as in Figure 1.7(b).

(C) as in Figure 1.7(c).

(D) as in Figure 1.7(d).

(E) zero.

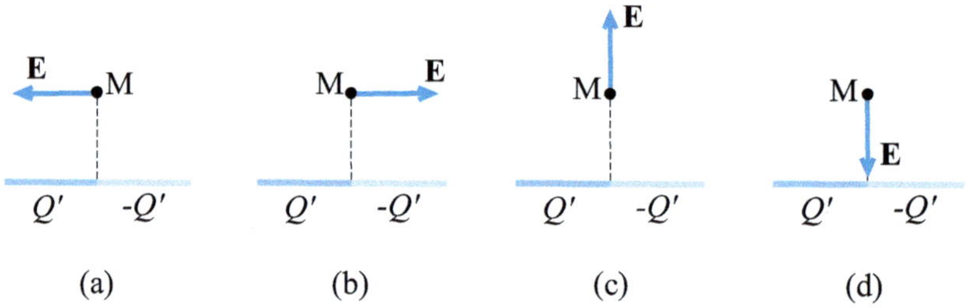

Figure 1.7 Half-positive, half-negative line charge of finite length; for Conceptual Question 1.4.

CONCEPTUAL QUESTION 1.5 **Two parallel strips with equal surface charge densities.** Two parallel, infinitely long strips of width a are uniformly charged with equal charge densities ρ_s ($\rho_s > 0$), and a cross section of the structure is shown in Figure 1.8. The ambient medium is air, and the separation between strips is d. The resultant electric field intensity vector $\mathbf{E}$ at the point M in the figure

(A) has a positive x-component only.

(B) has a negative x-component only.

(C) has a positive z-component only.

(D) has a negative z-component only.

(E) is zero.

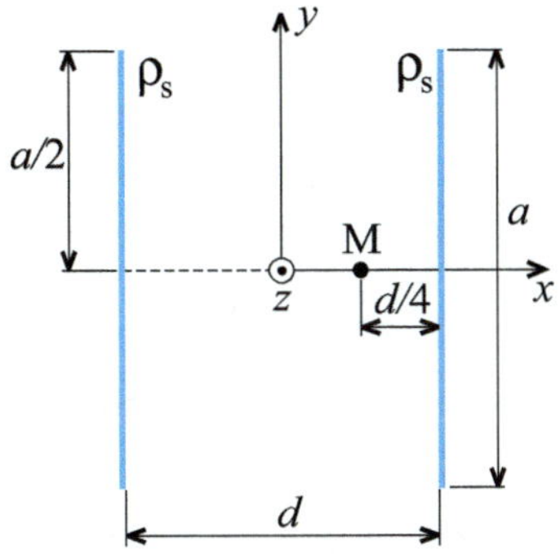

Figure 1.8 Cross section of two parallel, infinitely long strips with equal surface charge densities; for Conceptual Question 1.5.

CONCEPTUAL QUESTION 1.6 **Two parallel equally charged sheets.** Figure 1.9 shows a cross section of two parallel infinite sheets (infinitely long, infinitely wide strips, i.e., assuming that $a \to \infty$ in Figure 1.8) of charge with equal uniform (constant) positive densities ρ_s situated in air. The total electric field intensity vector $\mathbf{E}$ due to both sheets at the point M

(A) has a positive x-component only.

(B) has a negative x-component only.

(C) has a positive z-component only.

(D) has a negative z-component only.

(E) is zero.

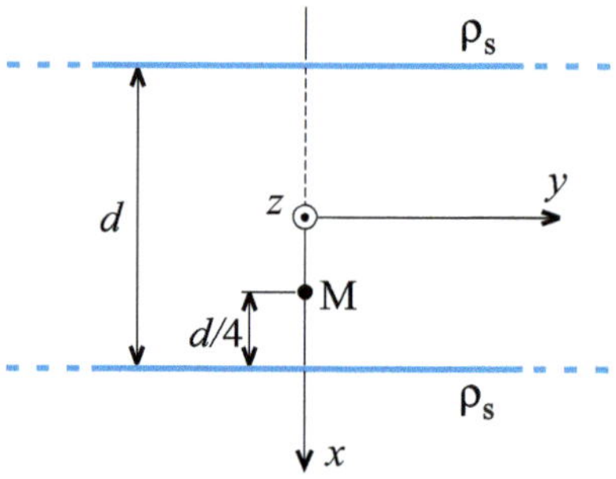

Figure 1.9 Cross section of two parallel infinite sheets of surface charge with equal densities; for Conceptual Question 1.6.

CONCEPTUAL QUESTION 1.7 **Electric field due to volume charge.** Consider an arbitrary (nonuniform) distribution of volume charges in a vacuum and the electric field intensity vector due to these charges at an arbitrary point in space where **E** is not zero. If the charge density, ρ, is doubled everywhere, so that $\rho_{\text{new}} = 2\rho$, the electric field vector considered

(A) becomes twice as large in magnitude and keeps the same direction.

(B) becomes larger in magnitude (not always twice) and keeps the same direction.

(C) becomes larger in magnitude and may change direction.

(D) becomes twice as large in magnitude and may change direction.

(E) may become larger or smaller in magnitude and may change direction.

1.3 Electric Scalar Potential

The electric scalar potential is a scalar quantity that can be used instead of the electric field intensity vector for the description of the electrostatic field. The potential, V, at a point P in an electric field is defined as the work W_e done by the field, that is, by the electric force, $\mathbf{F}_\text{e}$, in moving a test point charge, Q_p, from P to a reference point $\mathcal{R}$ (Figure 1.10),

$$W_\text{e} = \int_\text{P}^{\mathcal{R}} \mathbf{F}_\text{e} \cdot \text{d}\mathbf{l} = \int_\text{P}^{\mathcal{R}} F_\text{e}\, \text{d}l \cos\alpha \tag{1.10}$$

(the dot product of vectors **a** and **b** is a scalar given by $\mathbf{a}\cdot\mathbf{b} = |\mathbf{a}||\mathbf{b}|\cos\alpha$, α being the angle between **a** and **b**), divided by Q_p. Having in mind Eq. (1.3), this becomes

$$V = \frac{W_\mathrm{e}}{Q_\mathrm{p}} = \int_\mathrm{P}^{\mathcal{R}} \frac{\mathbf{F}_\mathrm{e}}{Q_\mathrm{p}} \cdot \mathrm{d}\mathbf{l} = \int_\mathrm{P}^{\mathcal{R}} \mathbf{E}\cdot\mathrm{d}\mathbf{l} \quad \text{(definition of the electric potential; unit: V)}\,, \tag{1.11}$$

namely, V equals the line integral of vector **E** from P to $\mathcal{R}$.[4] The unit for the potential is volt (abbreviated V). Note that Φ is also used to denote the electric potential.

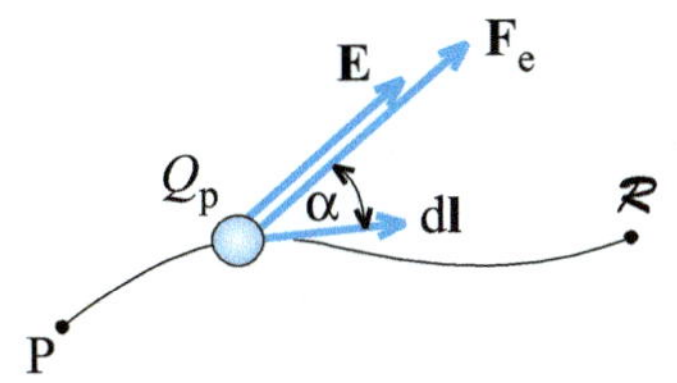

Figure 1.10 Displacement of a test charge in an electrostatic field.

From Eqs. (1.11) and (1.4), the electric scalar potential at a distance R from a point charge Q in free space with respect to the reference point at infinity is $V = Q/(4\pi\varepsilon_0 R)$. By the superposition principle, we then obtain the expressions for the resultant electric potential for the three characteristic continuous charge distributions in Figure 1.5. For instance, the potential expression corresponding to that in Eq. (1.9) for the field vector is given by

$$V = \frac{1}{4\pi\varepsilon_0}\int_l \frac{Q'\,\mathrm{d}l}{R} \quad \text{(potential due to line charge)}\,. \tag{1.12}$$

By the principle of conservation of energy, the net work done by the electrostatic field in moving Q_p from a point A to some point B and then moving it back to A along a different path is zero (because after the round trip, the system is the same as at the beginning). This means that the line integral of the electric field intensity vector along an arbitrary closed path (contour) is zero,

$$\oint_C \mathbf{E}\cdot\mathrm{d}\mathbf{l} = 0 \quad \text{(conservative nature of the electrostatic field)}\,, \tag{1.13}$$

which constitutes Maxwell's first equation for the electrostatic field.

[4] The line integral of a vector function (field) **a** along a line (curve) l, from a point A to a point B, is defined as $\int_l \mathbf{a}\cdot\mathrm{d}\mathbf{l} = \int_\mathrm{A}^\mathrm{B} \mathbf{a}\cdot\mathrm{d}\mathbf{l}$, where $\mathrm{d}\mathbf{l}$ is the differential length vector tangential to the curve (as in Figure 1.10) oriented from A toward B. If the line is closed (for example, a circle or a square), we call it contour (and usually mark it C), and the corresponding line integral, $\oint_C \mathbf{a}\cdot\mathrm{d}\mathbf{l}$, is termed the circulation of **a** along C. The reference direction of $\mathrm{d}\mathbf{l}$ coincides with the orientation of the contour.

By definition, the voltage between two points is the potential difference between them,

$$V_{AB} = V_A - V_B \quad \text{(voltage between points A and B)} , \tag{1.14}$$

where V_A and V_B are the potentials at point A and point B, respectively, with respect to the same reference point. Combining Eqs. (1.11) and (1.14), we get

$$V_{AB} = \int_A^B \mathbf{E} \cdot d\mathbf{l} \quad \text{(voltage via a line integral of } \mathbf{E}) . \tag{1.15}$$

CONCEPTUAL QUESTION 1.8 **Electric potential at the reference point.** The electric scalar potential (V) at the reference point for potential, $\mathcal{R}$, is

(A) zero.

(B) infinite.

(C) undefined.

CONCEPTUAL QUESTION 1.9 **Electric potential and field due to a charged ring.** A line charge of uniform charge density Q' is distributed around the circumference of a ring of radius a in air. Denoting by V and E, respectively, the electric scalar potential (with respect to the reference point at infinity) and field intensity due to this charge at the ring center, we have the following:

(A) $V = 0$ and $E = 0$.

(B) $V = 0$ and $E \neq 0$.

(C) $V \neq 0$ and $E = 0$.

(D) $V \neq 0$ and $E \neq 0$.

CONCEPTUAL QUESTION 1.10 **Electric potential at a point in a uniform electric field.** Consider a region with a unform (the same everywhere) electrostatic field of intensity E, as shown in Figure 1.11. If the electric scalar potential at the point A is zero ($V_A = 0$), the potential at the point B, given the notation in the figure, equals

(A) $V_{\mathrm{B}} = Ed$.

(B) $V_{\mathrm{B}} = Ed \sin \alpha$.

(C) $V_{\mathrm{B}} = -Ed \sin \alpha$.

(D) $V_{\mathrm{B}} = Ed \cos \alpha$.

(E) $V_{\mathrm{B}} = -Ed \cos \alpha$.

(F) $V_{\mathrm{B}} = 0$.

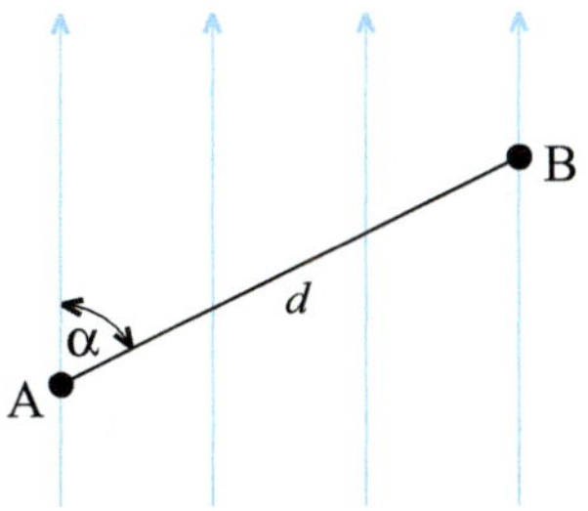

Figure 1.11 Points A and B in a uniform electric field; for Conceptual Question 1.10.

CONCEPTUAL QUESTION 1.11 **Contour in the field of a point charge.** A point charge Q is situated in free space. The line integral (circulation) of the electric field intensity vector $\mathbf{E}$ due to this charge along the contour C in Figure 1.12, composed of two circular parts of radii a and $2a$, respectively, and two radial parts of length a, amounts to

(A) $Q/(4\pi\varepsilon_0 a)$.

(B) $-Q/(4\pi\varepsilon_0 a)$.

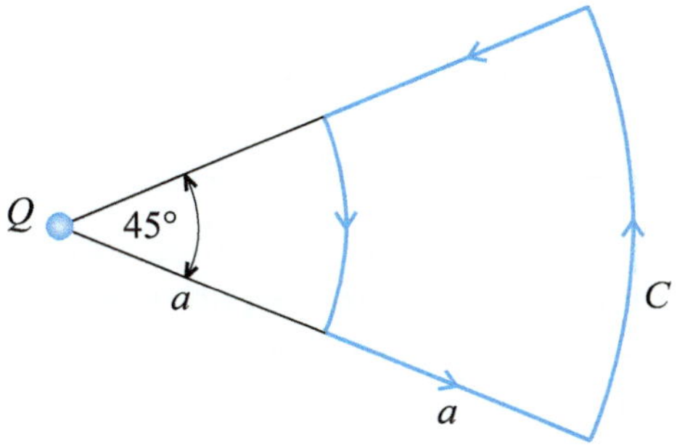

Figure 1.12 Contour C in the electric field of a point charge Q in free space; for Conceptual Question 1.11.

(C) $Q/(8\varepsilon_0 a)$.

(D) $-Q/(8\varepsilon_0 a)$.

(E) zero.

(ε_0 is the permittivity of a vacuum).

CONCEPTUAL QUESTION 1.12 **Potentials and voltages for a new reference point.** What happens to electric potentials and voltages in an electrostatic system after a new reference point is adopted for the potential?

(A) Both potentials and voltages change by the same value.

(B) Potentials become zero and voltages remain unchanged.

(C) Both potentials and voltages remain unchanged.

(D) Potentials change by the same value and voltages remain unchanged.

(E) Potentials remain unchanged and voltages change by the same value.

1.4 Differential Relationship between Field and Potential in Electrostatics, Gradient

Equation (1.11) represents an integral relationship between the electric field intensity vector and the potential in electrostatics, which enables us to determine V if we know $\mathbf{E}$. An equivalent, differential, relationship between these two quantities which can be used for evaluating $\mathbf{E}$ from V is given by

$$\mathbf{E} = E_x\,\hat{\mathbf{x}} + E_y\,\hat{\mathbf{y}} + E_z\,\hat{\mathbf{z}} = -\left(\frac{\partial V}{\partial x}\,\hat{\mathbf{x}} + \frac{\partial V}{\partial y}\,\hat{\mathbf{y}} + \frac{\partial V}{\partial z}\,\hat{\mathbf{z}}\right) = -\operatorname{grad} V = -\nabla V$$

$$(\mathbf{E}\text{ from } V \text{ in electrostatics, via gradient}), \quad (1.16)$$

where we have partial derivatives instead of ordinary ones because the potential is a function of all three coordinates (multivariable function), $V = V(x, y, z)$. The expression in the parentheses is called the gradient of the scalar function (V). It is sometimes written as $\operatorname{grad} V$, but much more frequently we write it as ∇V, using the so-called del operator or nabla operator, defined as

$$\nabla = \frac{\partial}{\partial x}\,\hat{\mathbf{x}} + \frac{\partial}{\partial y}\,\hat{\mathbf{y}} + \frac{\partial}{\partial z}\,\hat{\mathbf{z}} \quad \text{(del operator)}. \quad (1.17)$$

Similar formulas exist for computing the gradient (∇V) in cylindrical and spherical coordinate systems.

For a given scalar field f (not necessarily electrostatic potential), the magnitude of ∇f at a point in space equals the maximum space rate of change in the function f per unit distance [$|\nabla f| = (\mathrm{d}f/\mathrm{d}l)_{\mathrm{max}}$] and ∇f points in the direction of the maximum space rate of change in f. So, the gradient of f is a vector that provides us with both the direction in which f changes most rapidly and the magnitude of the maximum space rate of change.

CONCEPTUAL QUESTION 1.13 **Field maximum from a potential distribution.** The electrostatic potential V in a region is a function of the rectangular coordinate x only, and $V(x)$ is shown in Figure 1.13. Consider the electric field intensities at points A, B, C, D, and E. The largest field intensity is at point

(A) A.

(B) B.

(C) C.

(D) D.

(E) E.

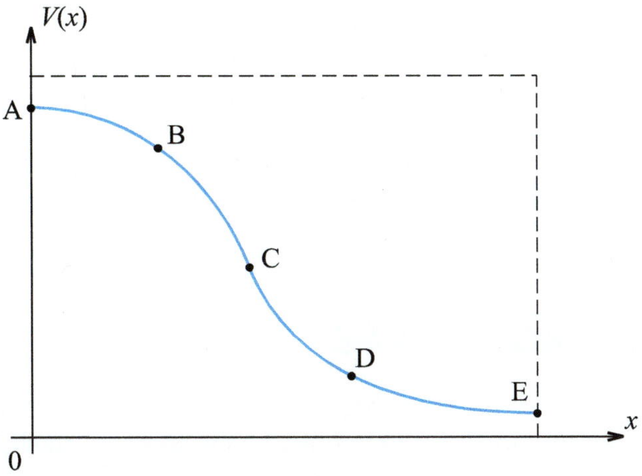

Figure 1.13 One-dimensional potential distribution; for Conceptual Question 1.13.

CONCEPTUAL QUESTION 1.14 **Zero potential and/or field at a point.** Consider an electrostatic field in a region of space and the following two statements: (a) If the electric scalar potential at a point in the region is zero, then the electric field vector at that point must be zero as well. (b) If the electric field vector at a point is zero, then the potential at the same point must be zero. Which of the statements is true?

(A) Statement (a) only.

(B) Statement (b) only.

(C) Both statements.

(D) Neither of the statements.

CONCEPTUAL QUESTION 1.15 **Direction of the steepest ascent.** The terrain elevation in a region is given by a function $h(x, y)$, where x and y are coordinates in the horizontal plane. The direction of the steepest ascent (maximum increase of the terrain elevation, h) at a location (x_0, y_0), where $\partial h/\partial x = 3$ and $\partial h/\partial y = 4$, is defined by the following unit vector $\hat{\mathbf{l}}$ ($|\hat{\mathbf{l}}| = 1$):

(A) $\hat{\mathbf{l}} = (3\,\hat{\mathbf{x}} + 4\,\hat{\mathbf{y}})/5$.

(B) $\hat{\mathbf{l}} = -(3\,\hat{\mathbf{x}} + 4\,\hat{\mathbf{y}})/5$.

(C) $\hat{\mathbf{l}} = (4\,\hat{\mathbf{x}} + 3\,\hat{\mathbf{y}})/5$.

(D) $\hat{\mathbf{l}} = -(4\,\hat{\mathbf{x}} + 3\,\hat{\mathbf{y}})/5$.

(E) $\hat{\mathbf{l}} = -(\hat{\mathbf{x}} + \hat{\mathbf{y}})/\sqrt{2}$.

(F) None of the above.

CONCEPTUAL QUESTION 1.16 **How steep is the maximum ascent?** Considering the terrain elevation function $h(x, y)$ (x and y being horizontal rectangular coordinates) and the location (x_0, y_0) at which $\partial h/\partial x = 3$ and $\partial h/\partial y = 4$, the steepest ascent at this location expressed as an angle α (in degrees) amounts to

(A) $\alpha = \arctan(3/4) = 36.87^\circ$ ($\arctan \equiv \tan^{-1}$).

(B) $\alpha = \arctan(4/3) = 53.13^\circ$.

(C) $\alpha = \arctan 5 = 78.69^\circ$.

(D) $\alpha = \arctan 2 = 63.43^\circ$.

(E) $\alpha = \arctan 1 = 45°$.

(F) None of the above.

CONCEPTUAL QUESTION 1.17 **Maximum increase in electrostatic potential.** The electrostatic field intensity vector in a region is given by $\mathbf{E}(x, y, z) = (\hat{\mathbf{x}} - z^2\,\hat{\mathbf{y}} + 2yz\,\hat{\mathbf{z}})$ V/m (x, y, z in m). The direction of the maximum increase in the electric scalar potential at a point (1 m, 1 m, 1 m) is determined by the unit vector

(A) $\hat{\mathbf{l}} = (-2\,\hat{\mathbf{y}} + 2\,\hat{\mathbf{z}})/\sqrt{8}$.

(B) $\hat{\mathbf{l}} = (2\,\hat{\mathbf{y}} - 2\,\hat{\mathbf{z}})/\sqrt{8}$.

(C) $\hat{\mathbf{l}} = (\hat{\mathbf{x}} - \hat{\mathbf{y}} + 2\,\hat{\mathbf{z}})/\sqrt{6}$.

(D) $\hat{\mathbf{l}} = (-\hat{\mathbf{x}} + \hat{\mathbf{y}} - 2\,\hat{\mathbf{z}})/\sqrt{6}$.

(E) $\hat{\mathbf{l}} = \hat{\mathbf{z}}$.

(F) None of the above.

1.5 Gauss' Law in Integral Form

Gauss' law (in integral form) states that the outward flux (surface integral) of the electric field intensity vector, $\mathbf{E}$, through any closed surface S in free space[5] is equal to the total charge enclosed by that surface, Q_S, divided by ε_0, namely,

$$\oint_S \mathbf{E} \cdot \mathrm{d}\mathbf{S} = \frac{Q_S}{\varepsilon_0} \quad \text{(Gauss' law)}\,. \tag{1.18}$$

The most general case of continuous charge distributions is the volume charge distribution (Figure 1.14), in terms of which Gauss' law can be written as

$$\oint_S \mathbf{E} \cdot \mathrm{d}\mathbf{S} = \frac{1}{\varepsilon_0} \int_v \rho\, \mathrm{d}v \quad \text{(Gauss' law for volume charge)}\,, \tag{1.19}$$

with v denoting the volume enclosed by the surface S and ρ the volume charge density. This particular form of Gauss' law is usually referred to as Maxwell's third equation for the electrostatic field in free space.

[5]The flux of a vector function $\mathbf{a}$ through an open or closed surface S is defined as $\int_S \mathbf{a} \cdot \mathrm{d}\mathbf{S}$, where $\mathrm{d}\mathbf{S}$ is the vector element of the surface perpendicular to it, and directed in accordance with the orientation of the surface.

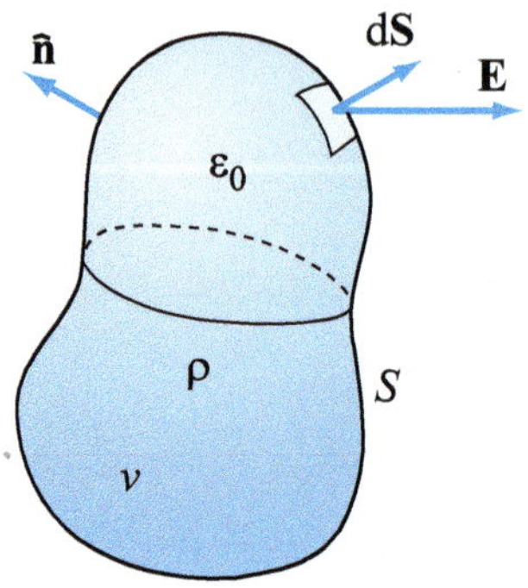

Figure 1.14 Arbitrary closed surface containing a volume charge distribution in free space.

CONCEPTUAL QUESTION 1.18 **Flux of the electric field vector through an infinite surface.** A point charge Q is situated in free space at a very small height h ($h \to 0$) above an imaginary (nonmaterial) infinite flat surface S, as depicted in Figure 1.15. The surface is oriented upward. The flux of the electric field intensity vector due to the charge Q through S (Ψ_E) comes out to be

(A) $\Psi_E = Q/(4\pi\varepsilon_0)$.

(B) $\Psi_E = Q/(2\varepsilon_0)$.

(C) $\Psi_E = -Q/(2\varepsilon_0)$.

(D) $\Psi_E = -Q/\varepsilon_0$.

(E) $\Psi_E = 0$.

(F) $\Psi_E \to \infty$.

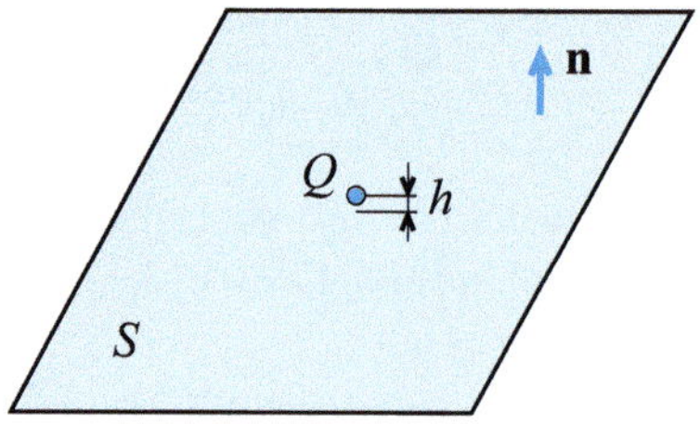

Figure 1.15 Point charge Q at a very small height $h \to 0$ above an infinite flat nonmaterial surface S; for Conceptual Question 1.18.

CONCEPTUAL QUESTION 1.19 **Flux through a cube side, charge at a vertex.** A point charge Q is located at one of the vertices of an imaginary cube in free space, as shown in Figure 1.16. The outward flux Ψ_E of the electric field intensity vector due to this charge through a cube side that does not contain the charge (e.g., the upper cube side in the figure) equals

(A) $\Psi_E = Q/\varepsilon_0$.

(B) $\Psi_E = Q$.

(C) $\Psi_E = Q/(2\varepsilon_0)$.

(D) $\Psi_E = Q/(6\varepsilon_0)$.

(E) $\Psi_E = Q/(24\varepsilon_0)$.

(F) $\Psi_E = 0$.

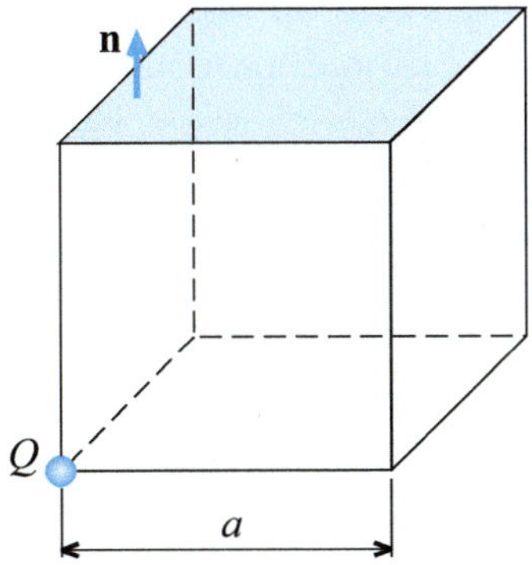

Figure 1.16 Point charge Q at a vertex of a cube; for Conceptual Question 1.19.

CONCEPTUAL QUESTION 1.20 **Gaussian surface inside a volume charge distribution.** A charge Q ($Q > 0$) is distributed uniformly throughout the volume of a sphere of radius a in free space. The outward flux of the electric field intensity vector $\mathbf{E}$ through the closed surface S shown in Figure 1.17 is

(A) Q/ε_0.

(B) $-Q/\varepsilon_0$.

(C) greater than Q/ε_0.

(D) positive and less than Q/ε_0.

(E) zero.

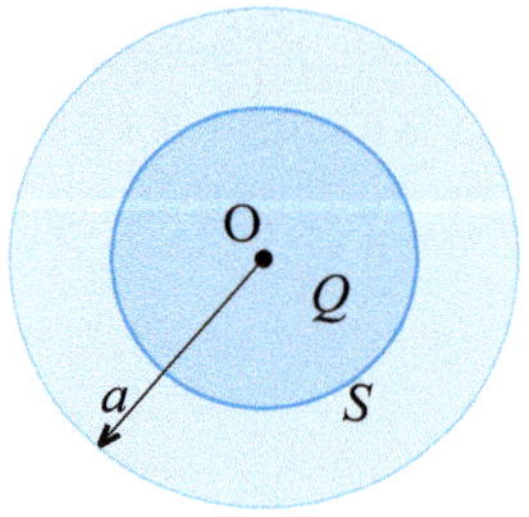

Figure 1.17 Closed surface S inside a uniform volume charge distribution; for Conceptual Question 1.20.

CONCEPTUAL QUESTION 1.21 **Electric dipole inside a spherical surface.** Figure 1.18 shows an electric dipole (an electrostatic system consisting of two point charges of equal magnitudes and opposite polarities, Q and $-Q$, separated by a distance d) characterized by a dipole moment $\mathbf{p} = Q\mathbf{d}$ ($Q > 0$), where $\mathbf{d}$ is the position vector of the positive charge with respect to the negative one, in free space. Consider a sphere of radius r, where $r \gg d$, centered at the center of the dipole (point P in Figure 1.18 belongs to the surface of the sphere). If $\mathbf{p}$ is doubled in magnitude, the outward flux of the electric field intensity vector through the surface of the sphere

(A) increases.

(B) decreases.

(C) remains the same.

(D) Need more information.

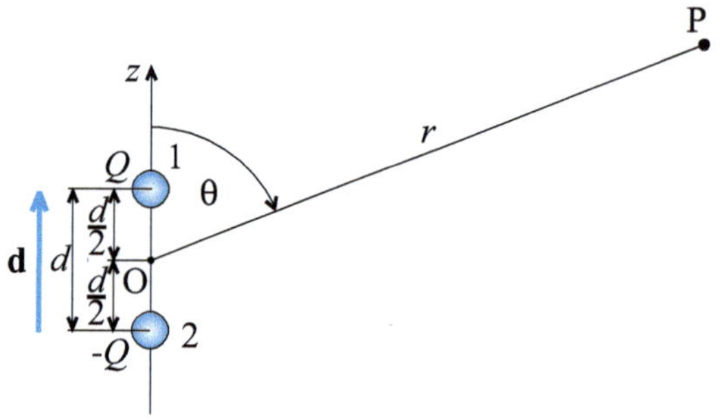

Figure 1.18 Electric dipole; for Conceptual Question 1.21.

CONCEPTUAL QUESTION 1.22 **Introducing a point charge near a Gaussian surface.** A spherical surface S is placed in free space concentrically with another sphere that is uniformly charged over its volume, and the radius of S is larger than that of the charged sphere. Then, a point charge Q, where Q equals the total charge of the sphere, is introduced in the system as in Figure 1.19. Let Ψ_E and E denote the outward flux of the electric field intensity vector through S and the electric field intensity at the point A in the figure, respectively. Which of the two quantities changes its value after the point charge is introduced?

(A) Ψ_E only.

(B) E only.

(C) both quantities.

(D) none of the quantities.

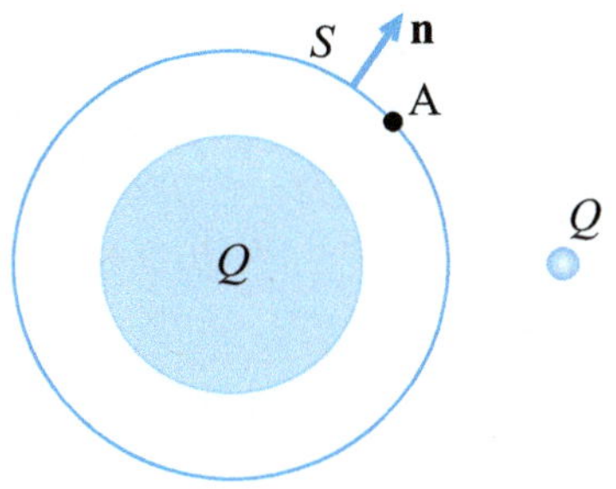

Figure 1.19 Closed concentric spherical surface S about a sphere with a uniform volume charge, and a point charge Q outside S; for Conceptual Question 1.22.

CONCEPTUAL QUESTION 1.23 **Electric field due to a uniformly charged spherical surface.** Consider a sphere of radius a that is uniformly charged over its surface with a total charge Q, and is situated in free space. The electric field intensity vector at a point whose radial distance from the sphere center is r is the same as $\mathbf{E}$ due to a point charge Q placed at the sphere center (and replacing the charged spherical surface) for the following range of values of r only:

(A) $a < r < \infty$.

(B) $r \gg a$.

(C) $0 < r < \infty$.

(D) $0 < r < a$.

(E) never (for none of the possible values of r).

CONCEPTUAL QUESTION 1.24 **Nonuniform surface charge over a sphere.** Compare a sphere of radius a that is nonuniformly charged over its surface (surface charge density is a function of the zenith angle, θ, in a spherical coordinate system) and a point charge at the sphere center replacing the surface charge and amounting to the total charge of the spherical surface, both in a vacuum. The electric field intensity vectors due to each of the systems, the surface charge and the point charge, are the same for the following radial distances, r, from the sphere center only:

(A) $a < r < \infty$.

(B) $r \gg a$.

(C) $0 < r < \infty$.

(D) $0 < r < a$.

(E) never (for none of the possible values of r).

1.6 Differential Form of Gauss' Law, Divergence

Gauss' law in Eq. (1.19) represents an integral relationship between the electric field intensity vector, $\mathbf{E}$, and the volume charge density, ρ. An equivalent, differential, relationship between $\mathbf{E}$ and ρ, that is, the differential form of Gauss' law, is given by

$$\operatorname{div}\mathbf{E} = \nabla\cdot\mathbf{E} = \frac{\partial E_x}{\partial x}+\frac{\partial E_y}{\partial y}+\frac{\partial E_z}{\partial z} = \frac{\rho}{\varepsilon_0} \quad \text{(Gauss' law in differential form)}\,. \tag{1.20}$$

The three-term expression with the partial derivatives of vector components of $\mathbf{E}$ is called the divergence of a vector function ($\mathbf{E}$), and is written as $\operatorname{div}\mathbf{E}$. Applying formally the formula for the dot product of two vectors in the Cartesian coordinate system to the del operator, Eq. (1.17), and vector $\mathbf{E}$, we get $\operatorname{div}\mathbf{E} = \nabla\cdot\mathbf{E}$. Similar formulas are used to calculate the divergence in cylindrical and spherical coordinates.

CONCEPTUAL QUESTION 1.25 **Plots of 1-D charge and field distributions.** The density of a volume charge in a region in free space depends on the Cartesian coordinate x only. Which of the following combinations of the two periodic functions $f_1(x)$ and $f_2(x)$ shown in Figure 1.20 can represent $\rho(x)$ and the associated electric field intensity, $E(x)$, in this region?

(A) $\rho(x) = f_1(x)$ and $E(x) = f_2(x)$.

(B) $\rho(x) = f_2(x)$ and $E(x) = f_1(x)$.

(C) $\rho(x) = f_1(x)$ and $E(x) = cf_1(x)$, where c is a constant.

(D) $\rho(x) = f_2(x)$ and $E(x) = cf_2(x)$ ($c =$ const).

(E) None of the above combinations is possible.

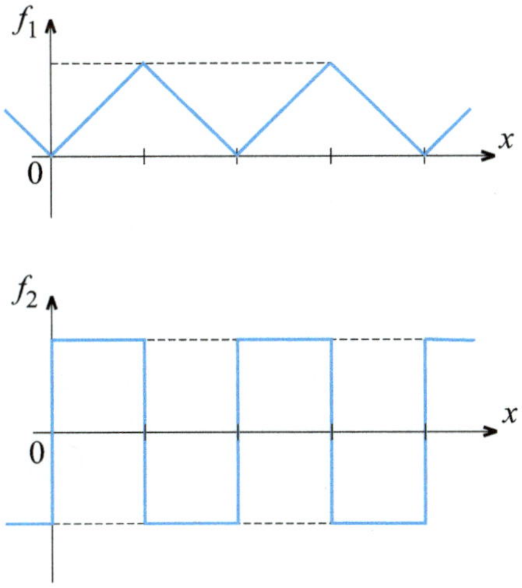

Figure 1.20 Two periodic functions of the Cartesian coordinate x in space; for Conceptual Question 1.25.

CONCEPTUAL QUESTION 1.26 **Divergence-free vector field.** Consider the field pattern (showing lines of a vector field $\mathbf{a}$ in a part of free space) in Figure 1.21(a) and that in Figure 1.21(b). Which of the fields is divergence-free ($\nabla \cdot \mathbf{a} = 0$)?

(A) field in Figure 1.21(a) only.

(B) field in Figure 1.21(b) only.

(C) both fields.

(D) neither of the fields.

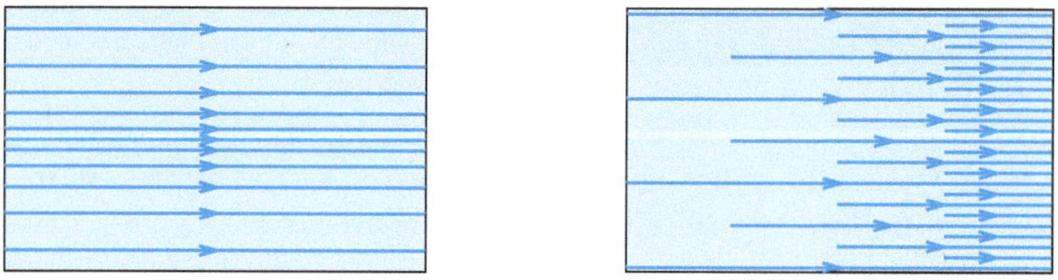

(a) (b)

Figure 1.21 Two patterns of vector fields in a part of space (the magnitude of the vector at a point is proportional to the density of field lines at that point); for Conceptual Question 1.26.

1.7 Conductors in the Electrostatic Field

Materials can broadly be classified in terms of their electrical properties as conductors (which conduct electric current) and dielectrics (insulators). In the rest of this chapter, we shall study the interaction of the electrostatic field with conductors, in which case essentially no theoretical modification is needed to the electrostatic equations, whereas the behavior of dielectrics in the electrostatic field will be discussed in the next chapter. Conductors have a large proportion of freely movable electric charges (free electrons and ions) that make the electric conductivity (ability to conduct electric current) of the material. Best conductors (with highest conductivity) are metals (such as silver, copper, gold, aluminum, etc.). In our studies of electrostatic fields, by conductor we normally mean a metallic conductor.

Under electrostatic conditions, after a transitional process, there cannot be electric field in a conductor,

$$\mathbf{E} = 0 \quad \text{(no electrostatic field inside a conductor)} \, . \tag{1.21}$$

According to Eqs. (1.21), (1.15), and (1.14), the voltage between any two points in the conductor, including points on its surface, is zero. This means that a conductor is an equipotential body, i.e., the potential is the same, $V = \text{const}$, everywhere in the conductor and on its surface. From Eq. (1.21), $\nabla \cdot \mathbf{E} = 0$ in a conductor, implying that [Eq. (1.20)] there cannot be surplus volume charges, $\rho = 0$, inside it. So, any locally surplus charge of a conductor, whether it is neutral (uncharged) as a whole or not, must be located at the surface of the conductor, as surface charge, of density $\rho_{\rm s}$. In addition, Eqs. (1.13) and (1.18) result in the following boundary conditions that the electric field must satisfy on a conductor surface:

$$E_{\rm t} = 0 \, , \quad E_{\rm n} = \frac{\rho_{\rm s}}{\varepsilon_0} \quad \text{(boundary conditions for a conductor surface)} \, , \tag{1.22}$$

where $E_{\rm t}$ is the tangential (on the boundary surface) component of $\mathbf{E}$ near the conductor in a vacuum (or air) and $E_{\rm n}$ is the normal component, defined with respect to the outward normal on the surface (directed from the conductor toward free space).

CONCEPTUAL QUESTION 1.27 Electric field vector near a metallic surface. A point charge Q ($Q < 0$) is located in air above a metallic screen. The electric field intensity vector $\mathbf{E}$ in air immediately above the surface of the screen is

(A) as in Figure 1.22(a).

(B) as in Figure 1.22(b).

(C) as in Figure 1.22(c).

(D) as in Figure 1.22(d).

(E) zero.

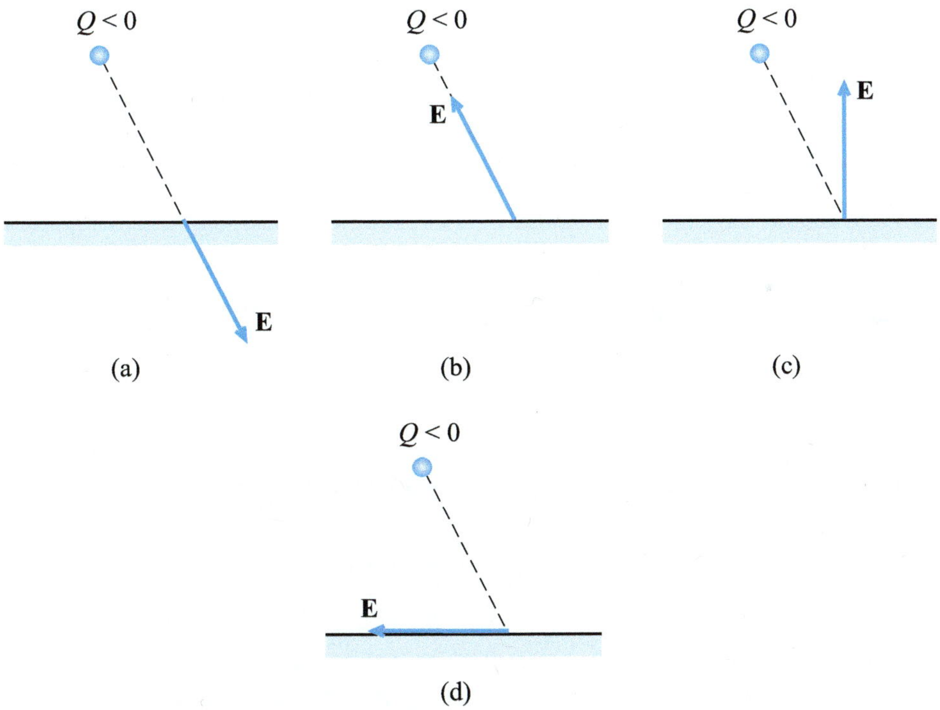

Figure 1.22 Point charge Q above a metallic screen, in air; for Conceptual Question 1.27.

CONCEPTUAL QUESTION 1.28 Introducing a metallic rod into an electric field. An uncharged thin metallic rod is introduced into a uniform electrostatic field, of intensity vector $\mathbf{E}_0$, in free space, such that it is either perpendicular or parallel to $\mathbf{E}_0$, as indicated in Figure 1.23. The rod affects the original field

(A) less in case (a).

(B) less in case (b).

(C) equally noticeably in both cases.

(D) negligibly in both cases.

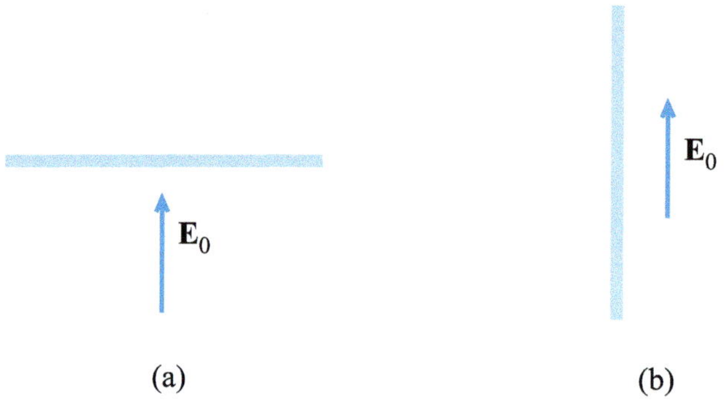

Figure 1.23 Introducing an uncharged thin metallic rod into a uniform electric field ($\mathbf{E}_0$ denotes the original field intensity vector, before a new electrostatic state is established); for Conceptual Question 1.28.

CONCEPTUAL QUESTION 1.29 **Insertion of a metallic slab in a uniform electric field.** A uniform electric field, of intensity vector $\mathbf{E}_0$, is established in the air-filled space between two metallic electrodes, as portrayed in Figure 1.24(a). If an uncharged (thick) metallic slab is then inserted in this space, without touching the electrodes, which gives the structure in Figure 1.24(b), the electric field intensity vector in region 3 (above the slab) in the new electrostatic state is

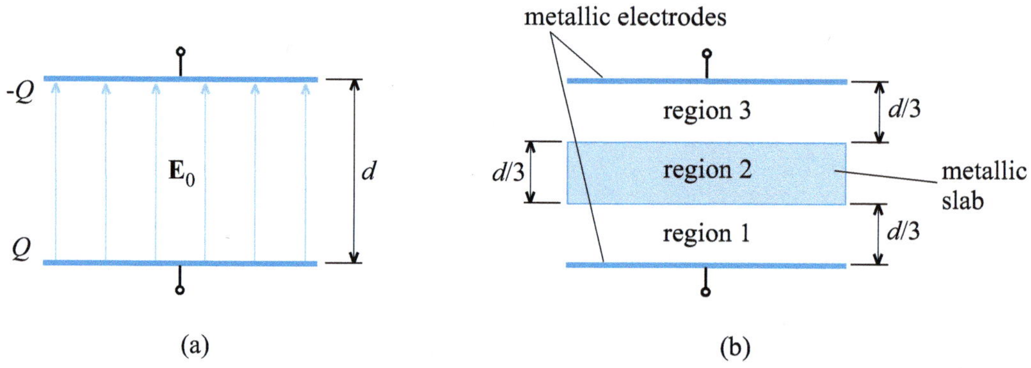

Figure 1.24 (a) Space with a uniform electric field and (b) insertion of an uncharged metallic slab in this field; for Conceptual Question 1.29.

(A) $\mathbf{E}_3 = 0$.

(B) $\mathbf{E}_3 = -\mathbf{E}_0$.

(C) $\mathbf{E}_3 = \mathbf{E}_0/3$.

(D) $\mathbf{E}_3 = 3\mathbf{E}_0/2$.

(E) $\mathbf{E}_3 = \mathbf{E}_0$.

1.8 Electrostatic Shielding

Let us consider an uncharged metallic sphere brought into a uniform electrostatic field, in free space. The field lines around the sphere after electrostatic equilibrium is reached are sketched in Figure 1.25(a). As a result of the electrostatic induction, there are induced surface charges on the sphere surface (creation of surplus charges in a conducting body caused by an external electrostatic field is called the electrostatic induction). Because the field due to the induced charges (this field exists both inside and outside the sphere) is superimposed to the external field, the total field inside the sphere becomes zero [Eq. (1.21)], and that outside it is not uniform any more. Negative induced charges are sinks of the field lines on the left-hand side of the sphere, whereas the positive induced charges are sources of the field lines on the right-hand side of the sphere. The field lines on both sides are normal to the sphere surface, and they therefore bend near the sphere. However, because there is no field throughout the sphere interior, we can remove it, without affecting the field outside the sphere. We thus obtain a domain with no field, bounded by a metallic shell, as shown in Figure 1.25(b). This means that the space inside the shell cavity is

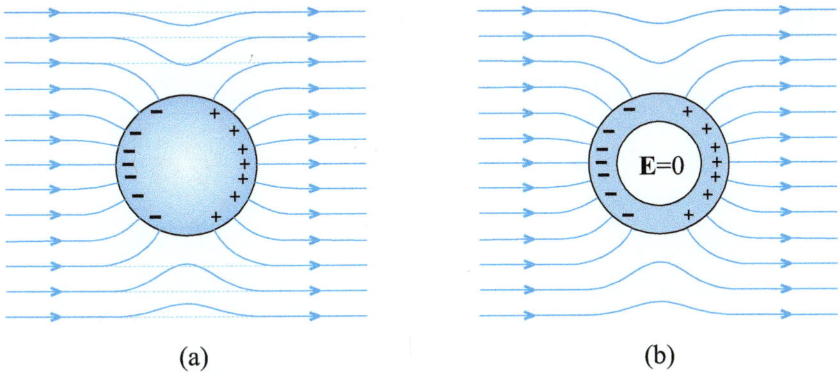

Figure 1.25 The principle of electrostatic shielding: (a) uncharged metallic sphere in a uniform external electrostatic field and (b) metallic shell in an electrostatic field – Faraday cage.

perfectly protected (isolated) from the external electrostatic field. The thickness of the shell can be arbitrary, and its shape does not need to be spherical. Hence, an arbitrary closed conducting shell represents a perfect electrostatic shield or screen for its interior domain. We call such a shield a Faraday cage. If the field outside the cage is changed, the charge on the cage walls will redistribute itself so that the field inside will remain zero. Charge redistribution is a nonelectrostatic transitional process, during which there is a nonzero time-varying electromagnetic field in the cavity; the process is very fast, practically instantaneous.

CONCEPTUAL QUESTION 1.30 **Negative point charge in a Faraday cage.** A negatively charged small body is situated inside an uncharged spherical metallic shell. The distribution of induced charges on the outer surface of the shell can be represented as in

(A) Figure 1.26(a).

(B) Figure 1.26(b).

(C) Figure 1.26(c).

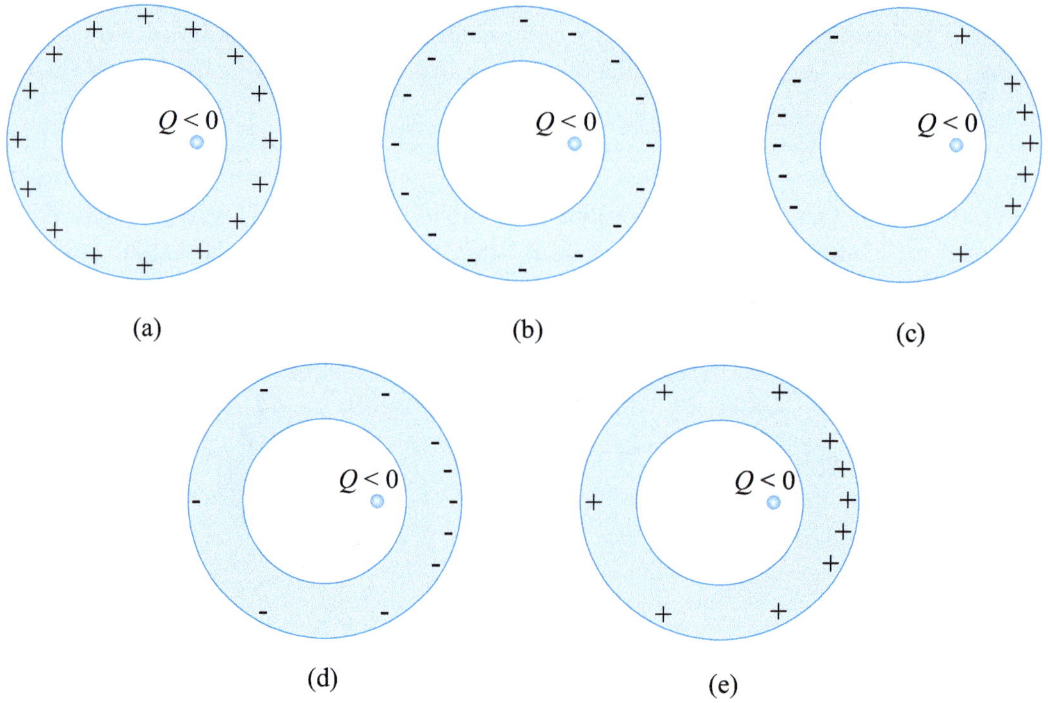

Figure 1.26 Negative point charge in an uncharged spherical metallic shell; for Conceptual Question 1.30.

(D) Figure 1.26(d).

(E) Figure 1.26(e).

CONCEPTUAL QUESTION 1.31 Electrostatic shielding – two bodies and a screen. In order to protect body B from the electrostatic field due to a charged body A, an ungrounded closed metallic screen is introduced (Figure 1.27). The protection is achieved for

(A) case (a) only.

(B) case (b) only.

(C) both cases.

(D) neither of the cases.

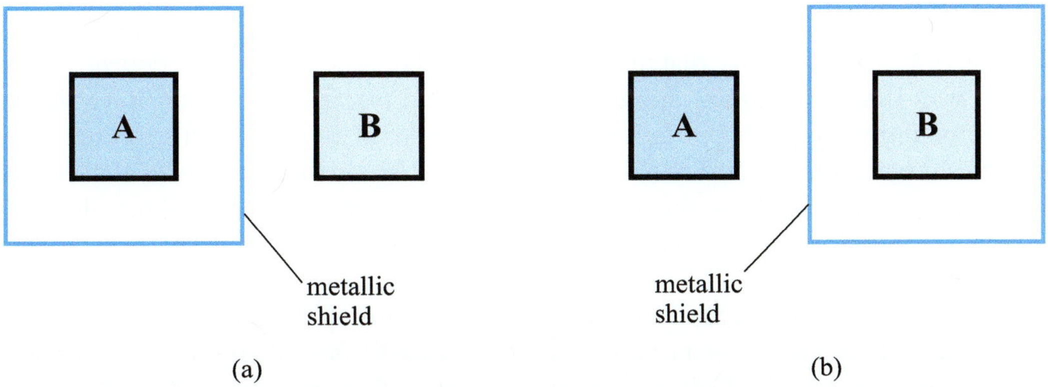

Figure 1.27 Two proposed configurations for electrostatic shielding; for Conceptual Question 1.31.

1.9 Charge Distribution on Metallic Bodies of Arbitrary Shapes

In the general case of a charged metallic body of an arbitrary shape, the charge distribution over the body surface is not uniform. The determination of this distribution for a given body with nonsymmetrical and/or nonsmooth surface is a rather complex problem. To get some qualitative insight about how the charge is distributed over the surface of an arbitrarily shaped isolated conducting body, consider a system composed of two charged metallic spheres of different radii, a and b, whose centers are a distance d apart, in free space. Let the spheres be connected by a very thin conductor, as shown in Figure 1.28. Assume, for simplicity, that $d \gg a, b$, so that the electric potential of each sphere can be evaluated as if the other one were not present. Hence, using Gauss' law in Eq. (1.18), the electric field due to the sphere with the total charge Q_a in Figure 1.28 comes out to be as in Eq. (1.4), and Eq. (1.11) results in the following expression for the potential of the sphere with respect to the reference point at infinity: $V_a = Q_a/(4\pi\varepsilon_0 a)$. Similarly, the potential of the other sphere is $V_b = Q_b/(4\pi\varepsilon_0 b)$, and since $V_a = V_b$ (the spheres are galvanically connected together, and thus represent a single conducting body, which must be equipotential), we obtain $Q_a/Q_b = a/b$. By means of Eq. (1.6), the sphere charges can be expressed in terms of the associated surface charge densities, $\rho_{sa} = Q_a/(4\pi a^2)$ and $\rho_{sb} = Q_b/(4\pi b^2)$, and then the second boundary condition in Eqs. (1.22) gives the corresponding relationship between the electric field intensities near the surfaces of spheres,

$$\frac{\rho_{sa}}{\rho_{sb}} = \frac{E_a}{E_b} = \frac{b}{a} \quad \text{(charge density and local field intensity} \propto \text{surface curvature)} , \tag{1.23}$$

We see that the charge is distributed between the two spheres in Figure 1.28 in such a way that the surface charge density on and electric field intensity near the surface of individual spheres is inversely proportional to the sphere radius. The surface charge is denser and the field stronger on the smaller sphere. However, the importance of Eq. (1.23) is much beyond the particular system in Figure 1.28. It implies a general conclusion that the surface charge density and the nearby field intensity at different parts of the surface of an arbitrarily shaped conducting body are approximately proportional to the local curvature of the surface, as long as it is convex.[6] This means, generally, that the largest concentration of charge and the strongest electric field are around sharp parts of conducting bodies. Note, for example, that this phenomenon is essential for the operation of lightning arresters.

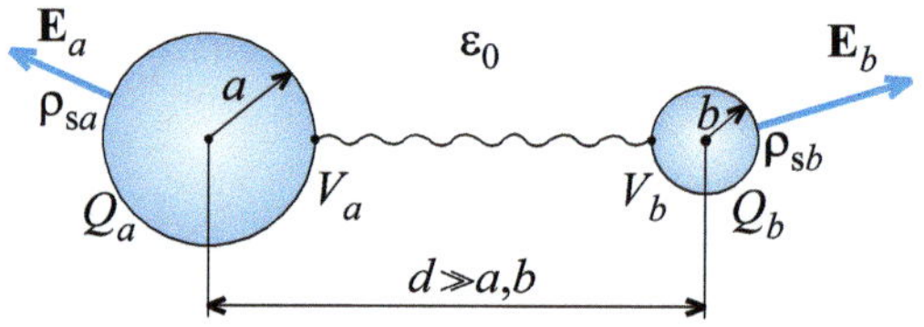

Figure 1.28 Two metallic spheres of different radii at the same potential.

[6]If the surface of a conducting body is concave (curved inward), the effect is just opposite; for a deep incurvature, we actually have a partial effect of a Faraday cage (cavity), Figure 1.25(b), and a decrease of the local field intensity.

CONCEPTUAL QUESTION 1.32 Measuring relative surface charge densities. Figure 1.29(a) shows a charged metallic body situated in free space. Consider the surface charge distribution on the body. A simple electrostatic device for measuring charge – an electroscope – is used; after touching the body at a point M by a small metallic ball on an isolated handle and then touching the electroscope by the same ball, the angle between the strips (leaves) of the electroscope indicates the magnitude of the surface charge density at the point M. Possible indications of the electroscope for points M_1, M_2, and M_3 on the body are those shown in

(A) Figure 1.29(b).

(B) Figure 1.29(c).

(C) Figure 1.29(d).

(D) Depends on whether the charge of the body is positive or negative.

(E) None of the cases shown is possible.

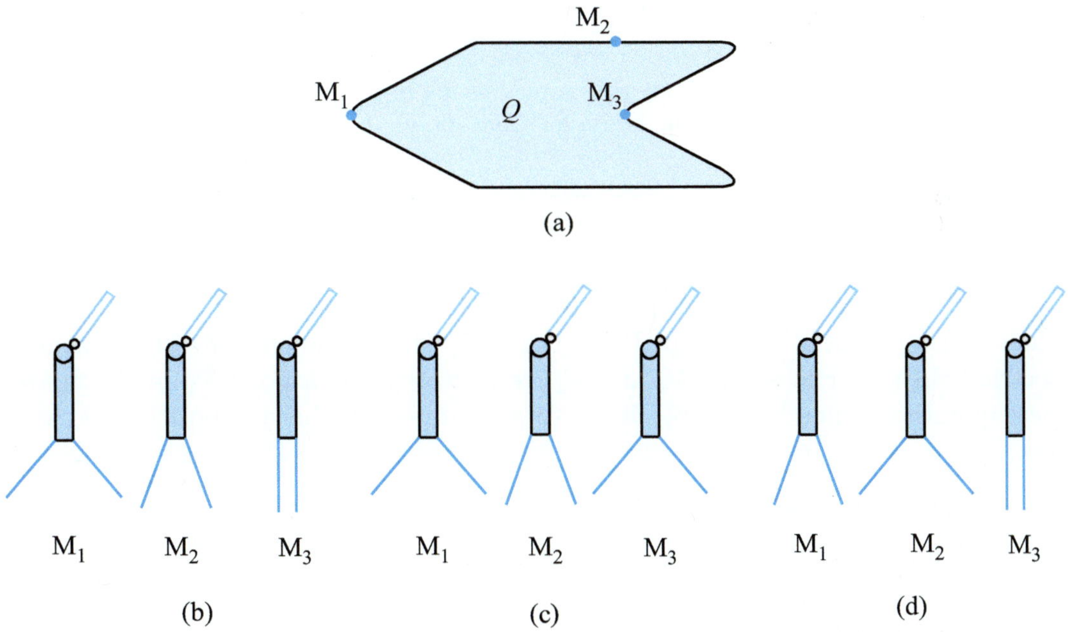

Figure 1.29 Measuring relative charge densities on different parts of a complex metallic surface – three offered sets of indications of an electroscope; for Conceptual Question 1.32.

CONCEPTUAL QUESTION 1.33 **Voltages between different points of a metallic body.** For the charged metallic body in Figure 1.30, compare the magnitudes of the voltage between points M_2 and M_3 and the voltage between points M_1 and M_3.

(A) The first one is larger.

(B) The first one is smaller.

(C) They are both zero.

(D) They are the same but nonzero.

(E) They are both infinite, so cannot be compared.

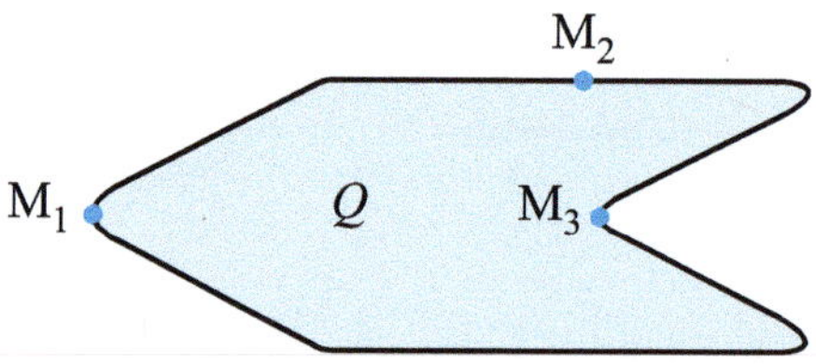

Figure 1.30 Charged metallic body of complex shape in free space, consideration of voltages between different points of the body; for Conceptual Question 1.33.

CONCEPTUAL QUESTION 1.34 **Electric field near curved charged surfaces.** A charged metallic body is situated in air, as shown in Figure 1.31. The electric potential at a point P_1 of the body and the magnitude of the nearby electric field vector in air are V_1 and E_1, respectively. At a point P_2 in the figure, these quantities equal V_2 and E_2. If the total charge of the body is positive, we have that

(A) $V_1 = V_2$ and $E_1 = E_2$.

(B) $V_1 = V_2$ and $E_1 < E_2$.

(C) $V_1 = V_2$ and $E_1 > E_2$.

(D) $V_1 < V_2$ and $E_1 < E_2$.

(E) $V_1 > V_2$ and $E_1 > E_2$.

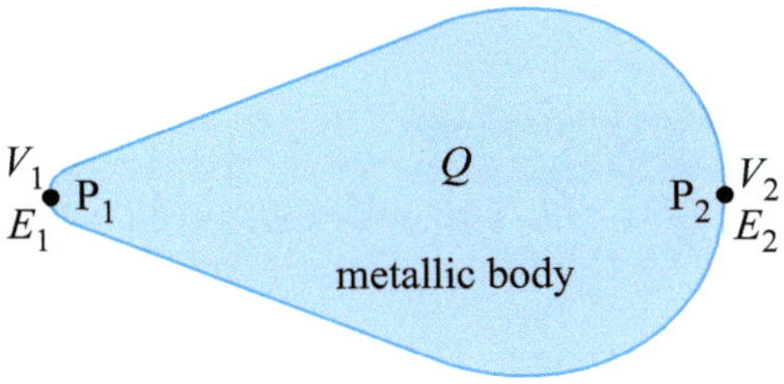

Figure 1.31 Charged metallic body; for Conceptual Question 1.34.

1.10 Image Theory

Often, electrostatic systems include charge configurations in the presence of grounded conducting planes. Examples are charged conductors near grounded metallic plates or large flat bodies, transmission lines in which one of the conductors is a ground plane (such as microstrip transmission lines), various charged objects above the earth's surface (power lines, charged clouds, charged airplanes, lightning rods, etc.), and so on. There is a very useful theory – illustrated in Figure 1.32 – by means of which we can remove the conducting plane from the system, in the analysis: namely, as far as the electrostatic field in the upper half-space is concerned, systems in Figures 1.32(a) and (b) are equivalent. This is so-called image theory, which, generalized to more than one point charge, i.e., to a (discrete or continuous) charge distribution, states that an arbitrary charge configuration above an infinite grounded conducting plane can be replaced by a new charge configuration in free space consisting of the original charge configuration itself and its negative image in the conducting plane. The equivalence is with respect to the electric field above the conducting plane, whose component due to the induced surface charge on the plane is equal to the field of the image.

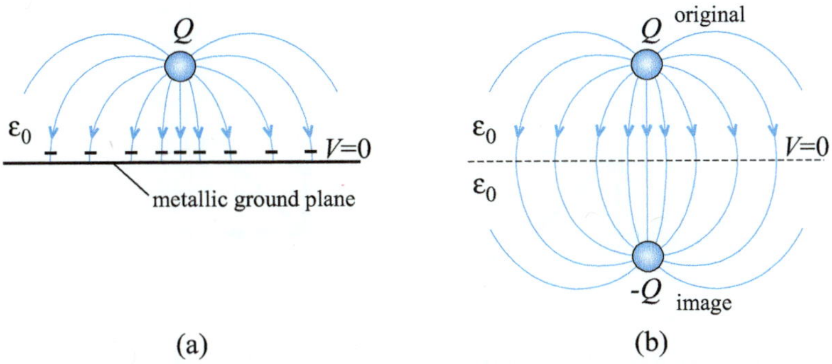

Figure 1.32 Image theory: systems (a) and (b) are equivalent with respect to the electric field in the upper half-space.

CONCEPTUAL QUESTION 1.35 **Force on a point charge above a conducting plane.** A point charge Q is situated in air at a height h above a grounded conducting plane. Relative to the plane, the electric force on this charge is

(A) always attractive (plane attracts the charge).

(B) always repulsive (plane repels the charge).

(C) attractive for Q positive and repulsive for Q negative.

(D) attractive for Q negative and repulsive for Q positive.

(E) always zero.

CONCEPTUAL QUESTION 1.36 **Image theory for electric dipoles above a conducting plane.** Consider an electric dipole (consisting of point charges $Q > 0$ and $-Q$ at a mutual distance d) with a moment $\mathbf{p} = Q\mathbf{d}$ ($\mathbf{d}$ being the position vector of Q with respect to $-Q$) placed in air above a horizontal conducting plane and its image in the plane such that the electric field in the upper half-space is the same in the original system and in the equivalent system using image theory. Images of three characteristic (vertical, horizontal, and oblique) dipoles are those shown in

(A) Figure 1.33(a).

(B) Figure 1.33(b).

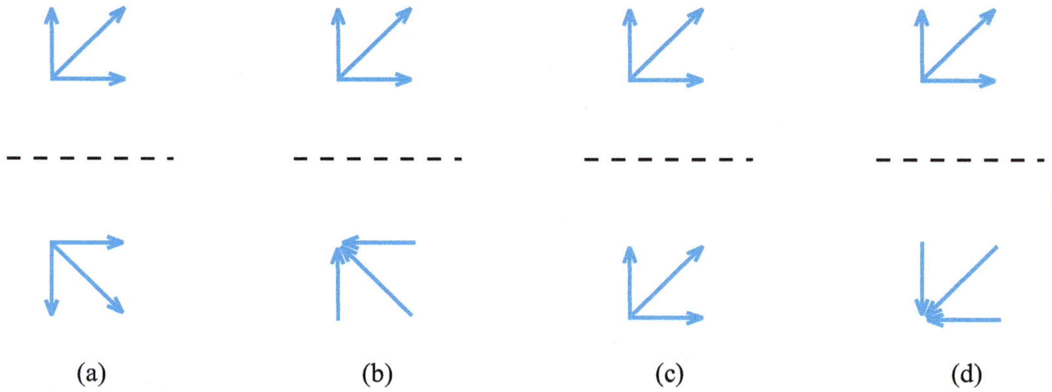

Figure 1.33 Image theory for three characteristic electric dipoles above a conducting plane: four offered sets of images (all vectors represent electric moments, $\mathbf{p}$, of dipoles); for Conceptual Question 1.36.

(C) Figure 1.33(c).

(D) Figure 1.33(d).

CONCEPTUAL QUESTION 1.37 **Image theory for a line charge parallel to a corner screen.** Illustrated in Figure 1.34 is the application of image theory to a line charge of density Q' in the presence of a 90° corner metallic screen in air. The densities of the three image line charges in the figure are

(A) $Q_1' = Q_2' = -Q'$ and $Q_3' = Q'$.

(B) $Q_1' = Q_2' = Q_3' = -Q'$.

(C) $Q_1' = Q_2' = Q'$ and $Q_3' = -Q'$.

(D) $Q_1' = Q_3' = -Q'$ and $Q_2' = Q'$.

(E) $Q_1' = Q_2' = -Q'$ and $Q_3' = 0$.

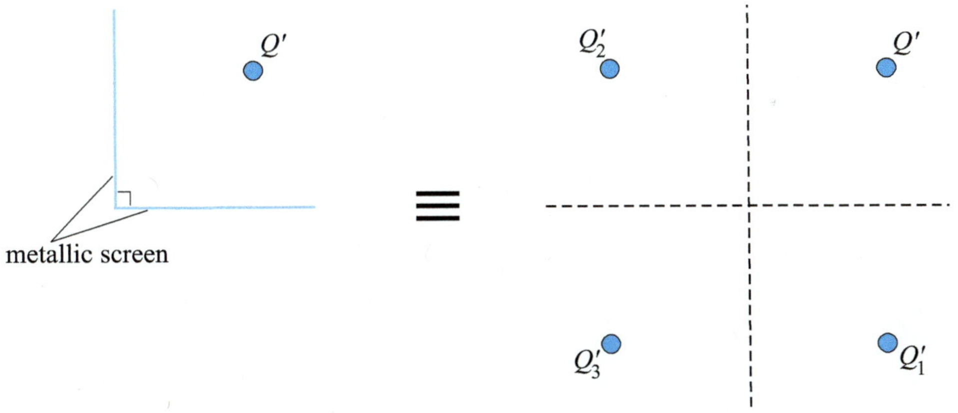

Figure 1.34 Image theory for a line charge of density Q' in the presence of a 90° corner metallic screen (cross section of the structure); for Conceptual Question 1.37.

---

2 ELECTROSTATIC FIELD IN DIELECTRICS

Introduction

Dielectrics or insulators are nonconducting materials, having very little free charges inside them (theoretically, perfect dielectrics have no free charges). However, another type of charge, called bound or polarization charges, exist in a polarized dielectric, as atoms and molecules in the dielectric behave like microscopic electric dipoles. In electrostatic systems containing both conductors and dielectrics, the equivalent electric-field sources are both free and bound charges, considered to reside in free space. By introducing the concept of dielectric permittivity, we are left, in turn, to deal with free charges in the system only, while the contribution of bound charges to the field is properly added through the permittivity. In continuation, we analyze capacitors and transmission lines, composed of both conductors and dielectrics, to evaluate and discuss their capacitance, energy, and breakdown characteristics, as a culmination of our study of the theory and applications of the electrostatic field.

2.1 Polarization of Dielectrics

Each atom or molecule in a dielectric is electrically neutral. For most dielectrics, centers of "gravity" of the positive and negative charges in an atom or molecule coincide – in the absence of the external electric field. When a dielectric is placed in an external field, of intensity $\mathbf{E}$, however, the positive and negative charges shift in opposite directions against their mutual attraction, and produce a small electric dipole. The moment of this equivalent dipole is given by $\mathbf{p} = Q\mathbf{d}$, where Q is the positive charge of the atom or molecule ($-Q$ is the negative charge), and $\mathbf{d}$ is the vector displacement of Q with respect to $-Q$. The charges are displaced from their equilibrium positions by forces [Eq. (1.3)] $\mathbf{F}_{e1} = Q\mathbf{E}$ and $\mathbf{F}_{e2} = -Q\mathbf{E}$, respectively. The displacement $\mathbf{d}$ is very small, on the order of the dimensions of atoms and molecules. The charges Q and $-Q$ are bound in place by atomic and molecular forces and can only shift positions slightly in response to the external field. So, the two charges in an equivalent small dipole cannot separate one from the other and migrate across the material in opposite directions run by the electric field. Hence, these charges are called bound charges (in contrast to free charges). Some dielectrics, such as water, have molecules with a permanent displacement between the centers of the positive and negative charge, so that they act as small electric dipoles even with no applied electric field. According to Figure 2.1, the torques (moments) of forces $\mathbf{F}_{e1}$ and $\mathbf{F}_{e2}$ with respect to the center of the dipole (point O) are $\mathbf{T}_1 = \mathbf{r}_1 \times \mathbf{F}_{e1}$ and $\mathbf{T}_2 = \mathbf{r}_2 \times \mathbf{F}_{e2}$, with $\mathbf{r}_1$ and $\mathbf{r}_2$ denoting the position vectors of Q and $-Q$ with respect to the dipole center. We notice that $\mathbf{r}_1 - \mathbf{r}_2 = \mathbf{d}$, and thus the resultant torque on the dipole turns out to be

$$\mathbf{T}_{\text{on dipole}} = \mathbf{T}_1 + \mathbf{T}_2 = Q(\mathbf{r}_1 - \mathbf{r}_2) \times \mathbf{E} = Q\mathbf{d} \times \mathbf{E}_{\text{ext}} = \mathbf{p} \times \mathbf{E} \,. \tag{2.1}$$

The process of making atoms and molecules in a dielectric behave as microscopic electric dipoles and orienting (rotating) the dipoles toward the direction of the external field, by means of the torque in Eq. (2.1), is termed the polarization of the dielectric, and bound charges are sometimes referred to as polarization charges. This process is extremely fast, practically instantaneous, and the dielectric in the new electrostatic state is said to be polarized or in the polarized state.

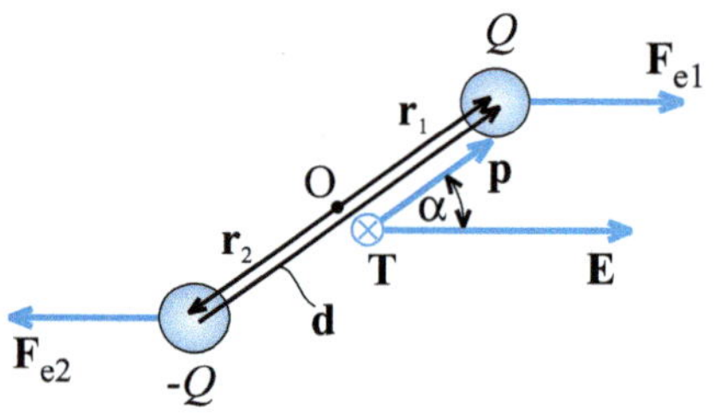

Figure 2.1 Small electric dipole representing an atom or molecule of a dielectric material in an external electric field.

To describe and analyze the polarized state of a dielectric, we introduce a macroscopic quantity called the polarization vector, defined by averaging dipole moments in an elementary volume dv as follows:

$$\mathbf{P} = \frac{(\sum \mathbf{p})_{\text{in } dv}}{dv} \quad \text{(polarization vector; unit: C/m}^2\text{)} \,. \tag{2.2}$$

Once the distribution of the vector $\mathbf{P}$ inside a dielectric body is known, the macroscopic distribution of volume and surface bound (polarization) charge densities [see Eqs. (1.5 and (1.6)], ρ_{p} and ρ_{ps}, throughout the body volume and over its surface, respectively, which represent macroscopic equivalents of all microscopic electric dipoles and their charges, are obtained as

$$\rho_{\text{p}} = -\operatorname{div} \mathbf{P} = -\nabla \cdot \mathbf{P} \,, \quad \rho_{\text{ps}} = \hat{\mathbf{n}} \cdot \mathbf{P} \quad (\hat{\mathbf{n}} \text{ outward normal})$$
$$\text{(bound volume and surface charge densities)} \,, \tag{2.3}$$

where $\hat{\mathbf{n}}$ stands for the normal unit vector on the surface oriented from the dielectric body outward. The electric field due to the polarized dielectric equals the field due to these volume and surface charges considered to be in a vacuum, as the rest of the material does not produce any field.

CONCEPTUAL QUESTION 2.1 **Torque on an electric dipole in a uniform electrostatic field.** An electric dipole, with charges Q and $-Q$, is placed in a uniform electrostatic field, whose intensity vector is $\mathbf{E}$, as shown in Figure 2.2. The field exerts a torque on the dipole for[1]

(A) cases (b) and (d) only.

(B) case (c) only.

(C) cases (b), (c), and (d) only.

(D) cases (a) and (e) only.

(E) cases (a), (b), (d), and (e) only.

(F) none of the cases shown.

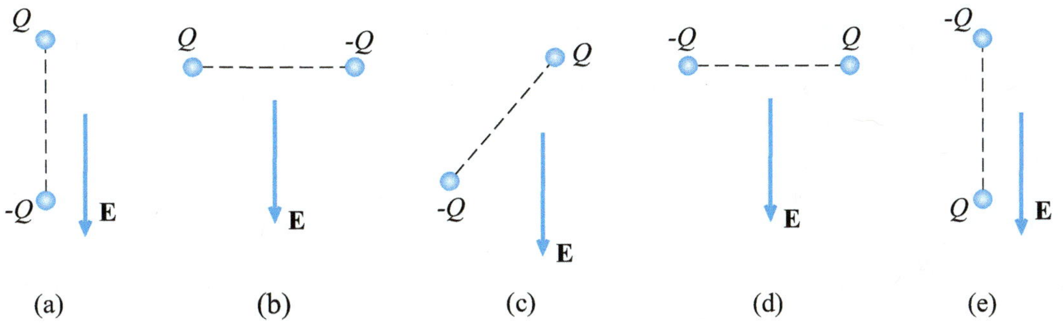

Figure 2.2 Electric dipole in a uniform electrostatic field of intensity $\mathbf{E}$ – five different positions of the dipole relative to the field lines; for Conceptual Question 2.1.

CONCEPTUAL QUESTION 2.2 **Electric dipole in a stable equilibrium.** How many positions of the dipole in Figure 2.2 (out of the five cases shown) represent stable equilibria?

(A) Zero.

(B) One.

(C) Two.

(D) Three.

[1] *For every conceptual question in this text, exactly one answer is correct.*

(E) Four.

CONCEPTUAL QUESTION 2.3 **Uniformly polarized dielectric.** If the polarization vector is the same at every point ($\mathbf{P}$ = const) inside a dielectric body (uniformly polarized dielectric), there is no macroscopic excess volume bound charge in the material.

(A) Always true.

(B) Sometimes true, sometimes false.

(C) Always false.

CONCEPTUAL QUESTION 2.4 **Nonuniformly polarized dielectric.** For a dielectric body with a polarization vector that varies ($\mathbf{P} \neq$ const) throughout the volume of the body (nonuniformly polarized dielectric), there is no macroscopic excess volume bound charge in the material.

(A) Always true.

(B) Sometimes true, sometimes false.

(C) Always false.

CONCEPTUAL QUESTION 2.5 **Excess surface bound charges.** On the entire surface of a polarized dielectric body, there always exist excess surface bound charges (there are ends of elementary electric dipoles pressed onto the surface that cannot be compensated by oppositely charged ends of neighboring dipoles), which are positive on some parts of the surface and negative on the other.

(A) True.

(B) False.

CONCEPTUAL QUESTION 2.6 **Total bound charge in a closed surface.** The total bound (polarization) charge $Q_{\mathrm{p}S}$ enclosed by an arbitrary imaginary closed surface S (as the one in Figure 1.14) that is situated (totally or partly) inside a polarized dielectric body

(A) is always positive.

(B) is always zero.

(C) is always negative.

(D) can be positive, negative, and zero.

CONCEPTUAL QUESTION 2.7 **Total bound charge of a dielectric body.** The total bound charge of a polarized dielectric body situated in free space is

(A) positive.

(B) zero.

(C) negative.

(D) Need more information.

CONCEPTUAL QUESTION 2.8 **Polarization in air.** There is no polarization ($\mathbf{P} = 0$) in air.

(A) True.

(B) False.

CONCEPTUAL QUESTION 2.9 **Uniformly polarized large dielectric slab, field inside.** An infinitely large dielectric slab of thickness $d = 2a$ is uniformly polarized throughout its volume such that the polarization vector, $\mathbf{P}$, is perpendicular to the faces (boundary surfaces) of the slab, as shown in Figure 2.3. The surrounding medium is air and there is no free charge in the dielectric. The electric field intensity vector (due to bound charges of the slab) at a point inside the slab is

(A) nonzero and has the same direction as $\mathbf{P}$.

(B) nonzero and is directed oppositely to $\mathbf{P}$.

(C) zero.

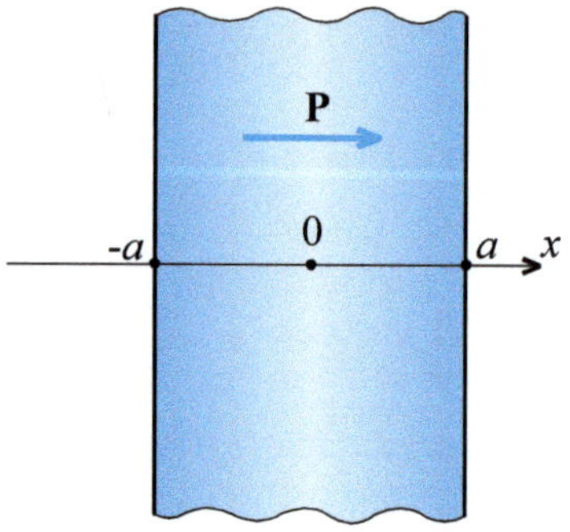

Figure 2.3 Infinitely large polarized dielectric slab, with a given polarization vector, $\mathbf{P}$, in air; for Conceptual Question 2.9.

CONCEPTUAL QUESTION 2.10 **Nonuniformly polarized slab, bound charge density.** If the polarization vector in the infinitely large dielectric slab shown in Figure 2.3 is perpendicular to the slab faces and given by $\mathbf{P} = P_0 x\,\hat{\mathbf{x}}/a$ $(-a < x < a)$, where P_0 is a positive constant, the bound volume charge density inside the slab, for $-a < x < a$, is

(A) a positive constant.

(B) a positive function of x.

(C) zero.

(D) a negative constant.

(E) a negative function of x.

(F) None of the above.

CONCEPTUAL QUESTION 2.11 **Nonuniformly polarized slab, field outside.** Assuming that the dielectric slab from Figure 2.3 is nonuniformly polarized such that the polarization vector is given by $\mathbf{P} = P(x)\,\hat{\mathbf{x}}$, $-a < x < a$, the electric field intensity vector outside the slab is zero for

(A) an odd function $P(x)$ only.

(B) an even function $P(x)$ only.

(C) any function $P(x)$.

(D) no function $P(x)$.

CONCEPTUAL QUESTION 2.12 **Flux of the electric field vector around a polarized body.** Consider a polarized dielectric body with no free charge, in free space. The outward flux of the electric field intensity vector, $\mathbf{E}$, through a closed surface S that completely encloses the body is

(A) positive.

(B) zero.

(C) negative.

(D) Depends on whether the dielectric is homogeneous or not.

(E) Depends on the distribution of the polarization vector in the body.

2.2 Generalized Gauss' Law and Permittivity

We now consider the most general electrostatic system containing both conductors and dielectrics. The equivalent field sources are now both free and bound charges, in free space, and we define a new vector quantity,

$$\mathbf{D} = \varepsilon_0 \mathbf{E} + \mathbf{P} \quad \text{(electric flux density vector; unit: C/m}^2\text{)} , \tag{2.4}$$

which is called the electric flux density vector (also known as the electric displacement vector or electric induction vector) and with which we have, in place of Eqs. (1.18)–(1.20), the corresponding forms of the generalized Gauss' law:

$$\oint_S \mathbf{D} \cdot \mathrm{d}\mathbf{S} = Q_S \, , \quad \oint_S \mathbf{D} \cdot \mathrm{d}\mathbf{S} = \int_v \rho \, \mathrm{d}v \, , \quad \nabla \cdot \mathbf{D} = \rho \quad \text{(generalized Gauss' law)} , \tag{2.5}$$

where Q_S is the total free charge enclosed by an arbitrary closed surface S, and ρ is the free charge density.

For linear dielectrics,

$$\mathbf{D} = \varepsilon \mathbf{E} \, , \quad \varepsilon = \varepsilon_\mathrm{r} \varepsilon_0 \quad \text{(permittivity of a linear dielectric; unit: F/m)} , \tag{2.6}$$

with ε being the permittivity and ε_r the relative permittivity of the medium ($\varepsilon_\mathrm{r} \geq 1$). The unit for ε is farad per meter (F/m), while ε_r is dimensionless. For nonlinear dielectrics, the relation between $\mathbf{D}$ and $\mathbf{E}$, $\mathbf{D} = \mathbf{D}(\mathbf{E})$, is nonlinear. In addition, a material is said to be homogeneous when its properties do not change from point to point in the region being considered, so in a linear homogeneous dielectric, ε is a constant independent of spatial coordinates. Otherwise, the material is inhomogeneous [e.g., $\varepsilon = \varepsilon(x, y, z)$ in the region].

CONCEPTUAL QUESTION 2.13 **Flux of the electric field intensity vector.** The polarization vector, $\mathbf{P}$, and free volume charge density, ρ, are known at every point of a dielectric body. The expression for the flux of the electric field intensity vector through a closed surface S situated entirely inside the body is given by

(A) $\oint_S \mathbf{E} \cdot \mathrm{d}\mathbf{S} = \int_v \rho \, \mathrm{d}v$, where v denotes the volume enclosed by S.

(B) $\oint_S \mathbf{E} \cdot \mathrm{d}\mathbf{S} = \left(\int_v \rho \, \mathrm{d}v\right) / \varepsilon_0$.

(C) $\oint_S \mathbf{E} \cdot \mathrm{d}\mathbf{S} = \varepsilon_0 \oint_S \mathbf{P} \cdot \mathrm{d}\mathbf{S}$.

(D) $\oint_S \mathbf{E} \cdot \mathrm{d}\mathbf{S} = \left(\int_v \rho \, \mathrm{d}v - \oint_S \mathbf{P} \cdot \mathrm{d}\mathbf{S}\right) / \varepsilon_0$.

(E) $\oint_S \mathbf{E} \cdot \mathrm{d}\mathbf{S} = 0$.

CONCEPTUAL QUESTION 2.14 **Uniform field in a dielectric.** There is a uniform electric field ($\mathbf{E} = \text{const}$) in a certain dielectric region. The free volume charge density is ρ. The bound volume charge density amounts to

(A) $\rho_{\mathrm{p}} = 0$.

(B) $\rho_{\mathrm{p}} = \rho$.

(C) $\rho_{\mathrm{p}} = -\rho$.

(D) $\rho_{\mathrm{p}} = \varepsilon_0 \nabla \cdot \mathbf{E}$.

(E) $\rho_{\mathrm{p}} = \varepsilon_0 \nabla \cdot \mathbf{E} + \rho$.

CONCEPTUAL QUESTION 2.15 **Volume charges in a homogeneous linear dielectric.** In a homogeneous linear dielectric of relative permittivity ε_{r}, the free volume charge density is ρ. The bound volume charge density equals

(A) $\rho_{\mathrm{p}} = 0$.

(B) $\rho_\mathrm{p} = \rho$.

(C) $\rho_\mathrm{p} = \rho/\varepsilon_\mathrm{r}$.

(D) $\rho_\mathrm{p} = -\rho/\varepsilon_\mathrm{r}$.

(E) $\rho_\mathrm{p} = -(\varepsilon_\mathrm{r} - 1)\rho/\varepsilon_\mathrm{r}$.

CONCEPTUAL QUESTION 2.16 **Charge-free homogeneous medium.** In a homogeneous linear medium with no free volume charge, there is no bound volume charge either.

(A) True.

(B) False.

CONCEPTUAL QUESTION 2.17 **Poisson's equation for an inhomogeneous medium.** For an inhomogeneous dielectric region of permittivity ε, the following second-order differential equation (Poisson's equation) relating the electric potential, V, to the free volume charge density, ρ, holds true:

(A) $\nabla^2 V = \rho/\varepsilon$.

(B) $\nabla^2 V = -\rho/\varepsilon$.

(C) $\nabla \cdot (\nabla V) = \rho$.

(D) $\nabla \cdot (\varepsilon \nabla V) = -\rho$.

(E) $\nabla V = -\rho/\varepsilon$.

CONCEPTUAL QUESTION 2.18 **Charge density from 1-D electric potential distribution.** In an electrostatic system filled with a homogeneous dielectric, of permittivity ε, the potential V is a quadratic function of x (and does not depend on other coordinates). The charge density, ρ, in the system is

(A) a linear function of x.

(B) a quadratic function of x.

(C) a cubic function of x.

(D) a nonzero constant.

(E) zero.

CONCEPTUAL QUESTION 2.19 **Permittivity versus field intensity in a nonlinear dielectric.** The polarization properties and permittivity, ε, of a nonlinear dielectric material depend on the electric field intensity, E, in the material.

(A) True.

(B) False.

CONCEPTUAL QUESTION 2.20 **Some vector relationships for a nonlinear dielectric.** Consider a metallic structure filled with a nonlinear dielectric. The electric field intensity, electric flux density, and polarization vectors in the dielectric are $\mathbf{E}$, $\mathbf{D}$, and $\mathbf{P}$, respectively. At any point in the dielectric and for any field intensity, the following vectors are linearly proportional to $\mathbf{E}$:

(A) $\mathbf{D}$ only.

(B) $\mathbf{P}$ only.

(C) Both $\mathbf{D}$ and $\mathbf{P}$.

(D) $\mathbf{D} - \mathbf{P}$.

(E) None of the above.

2.3 Dielectric–Dielectric Boundary Conditions

Let us consider a dielectric–dielectric boundary surface, shown in Figure 2.4. Let $\mathbf{E}_1$ and $\mathbf{D}_1$ be, respectively, the electric field intensity vector and electric flux density vector close to the boundary in medium 1, whereas $\mathbf{E}_2$ and $\mathbf{D}_2$ stand for the same quantities in medium 2. Equations (1.13) and (2.5) result in the following boundary conditions for tangential components of **E** and normal components of **D** on the boundary:

$$E_{1t} = E_{2t} , \quad D_{1n} - D_{2n} = \rho_s \quad \text{(dielectric–dielectric boundary conditions)} , \tag{2.7}$$

where the normal components are defined with respect to the unit normal $\hat{\mathbf{n}}$ directed from region 2 to region 1 (Figure 2.4), and ρ_s is the free surface charge density that may exist on the surface. In the absence of charge,

$$D_{1n} = D_{2n} \quad \longrightarrow \quad \varepsilon_1 E_{1n} = \varepsilon_2 E_{2n} \quad (\rho_s = 0) . \tag{2.8}$$

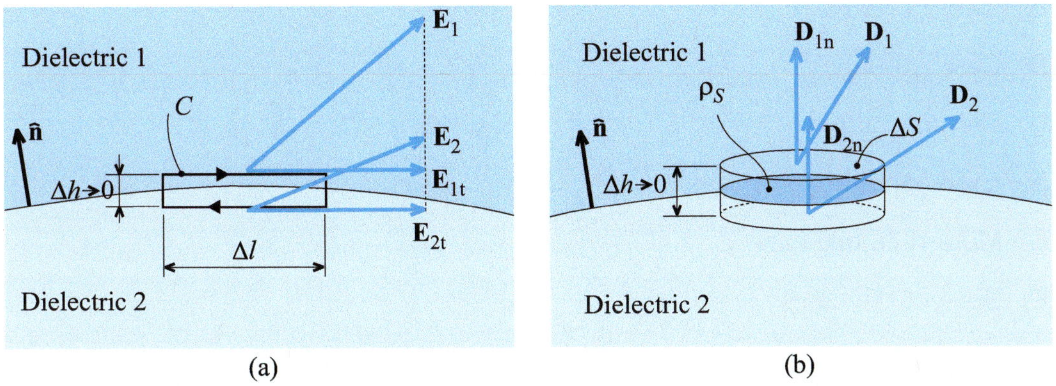

Figure 2.4 Dielectric–dielectric boundary surface: boundary conditions for (a) tangential components of **E** and (b) normal components of **D**.

With α_1 and α_2 denoting the angles that field lines in region 1 and region 2 make with the normal to the boundary interface, **n**, as shown in Figure 2.5, we have $\tan\alpha_1 = E_{1t}/E_{1n}$ and $\tan\alpha_2 = E_{2t}/E_{2n}$, and then dividing these tangents and using Eqs. (2.7) and (2.8) we obtain the law of refraction of the electric field lines

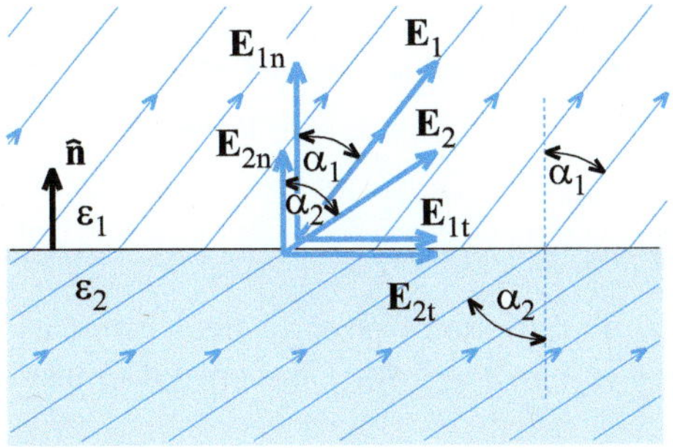

Figure 2.5 Refraction of electric field lines at a dielectric–dielectric interface.

at a dielectric–dielectric boundary that is free of charge ($\rho_s = 0$):

$$\frac{\tan\alpha_1}{\tan\alpha_2} = \frac{\varepsilon_1}{\varepsilon_2} \quad \text{(law of refraction of electric field lines)}\,. \tag{2.9}$$

CONCEPTUAL QUESTION 2.21 **Boundary conditions at a dielectric–dielectric interface.** Consider a boundary surface between two dielectric media, with relative permittivities $\varepsilon_{r1} = 4$ and $\varepsilon_{r2} = 2$, respectively. Assuming that there is no surface charge on the boundary, which of the cases shown in Figure 2.6 represent possible electric field intensity vectors on the two sides of the boundary?

(A) Case (a) only.

(B) Case (b) only.

(C) Case (c) only.

(D) Case (d) only.

(E) More than one case.

(F) None of the cases.

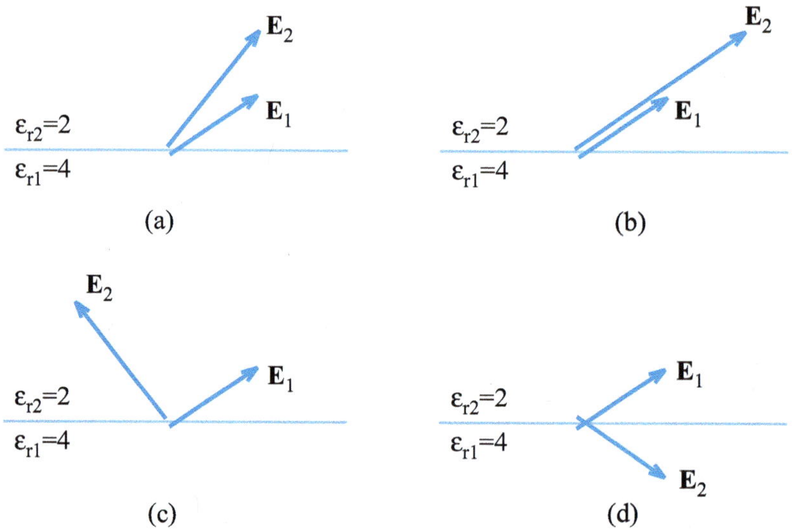

Figure 2.6 Four offered combinations of electric field intensity vectors on two sides of a dielectric–dielectric interface ($\varepsilon_{r1} = 2\varepsilon_{r2}$); for Conceptual Question 2.21.

CONCEPTUAL QUESTION 2.22 **Refraction of electrostatic field lines.** Figure 2.7 shows lines of an electrostatic field near a dielectric–dielectric boundary that is free of charge ($\rho_s = 0$). Which of the following is a possible combination of the two media?

(A) Medium 1 is PVC ($\varepsilon_{r1} = 2.7$) and medium 2 is mica ($\varepsilon_{r2} = 5.4$).

(B) Medium 1 is mica ($\varepsilon_{r1} = 5.4$) and medium 2 is PVC ($\varepsilon_{r2} = 2.7$).

(C) Medium 1 is water ($\varepsilon_{r1} = 81$) and medium 2 is air ($\varepsilon_{r2} = 1$).

(D) Medium 1 is air ($\varepsilon_{r1} = 1$) and medium 2 is water ($\varepsilon_{r2} = 81$).

(E) Both combinations (A) and (B) above.

(F) Both combinations (C) and (D) above.

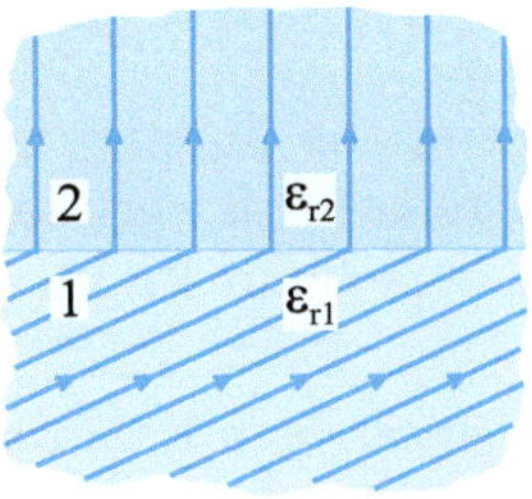

Figure 2.7 Refraction of electric field lines at an interface between two dielectric media; for Conceptual Question 2.22.

2.4 Analysis of Capacitors with Homogeneous Dielectrics

Figure 2.8 shows a capacitor – consisting of two metallic bodies (electrodes) embedded in a dielectric, and charged with equal charges of opposite polarities, Q and $-Q$. In linear capacitors (filled with linear dielectrics), Q is linearly proportional to the capacitor voltage, which, in turn, is evaluated as (Figure 2.8)

$$V = \int_1^2 \mathbf{E} \cdot \mathrm{d}\mathbf{l} \, . \tag{2.10}$$

Based on this proportionality, the capacitance of the capacitor is defined as

$$C = \frac{Q}{V} \quad \text{(capacitance of a capacitor; unit: F)} \, . \tag{2.11}$$

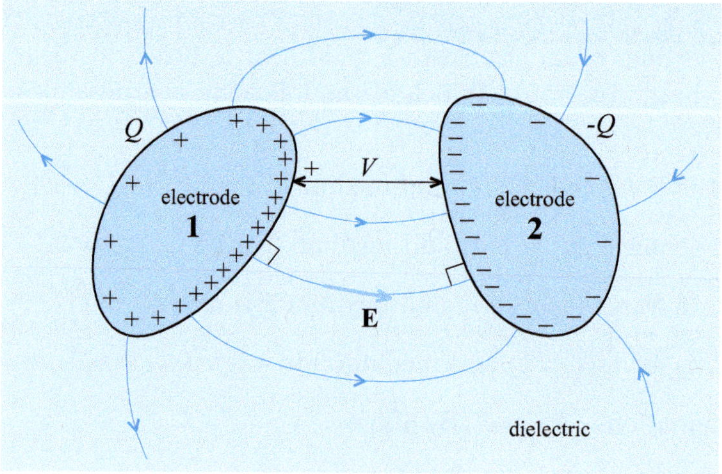

Figure 2.8 Capacitor.

It is always positive ($C > 0$), and the unit is the farad (F). For two-conductor transmission lines (two-body systems with very long conductors of uniform cross section), we define the capacitance per unit length of the line,

$$C' = C_{\text{p.u.l.}} = \frac{C}{l} = \frac{Q'}{V} \quad \text{(capacitance p.u.l. of a transmission line; unit: F/m)} \,, \tag{2.12}$$

where C, l, and Q' are the total capacitance, length, and charge per unit length of the structure [see Eq. (1.8)].

Shown in Figure 2.9 are some of the most frequently used capacitors and transmission lines. For a spherical capacitor, Figure 2.9(a), applying the generalized Gauss' law in integral form, Eqs. (2.5), and using Eqs. (2.10) and (2.11), we obtain

$$E(r) = \frac{Q}{4\pi\varepsilon r^2} \quad (a < r < b) \quad \longrightarrow \quad V = \int_a^b E(r)\,\mathrm{d}r = \frac{Q}{4\pi\varepsilon}\left(\frac{1}{a} - \frac{1}{b}\right)$$

$$\longrightarrow \quad C = \frac{Q}{V} = \frac{4\pi\varepsilon ab}{b-a} \quad \text{(capacitance of a spherical capacitor)} \,. \tag{2.13}$$

In an analogous fashion, the capacitance per unit length [Eq. (2.12)] of a coaxial cable, Figure 2.9(b), is found to be $C' = 2\pi\varepsilon/\ln(b/a)$, while that of a thin symmetrical two-wire transmission line (with $d \gg a$), Figure 2.9(c), equals $C' = \pi\varepsilon/\ln(d/a)$.

To analyze a parallel-plate capacitor with the fringing effects neglected (we assume that the electric field in the dielectric is uniform and that there is no field outside the dielectric), we apply the generalized Gauss' law to the rectangular Gaussian surface shown in Figure 2.9(d), which results in

$$C = \varepsilon\frac{S}{d} \quad \text{(capacitance of a parallel-plate capacitor, fringing neglected)} \,. \tag{2.14}$$

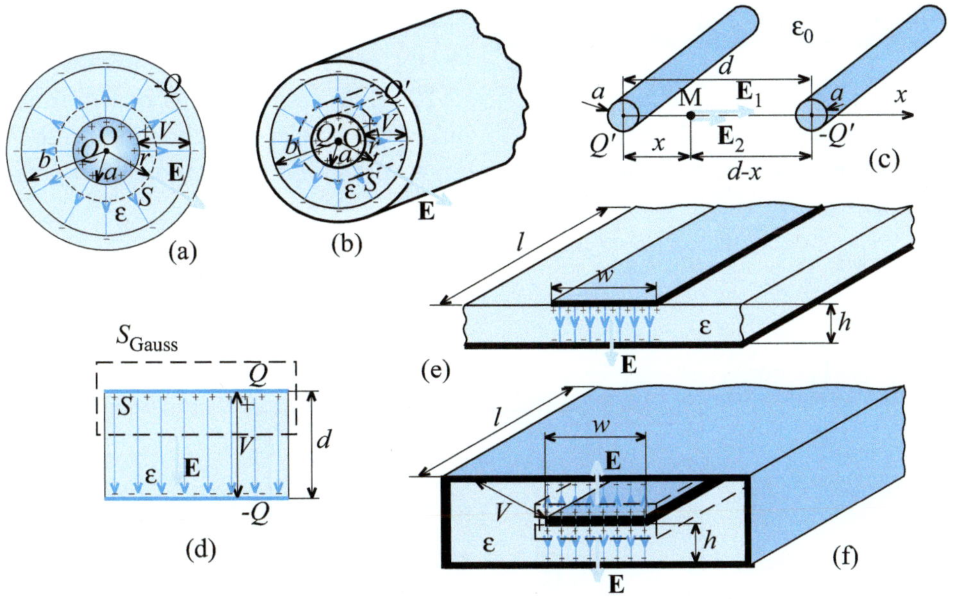

Figure 2.9 Examples of capacitors and transmission lines: (a) spherical capacitor, (b) coaxial cable, (c) thin two-wire transmission line, (d) parallel-plate capacitor, (e) microstrip transmission line, and (f) strip transmission line; in cases (d)–(f), fringing effects neglected.

By a similar token, the capacitance p.u.l. of a microstrip transmission line, Figure 2.9(e), comes out to be $C' = \varepsilon w/h$, and that of a strip transmission line, Figure 2.9(f), amounts to $C' = 2\varepsilon w/h$ (with fringing neglected in both cases).

CONCEPTUAL QUESTION 2.23 **Change of voltage due to a change of flux density.** The voltage between the terminals of a charged capacitor with a linear dielectric equals V. If the electric flux density at every point in the dielectric is doubled, the voltage of the capacitor in the new electrostatic state equals

(A) $V/2$.

(B) V.

(C) $2V$.

(D) $4V$.

(E) None of the above.

CONCEPTUAL QUESTION 2.24 **Capacitance of a "cubical" capacitor.** Consider a "cubical" capacitor, which consists of two concentric hollow metallic cubes with thin walls, as shown in Figure 2.10. The edge lengths of the inner and outer conductors are $a = 5$ cm and $b = 15$ cm, respectively, and the medium between the conductors is air. If a is made twice as large and b kept the same, the capacitance of the capacitor

(A) increases.

(B) decreases.

(C) remains the same.

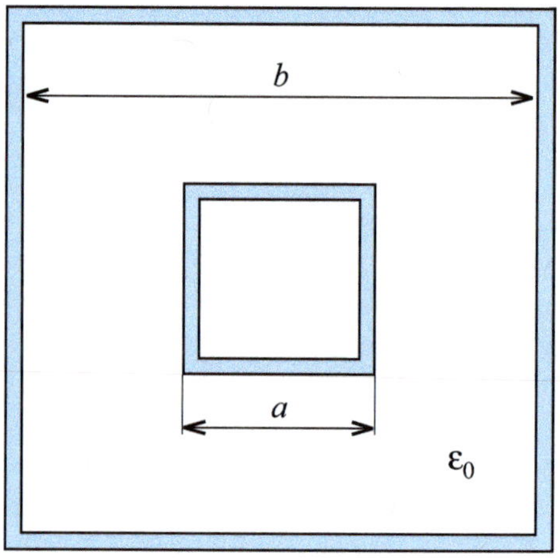

Figure 2.10 "Cubical" air-filled capacitor; for Conceptual Question 2.24.

CONCEPTUAL QUESTION 2.25 **"Cubical" capacitor versus isolated cube.** The capacitance of the "cubical" capacitor in Figure 2.10 is

(A) larger than

(B) the same as

(C) smaller than

the capacitance of an isolated cubical conductor of edge length a in air (which can be regarded as the inner electrode of a "cubical" capacitor with $b \to \infty$).

CONCEPTUAL QUESTION 2.26 **Change of field intensity due to a change of dielectric.** An air-filled parallel-plate capacitor is charged and its terminals left open. A dielectric slab with relative permittivity $\varepsilon_r = 2$ is then inserted so as to just fill the space between the plates, without touching the plates by hands or any other conducting body. As a result, the electric field intensity between the plates

(A) increases.

(B) decreases.

(C) remains the same.

CONCEPTUAL QUESTION 2.27 Change of dielectric with a source connected. An air-filled parallel-plate capacitor is attached to a voltage source. While the source is still connected, the space between the plates is completely filled by a dielectric slab ($\varepsilon_r = 2$). In the new electrostatic state, the electric field intensity is

(A) larger than

(B) the same as

(C) smaller than

that in the air-filled capacitor.

CONCEPTUAL QUESTION 2.28 Arbitrarily shaped capacitor with and without dielectric. A capacitor with electrodes of arbitrary shapes has a homogeneous dielectric of relative permittivity ε_r ($\varepsilon_r > 1$). If the dielectric is removed (without changing the shapes of the electrodes), the capacitance of the capacitor

(A) increases.

(B) decreases.

(C) remains the same.

(D) The answer depends on the actual shapes of the electrodes.

CONCEPTUAL QUESTION 2.29 Parallel-plate capacitor with an inserted metallic slab. The charges of the plates of an air-filled parallel-plate capacitor are Q and $-Q$. The capacitor terminals are open and the fringing effects can be neglected. An uncharged metallic slab, the thickness of which is smaller than the plate separation, is next inserted between the plates, as shown in Figure 2.11. The voltage between the capacitor plates is now

(A) larger than

(B) the same as

(C) smaller than

before the slab was inserted.

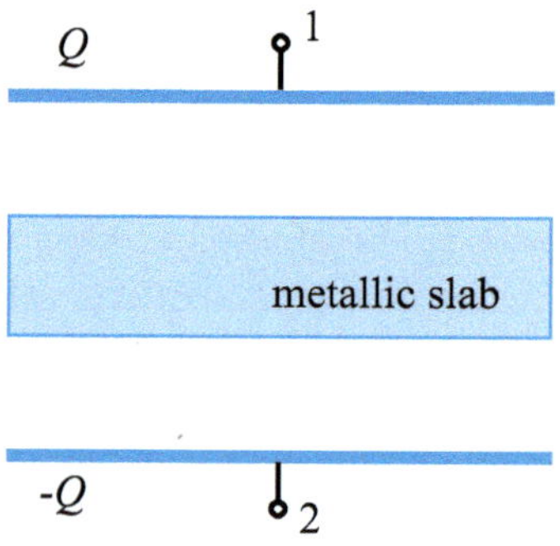

Figure 2.11 Parallel-plate capacitor with a metallic slab inserted between the plates; for Conceptual Question 2.29.

CONCEPTUAL QUESTION 2.30 **Capacitor with a slab in two configurations.** Consider the capacitor in Figure 2.11 with the following two modifications.

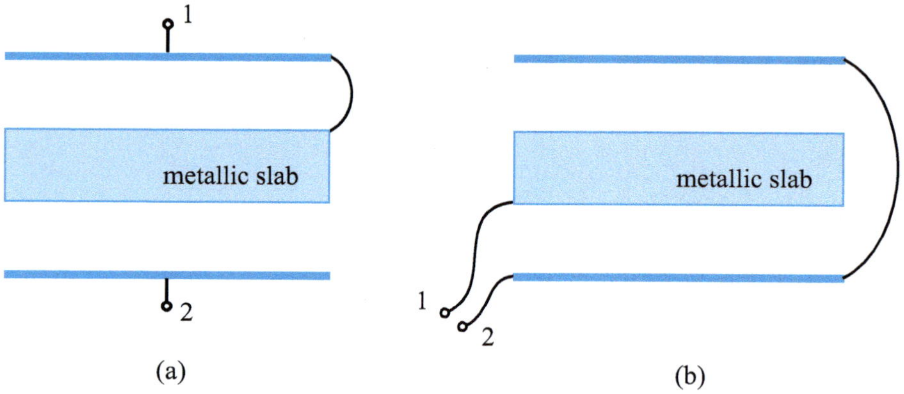

Figure 2.12 The capacitor from Figure 2.11 with the slab galvanically connected to the upper plate (a) and the plates galvanically connected together (b); for Conceptual Question 2.30.

In case (a), the slab is galvanically connected to the upper plate [Figure 2.12(a)]. In case (b), the plates are galvanically connected together [Figure 2.12(b)]. The capacitance between the terminals 1 and 2 is higher for

(A) case (a).

(B) case (b).

(C) The two capacitances are equal.

CONCEPTUAL QUESTION 2.31 **Capacitor consisting of seven parallel plates.** The capacitor shown in Figure 2.13 consists of seven parallel square metallic plates of edge lengths a and separations between all adjacent plates d ($d \ll a$). The medium is air. With C designating the capacitance of an air-filled parallel-plate capacitor of plate area a^2 and plate separation d, the capacitance of the capacitor in Figure 2.13 equals

(A) $C/6$.

(B) $C/3$.

(C) C.

(D) $3C$.

(E) $6C$.

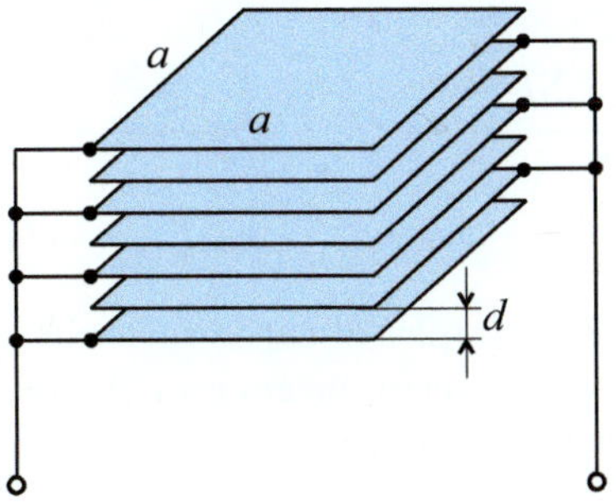

Figure 2.13 Capacitor consisting of seven parallel metallic plates in air; for Conceptual Question 2.31.

2.5 Analysis of Capacitors with Inhomogeneous Dielectrics

Often we deal with capacitors and transmission lines containing inhomogeneous dielectrics. To illustrate the analysis of such electrostatic systems, consider the two parallel-plate capacitors with piece-wise homogeneous dielectrics shown in Figure 2.14 and assume that the fringing effects are negligible in both cases. The field analysis of the capacitor with two dielectric layers in Figure 2.14(a), based on applying the generalized Gauss' law [Eqs. (2.5)] to a rectangular closed surface enclosing the plate charged with Q, with the right-hand side positioned in either one of the dielectrics, gives

$$D_1 = D_2 = D = \frac{Q}{S} \quad \longrightarrow \quad E_1 = \frac{D}{\varepsilon_1} = \frac{Q}{\varepsilon_1 S} \quad \text{and} \quad E_2 = \frac{D}{\varepsilon_2} = \frac{Q}{\varepsilon_2 S}$$

$$\longrightarrow \quad V = E_1 d_1 + E_2 d_2 = \frac{Q}{S}\left(\frac{d_1}{\varepsilon_1} + \frac{d_2}{\varepsilon_2}\right) \quad \longrightarrow \quad C_\mathrm{a} = \frac{Q}{V} = \frac{\varepsilon_1 \varepsilon_2 S}{\varepsilon_2 d_1 + \varepsilon_1 d_2} \,. \tag{2.15}$$

Alternatively, C_a can be obtained as the equivalent capacitance of two capacitors (with homogeneous dielectrics) in series, using Eq. (2.14) twice,

$$C_\mathrm{a} = \frac{C_1 C_2}{C_1 + C_2}, \quad \text{where} \quad C_1 = \varepsilon_1 \frac{S}{d_1} \quad \text{and} \quad C_2 = \varepsilon_2 \frac{S}{d_2} \,. \tag{2.16}$$

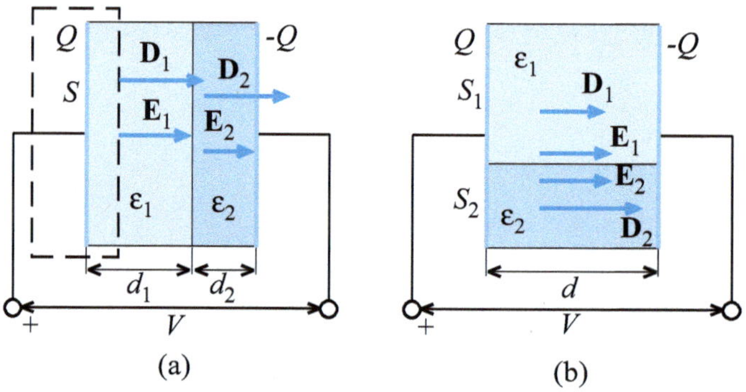

Figure 2.14 Two characteristic examples of capacitors with inhomogeneous dielectrics: (a) parallel-plate capacitor with two dielectric layers and (b) parallel-plate capacitor with two dielectric sectors (fringing neglected in both cases).

On the other side, the field analysis of the capacitor in Figure 2.14(b) can be carried out as

$$E_1 = E_2 = E = \frac{V}{d} \quad \longrightarrow \quad D_1 = \varepsilon_1 E = \frac{\varepsilon_1 V}{d} \quad \text{and} \quad D_2 = \varepsilon_2 E = \frac{\varepsilon_2 V}{d}$$

$$\longrightarrow \quad Q = D_1S_1 + D_2S_2 = (\varepsilon_1 S_1 + \varepsilon_2 S_2)\,\frac{V}{d} \quad \longrightarrow \quad C_\mathrm{b} = \frac{Q}{V} = \frac{\varepsilon_1 S_1 + \varepsilon_2 S_2}{d}\,, \tag{2.17}$$

whereas the alternative approach using the concept of the equivalent capacitance of two capacitors in parallel results in

$$C_\mathrm{b} = C_1 + C_2\,, \quad \text{where} \quad C_1 = \varepsilon_1\frac{S_1}{d} \quad \text{and} \quad C_2 = \varepsilon_2\frac{S_2}{d}\,. \tag{2.18}$$

CONCEPTUAL QUESTION 2.32 **Parallel-plate capacitor with an inserted dielectric slab.** An unpolarized dielectric slab of permittivity ε ($\varepsilon > \varepsilon_0$) is inserted between the plates of an open-circuited air-filled parallel-plate capacitor, which were previously charged with Q and $-Q$, respectively, as shown in Figure 2.15. The fringing effects can be neglected. The voltage between the capacitor plates is now

(A) larger than

(B) the same as

(C) smaller than

before the slab was inserted.

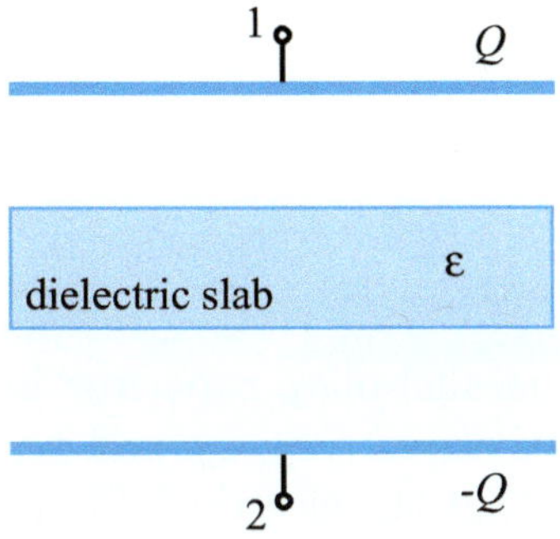

Figure 2.15 Parallel-plate capacitor with a dielectric slab inserted between the plates; for Conceptual Question 2.32.

CONCEPTUAL QUESTION 2.33 **Electric field intensity in a spherical capacitor.** Consider a charged spherical capacitor with a linear dielectric. Designating by r the radial distance of an arbitrary point from the capacitor center, the magnitude of the electric field intensity vector, $\mathbf{E}$, between the capacitor electrodes is

(A) inversely proportional to r.

(B) inversely proportional to r^2.

(C) inversely proportional to $\ln r$.

(D) uniform.

(E) Need more information.

CONCEPTUAL QUESTION 2.34 **Electric flux density in a spherical capacitor.** The magnitude of the electric flux density vector, $\mathbf{D}$, between the electrodes of a charged spherical capacitor with a linear dielectric and with r standing for the radial distance from the capacitor center is

(A) inversely proportional to r.

(B) inversely proportional to r^2.

(C) inversely proportional to $\ln r$.

(D) uniform.

(E) Need more information.

CONCEPTUAL QUESTION 2.35 **Parallel-plate capacitor with a dielectric in four parts.** A parallel-plate capacitor is filled with a dielectric composed of four parts, of permittivities ε_1, ε_2, ε_3, and ε_4, as in Figure 2.16. Assuming that the capacitor is charged, that the electric field in each of the pieces is uniform, and that no surface free charges exist on dielectric–dielectric boundaries, consider the following four statements: (a) If $\varepsilon_1 = \varepsilon_2$ and $\varepsilon_3 = \varepsilon_4$, then vector $\mathbf{E}$ is the same in all the pieces. (b) If $\varepsilon_1 = \varepsilon_2$ and $\varepsilon_3 = \varepsilon_4$, then vector $\mathbf{D}$ is the same in all the pieces. (c) If $\varepsilon_1 = \varepsilon_3$ and $\varepsilon_2 = \varepsilon_4$, then vector $\mathbf{E}$ is the same in all the pieces. (d) If $\varepsilon_1 = \varepsilon_3$ and $\varepsilon_2 = \varepsilon_4$, then vector $\mathbf{D}$ is the same in all the pieces. Which of the above statements are true?

(A) Statements (a) and (b).

(B) Statements (c) and (d).

(C) Statements (a) and (d).

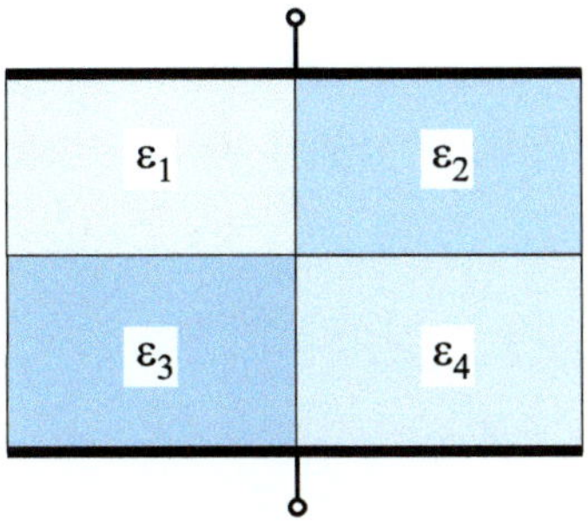

Figure 2.16 Parallel-plate capacitor filled with four dielectric pieces; for Conceptual Question 2.35.

(D) Statements (b) and (c).

(E) None of the statements.

CONCEPTUAL QUESTION 2.36 Coaxial cable with a radial variation of permittivity. A coaxial cable is filled with a continuously inhomogeneous dielectric and connected to a voltage source. The permittivity of the dielectric is a function of the radial distance r from the cable axis and no other coordinates. Consider vectors $\mathbf{D}$ and $\mathbf{E}$ in the cable. The way in which each of the vectors varies throughout the dielectric is the same as in the same cable if air-filled for

(A) both vectors.

(B) vector $\mathbf{D}$ only.

(C) vector $\mathbf{E}$ only.

(D) none of the vectors.

CONCEPTUAL QUESTION 2.37 Coaxial cable with four dielectric sectors. Consider a coaxial cable with a dielectric in the form of four 90° sectors with different permittivities, the cross section of which is shown in Figure 2.17, connected to a voltage source. The way in which each of the vectors $\mathbf{D}$ and $\mathbf{E}$ varies throughout the dielectric of the cable is the same as in the same cable if air-filled for

(A) both vectors.

(B) vector **D** only.

(C) vector **E** only.

(D) none of the vectors.

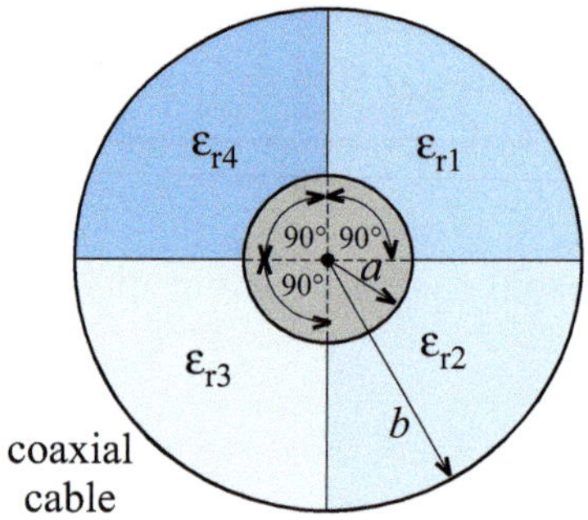

Figure 2.17 Cross section of a coaxial cable with a four-piece dielectric; for Conceptual Question 2.37.

CONCEPTUAL QUESTION 2.38 **One hundred capacitors connected in series/parallel.** We have a set of 100 capacitors of arbitrary geometries, with different capacitances $C_1, C_2, \ldots, C_{100}$. Let C_{series} and C_{parallel} be the equivalent total capacitances of the capacitors connected in series [Figure 2.18(a)] and parallel [Figure 2.18(b)], respectively. Comparing these two equivalent capacitances, we have

(A) $C_{\text{series}} < C_{\text{parallel}}$.

(B) $C_{\text{series}} = C_{\text{parallel}}$.

(C) $C_{\text{series}} > C_{\text{parallel}}$.

(D) Depends on the geometries of the individual capacitors and nothing else.

(E) Depends on the relationships between particular values of $C_1, C_2, \ldots, C_{100}$.

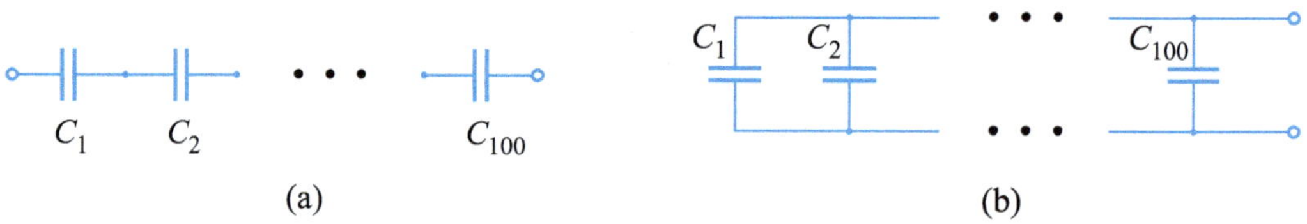

Figure 2.18 Set of 100 arbitrary capacitors connected in series (a) and parallel (b); for Conceptual Question 2.38.

2.6 Energy of an Electrostatic System

Every charged capacitor and every system of charged conducting bodies contains a certain amount of energy, which, by the principle of conservation of energy, equals the work done in the process of charging the system. This energy is called the electric energy and is related to the charges and potentials of the conducting bodies in the system. For a linear capacitor (one filled with a linear dielectric) of capacitance C, charge Q, and voltage V [see Figure 2.8 and Eq. (2.11)], the energy is given by

$$W_{\mathrm{e}} = \frac{1}{2}CV^2 = \frac{1}{2}QV = \frac{Q^2}{2C} \quad \text{(energy of a capacitor; unit: J)} . \tag{2.19}$$

The unit for the electric energy is the joule (J).

As the charges in an electrostatic system are sources of the electric field, it turns out that the energy of the system can be expressed also in terms of the electric field intensity throughout the system. This leads to an assumption that the electric energy is actually localized in the electric field, and therefore in the dielectric (which can be air and a vacuum) between the conductors of an electrostatic system (e.g., a capacitor). Quantitatively, the concentration (density) of energy at specific locations in the dielectric is proportional to the local electric field intensity, E, squared and can be determined as [see Eq. (2.6)]

$$w_{\mathrm{e}} = \frac{1}{2}\varepsilon E^2 = \frac{1}{2}ED = \frac{D^2}{2\varepsilon} \quad \text{(electric energy density; unit: J/m}^3\text{)} . \tag{2.20}$$

The total electric energy of the system is now obtained as $W_{\mathrm{e}} = \int_v w_{\mathrm{e}}\,\mathrm{d}v$, where v denotes the volume of the system dielectric.

CONCEPTUAL QUESTION 2.39 **Change of capacitor energy due to a change of dielectric.** The dielectric in a spherical capacitor [Figure 2.9(a)] is oil. The capacitor is connected to a voltage source. The source is then disconnected and the oil is drained from the capacitor. The energy of the capacitor in the final electrostatic state is

(A) larger than

(B) the same as

(C) smaller than

before the source was disconnected.

CONCEPTUAL QUESTION 2.40 **Dielectric drain under different circumstances.** The oil dielectric in a spherical capacitor is completely drained while the voltage source is still connected to the capacitor. As a result, the energy of the capacitor

(A) increases.

(B) decreases.

(C) remains the same.

CONCEPTUAL QUESTION 2.41 **Energies of isolated metallic spheres of different sizes.** Consider two isolated metallic spheres with the same charges and different radii in air, and compare their energies (note that an isolated sphere can be regarded as the inner electrode of a spherical capacitor whose outer electrode has an infinite radius). The larger energy is that of

(A) the larger sphere.

(B) the smaller sphere.

(C) The energies are the same.

(D) Need more information.

CONCEPTUAL QUESTION 2.42 **Change of field intensity/flux density and energy.** Two capacitors contain the same amounts of electric energy. If the electric field intensity (E) at every point in the first capacitor becomes twice as large, while the electric flux density (D) at every point in the second capacitor is halved, the energy stored in the first capacitor in the new electrostatic state is

(A) 1/4 of

(B) 1/16 of

(C) 4 times

(D) 16 times

(E) the same as

that stored in the second capacitor.

CONCEPTUAL QUESTION 2.43 **Energy densities in a half-filled capacitor.** The space between the electrodes of a capacitor is half filled with a dielectric of relative permittivity $\varepsilon_r = 2$ and half filled with air. The electric field in the entire space is unform ($\mathbf{E} = \text{const}$). Compared with the electric energy density in the dielectric, that in air is

(A) larger.

(B) smaller.

(C) the same.

(D) zero.

(E) Need more information.

2.7 Dielectric Breakdown in Electrostatic Systems

The electric field intensity, E, in the dielectric of an electrostatic system cannot be increased indefinitely: if a certain value is exceeded, the dielectric becomes conducting; it temporarily or permanently loses its insulating property, and is said to break down. The breaking field value, i.e., the maximum electric field intensity that an individual dielectric material can withstand without breakdown, is termed the dielectric strength of the material and is denoted by E_{cr} (critical field intensity). For air, $E_{cr0} = 3$ MV/m, while $E_{cr} > E_{cr0}$ for other dielectrics (for example, $E_{cr} = 47$ MV/m for polyethylene). We shall now study capacitors and transmission lines in high-voltage applications, i.e., in situations where the electric field in the dielectric is so strong that there is a danger of dielectric breakdown in the structure. Breakdown occurs when the largest local field intensity in the dielectric reaches E_{cr} for that particular material. In structures with nonuniform electric field distributions, the principal task is to identify the most vulnerable spot for breakdown, and this task is more complex for structures containing multiple dielectric regions.

CONCEPTUAL QUESTION 2.44 **Breakdown in a vacuum.** The dielectric strength of a vacuum is

(A) larger than

(B) the same as

(C) smaller than

that of air ($E_{\mathrm{cr0}} = 3$ MV/m).

CONCEPTUAL QUESTION 2.45 **Breakdown of a nonsymmetrical two-wire line in air.** A nonsymmetrical thin two-wire transmission line [the same as in Figure 2.9(c) but with wires of different radii] is placed in air and charged. The wire with a positive charge ($Q' > 0$ per unit length) has a radius a. The radius of the other wire (charged by $-Q'$ per unit length) is $b = a/2$. The distance between the wire axes is $d = 100a$. If Q' is made sufficiently large for the dielectric breakdown to occur, this will be

(A) next to the surface of the positively charged wire.

(B) next to the surface of the negatively charged wire.

(C) close to the surfaces of both wires (simultaneously).

(D) approximately halfway between the wires, in the plane containing wire axes.

(E) Need more information.

CONCEPTUAL QUESTION 2.46 **Breakdown in a capacitor with two dielectric parts.** The dielectric of a parallel-plate capacitor has two homogeneous parts, referred to as dielectric 1 and dielectric 2. The boundary surface between the parts is flat and parallel to the capacitor plates, as in Figure 2.19. The dielectric strength of dielectric 1 is twice that of dielectric 2. If a voltage larger than the breakdown voltage of the capacitor [the highest possible voltage that can be applied to the capacitor (before it breaks down)] is applied across the capacitor terminals, the breakdown occurs in

(A) dielectric 1.

(B) dielectric 2.

(C) both dielectrics simultaneously.

(D) Need more information.

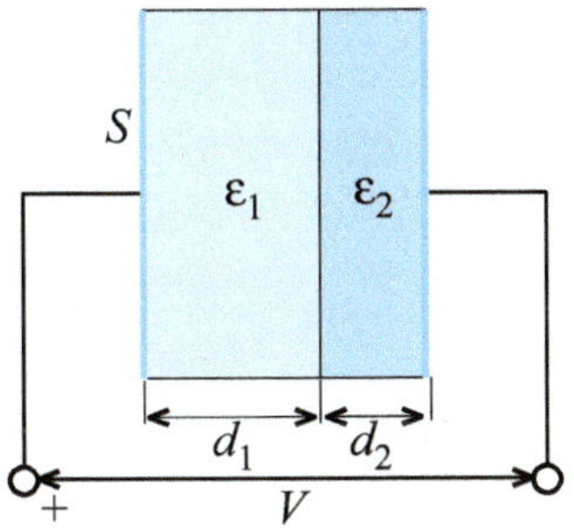

Figure 2.19 Dielectric breakdown analysis for a parallel-plate capacitor with two dielectric layers; for Conceptual Question 2.46.

CONCEPTUAL QUESTION 2.47 **Breakdown for a different position of dielectric parts.** Consider a parallel-plate capacitor with two dielectric sectors; namely, the boundary surface between the two homogeneous material parts constituting the dielectric of the capacitor is perpendicular to the capacitor plates, as in Figure 2.20. Assume that the dielectric strength of dielectric 1 is twice that of dielectric 2. If the applied voltage is larger than the breakdown voltage of the capacitor, the breakdown occurs in

(A) dielectric 1.

(B) dielectric 2.

(C) both dielectrics simultaneously.

(D) Need more information.

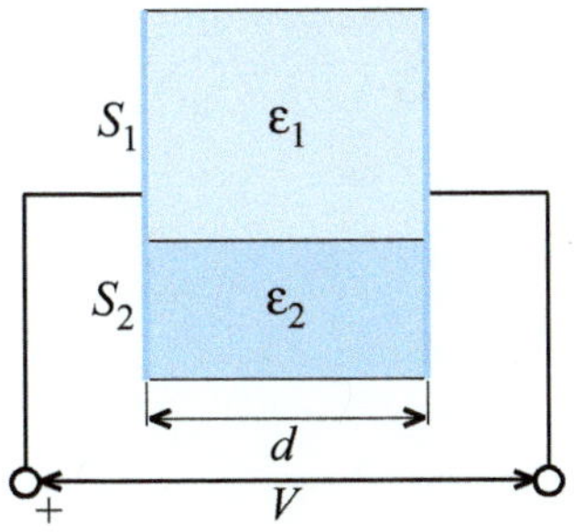

Figure 2.20 Breakdown analysis for a parallel-plate capacitor with two dielectric sectors; for Conceptual Question 2.47.

---

3 STEADY ELECTRIC CURRENTS

Introduction

So far, we have dealt with electrostatic fields, namely, the fields associated with time-invariant charges at rest. We now consider the charges in an organized macroscopic motion, which constitute an electric current. Our focus in this chapter is on the steady flow of free charges in conducting materials, i.e., on steady (time-invariant) electric currents, whose macroscopic characteristics (like the amount of current through a wire conductor) do not vary with time. Steady currents are also called direct currents, abbreviated dc.

3.1 Continuity Equation, Conductivity, and Ohm's and Joule's Laws in Local Form

The current intensity, I, is defined as a rate of movement of charge passing through a surface (e.g., cross section of a cylindrical conductor),

$$I = \frac{\mathrm{d}Q}{\mathrm{d}t} \quad \text{(current intensity or, simply, current; unit: A)} , \tag{3.1}$$

i.e., I equals the total amount of charge that flows through the surface during an elementary time $\mathrm{d}t$, divided by $\mathrm{d}t$. The unit for current intensity, which is usually referred to as, simply, current, is ampere or amp (A), equal to C/s. The current density vector, $\mathbf{J}$, is a vector that is directed along the current lines and whose magnitude, with reference to Figure 3.1(a), is given by

$$J = \frac{\mathrm{d}I}{\mathrm{d}S} \quad \text{(current density; unit: A/m}^2\text{)} , \tag{3.2}$$

where $\mathrm{d}I$ is the current flowing through an elementary surface of area $\mathrm{d}S$.

In many situations, current flow is localized in a very thin (theoretically infinitely thin) film over a surface, as shown in Figure 3.1(b). This is so-called surface current, described by the surface current density vector, $\mathbf{J}_\mathrm{s}$, which is defined as

$$J_\mathrm{s} = \frac{\mathrm{d}I}{\mathrm{d}l} \quad \text{(surface current density; unit: A/m)} , \tag{3.3}$$

where $\mathrm{d}I$ is the current flowing across a line element $\mathrm{d}l$ set normal to the current flow [Figure 3.1(b)]. Note that the surface current density vector is sometimes denoted as $\mathbf{K}$.

By the continuity equation, the total current of any time dependence leaving a domain v through a closed surface S, that is, the total outward flux of the current

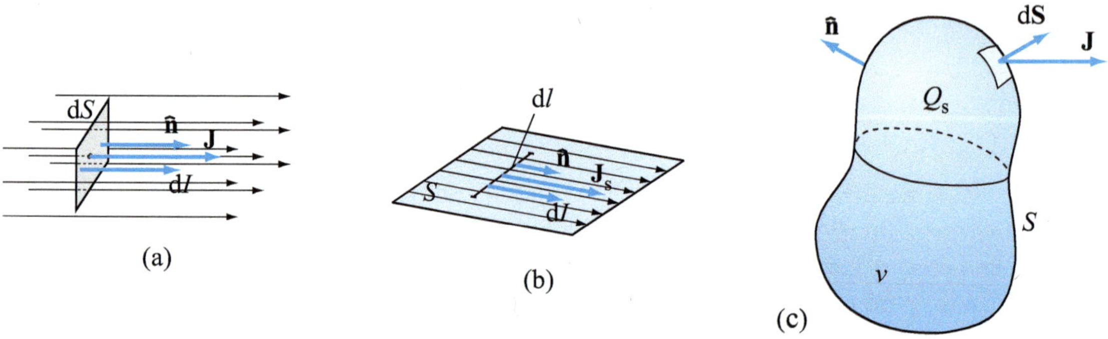

Figure 3.1 (a) Current density vector ($\mathbf{J}$). (b) Surface current density vector ($\mathbf{J}_s$). (c) Arbitrary closed surface in a region with currents.

density vector through S, Figure 3.1(c), is equal to the negative derivative in time of the total charge enclosed by S,

$$\oint_S \mathbf{J} \cdot d\mathbf{S} = -\frac{dQ_S}{dt} \quad \text{(continuity equation, for currents of any time dependence)} . \tag{3.4}$$

By analogy to the integral and differential forms of the generalized Gauss' law, Eqs. (2.5), we get the differential form of the continuity equation:

$$\nabla \cdot \mathbf{J} = -\frac{\partial \rho}{\partial t} \quad \text{(continuity equation in differential form)} . \tag{3.5}$$

For steady (time-invariant) currents, Eqs. (3.4) and (3.5) reduce to

$$\oint_S \mathbf{J} \cdot d\mathbf{S} = 0 \,, \quad \nabla \cdot \mathbf{J} = 0 \quad \text{(continuity equation for steady currents)} , \tag{3.6}$$

where the integral equation represents a generalization of Kirchhoff's current law in circuit theory.

In linear conducting media, $\mathbf{J}$ is linearly proportional to the electric field intensity vector, $\mathbf{E}$,

$$\mathbf{J} = \sigma \mathbf{E} \quad \text{(Ohm's law in local form)} , \tag{3.7}$$

where σ is the conductivity [unit: siemens per meter (S/m)] of the medium. This relationship is known as Ohm's law in local or point form. In addition, in studying steady current fields we always have in mind that time-invariant currents in a conductor are produced by a static electric field, which is a conservative field, meaning that the line integral of $\mathbf{E}$ along an arbitrary closed path is zero, Eq. (1.13).

Copper (Cu), the most commonly used metallic conductor, has a conductivity of $\sigma_{\text{Cu}} = 58$ MS/m at room temperature (20°C). In many applications, we consider copper and other metallic conductors as perfect electric conductors (PEC), with

$$\sigma \to \infty \quad \text{(perfect electric conductor – PEC)} . \tag{3.8}$$

In a conductor with electric current, electric energy is constantly converted into heat, and the rate (power) of this energy transformation is described by Joule's law in local (point) form, which states that the volume density of the power of Joule's (ohmic) losses at a point in the material equals

$$p_J = JE = \frac{J^2}{\sigma} = \sigma E^2 \quad \text{(Joule's law in local form)} \ . \tag{3.9}$$

The unit for power is watt (W), and hence the unit for p_J is W/m^3. The total power of Joule's losses (the electric power that is lost to heat) in a domain of volume v (e.g., in the entire conducting body) is obtained as $P_J = \int_v p_J \, dv$.

CONCEPTUAL QUESTION 3.1 **Change of current due to a change of field intensity.** A metallic conductor carries a steady current of intensity I. If the electric field intensity at every point in the conductor is doubled, the current intensity of the conductor in the new steady state equals[1]

(A) $I/2$.

(B) I.

(C) $2I$.

(D) $4I$.

(E) None of the above.

CONCEPTUAL QUESTION 3.2 **Conductor of variable cross section.** A steady current flows through a homogeneous metallic conductor of variable cross section, shown in Figure 3.2. The electric field intensities E_1 and E_2 ($E_1, E_2 > 0$) in the two long parts of the conductor (see the figure) are related as

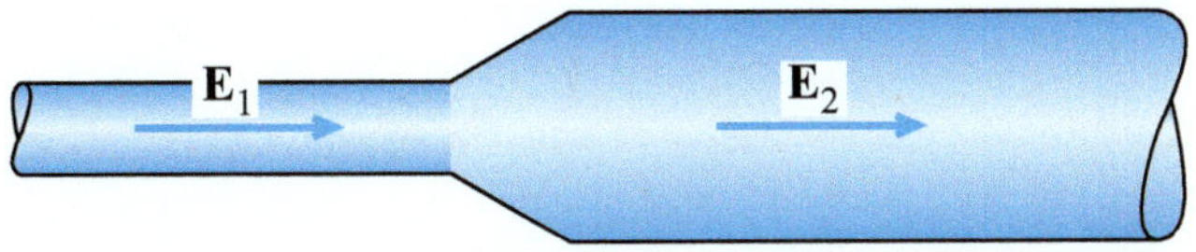

Figure 3.2 Metallic conductor of variable cross section with a steady current; for Conceptual Question 3.2.

[1] *For every conceptual question in this text, exactly one answer is correct.*

(A) $E_1 < E_2$.

(B) $E_1 = E_2$.

(C) $E_1 > E_2$.

CONCEPTUAL QUESTION 3.3 **Volume current density of a thin plate.** A very thin aluminum plate of length l, width w, and thickness d ($d \ll w$) carries a current I that is uniformly distributed across the plate cross section (of dimensions w and d). The current lines are parallel to the plate length. If this current is considered as a volume current, the current density is given by

(A) $J = I/w$.

(B) $J = I/l$.

(C) $J = I/d$.

(D) $J = I/(wl)$.

(E) $J = I/(wd)$.

(F) $J = I/(wld)$.

CONCEPTUAL QUESTION 3.4 **Surface current density of a thin plate.** If the current I flowing through a very thin aluminum plate of length l, width w, and thickness d ($d \ll w$) (current flow is along the plate length and is uniform across the plate cross section) is considered as a surface current, the surface current density of the plate amounts to

(A) $J_s = I/w$.

(B) $J_s = I/l$.

(C) $J_s = I/d$.

(D) $J_s = I/(wl)$.

(E) $J_s = I/(wd)$.

(F) $J_s = I/(wld)$.

CONCEPTUAL QUESTION 3.5 Possible distributions of steady currents. Consider each of the following four vectors in the Cartesian coordinate system: $\mathbf{J}_1 = J_0\,\hat{\mathbf{x}}$, $\mathbf{J}_2 = J_0(x/a)\,\hat{\mathbf{x}}$, $\mathbf{J}_3 = J_0(xy/a^2)\,\hat{\mathbf{x}}$, and $\mathbf{J}_4 = J_0(y/a)^2\,\hat{\mathbf{x}}$, where J_0 and a are constants. Which of them can be the density vector of a steady current in a conducting medium?

(A) $\mathbf{J}_1$ only.

(B) $\mathbf{J}_1$ and $\mathbf{J}_2$ only.

(C) $\mathbf{J}_1$ and $\mathbf{J}_4$ only.

(D) $\mathbf{J}_4$ only.

(E) All of the vectors.

(F) None of the vectors.

CONCEPTUAL QUESTION 3.6 Possible distributions of time-varying currents. Which of the four vectors, $\mathbf{J}_1 = J_0\,\hat{\mathbf{x}}$, $\mathbf{J}_2 = J_0(x/a)\,\hat{\mathbf{x}}$, $\mathbf{J}_3 = J_0(xy/a^2)\,\hat{\mathbf{x}}$, and $\mathbf{J}_4 = J_0(y/a)^2\,\hat{\mathbf{x}}$ (J_0 and a are constants), can be the density vector of a time-varying current in a conducting medium?

(A) $\mathbf{J}_1$ only.

(B) $\mathbf{J}_1$ and $\mathbf{J}_2$ only.

(C) $\mathbf{J}_1$ and $\mathbf{J}_4$ only.

(D) $\mathbf{J}_4$ only.

(E) All of the vectors.

(F) None of the vectors.

CONCEPTUAL QUESTION 3.7 Circulation of the current density vector. Consider a distribution of steady currents in a conducting medium. The line integral of the current density vector, $\mathbf{J}$, along an arbitrary closed path in this medium is

(A) zero.

(B) nonzero.

(C) Need more information.

CONCEPTUAL QUESTION 3.8 **Conductor with a uniform cross section of complex shape.** Figure 3.3 shows the cross section of a long homogeneous metallic conductor carrying a steady current. The current densities J_1 and J_2 in the two parts of the conductor (see the figure) are related as

(A) $J_1 < J_2$.

(B) $J_1 = J_2$.

(C) $J_1 > J_2$.

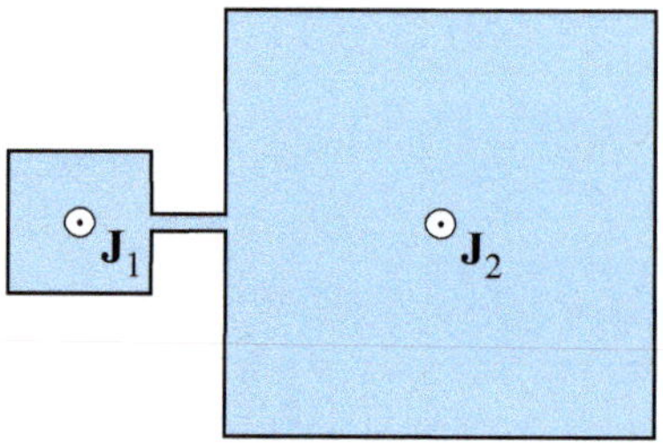

Figure 3.3 Cross section of a homogeneous conductor with a steady current; for Conceptual Question 3.8.

CONCEPTUAL QUESTION 3.9 **Conductor with two parts of different conductivities.** A conductor is composed from two homogeneous pieces of the same size but of different conductivities ($\sigma_1 \neq \sigma_2$). In case (a), the pieces are connected one along the other [Figure 3.4(a)]. In case (b), they are connected one behind the other [Figure 3.4(b)]. If a time-invariant current of intensity I is made to flow through the conductor, the current density vectors in the two pieces are the same ($\mathbf{J}_1 = \mathbf{J}_2$) for

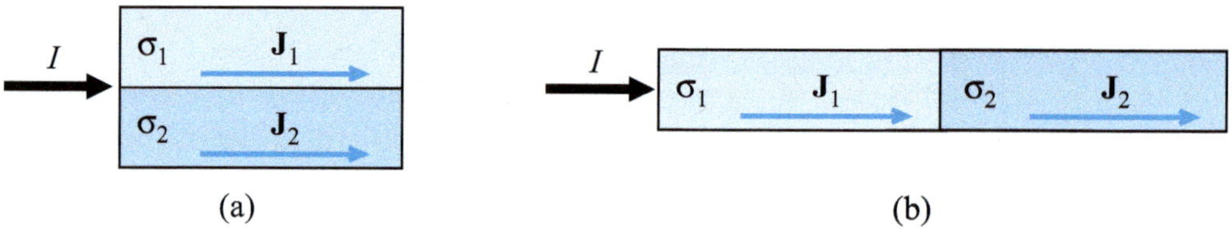

Figure 3.4 Current conductor made of two pieces with different conductivities connected one along the other (a) and one behind the other (b); for Conceptual Question 3.9.

(A) case (a) only.

(B) case (b) only.

(C) both cases.

(D) neither of the cases.

(E) Depends on other material parameters.

CONCEPTUAL QUESTION 3.10 **Electric field intensities in two conducting pieces.** Consider the electric field intensity vectors in the two parts of the two conductors in Figure 3.5, and determine in which of the cases $\mathbf{E}_1 = \mathbf{E}_2$:

(A) case (a) only.

(B) case (b) only.

(C) both cases.

(D) neither of the cases.

(E) Depends on other material parameters.

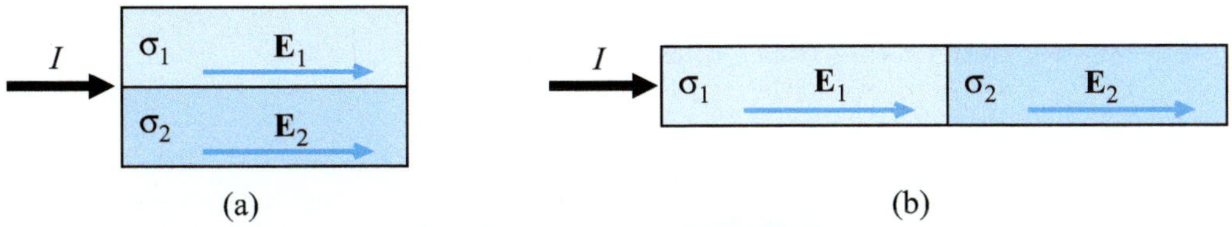

Figure 3.5 Analysis of the electric field intensity vectors in two current conductors made of two pieces with different conductivities ($\sigma_1 \neq \sigma_2$); for Conceptual Question 3.10.

CONCEPTUAL QUESTION 3.11 **Electric flux densities in two conducting pieces.** With reference to Figure 3.6, the electric flux density vectors in the two conducting pieces are the same, $\mathbf{D}_1 = \mathbf{D}_2$, in

(A) case (a) only.

(B) case (b) only.

(C) both cases.

(D) neither of the cases.

(E) Depends on other material parameters.

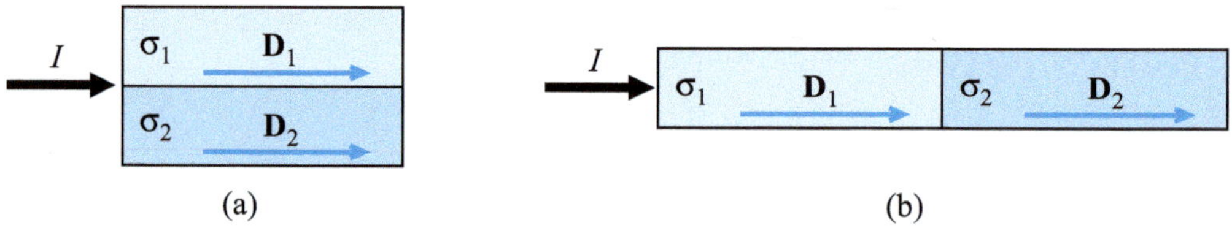

Figure 3.6 Analysis of the electric flux density vectors in conductors made of pieces with $\sigma_1 \neq \sigma_2$; for Conceptual Question 3.11.

CONCEPTUAL QUESTION 3.12 **Power of Joule's losses in two conducting pieces.** For the current-carrying conductor with two parts of different conductivities in the arrangement in Figure 3.7, assume that $\sigma_1 > \sigma_2$ and consider the power of Joule's losses in each of the pieces. This power

(A) is larger in the piece with larger conductivity.

(B) is larger in the piece with smaller conductivity.

(C) is the same in both pieces.

(D) The answer depends on other material parameters.

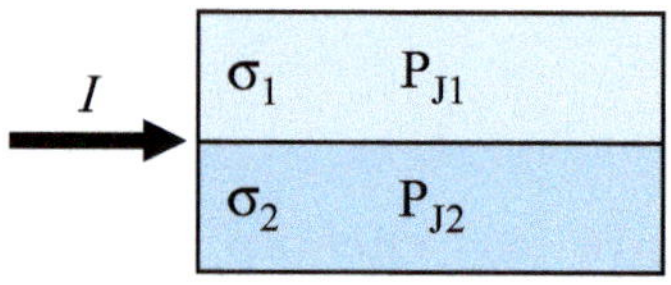

Figure 3.7 Power of Joule's losses in a current conductor made of two pieces with $\sigma_1 > \sigma_2$ connected one along the other; for Conceptual Question 3.12.

CONCEPTUAL QUESTION 3.13 **Joule's losses for a different arrangement of pieces.** If the two material pieces constituting the current-carrying conductor are arranged as in Figure 3.8 and $\sigma_1 > \sigma_2$, the power of Joule's losses

(A) is larger in the piece with σ_1.

(B) is larger in the piece with σ_2.

(C) is the same in both pieces.

(D) The answer depends on other material parameters.

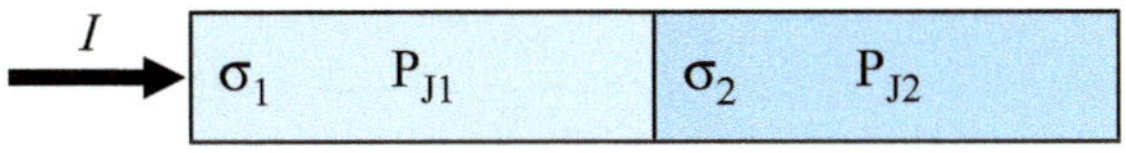

Figure 3.8 Power of Joule's losses for two conducting pieces with $\sigma_1 > \sigma_2$ connected one behind the other; for Conceptual Question 3.13.

CONCEPTUAL QUESTION 3.14 Power of Joule's losses in two steady states. The power of Joule's losses in two conductors appears to be the same. If the current density at every point in the first conductor becomes twice as large, while the electric field intensity at every point in the second conductor is halved, the power of Joule's losses in the first conductor in the new steady state is

(A) 1/4 of

(B) 1/16 of

(C) 4 times

(D) 16 times

(E) the same as

that in the second conductor.

3.2 Resistance, Conductance, and Ohm's Law

A conductor with two terminals and a (substantial) resistance, R, is usually referred to as a resistor. The relation between the voltage, current, and resistance of a resistor is known as Ohm's law:

$$V = RI \quad \text{or} \quad I = GV \quad \left(G = \frac{1}{R}\right) \quad \text{(Ohm's law)}\,. \tag{3.10}$$

The resistance is always nonnegative ($R \geq 0$), and the unit is the ohm (Ω), equal to V/A. The value of R depends on the shape and size of the conductor (resistor), and on the conductivity σ (or resistivity ρ) of the material. The reciprocal of resistance is called the conductance and symbolized by G. Its unit is the siemens (S), where

$S = \Omega^{-1} = A/V$. Note that sometimes the mho (ohm spelled backwards) is used instead of the siemens.

As an example, the resistance of a homogeneous resistor with a uniform cross section of an arbitrary shape and surface area S can be found as follows:

$$J = \frac{I}{S} \quad \longrightarrow \quad V = El = \frac{J}{\sigma} l = \frac{Il}{\sigma S} \quad \longrightarrow \quad R = \frac{V}{I} = \frac{l}{\sigma S}, \tag{3.11}$$

where I, J, E, l, and V are the current intensity, current density, electric field intensity, length, and voltage of the resistor and σ is the conductivity of its material.

CONCEPTUAL QUESTION 3.15 **Resistor with two cuboidal parts.** A resistor is formed from two rectangular cuboids of the same size, with sides a, b, and c, made from different resistive materials, with conductivities σ_1 and σ_2 ($\sigma_1 \neq \sigma_2$). If the two cuboids are connected as in Figure 3.9 and the voltage between the resistor terminals is V, the current intensity of the resistor amounts to

(A) $I = (\sigma_1 + \sigma_2)aV/(2bc)$.

(B) $I = (\sigma_1 + \sigma_2)bcV/(2a)$.

(C) $I = (\sigma_1 + \sigma_2)aV/(\sigma_1\sigma_2 bc)$.

(D) $I = \sigma_1\sigma_2 bcV/[(\sigma_1 + \sigma_2)a]$.

(E) $I = \sigma_1\sigma_2 aV/[(\sigma_1 + \sigma_2)bc]$.

(F) $I = 2aV/[(\sigma_1 + \sigma_2)bc]$.

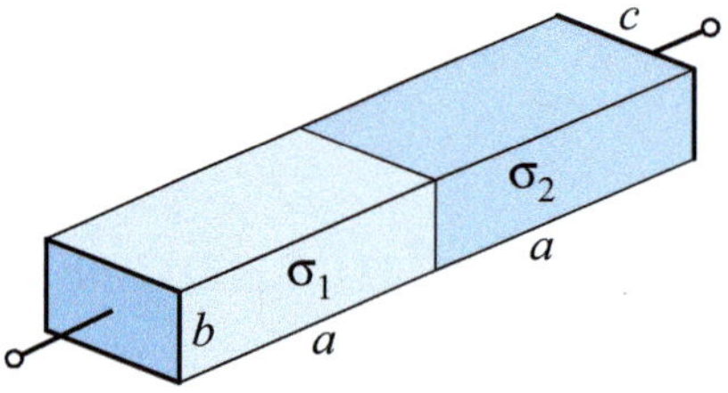

Figure 3.9 Two rectangular cuboids made from different resistive materials connected one behind the other; for Conceptual Question 3.15.

CONCEPTUAL QUESTION 3.16 **Resistor with two parts in a different connection.** If a resistor is formed from two cuboids with different conductivities connected as in Figure 3.10 and the current intensity through the resistor is I, the voltage between the resistor terminals equals

(A) $V = (\sigma_1 + \sigma_2)bcI/a$.

(B) $V = aI/[(\sigma_1 + \sigma_2)bc]$.

(C) $V = (\sigma_1 + \sigma_2)bcI/(2a)$.

(D) $V = 2aI/[(\sigma_1 + \sigma_2)bc]$.

(E) $V = \sigma_1\sigma_2 aI/[(\sigma_1 + \sigma_2)bc]$.

(F) $V = \sigma_1\sigma_2 bcI/[(\sigma_1 + \sigma_2)a]$.

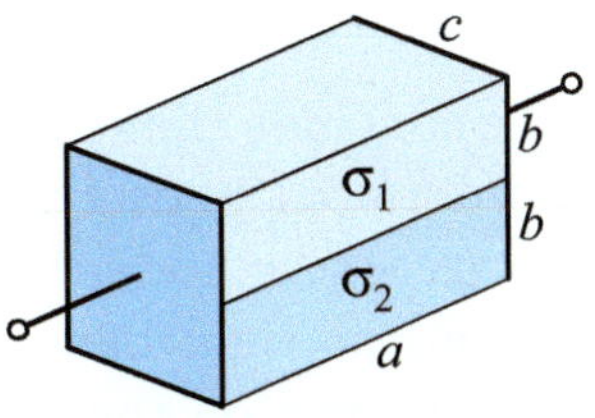

Figure 3.10 Two cuboids with $\sigma_1 \neq \sigma_2$ connected one along the other; for Conceptual Question 3.16.

CONCEPTUAL QUESTION 3.17 One hundred resistors connected in series/parallel. One hundred resistors of arbitrary geometries, with different resistances $R_1, R_2, \ldots, R_{100}$, are connected first in series [Figure 3.11(a)] and then in parallel [Figure 3.11(b)]. Which combination has greater resistance?

(A) The connection in series.

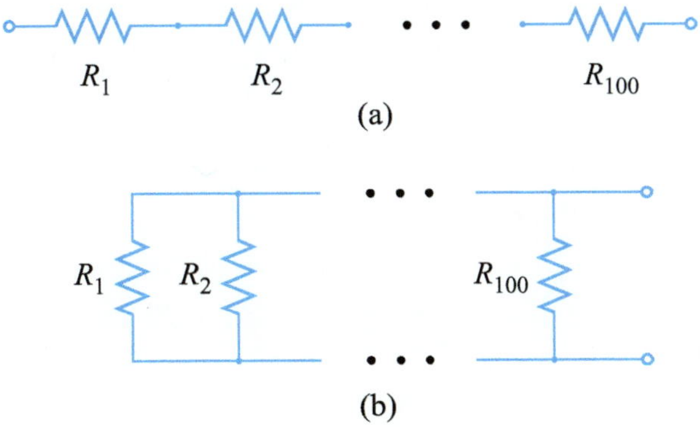

Figure 3.11 One hundred resistors of arbitrary geometries and resistances connected in series (a) and parallel (b); for Conceptual Question 3.17.

(B) The connection in parallel.

(C) The two connections have the same resistance.

(D) Depends on the geometries of the individual resistors and nothing else.

(E) Depends on the relationships between particular values of $R_1, R_2, \ldots, R_{100}$.

3.3 Boundary Conditions for Steady Currents

Comparing Eqs. (3.4) and (2.5), we conclude that the boundary condition for normal components of the vector **J** at interfaces between conducting media of different conductivity is of the same form as the boundary condition for the vector **D**, in Eqs. (2.7). The only difference is on the right-hand side of the equation, where ρ_s (the surface charge density that may exist on the surface) is replaced by $-\partial\rho_s/\partial t$. For steady currents, $-\partial\rho_s/\partial t = 0$, and the boundary condition that corresponds to Eq. (1.13) is the same as for the electrostatic field, as in Eqs. (2.7); hence, the complete set of boundary conditions for steady currents is given by

$$E_{1t} = E_{2t}\,, \quad J_{1n} = J_{2n} \quad \text{(boundary conditions for steady currents)}\,. \tag{3.12}$$

From these conditions and Eq. (3.7), we obtain the law of refraction of the current density lines at a boundary interface, Figure 3.12, analogous to that in Eq. (2.9),

$$\frac{\tan\alpha_1}{\tan\alpha_2} = \frac{\sigma_1}{\sigma_2} \quad \text{(law of refraction of current streamlines)}\,. \tag{3.13}$$

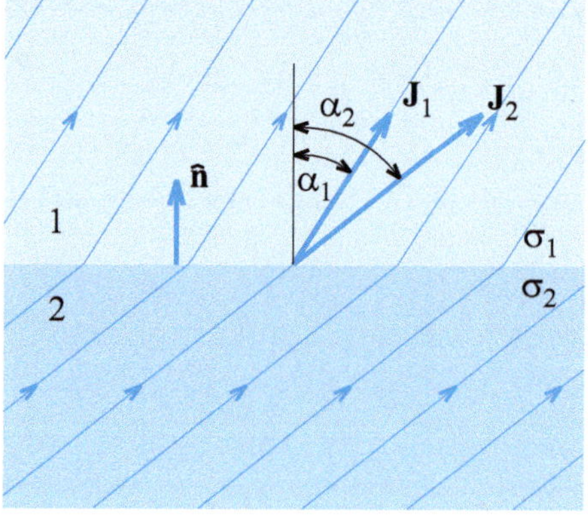

Figure 3.12 Refraction of steady current lines at a conductor–conductor interface.

CONCEPTUAL QUESTION 3.18 Boundary conditions at a conductor–conductor interface. Consider a boundary surface between two conducting media of conductivities σ_1 and σ_2, where $\sigma_1 = 2\sigma_2$. Which of the cases shown in Figure 3.13 represent possible time-invariant current density vectors on the two sides of the boundary?

(A) Case (a) only.

(B) Case (b) only.

(C) Case (c) only.

(D) Case (d) only.

(E) More than one case.

(F) None of the cases.

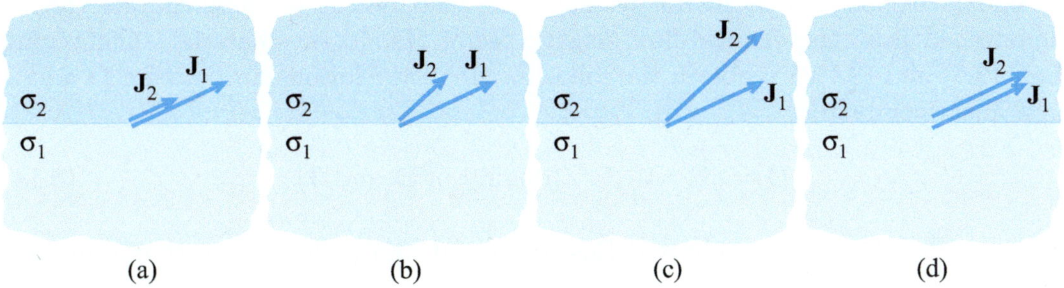

Figure 3.13 Interface between two conducting media ($\sigma_1 = 2\sigma_2$) – four cases with different combinations (not all necessarily physically meaningful) of vectors $\mathbf{J}_1$ and $\mathbf{J}_2$ on the two sides of the boundary; for Conceptual Question 3.18.

CONCEPTUAL QUESTION 3.19 Refraction of steady current lines. Figure 3.14 shows steady current density lines near a boundary between two conducting media. Which of the two media is a better conductor?

(A) Medium 1.

(B) Medium 2.

(C) Need more information.

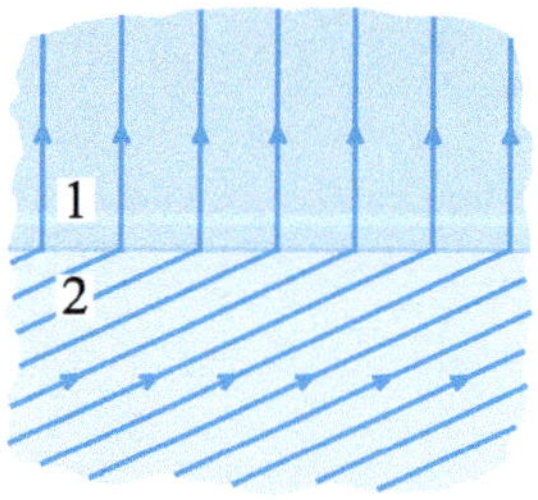

Figure 3.14 Refraction of steady current density lines at an interface between two conducting media; for Conceptual Question 3.19.

3.4 Duality Relationships in the Steady Current Field

The electric field intensity vector, $\mathbf{E}$, in the steady current field is produced by stationary excess charges in the system. The distribution of these charges can be determined from the electric flux density vector, $\mathbf{D}$, in the material. Combining Eqs. (2.6) and (3.7), we obtain the following duality relationship between $\mathbf{D}$ and $\mathbf{J}$ in a linear conductor:

$$\mathbf{D} = \varepsilon \mathbf{E} = \frac{\varepsilon}{\sigma}\mathbf{J} \quad \text{(duality of } \mathbf{D} \text{ and } \mathbf{J}) \, . \tag{3.14}$$

By means of this relationship and Eqs. (2.5) and (3.6), the volume charge density in the conductor is

$$\rho = \nabla \cdot \mathbf{D} = \nabla \cdot \left(\frac{\varepsilon}{\sigma}\mathbf{J}\right) = \left[\nabla \left(\frac{\varepsilon}{\sigma}\right)\right] \cdot \mathbf{J} + \frac{\varepsilon}{\sigma}(\nabla \cdot \mathbf{J}) = \mathbf{J} \cdot \nabla \left(\frac{\varepsilon}{\sigma}\right) . \tag{3.15}$$

Next, consider a pair of metallic bodies (electrodes) placed in a homogeneous conducting medium of conductivity σ and permittivity ε, as shown in Figure 3.15. We now use Eq. (3.14) to relate the conductance, G, and capacitance, C, between the electrodes. From Eq. (3.15), there are no volume charges ($\rho = 0$) in the medium (σ and ε are constants). Applying the integral forms of the continuity equation for steady currents, Eqs. (3.6), and generalized Gauss' law, Eqs. (2.5), to an arbitrary surface S completely enclosing the positive electrode (Figure 3.15) gives $I = \oint_S \mathbf{J} \cdot \mathrm{d}\mathbf{S} = (\sigma/\varepsilon) \oint_S \mathbf{D} \cdot \mathrm{d}\mathbf{S} = \sigma Q/\varepsilon$. Dividing this equation by V and having in mind Eqs. (3.10) and (2.11), the duality relationship between G and C turns out to be

$$G = \frac{\sigma}{\varepsilon} C \quad \text{(duality of } G \text{ and } C) \, . \tag{3.16}$$

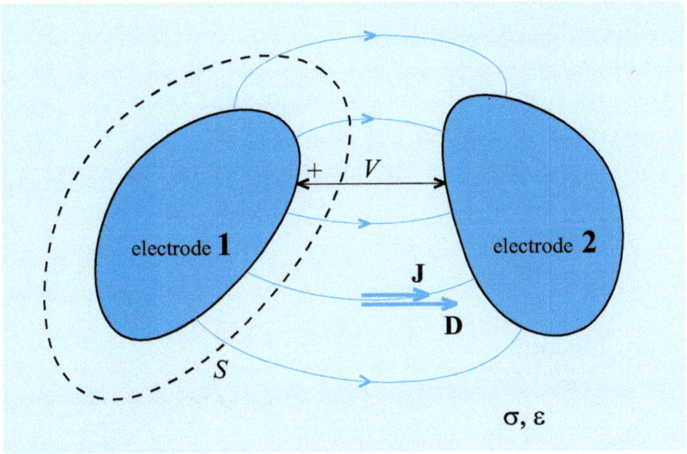

Figure 3.15 Two metallic electrodes in a homogeneous conducting medium.

CONCEPTUAL QUESTION 3.20 **Excess volume charge in a resistor.** There cannot be an excess free charge distributed throughout the volume of a resistor carrying a steady current.

(A) True.

(B) False.

CONCEPTUAL QUESTION 3.21 **Excess volume charge in a lossy dielectric.** If a capacitor with an imperfect dielectric is connected to a time-invariant voltage source, an excess volume charge may be accumulated in the dielectric.

(A) True.

(B) False.

CONCEPTUAL QUESTION 3.22 **Excess volume charge in a homogeneous medium.** There cannot be volume excess charges ($\rho = 0$) inside homogeneous media with steady currents.

(A) True.

(B) False.

CONCEPTUAL QUESTION 3.23 **Same system in a lossy medium and in air.** Two metallic spheres are placed in a homogeneous lossy medium of conductivity σ and they do not touch. The conductance between these spheres can be found from the capacitance between the same spheres at the same mutual distance in air (although the conductance of the air-filled system is zero).

(A) True.

(B) False.

(C) Need more information.

CONCEPTUAL QUESTION 3.24 **Conductance versus capacitance for two capacitors.** Consider two capacitors filled with the same homogeneous imperfect dielectric. If the capacitance of the first capacitor is greater than the capacitance of the second one, the same is true for their conductances.

(A) True.

(B) False.

(C) Need more information.

CONCEPTUAL QUESTION 3.25 **Two capacitors with different lossy dielectrics.** Consider capacitances and conductances of two capacitors with different homogeneous imperfect dielectrics. If $C_1 > C_2$, then $G_1 > G_2$ as well.

(A) True.

(B) False.

(C) Need more information.

CONCEPTUAL QUESTION 3.26 **Conductance vs. capacitance for inhomogeneous dielectric.** A capacitor is filled with an inhomogeneous imperfect dielectric of known parameters. Can the conductance of this capacitor be found from its capacitance?

(A) Never.

(B) Always.

(C) Sometimes.

CONCEPTUAL QUESTION 3.27 Circuit with ideal resistors and capacitors. For a dc circuit with an ideal voltage generator, two ideal resistors, and two ideal capacitors, shown in Figure 3.16, $\mathcal{E} = 10$ V, $R_1 = R_2 = 3\ \Omega$, $C_1 = 4$ nF, and $C_2 = 6$ nF. The voltage across the second capacitor is

(A) $V = 0$.

(B) $V = 3$ V.

(C) $V = 4$ V.

(D) $V = 5$ V.

(E) $V = 6$ V.

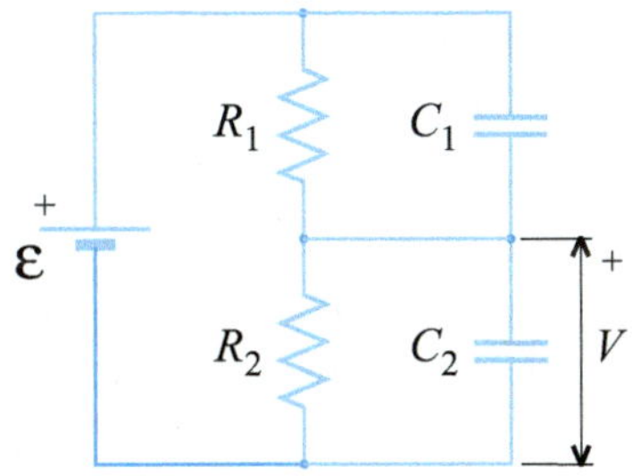

Figure 3.16 Circuit with ideal elements in a dc regime; for Conceptual Question 3.27.

CONCEPTUAL QUESTION 3.28 Circuit with one nonideal capacitor. Consider the dc circuit in Figure 3.16 ($\mathcal{E} = 10$ V, $R_1 = R_2 = 3\ \Omega$, $C_1 = 4$ nF, $C_2 = 6$ nF), and assume that the first capacitor has an imperfect dielectric. With this change, which of the capacitors will have a different voltage as compared to the respective voltage in the circuit with all ideal elements?

(A) The first capacitor only.

(B) The second capacitor only.

(C) Both capacitors.

(D) Neither of the capacitors.

3.5 Lossy Transmission Lines with Steady Currents

In this section, we study two-conductor transmission lines [e.g., lines in Figures 2.9(b), (c), (e), and (f)] with losses in a time-invariant (dc) regime. We subdivide the line into short sections, of length Δz, and represent each such section by a circuit cell consisting of a series resistor of resistance $\Delta R = R'\Delta z$ and a shunt (parallel) resistor of conductance $\Delta G = G'\Delta z$, as indicated in Figure 3.17, where R' and G' are the resistance and leakage conductance per unit length of the line, modeling losses in line conductors and dielectric, respectively. With this, a transmission line is said to be a circuit with distributed parameters (parameters per unit length). Using Kirchhoff's voltage and current laws and Ohm's law for the cells of the circuit (Figure 3.17), we obtain $V(z+\Delta z)-V(z) = -\Delta RI$ and $I(z+\Delta z)-I(z) = -\Delta GV$, and these equations, in the limit of $\Delta z \to 0$, become

$$\frac{\mathrm{d}V}{\mathrm{d}z} = -R'I\,, \quad \frac{\mathrm{d}I}{\mathrm{d}z} = -G'V \quad \text{(telegrapher's equations in a dc regime)}\,, \qquad (3.17)$$

the so-called telegrapher's equations for time-invariant currents and voltages on transmission lines. For a line with both $R' \neq 0$ and $G' \neq 0$, these are coupled differential equations, whose solutions are exponential functions for $V(z)$ and $I(z)$. If $R' = 0$ and $G' \neq 0$, $V =$ const and $I(z)$ is a linear function along the line. Conversely, if $G' = 0$ and $R' \neq 0$, $I =$ const and $V(z)$ is a linear function. Finally, if both $G' = 0$ and $R' = 0$, both current and voltage do not vary along the line.

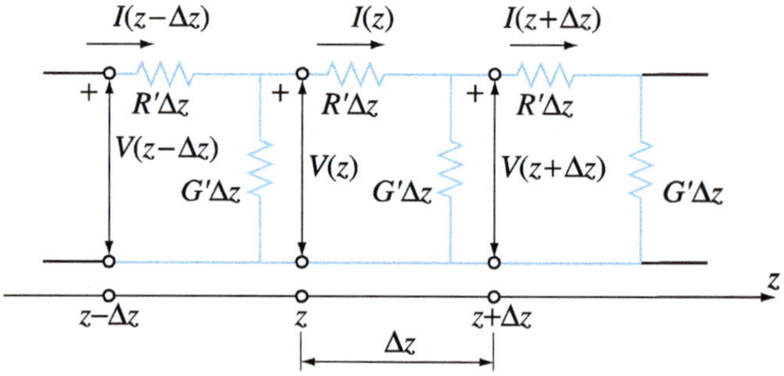

Figure 3.17 Circuit-theory representation of a transmission line with losses in conductors and dielectric in a dc regime.

CONCEPTUAL QUESTION 3.29 **Infinite conductivity of line conductors.** In a transmission line, the conductivity of conductors is $\sigma_\mathrm{c} \to \infty$, while that of the dielectric, σ_d, is finite and nonzero. For this line, we have the following in the dc circuit model in Figure 3.18:

(A) $R' \to \infty$.

(B) $G' \to \infty$.

(C) $R' = 0$.

(D) $G' = 0$.

(E) $R' \to \infty$ and $G' = 0$.

(F) $R' = 0$ and $G' \to \infty$.

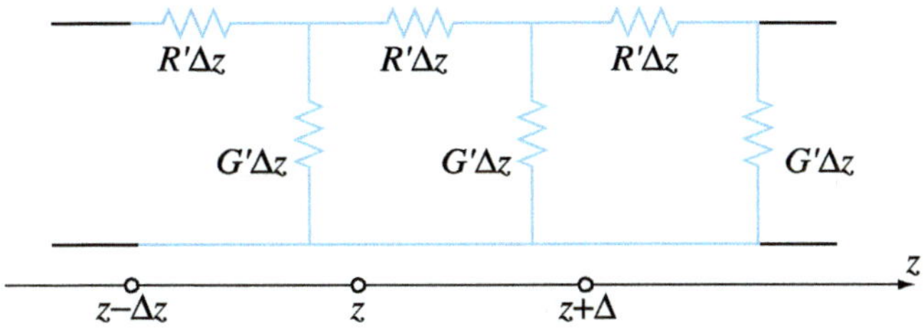

Figure 3.18 General dc circuit model of a lossy transmission line; for Conceptual Question 3.29.

CONCEPTUAL QUESTION 3.30 **Zero conductivity of line dielectric.** In the dc circuit model (Figure 3.18) of a transmission line with $\sigma_\mathrm{d} = 0$ and σ_c finite and nonzero,

(A) $R' \to \infty$.

(B) $G' \to \infty$.

(C) $R' = 0$.

(D) $G' = 0$.

(E) $R' \to \infty$ and $G' = 0$.

(F) $R' = 0$ and $G' \to \infty$.

CONCEPTUAL QUESTION 3.31 **P.u.l. resistance and conductance of a transmission line.** Consider the per-unit-length resistance (R') and leakage conductance (G') of a lossy transmission line in a dc regime, Figure 3.18. The total length of the line is l. Which one of the following relationships is always satisfied?

(A) $R' = 1/G'$.

(B) $R' = l^2/G'$.

(C) $R'l^2 = 1/G'$.

(D) None.

CONCEPTUAL QUESTION 3.32 **Current along a lossy transmission line.** A transmission line with an imperfect dielectric is connected to an ideal time-invariant voltage generator. The other end of the line is open. The current intensity through the line conductors

(A) decreases linearly with the distance from the generator.

(B) increases linearly with the distance from the generator.

(C) varies nonlinearly with the distance from the generator.

(D) does not vary with the distance from the generator, but is not zero.

(E) is zero at any distance from the generator.

(F) Need more information.

CONCEPTUAL QUESTION 3.33 **Coaxial cable with a lossy dielectric.** A coaxial cable [Figure 2.9(b)] with copper conductors is filled with a liquid dielectric. The cable is connected at one end to an ideal time-invariant voltage source, while its other end is terminated in a purely resistive load. The losses in the cable conductors are negligible, whereas the losses in the dielectric have to be taken into account. Considering the voltage between the conductors and the current through them, the following of the two quantities is (are) constant along the cable:

(A) voltage only.

(B) current only.

(C) both.

(D) neither.

CONCEPTUAL QUESTION 3.34 Two-wire line with a lossy dielectric. A two-wire transmission line [Figure 2.9(c)] has lossless conductors and a lossy dielectric. The line is fed by a time-invariant voltage generator. The current density in the wires and that in the dielectric vary in the same way with the distance from the generator.

(A) True.

(B) False.

4 MAGNETOSTATIC FIELD IN FREE SPACE

Introduction

We now introduce a series of new phenomena associated with steady electric currents, which are essentially the consequence of a new simple experimental fact – that conductors with currents exert forces on one another. These forces are called magnetic forces, and the field due to one current conductor in which the other conductor is situated and which causes the force on it is called the magnetic field. Any motion of electric charges and any electric current are followed by the magnetic field. The magnetic field due to steady electric currents is termed the steady (static) magnetic field or magnetostatic field. The theory of the magnetostatic field, the magnetostatics, restricted to a vacuum and nonmagnetic media is the subject of this chapter. The magnetic materials will be studied in the following chapter.

4.1 Magnetic Force and Magnetic Flux Density Vector

To quantitatively describe the magnetic field, we introduce a vector quantity called the magnetic flux density vector, **B**. It is defined analogously to the electric field intensity vector, **E**, in electrostatics [Eq. (1.3)] through the force, magnetic force, on a small probe point charge Q_p moving at a velocity **v** in the field, which equals the cross product[1] of vectors $Q_\mathrm{p}\mathbf{v}$ and **B**,

$$\mathbf{F}_\mathrm{m} = Q_\mathrm{p}\mathbf{v} \times \mathbf{B} \quad (Q_\mathrm{p} \to 0) \quad \text{(definition of the magnetic flux density } \mathbf{B}\text{; unit: T)} . \tag{4.1}$$

The unit for **B** is tesla (abbreviated T).

The magnetic equivalent of Coulomb's law, Eq. (1.1), states that the magnetic force on a point charge Q_2 that moves at a velocity $\mathbf{v}_2$ in the magnetic field due to a point charge Q_1 moving with a velocity $\mathbf{v}_1$ in a vacuum (or air), Figure 4.1(a), is given by

$$\mathbf{F}_{\mathrm{m}12} = \frac{\mu_0}{4\pi} \frac{Q_2\mathbf{v}_2 \times \left(Q_1\mathbf{v}_1 \times \hat{\mathbf{R}}_{12}\right)}{R^2} \quad \text{(magnetic force)} , \tag{4.2}$$

[1]The cross product of vectors **a** and **b**, $\mathbf{a} \times \mathbf{b}$, is a vector whose magnitude is given by $|\mathbf{a} \times \mathbf{b}| = |\mathbf{a}||\mathbf{b}| \sin\alpha$, where α is the angle between the two vectors in the product. It is perpendicular to the plane defined by the vectors **a** and **b**, and its direction (orientation) is determined by the right-hand rule when the first vector (**a**) is rotated by the shortest route toward the second vector (**b**). In this rule, the direction of rotation is defined by the fingers of the right hand when the thumb points in the direction of the cross product.

where μ_0 is the permeability of a vacuum (free space),

$$\mu_0 = 4\pi \cdot 10^{-7}\ \text{H/m} \quad \text{(permeability of a vacuum)}\,. \tag{4.3}$$

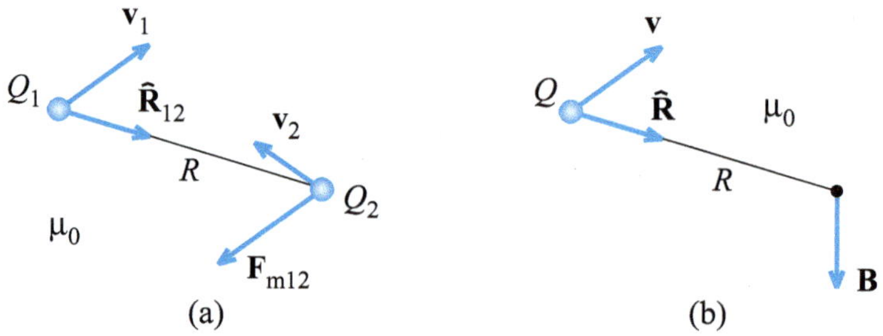

Figure 4.1 (a) Magnetic force between two point charges moving in a vacuum, given by Eq. (4.2). (b) Magnetic flux density vector due to a moving point charge, Eq. (4.4).

Combining Eqs. (4.1) and (4.2), we can identify the expression for the magnetic flux density vector of a point charge Q moving with a velocity $\mathbf{v}$ [Figure 4.1(b)]:

$$\mathbf{B} = \frac{\mu_0}{4\pi}\frac{Q\mathbf{v}\times\hat{\mathbf{R}}}{R^2} \quad \text{(}\mathbf{B}\text{ due to a moving point charge in free space)}\,. \tag{4.4}$$

Finally, note that, from Eq. (4.1) and the superposition principle, the magnetic force on a straight wire conductor of length l with a steady current of intensity I placed in a uniform magnetic field of flux density $\mathbf{B}$ comes out to be

$$\mathbf{F}_\text{m} = I\mathbf{l}\times\mathbf{B} \quad \text{(}\mathbf{F}_\text{m}\text{ on a straight conductor in a uniform magnetic field)}\,, \tag{4.5}$$

where the direction of the vector $\mathbf{l}$ is the same as the direction of the current flow along the conductor and $|\mathbf{l}| = l$.

CONCEPTUAL QUESTION 4.1 **Magnetic force between two moving point charges.** Consider point charges Q_1 and Q_2 that move at velocities $\mathbf{v}_1$ and $\mathbf{v}_2$, respectively, in free space. In particular, consider the situation depicted in Figure 4.2, where vectors $Q_1\mathbf{v}_1$ and $Q_2\mathbf{v}_2$ are parallel to each other and normal to the line joining them. For the two cases shown, the magnetic force between the charges is[2]

(A) attractive in case (a) and repulsive in case (b).

(B) repulsive in case (a) and attractive in case (b).

[2] *For every conceptual question in this text, exactly one answer is correct.*

(C) attractive in both cases.

(D) repulsive in both cases.

(E) zero in both cases.

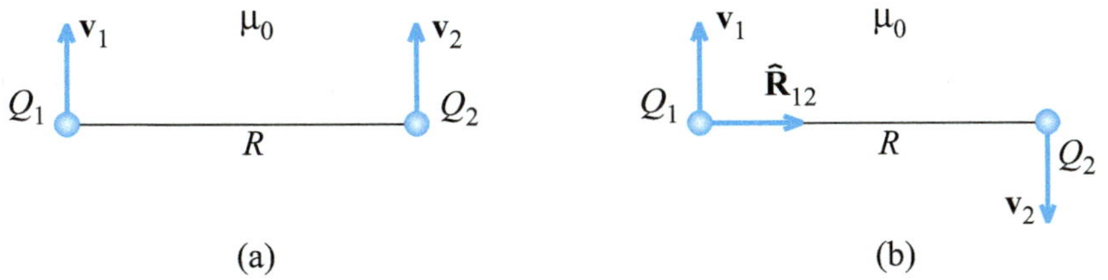

Figure 4.2 Two charges moving parallel to each other in the same direction (a) and in opposite directions (b) in a vacuum; for Conceptual Question 4.1.

CONCEPTUAL QUESTION 4.2 Point charge moving near a wire current conductor. A point charge Q is moving in air with a velocity $\mathbf{v}$ near a straight wire conductor with a time-invariant current of intensity I. Referring to three cases with different directions of $\mathbf{v}$ shown in Figure 4.3, the magnetic force on Q is zero for

(A) case (a) only.

(B) case (b) only.

(C) case (c) only.

(D) at least two of the cases.

(E) none of the cases.

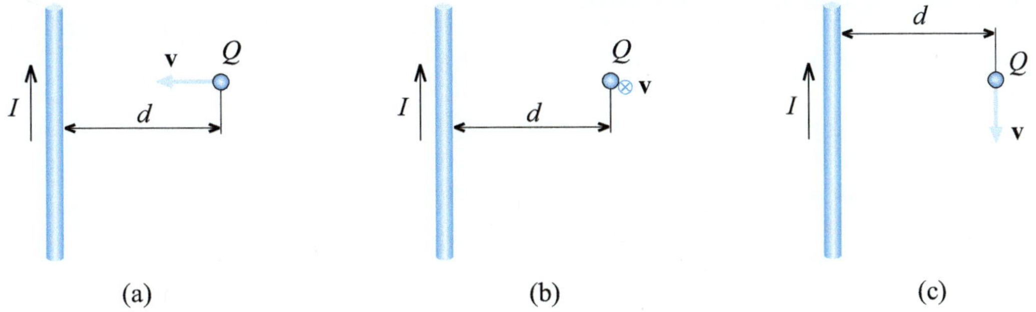

Figure 4.3 Point charge moving with a velocity $\mathbf{v}$ near a wire conductor with a steady current – three cases with different directions of $\mathbf{v}$; for Conceptual Question 4.2.

CONCEPTUAL QUESTION 4.3 **Comparing B and E fields due to point sources.** Comparing the magnetic flux density, B, due to a product $Q\mathbf{v}$ (point charge Q moving at a velocity $\mathbf{v}$) and the electric field intensity, E, due to a point charge Q at rest, both in free space, there is the same dependence on the amount of sources ($Q|\mathbf{v}|$ and Q) and on distance R, while the constant μ_0 corresponds to the constant $1/\varepsilon_0$.

(A) True.

(B) False.

CONCEPTUAL QUESTION 4.4 **Magnetic field lines due to a moving charge.** The lines of the magnetic flux density vector, $\mathbf{B}$, due to a point charge Q moving with a velocity $\mathbf{v}$ in free space are

(A) radials starting at Q.

(B) straight lines parallel to the vector $Q\mathbf{v}$.

(C) straight lines perpendicular to the vector $Q\mathbf{v}$.

(D) circles tangential to the line containing the vector $Q\mathbf{v}$.

(E) circles centered on the line containing the vector $Q\mathbf{v}$.

CONCEPTUAL QUESTION 4.5 **Acceleration and/or deflection of a charged particle.** A charged particle moves with a velocity $\mathbf{v}$ in a vacuum. An applied magnetic field of flux density $\mathbf{B}$ can change

(A) both the magnitude of $\mathbf{v}$ and its direction.

(B) the magnitude of $\mathbf{v}$ but not its direction.

(C) the direction of $\mathbf{v}$ but not its magnitude.

(D) neither the magnitude of $\mathbf{v}$ nor its direction.

4.2 Biot–Savart Law

Generalizing, by the principle of superposition, the expression in Eq. (4.4), we obtain the expression for the resultant magnetic flux density vector due to a current of intensity I flowing along a line (wire) l, Figure 4.4(a),

$$\mathbf{B} = \frac{\mu_0}{4\pi} \int_l \frac{I\,\mathrm{d}\mathbf{l} \times \hat{\mathbf{R}}}{R^2} \quad \text{(Biot–Savart law)} , \tag{4.6}$$

which is known as the Biot–Savart law.

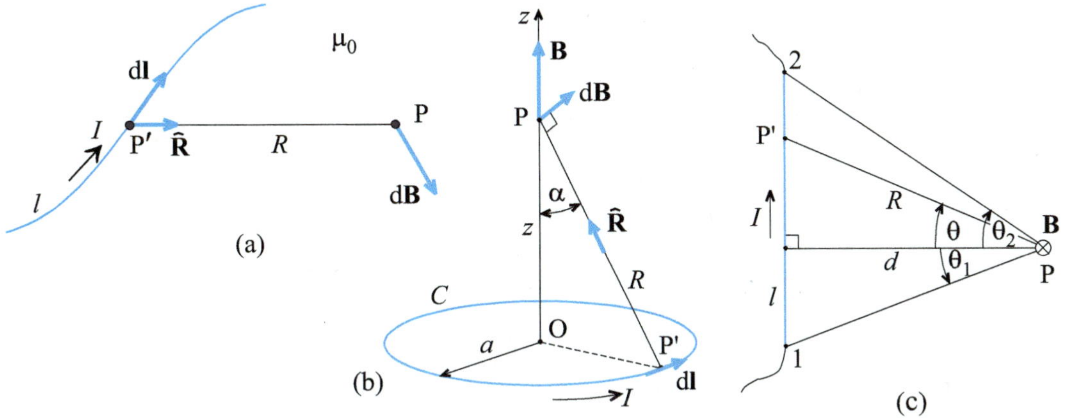

Figure 4.4 Evaluation of the magnetic flux density vector due to (a) an arbitrary line current, (b) a circular current loop, and (c) a finite straight wire conductor in free space.

Equation (4.6), along with its versions for volume and surface currents (see Figure 3.1), is a general means for evaluating (by superposition and integration) the field $\mathbf{B}$ due to given current distributions in free space or any nonmagnetic medium [versions for volume and surface currents have $\mathbf{J}\,\mathrm{d}v$ and $\mathbf{J}_\mathrm{s}\,\mathrm{d}S$, respectively, as the current element in place of $I\,\mathrm{d}\mathbf{l}$ in Eq. (4.6)]. For example, $\mathbf{B}$ due to a circular current loop at its axis normal to its plane, Figure 4.4(b), is found as follows:

$$\mathbf{B} = \oint_C \mathrm{d}B_z\,\hat{\mathbf{z}} = \oint_C \mathrm{d}B \sin\alpha\,\hat{\mathbf{z}} = \oint_C \frac{\mu_0 I\,\mathrm{d}l}{4\pi R^2}\frac{a}{R}\,\hat{\mathbf{z}} = \frac{\mu_0 I a}{4\pi R^3}\,\hat{\mathbf{z}} \oint_C \mathrm{d}l = \frac{\mu_0 I a^2}{2(z^2+a^2)^{3/2}}\,\hat{\mathbf{z}}$$

$$(\mathbf{B} \text{ due to a circular current loop}) . \tag{4.7}$$

Similarly, the magnetic flux density at an arbitrary point in space due to a steady current in a straight wire conductor of finite length, Figure 4.4(c), comes out to be

$$B = \frac{\mu_0 I}{4\pi d}(\sin\theta_2 - \sin\theta_1) \quad (B \text{ due to a straight wire conductor of finite length}) , \tag{4.8}$$

with the vector **B** shown in Figure 4.4(c). By taking $\theta_1 = -\pi/2$ and $\theta_2 = \pi/2$ in Figure 4.4(c), Eq. (4.8) gives the expression for B due to an infinitely long straight wire conductor carrying a current I, which, with a notation $d = r$, becomes

$$B = \frac{\mu_0 I}{2\pi r} \quad (B \text{ due to an infinite wire conductor}) . \tag{4.9}$$

CONCEPTUAL QUESTION 4.6 **Magnetic field due to volume currents.** Consider an arbitrary distribution of volume currents in a vacuum and the magnetic flux density vector due to these currents at an arbitrary point in space. If the magnitude of the current density vector is doubled everywhere, the magnetic flux density vector considered

(A) becomes twice as large in magnitude and keeps the same direction.

(B) becomes larger in magnitude (not always twice) and keeps the same direction.

(C) becomes larger in magnitude and may change direction.

(D) becomes twice as large in magnitude and may change direction.

(E) may become larger or smaller in magnitude and may change direction.

CONCEPTUAL QUESTION 4.7 **Current loop with two circular and two linear parts.** A steady current of intensity I ($I > 0$) flows along a planar loop consisting of two circular and two straight wire conductors, as in Figure 4.5. The medium is air. The magnetic flux density vector at the center of the circles (point O)

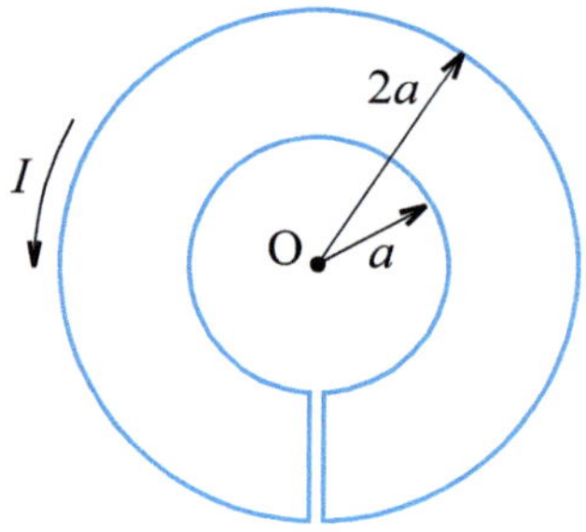

Figure 4.5 Current loop consisting of two circular and two straight wire conductors; for Conceptual Question 4.7.

(A) is directed into the plane of the drawing.

(B) is directed out of the plane of the drawing.

(C) lies in the plane of the drawing.

(D) is zero.

(E) Need more information.

CONCEPTUAL QUESTION 4.8 **Magnetic field of a rectangular current loop.** A rectangular wire loop of edge lengths a and b in air carries a steady current of intensity I ($I > 0$), as shown in Figure 4.6. The magnetic flux density vector $\mathbf{B}$ at the point M in the figure can be represented as

(A) $\mathbf{B} = B_x\,\hat{\mathbf{x}}$, where $B_x > 0$.

(B) $\mathbf{B} = B_x\,\hat{\mathbf{x}}$, where $B_x < 0$.

(C) $\mathbf{B} = B_z\,\hat{\mathbf{z}}$, where $B_z > 0$.

(D) $\mathbf{B} = B_z\,\hat{\mathbf{z}}$, where $B_z < 0$.

(E) $\mathbf{B} = B_x\,\hat{\mathbf{x}} + B_y\,\hat{\mathbf{y}}$, where $B_x \neq 0$ and $B_y \neq 0$.

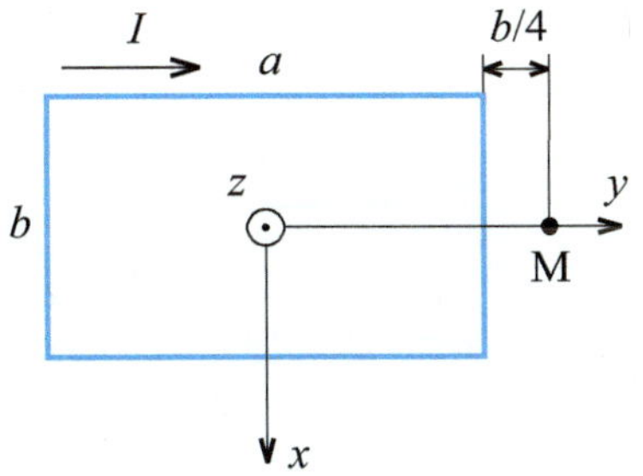

Figure 4.6 Rectangular wire loop with a steady current; for Conceptual Question 4.8.

CONCEPTUAL QUESTION 4.9 **Magnetic field due to three solenoidal coils.** Three identical solenoidal coils, wound uniformly and densely on cylindrical nonmagnetic supports with N turns of insulated thin wire in one layer [every wire turn of each of the solenoids can be regarded as a circular current loop, as the one in Figure 4.4(b)], are positioned in space as shown in Figure 4.7. The axes of coils lie in the same plane and the permeability everywhere is μ_0. Let I_1, I_2, and I_3 denote the intensities of time-invariant currents in the coils. Consider the following two cases: (a) $I_1 = I_2 = I_3 = I$ and (b) $I_1 = I$, $I_2 = I_3 = 0$. If $I > 0$, the magnetic flux density at the center of the system (the point P) for case (a) is

(A) larger than

(B) the same as

(C) smaller than

the magnetic flux density at the same point for case (b).

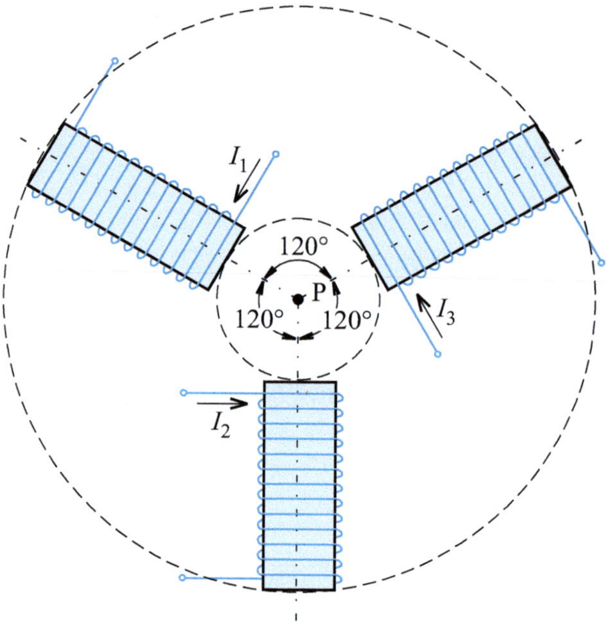

Figure 4.7 Three solenoidal coils with steady currents; for Conceptual Question 4.9.

CONCEPTUAL QUESTION 4.10 Rectangular loop and long straight wire – forces. Two conductors, a rectangular loop and a long straight wire, lie in the same plane in free space and carry steady currents of the same intensity, I, as shown in Figure 4.8. For the directions of currents indicated in the figure, the magnetic force on the straight wire conductor is directed

(A) into the plane of the drawing.

(B) out of the plane of the drawing.

(C) toward the loop.

(D) away from the loop.

(E) The force is zero.

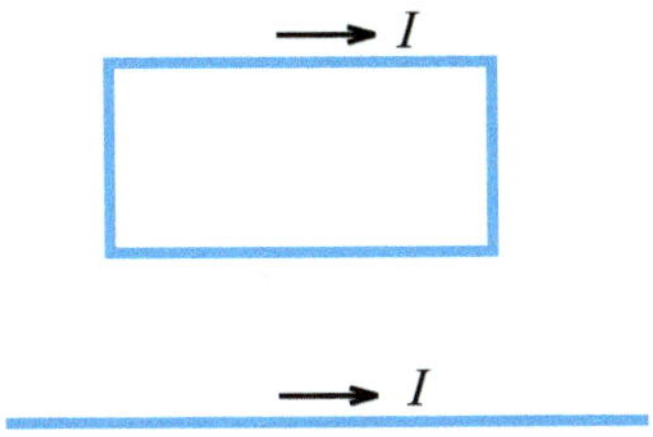

Figure 4.8 Rectangular loop and long straight wire carrying steady currents of the same intensity; for Conceptual Question 4.10.

CONCEPTUAL QUESTION 4.11 Different positions of a current loop in a magnetic field. A rectangular loop carrying a steady current of intensity I is placed in a steady uniform magnetic field of flux density $\mathbf{B}$. Figure 4.9 shows five cases with different positions of the loop with respect to the magnetic field lines. The field exerts a torque on the loop for

(A) cases (a) and (b) only.

(B) case (c) only.

(C) cases (a), (b), and (c) only.

(D) cases (d) and (e) only.

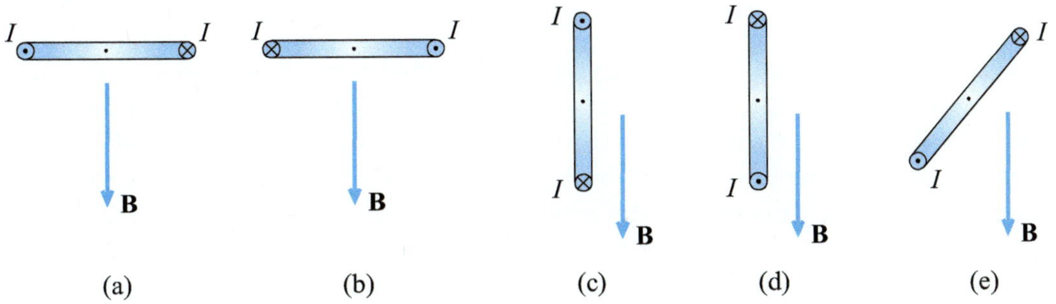

Figure 4.9 Rectangular current loop in a uniform magnetic field of flux density $\mathbf{B}$ – five different positions of the loop with respect to the field lines; for Conceptual Question 4.11.

(E) cases (c), (d), and (e) only.

(F) none of the cases shown.

CONCEPTUAL QUESTION 4.12 **Current loop in a stable equilibrium.** How many positions of the current loop in Figure 4.9 (out of the five positions shown) represent stable equilibria?

(A) Zero.

(B) One.

(C) Two.

(D) Three.

(E) Four.

4.3 Ampère's Law in Integral Form

In magnetostatics, the law that helps us evaluate the magnetic field due to highly symmetrical current distributions in free space more easily than the Biot–Savart law is Ampère's law. It states that the line integral (circulation) of the magnetic flux density vector around any contour (C) in a vacuum (free space), Figure 4.10(a), is equal to μ_0 times the total current enclosed by that contour, I_C,

$$\oint_C \mathbf{B} \cdot \mathrm{d}\mathbf{l} = \mu_0 I_C \,, \quad \oint_C \mathbf{B} \cdot \mathrm{d}\mathbf{l} = \mu_0 \int_S \mathbf{J} \cdot \mathrm{d}\mathbf{S} \quad \text{(Ampère's law)} \,, \tag{4.10}$$

with S being a surface of arbitrary shape spanned over (bounded by) C. The reference direction of the current flow, that is, the orientation of the surface S, is related to the reference direction of the contour by means of the right-hand rule: the current is in the direction defined by the thumb of the right hand when the other fingers point in the direction of the contour, as shown in Figure 4.10(a). Equation (4.10) represents Maxwell's second equation for static fields in free space.

As an illustration of the application of Ampère's law, consider an infinitely long cylindrical copper conductor of radius a in air, carrying a steady current of intensity I. Because of symmetry, the lines of the magnetic field due to the conductor current are circles centered at the conductor axis, i.e., the vector $\mathbf{B}$ at an arbitrary point

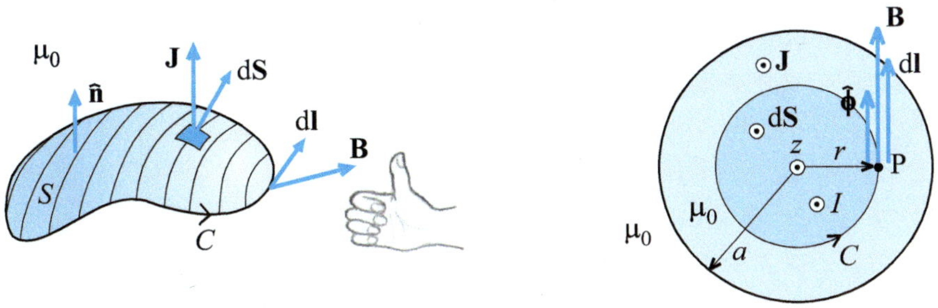

(a) (b)

Figure 4.10 (a) Arbitrary contour in a magnetostatic field – for the formulation of Ampère's law. (b) Cross section of a cylindrical conductor with a steady current I: application of Ampère's law to find the field $\mathbf{B}$.

P either inside or outside the conductor is of the form $\mathbf{B} = B(r)\,\hat{\boldsymbol{\phi}}$, as shown in Figure 4.10(b). From Eq. (4.10) and Figure 4.10(b),

$$\oint_C \mathbf{B}\cdot \mathrm{d}\mathbf{l} = \oint_C B(r)\,\mathrm{d}l = B(r)\oint_C \mathrm{d}l = B(r)\,l = B(r)\,2\pi r = \mu_0 \begin{cases} J\pi r^2 & \text{for } r < a \\ I & \text{for } r \geq a \end{cases}, \tag{4.11}$$

where $J = I/(\pi a^2)$ is the current density of the conductor, and hence

$$B = \begin{cases} \mu_0 I r/(2\pi a^2) & \text{for } r < a \\ \mu_0 I/(2\pi r) & \text{for } r \geq a \end{cases} \quad (B \text{ due to a thick cylindrical conductor})\,. \tag{4.12}$$

CONCEPTUAL QUESTION 4.13 **Algebraic total enclosed current.** The circulation of the magnetic flux density vector along the contour C in Figure 4.11 (the medium is air), $\oint_C \mathbf{B}\cdot \mathrm{d}\mathbf{l}$, equals

(A) $\mu_0\,(I_4 - I_5)$.

(B) $\mu_0\,(-I_4 + I_5)$.

(C) $\mu_0\,(I_1 - I_4 - I_7)$.

(D) $\mu_0\,(-I_1 + I_4 + I_7)$.

(E) $\mu_0\,(I_1 + 4I_2 - I_4 + I_5 - I_7)$.

(F) $\mu_0\,(I_1 + I_2 + I_3 + I_4 + I_5 + I_6 + I_7)$.

(μ_0 is the permeability of a vacuum).

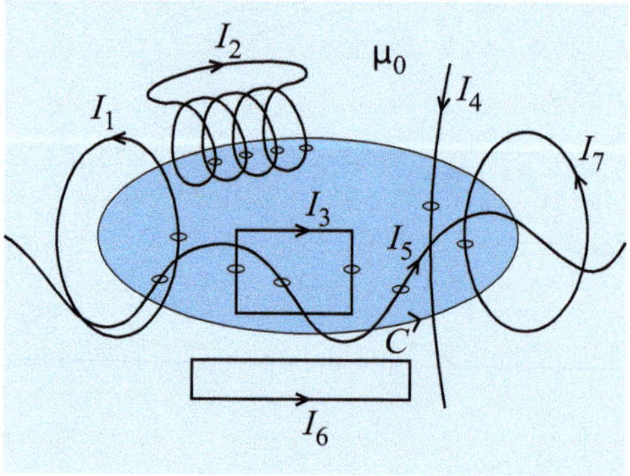

Figure 4.11 Closed path and seven line currents in air; for Conceptual Question 4.13.

CONCEPTUAL QUESTION 4.14 Amperian contour outside a current conductor. A time-invariant current of intensity I ($I > 0$) is established in a cylindrical copper conductor. The conductor is situated in air. The circulation (line integral) of the magnetic flux density vector, $\mathbf{B}$, along a contour C composed of two circular and two radial parts and positioned outside the conductor, as shown in Figure 4.12, is

(A) $\mu_0 I$.

(B) $-\mu_0 I$.

(C) greater than $\mu_0 I$.

(D) positive and less than $\mu_0 I$.

(E) zero.

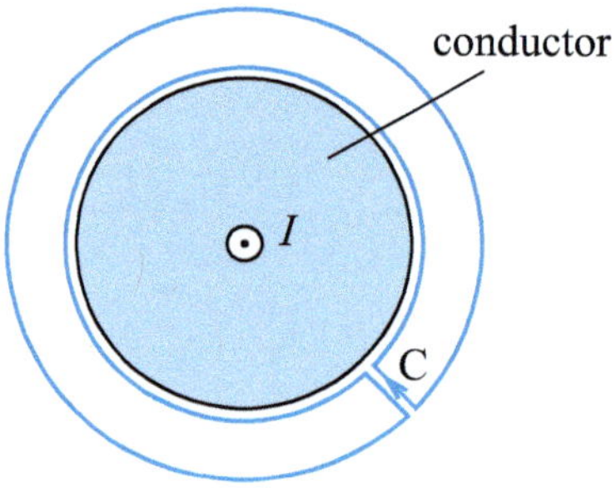

Figure 4.12 Amperian contour outside a conductor with a steady current; for Conceptual Question 4.14.

CONCEPTUAL QUESTION 4.15 Amperian contour inside a conductor. The line integral of the vector **B** along a circular contour C positioned inside a cylindrical copper conductor (in air) carrying a dc current I ($I > 0$) (the contour radius is smaller than the conductor radius), as shown in Figure 4.13, is

(A) $\mu_0 I$.

(B) $-\mu_0 I$.

(C) greater than $\mu_0 I$.

(D) positive and less than $\mu_0 I$.

(E) zero.

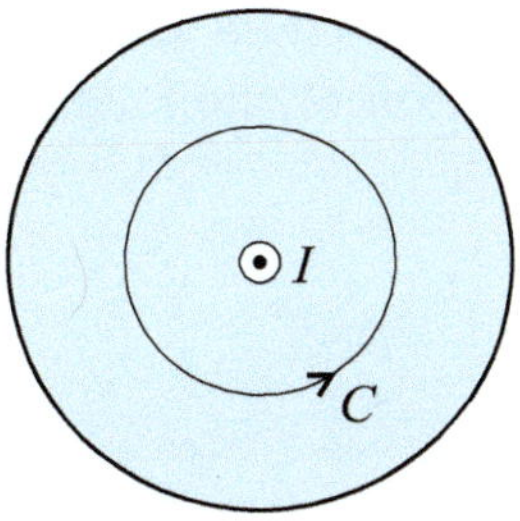

Figure 4.13 Contour inside a conductor with a dc current; for Conceptual Question 4.15.

CONCEPTUAL QUESTION 4.16 Same contour inside different conductors. Consider two very long metallic conductors, one of a circular and the other of square cross section. Both conductors carry steady currents of the same density. If the same circular contour C is positioned inside each of the conductors, as in Figure 4.14, the circulation of the magnetic flux density vector along C in the circular conductor is

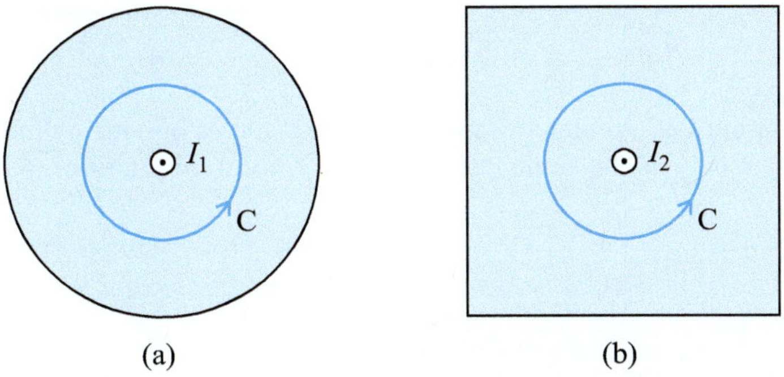

Figure 4.14 Very long conductors of circular (a) and square (b) cross sections with steady currents of the same density and the same circular Amperian contour positioned inside conductors; for Conceptual Question 4.16.

(A) larger than

(B) the same as

(C) smaller than

the circulation in the square conductor.

CONCEPTUAL QUESTION 4.17 **Different contours around the same conductor.** Consider two identical cylindrical metallic conductors carrying steady currents of the same intensity and Amperian contours positioned around each of them. The first contour is circular, while the other one has a square shape, as shown in Figure 4.15. The circulation of the magnetic flux density vector along the circular contour is

(A) larger than

(B) the same as

(C) smaller than

that along the square contour.

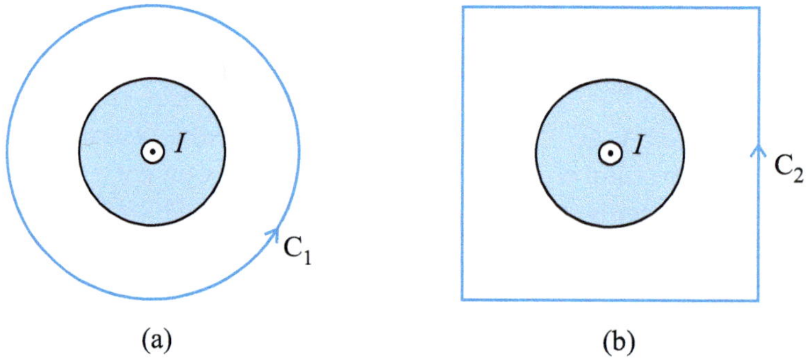

Figure 4.15 Contours of circular (a) and square (b) shapes outside cylindrical conductors carrying steady currents of the same intensity; for Conceptual Question 4.17.

CONCEPTUAL QUESTION 4.18 **Line integral along a part of a contour.** A contour composed of eight straight segments is positioned in air near a very long wire conductor with a steady current of intensity I (Figure 4.16). The line integral of the magnetic flux density vector due to this current along the part of the contour between points M and Q, via N and P, equals

(A) $\mu_0 I$.

(B) $\mu_0 I/2$.

(C) zero.

(D) $-\mu_0 I$.

(E) none of the above.

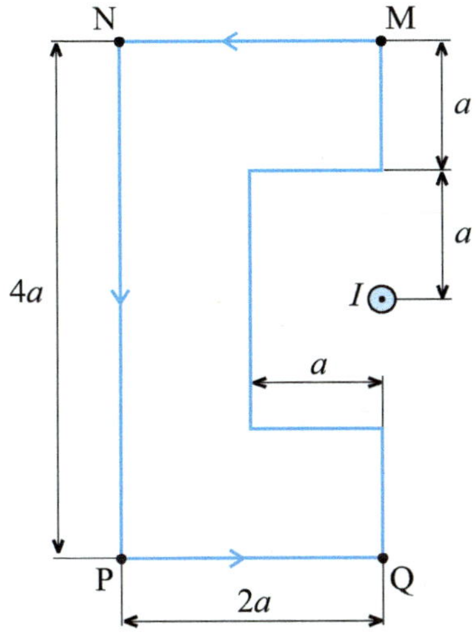

Figure 4.16 Contour composed of eight straight segments in the magnetic field of a very long current conductor; for Conceptual Question 4.18.

CONCEPTUAL QUESTION 4.19 Magnetic field in the outer conductor of a coaxial cable. The inner and outer conductors of a coaxial cable carry steady currents of the same intensity and opposite directions, as shown in Figure 4.17. The cable dielectric and conductors are nonmagnetic, and the surrounding medium is air. If the current intensity in the inner conductor is increased, while keeping the current in the outer conductor unchanged, the magnetic flux density at every point of the outer conductor

(A) increases.

(B) decreases.

(C) remains the same.

(D) Need more information.

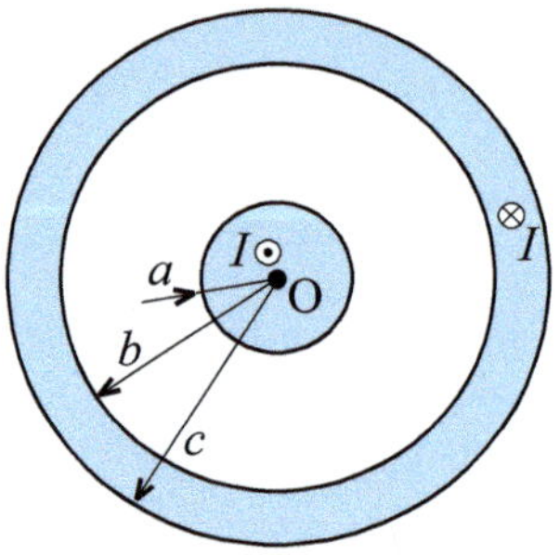

Figure 4.17 Cross section of a coaxial cable with a steady current; for Conceptual Question 4.19.

CONCEPTUAL QUESTION 4.20 Magnetic field in the inner conductor of a coaxial cable. For the coaxial cable in Figure 4.17, assume that the current intensity in the outer conductor is decreased, so the new current is $I_2 < I$, while keeping the current in the inner conductor unchanged, $I_1 = I$. As a result, the magnetic flux density at every point of the inner conductor (not considering the points at the conductor axis)

(A) increases.

(B) decreases.

(C) remains the same.

(D) Need more information.

CONCEPTUAL QUESTION 4.21 Folded metallic strip with current. An insulated metallic strip folded as in Figure 4.18 carries a steady current of intensity I. The width of the strip is $a = 20d$, where d is the diameter of the cylindrical cavity formed by the strip. With this, the magnitude of the **B** field at a point P inside the cavity (see the figure) is

(A) considerably larger than

(B) considerably smaller than

(C) practically the same as

that at a point Q outside the cavity.

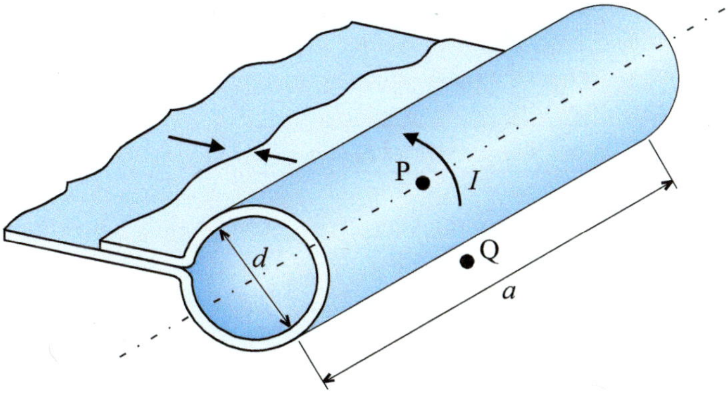

Figure 4.18 Folded metallic strip carrying a steady current; for Conceptual Question 4.21.

CONCEPTUAL QUESTION 4.22 Wider strip conductor (longer cavity). Consider the folded strip conductor in Figure 4.18, and assume that the width of the strip, a, is made twice as large, while keeping the strip current, I, and the cavity diameter, d, the same. The magnitude of the $\mathbf{B}$ field at the point P (inside the cavity) in the modified structure is

(A) noticeably larger than

(B) noticeably smaller than

(C) practically the same as

in the original structure.

CONCEPTUAL QUESTION 4.23 Two parallel large strip conductors. Two very long and wide conducting strips are placed in air parallel and very close to each other. Steady currents of the same intensity, I, and opposite directions flow through the strips, as shown in Figure 4.19. If the current direction is reversed in one of the strips, the magnetic field between the strips, away from the strip edges

(A) becomes noticeably stronger.

(B) becomes noticeably weaker.

(C) remains practically the same.

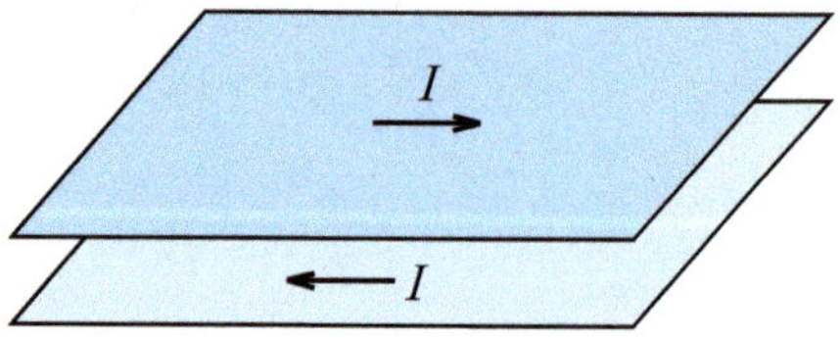

Figure 4.19 Two parallel large conducting strips with steady currents; for Conceptual Question 4.23.

4.4 Differential Form of Ampère's Law, Curl

The counterpart of Eq. (4.10) in differential notation reads

$$\left(\frac{\partial B_z}{\partial y} - \frac{\partial B_y}{\partial z}\right)\hat{\mathbf{x}} + \left(\frac{\partial B_x}{\partial z} - \frac{\partial B_z}{\partial x}\right)\hat{\mathbf{y}} + \left(\frac{\partial B_y}{\partial x} - \frac{\partial B_x}{\partial y}\right)\hat{\mathbf{z}} = \mu_0 \mathbf{J} \,, \tag{4.13}$$

with the expression on the left-hand side of the equation being the so-called curl of a vector function ($\mathbf{B}$), written as $\operatorname{curl}\mathbf{B}$. We notice that applying formally the formula for the cross product of two vectors in the Cartesian coordinate system to $\nabla \times \mathbf{B}$, where the del operator is given by Eq. (1.17), we obtain exactly $\operatorname{curl}\mathbf{B}$. Hence, the differential Ampère's law can be written in a short form as

$$\operatorname{curl}\mathbf{B} = \nabla \times \mathbf{B} = \mu_0 \mathbf{J} \quad \text{(Ampère's law in differential form)} \,. \tag{4.14}$$

Similar formulas to that in Eq. (4.13) are used to calculate the curl, namely, $\nabla \times \mathbf{B}$ in Eq. (4.14), in cylindrical and spherical coordinate systems.

CONCEPTUAL QUESTION 4.24 **Uniform magnetic field.** In a certain region in free space, there is a uniform magnetic field, with flux density $\mathbf{B}_0$. The volume current density in that region

(A) does not vary with spatial coordinates and is nonzero.

(B) is zero.

(C) varies with spatial coordinates (in a certain fashion).

(D) Need more information.

CONCEPTUAL QUESTION 4.25 **Plots of 1-D current and magnetic field distributions.** In a region in free space, the time-invariant current density vector, **J**, has a single Cartesian (x, y, or z) component, and the same is true for the associated magnetic flux density vector, **B**, in the region. Both **J** and **B** depend on the Cartesian coordinate x only, and their nonzero components are represented by a combination of the two periodic functions $f_1(x)$ and $f_2(x)$, shown in Figure 4.20. Which one is a possible combination?

(A) $J_y(x) = f_1(x)$ and $B_x(x) = f_2(x)$.

(B) $J_y(x) = f_2(x)$ and $B_x(x) = f_1(x)$.

(C) $J_z(x) = f_1(x)$ and $B_y(x) = f_2(x)$.

(D) $J_z(x) = f_2(x)$ and $B_y(x) = f_1(x)$.

(E) $J_x(x) = f_1(x)$ and $B_x(x) = f_2(x)$.

(F) $J_x(x) = f_2(x)$ and $B_x(x) = f_1(x)$.

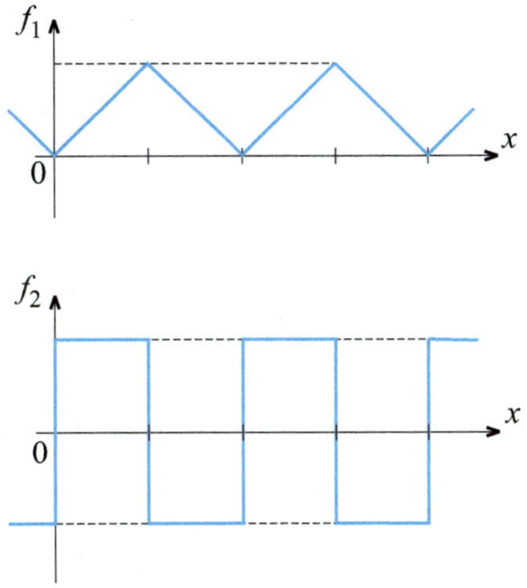

Figure 4.20 Two periodic 1-D spatial functions; for Conceptual Question 4.25.

CONCEPTUAL QUESTION 4.26 **Curl-free vector field.** Which of the field patterns in Figure 4.21 represent a curl-free ($\nabla \times \mathbf{a} = 0$) vector field **a**?

(A) Field in Figure 4.21(a) only.

(B) Field in Figure 4.21(b) only.

(C) Both fields.

(D) Neither of the fields.

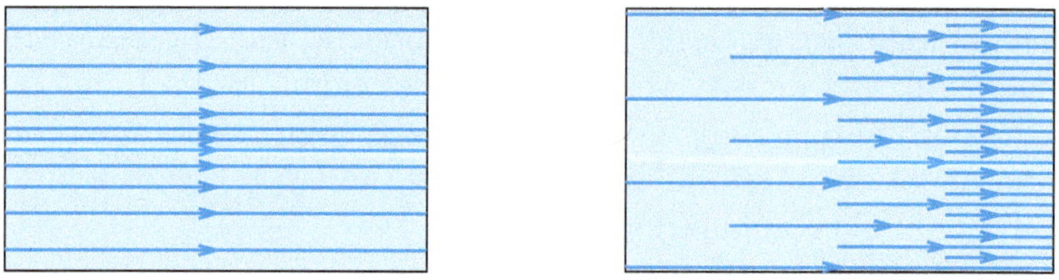

(a) (b)

Figure 4.21 Two field patterns (showing lines of a vector field **a** in a part of free space, with the magnitude of **a** at a point being proportional to the density of field lines at that point); for Conceptual Question 4.26.

4.5 Law of Conservation of Magnetic Flux

As vector **B** is called the magnetic flux density vector, its flux through a surface S is called, accordingly, the magnetic flux. It is denoted as Φ,

$$\Phi = \int_S \mathbf{B} \cdot \mathrm{d}\mathbf{S} \quad \text{(magnetic flux; unit: Wb)}\,, \tag{4.15}$$

and measured in webers (Wb). The net outward magnetic flux through a closed surface (S), like the one in Figure 1.14, is always zero, and thus [by analogy with Eqs. (3.6)] the **B** field is a divergenceless (divergence-free) or solenoidal vector field,

$$\oint_S \mathbf{B} \cdot \mathrm{d}\mathbf{S} = 0\,, \quad \nabla \cdot \mathbf{B} = 0 \quad \text{(law of conservation of magnetic flux)}\,. \tag{4.16}$$

These relations are known as the law of conservation of magnetic flux, and also as Maxwell's fourth equation (in integral and differential forms). For obvious reasons, it is sometimes referred to as Gauss' law for the magnetic field. Essentially, this law tells us that there exist no positive or negative "magnetic charges" and that the magnetic field lines must close upon themselves.

CONCEPTUAL QUESTION 4.27 **Magnetic flux through a spherical surface.** A sphere of radius a is placed in free space near a very long, straight wire carrying a steady current of intensity I. The distance of the sphere center from the wire axis is d. The magnetic flux through the sphere surface depends on

(A) a only.

(B) I only.

(C) a and I only.

(D) all three parameters (a, I, and d).

(E) none of the above parameters.

CONCEPTUAL QUESTION 4.28 Magnetic flux through different surfaces bounded by a contour. Consider an arbitrary contour C and three different open surfaces, S_1, S_2, and S_3, that are all bounded by the contour and oriented in the same way – according to the right-hand rule with respect to the orientation of the contour, as shown in Figure 4.22. Denoting the magnetic fluxes through these surfaces by Φ_1, Φ_2, and Φ_3, respectively, we have the following:

(A) $\Phi_1 = \Phi_2 = \Phi_3$.

(B) $\Phi_1 = \Phi_2 = -\Phi_3$.

(C) $\Phi_1 = -\Phi_2 = -\Phi_3$.

(D) $\Phi_1 + \Phi_2 + \Phi_3 = 0$.

(E) $\Phi_1 \neq \Phi_2$, $\Phi_2 \neq \Phi_3$, and $\Phi_1 \neq \Phi_3$.

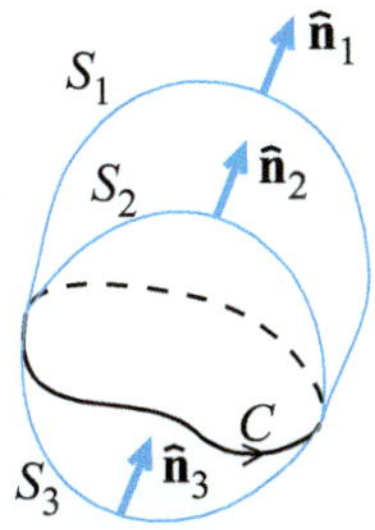

Figure 4.22 Evaluating the magnetic flux through three different open surfaces (S_1, S_2, and S_3) bounded by a contour C; for Conceptual Question 4.28.

CONCEPTUAL QUESTION 4.29 Magnetic flux through a conical surface. Consider an imaginary open conical surface in a uniform steady magnetic field of flux density $B = 1$ T. The height (length) of the cone is $h = 20$ cm and the radius of its opening is $a = 10$ cm. The vector $\mathbf{B}$ makes an angle $\alpha = 45°$ with the cone axis, as in Figure 4.23. If h is doubled (without changing a, B, and α), the magnetic flux through the conical surface (oriented downward)

(A) increases.

(B) decreases.

(C) remains the same.

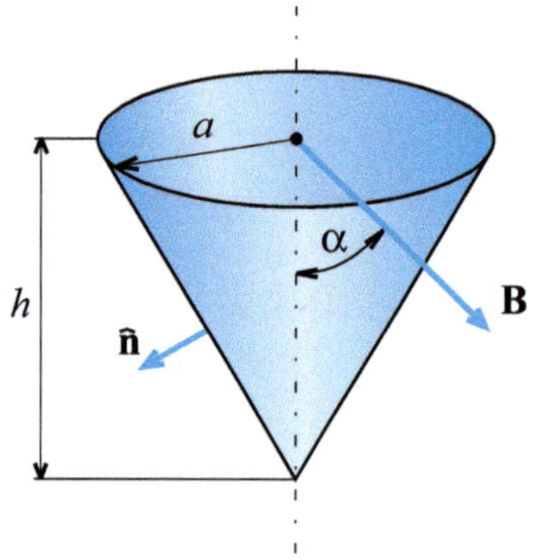

Figure 4.23 Open conical surface in a uniform magnetic field; for Conceptual Question 4.29.

CONCEPTUAL QUESTION 4.30 **Magnetic flux through a cylindrical surface.** Consider an imaginary (nonmaterial) cylinder of radius a and height h in a uniform magnetic field of flux density B. The vector $\mathbf{B}$ makes an angle α with the cylinder axis. The outward magnetic flux through the lateral surface of the cylinder equals

(A) $\Phi = 2\pi Bah \sin\alpha$.

(B) $\Phi = -2\pi Bah \sin\alpha$.

(C) $\Phi = 2\pi Ba^2 \cos\alpha$.

(D) $\Phi = -2\pi Ba^2 \cos\alpha$.

(E) $\Phi = \pi Ba^2 \cos\alpha$.

(F) $\Phi = 0$.

CONCEPTUAL QUESTION 4.31 **Magnetic flux through cylinder bases.** Consider a vertical cylinder in a steady magnetic field in free space. If Φ_1 denotes the magnetic flux through the lower basis of the cylinder and Φ_2 that through the upper basis with both surfaces oriented in the same way (upward), we have that

(A) $\Phi_1 = \Phi_2$.

(B) $\Phi_1 \neq \Phi_2$.

(C) Need more information.

CONCEPTUAL QUESTION 4.32 **Possible magnetostatic field patterns.** Which of the two field patterns in Figure 4.24 represent(s) a possible steady magnetic field **B**?

(A) Pattern in Figure 4.24(a) only.

(B) Pattern in Figure 4.24(b) only.

(C) Both patterns.

(D) Neither.

Figure 4.24 Two vector field patterns (vector magnitude is proportional to the field lines density); for Conceptual Question 4.32.

CONCEPTUAL QUESTION 4.33 **Possible electrostatic field patterns.** Which of the two patterns in Figure 4.24 represent(s) a possible static electric field vector **E**?

(A) Pattern in Figure 4.24(a) only.

(B) Pattern in Figure 4.24(b) only.

(C) Both patterns.

(D) Neither.

CONCEPTUAL QUESTION 4.34 **Vector with zero curl and divergence at a point.** If both the curl and divergence of a vector **a** at a point are zero, then **a** must be zero at that point.

(A) True.

(B) False.

CONCEPTUAL QUESTION 4.35 **Zero vector at a point.** If a vector **a** at a point is zero, then both the curl and divergence of **a** must be zero at that point.

(A) True.

(B) False.

4.6 Magnetic Vector Potential

In analogy to the computation of the electric scalar potential (V) in Eq. (1.12) and the electric field intensity vector ($\mathbf{E}$) from V in Eq. (1.16), in magnetostatics we have

$$\mathbf{A} = \frac{\mu_0}{4\pi}\int_l \frac{I\,\mathrm{d}\mathbf{l}}{R} \quad \longrightarrow \quad \mathbf{B} = \nabla \times \mathbf{A} \quad \text{(magnetic flux density from potential)}\,, \tag{4.17}$$

where the vector $\mathbf{A}$ is called the magnetic vector potential, and its unit is $\mathrm{T \cdot m}$.

CONCEPTUAL QUESTION 4.36 **Magnetic vector potential of a straight wire conductor.** A steady current is established in a straight metallic wire conductor in a nonmagnetic medium. The magnetic vector potential due to this current at an arbitrary point in space that is not on the wire axis is

(A) parallel to the wire.

(B) perpendicular to the wire.

(C) Neither of the above.

(D) Depends on the particular position of the point with respect to the wire.

CONCEPTUAL QUESTION 4.37 **Magnetic vector potential of a square current loop.** Consider a square loop with a steady current of intensity I, in free space. Let A_1 denote the magnitude of the magnetic vector potential at the loop center due to the current along one of the square sides. The magnitude of the total magnetic vector potential at the center equals

(A) $4A_1$.

(B) $2A_1$.

(C) zero.

(D) None of the above.

CONCEPTUAL QUESTION 4.38 **Magnetic vector potential of a circular current loop.** The magnetic vector potential, **A**, due to a circular wire loop with a steady current is zero at any point along the axis of the loop normal to the loop plane [z-axis in Figure 4.4(b)].

(A) True.

(B) False.

CONCEPTUAL QUESTION 4.39 **Field from potential, circular current loop.** Can the expression for the magnetic flux density vector, **B**, due to a circular current loop at its axis normal to its plane [expression for **B** in Eq. (4.7)] be obtained from the expression for the vector potential **A** due to the loop along the same axis [z-axis in Figure 4.4(b)]?

(A) Yes.

(B) No.

---

5 MAGNETOSTATIC FIELD IN MATERIAL MEDIA

Introduction

In analysis of the magnetostatic field in the presence of magnetic materials, many basic concepts, physical laws, and mathematical techniques are entirely analogous to the corresponding concepts, laws, and techniques in electrostatics (Chapter 2). Magnetized materials can be represented by vast collections of tiny atomic current loops, i.e., magnetic dipoles, while the concept of permeability of a medium allows for macroscopic characterization of materials and their fields. The most important difference, however, with respect to the analysis of dielectric materials is the inherent nonlinear behavior of the most important class of magnetic materials, called ferromagnetics. This is a class of materials with striking magnetic properties (many orders of magnitude stronger than in other materials), with iron as a typical example. Analysis of magnetic circuits (consisting of ferromagnetic cores of different shapes with current-carrying windings), which is a culmination of our study of the magnetostatic field in material media, thus essentially resembles the dc analysis of nonlinear electric circuits. In addition, configurations of current-carrying conductors store magnetic energy. For systems that contain ferromagnetic materials with pronounced nonlinearity, the energy balance in the system includes so-called hysteresis losses in the material.

5.1 Magnetization Current

According to the elementary atomic model of matter, all materials are composed of atoms, each with a central fixed positively charged nucleus and a number of negatively charged electrons circulating around the nucleus in various orbits. Both these orbital motions and the inherent spins of the electrons about their own axes can be represented by small current loops, also referred to as magnetic dipoles. The magnetic moment of each loop is given by $\mathbf{m} = I\mathbf{S}$, where I is the current intensity of the loop and $\mathbf{S}$ is the loop surface area vector, oriented in accordance to the right-hand rule with respect to the reference direction of the current. With an applied magnetic field, these current loops experience torques, which lead to a net alignment of microscopic magnetic dipole moments with the external field and a net magnetic moment in the material, and this process is known as the magnetization of the material. However, instead of analyzing every single atom and all microscopic magnetic dipole moments, we introduce a macroscopic quantity termed the magnetization vector and defined as [note the analogy with Eq. (2.2)]

$$\mathbf{M} = \frac{(\sum \mathbf{m})_{\text{in } dv}}{dv} \quad \text{(magnetization vector; unit: A/m)} . \tag{5.1}$$

The macroscopic current equivalent to atomic current loops (magnetic dipoles) is named the magnetization current, and the corresponding volume and surface current densities [see Figure 3.1 and Eqs. (3.2) and (3.3)], $\mathbf{J}_{\rm m}$ and $\mathbf{J}_{\rm ms}$, throughout the volume and over the surface, respectively, of the magnetized body are evaluated as

$$\mathbf{J}_{\rm m} = \operatorname{curl}\mathbf{M} = \nabla\times\mathbf{M}\,, \quad \mathbf{J}_{\rm ms} = \mathbf{M}\times\hat{\mathbf{n}} \quad (\hat{\mathbf{n}} \text{ outward normal})$$
$$\text{(magnetization volume and surface current density vectors)}\,. \quad (5.2)$$

We can then, considering these macroscopic currents to reside in a vacuum, calculate the magnetic flux density vector, **B**, due to the magnetized body (and any other related quantity of interest) using the appropriate free-space equations (e.g., various forms of the Biot–Savart law and Ampère's law).

CONCEPTUAL QUESTION 5.1 **Uniformly magnetized material.** In a uniformly magnetized material, with a magnetization vector that does not change spatially ($\mathbf{M} = \text{const}$), there is no net volume macroscopic magnetization current.[1]

(A) Always true.

(B) Sometimes true, sometimes false.

(C) Always false.

CONCEPTUAL QUESTION 5.2 **Nonuniformly magnetized material.** Consider a magnetic body with a magnetization vector varying ($\mathbf{M} \neq \text{const}$) throughout the volume of the material (nonuniformly magnetized material). In such a material, there is no net volume macroscopic magnetization current.

(A) Always true.

(B) Sometimes true, sometimes false.

(C) Always false.

CONCEPTUAL QUESTION 5.3 **Surface magnetization current.** On the entire surface of a magnetized magnetic body, there always exists surface macroscopic magnetization current (there are parts of atomic current loops pressed onto the surface that cannot be compensated by oppositely flowing currents of neighboring loops).

[1] *For every conceptual question in this text, exactly one answer is correct.*

(A) True.

(B) False.

CONCEPTUAL QUESTION 5.4 **Infinitely long cylindrical bar magnet.** An infinitely long cylindrical bar magnet of radius a, in air, is permanently magnetized with a uniform magnetization, and the magnetization vector, of magnitude M, is parallel to the bar axis. Magnitudes of the magnetization volume and surface current density vectors, $\mathbf{J}_{\rm m}$ and $\mathbf{J}_{\rm ms}$, over the volume and surface of the magnet, respectively, are

(A) $J_{\rm m} = M$ and $J_{\rm ms} = 0$.

(B) $J_{\rm m} = 0$ and $J_{\rm ms} = M$.

(C) $J_{\rm m} = \mu_0 M$ and $J_{\rm ms} = M/a$.

(D) $J_{\rm m} = M/a$ and $J_{\rm ms} = M$.

(E) $J_{\rm m} = 0$ and $J_{\rm ms} = 0$.

(μ_0 is the permeability of a vacuum).

CONCEPTUAL QUESTION 5.5 **Field due to a uniformly magnetized ferromagnetic disk.** A thin ferromagnetic disk of radius a and thickness d ($d \ll a$) is situated in air. The disk is uniformly magnetized throughout its volume. The magnetization vector is normal to disk bases and its magnitude is M. The magnetic flux density vector, $\mathbf{B}$, due to the magnetized disk is equal to $\mathbf{B}$ due to an equivalent circular current loop with radius a and current intensity $I = Md$ placed along the circumference of the disk and assumed to be in a vacuum – if observed (i.e., if $\mathbf{B}$ is computed) at

(A) an arbitrary point along the disk axis normal to its bases (and no other point).

(B) the center of the disk (and no other point).

(C) an arbitrary point inside the disk (and no other point).

(D) an arbitrary point outside the disk (and no other point).

(E) an arbitrary point in space.

(F) no point.

CONCEPTUAL QUESTION 5.6 **Nonuniformly magnetized ferromagnetic body.** The magnetization vector in a ferromagnetic body is given by $\mathbf{M} = M(x)\,\hat{\mathbf{y}}$. The magnetization volume current density vector inside this body can be represented as

(A) $\mathbf{J}_{\rm m} = J_{\rm m}(x)\,\hat{\mathbf{x}}$.

(B) $\mathbf{J}_{\rm m} = J_{\rm m}(x)\,\hat{\mathbf{y}}$.

(C) $\mathbf{J}_{\rm m} = J_{\rm m}(x)\,\hat{\mathbf{z}}$.

(D) $\mathbf{J}_{\rm m} = J_{\rm m}(y)\,\hat{\mathbf{x}}$.

(E) $\mathbf{J}_{\rm m} = J_{\rm m}(z)\,\hat{\mathbf{x}}$.

(F) $\mathbf{J}_{\rm m} = 0$.

5.2 Generalized Ampère's Law and Permeability

For a general magnetostatic system where we have conduction currents flowing through conductors (including conducting magnetic materials) and magnetization currents inside magnetic bodies and over their surfaces, inclusion of magnetization properties of materials in Eqs. (4.10) and (4.14) leads to the corresponding forms of the generalized Ampère's law:

$$\oint_C \mathbf{H}\cdot d\mathbf{l} = I_C\,, \qquad \nabla\times\mathbf{H} = \mathbf{J} \qquad \text{(generalized Ampère's law)}\,, \tag{5.3}$$

where, as in Eqs. (4.10) and (4.14), I_C is the total conduction current enclosed by an arbitrary contour C and $\mathbf{J}$ is the conduction current density vector, while $\mathbf{H}$ is a new quantity called the magnetic field intensity vector and defined as

$$\mathbf{H} = \frac{\mathbf{B}}{\mu_0} - \mathbf{M} \qquad \text{(magnetic field intensity vector; unit: A/m)}\,. \tag{5.4}$$

For linear magnetic materials, the magnetization vector, $\mathbf{M}$, is linearly proportional to $\mathbf{B}$, so that, from Eq. (5.4), $\mathbf{B}$ is linearly proportional to $\mathbf{H}$, and the proportionality constant in the latter case is the permeability of the medium, μ,

$$\mathbf{B} = \mu\mathbf{H}\,, \qquad \mu = \mu_{\rm r}\mu_0 \qquad \text{(permeability; unit: H/m)}\,, \tag{5.5}$$

with the unit for μ being henry per meter (H/m) and $\mu_{\rm r}$ denoting the relative permeability of the material (dimensionless).

In ferromagnetics, the function $B = B(H)$ is in general nonlinear and has multiple branches. The value of μ generally is not unique. Typically, the maximum value of μ_r is around 250 for cobalt, 600 for nickel, and 5,000 for iron (with 0.4% impurity), whereas it is as high as about 200,000 for purified iron (0.04% impurity). In many applications involving ferromagnetics, we assume

$$\mu_r \to \infty \quad \text{(perfect magnetic conductor – PMC)} , \tag{5.6}$$

and such media are customarily referred to as perfect magnetic conductors (PMCs).

CONCEPTUAL QUESTION 5.7 Magnetic region with no conduction currents. If in a certain magnetic region there are no conduction volume currents ($\mathbf{J} = 0$) and the magnetic flux density vector does not vary with spatial coordinates ($\mathbf{B} = \text{const}$), the magnetization volume current density vector, $\mathbf{J}_m$, in the region is

(A) always zero.

(B) generally nonzero, but not a function of spatial coordinates ($\mathbf{J}_m = \text{const}$).

(C) generally a (nonzero) function of spatial coordinates.

CONCEPTUAL QUESTION 5.8 Magnetic field intensity in a toroidal magnetic core. A uniform and dense winding (coil) with N turns of wire is placed over a toroidal magnetic core, as shown in Figure 5.1. There is a steady current of intensity I through the winding. The magnetic field intensity vector, $\mathbf{H}$, in the core is the same as inside the same coil when air-filled – under the following condition:

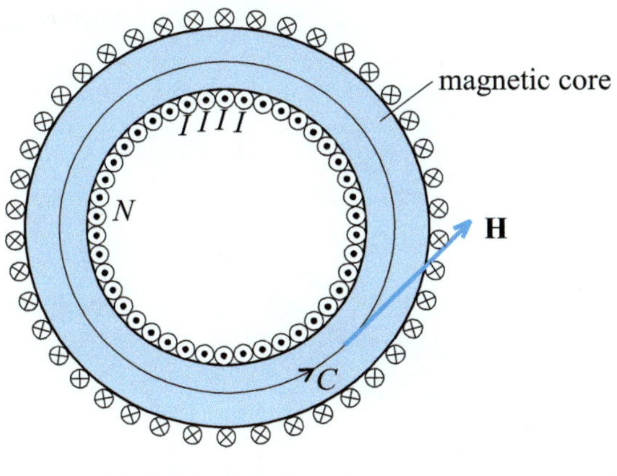

Figure 5.1 Evaluation of the magnetic field intensity vector in a toroidal coil with a magnetic core; for Conceptual Question 5.8.

(A) if the magnetic material of the core is linear.

(B) if the toroid is thin.

(C) if the core material has not reached the state of saturation.

(D) if the core material is in the state of saturation.

(E) always.

(F) never.

CONCEPTUAL QUESTION 5.9 **Coaxial cable partly filled with a ferromagnetic layer.** A coaxial cable carries a time-invariant current I. A thin layer of a ferromagnetic material is placed near the outer conductor, and the rest of the space between the conductors is air-filled. With respect to the notation in Figure 5.2 showing the cable cross section, the magnetic field exists only in

(A) region 3.

(B) regions 2 and 3.

(C) regions 1, 3, and 4.

(D) regions 1, 2, 3, and 4.

(E) regions 1, 2, 3, 4, and 5.

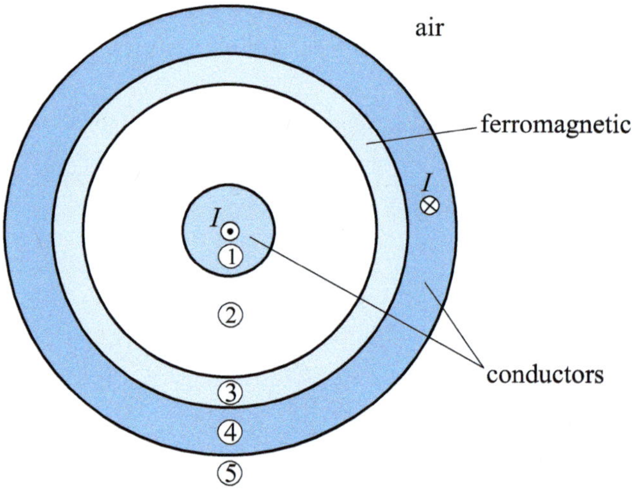

Figure 5.2 Cross section of a coaxial cable partly filled with a ferromagnetic material; for Conceptual Question 5.9.

CONCEPTUAL QUESTION 5.10 Homogeneous linear magnetic material. The conduction volume current density vector at a point in a homogeneous and linear magnetic material of relative permeability μ_r is $\mathbf{J}$. The curl of the magnetic flux density vector, $\nabla \times \mathbf{B}$, at that point equals

(A) $\mu_0 \mathbf{J}$.

(B) $\mu_r \mu_0 \mathbf{J}$.

(C) $\mu_r \mathbf{J}$.

(D) $(\mu_r - 1)\mathbf{J}$.

(E) $\mathbf{J}$.

(F) 0.

CONCEPTUAL QUESTION 5.11 Magnetization current in a homogeneous linear material. For a homogeneous and linear magnetic material of relative permeability μ_r, the magnetization volume current density vector at a point in the material at which the conduction volume current density vector is $\mathbf{J}$ amounts to

(A) $\mathbf{J}_m = 0$.

(B) $\mathbf{J}_m = \mathbf{J}$.

(C) $\mathbf{J}_m = \mu_r \mathbf{J}$.

(D) $\mathbf{J}_m = -\mu_r \mathbf{J}$.

(E) $\mathbf{J}_m = (\mu_r - 1)\mathbf{J}$.

CONCEPTUAL QUESTION 5.12 Current-free homogeneous medium. There cannot be magnetization volume current ($\mathbf{J}_m = 0$) in a homogeneous linear magnetic medium with no conduction current ($\mathbf{J} = 0$).

(A) True.

(B) False.

CONCEPTUAL QUESTION 5.13 **Permeability versus field intensity in a nonlinear ferromagnetic.** The permeability, μ, of a nonlinear ferromagnetic material is a function of the applied magnetic field intensity, H, in the material.

(A) True.

(B) False.

5.3 Boundary Conditions for the Magnetic Field

From the integral forms of the generalized Ampère's law, Eq. (5.3), and the law of conservation of magnetic flux, Eq. (4.16), we obtain the following set of boundary conditions for the magnetic field on the boundary surface between two arbitrary media, analogous to those in Eqs. (2.7):

$$H_{1\mathrm{t}} - H_{2\mathrm{t}} = J_\mathrm{s} \,, \quad B_{1\mathrm{n}} = B_{2\mathrm{n}} \quad \text{(magnetic–magnetic boundary conditions)} \,, \tag{5.7}$$

where J_s is the density of a surface conduction current that may exist on the boundary, Figure 5.3(a). In the case of linear magnetic media of permeabilities μ_1 and μ_2 and $J_\mathrm{s} = 0$, Figure 5.3(b), the law of refraction of the magnetic field lines holds, of the same form as in Eqs. (2.9) and (3.13),

$$\frac{\tan\alpha_1}{\tan\alpha_2} = \frac{\mu_1}{\mu_2} \quad \text{(law of refraction of magnetic field lines)} \,. \tag{5.8}$$

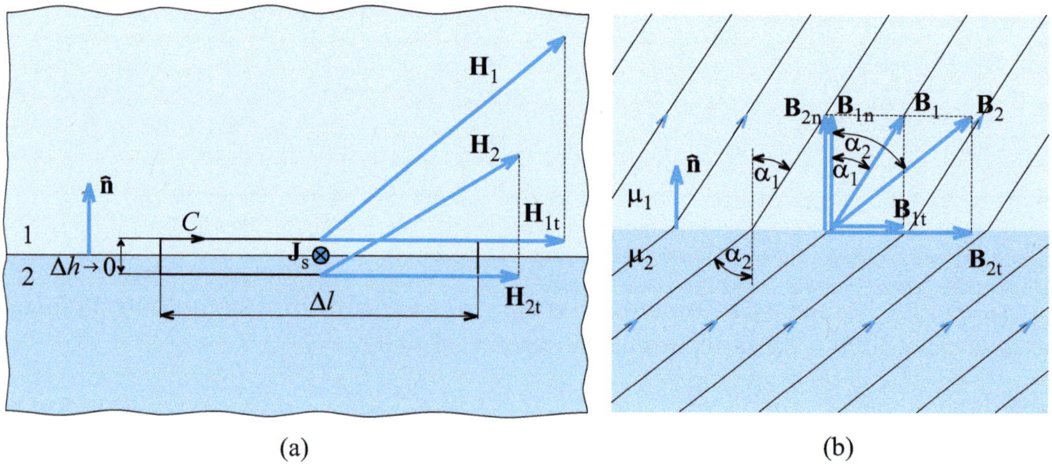

Figure 5.3 (a) Boundary condition for tangential components of vector **H** at an arbitrary magnetic–magnetic interface. (b) Refraction of magnetic field lines at a boundary surface between two linear magnetic media with $\mathbf{J}_\mathrm{s} = 0$.

CONCEPTUAL QUESTION 5.14 Boundary conditions at a magnetic–magnetic interface. Consider a boundary surface between two magnetic media, with relative permeabilities $\mu_{r1} = 600$ and $\mu_{r2} = 300$, respectively. Assuming that no conduction current exists on the boundary ($\mathbf{J}_s = 0$), which of the cases shown in Figure 5.4 represent possible magnetic field intensity vectors on the two sides of the boundary?

(A) Case (a) only.

(B) Case (b) only.

(C) Case (c) only.

(D) Case (d) only.

(E) More than one case.

(F) None of the cases.

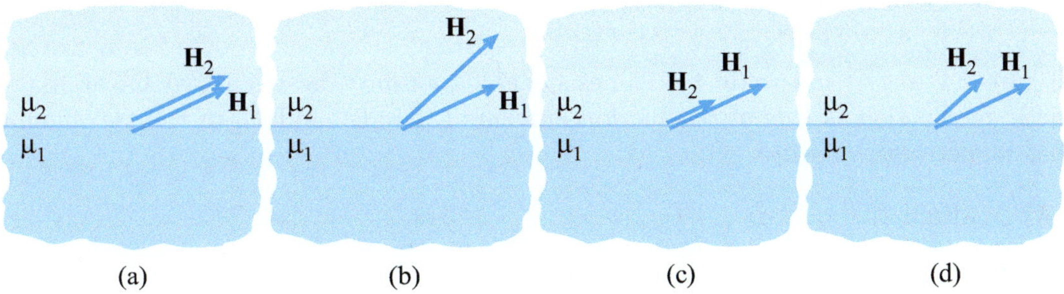

Figure 5.4 Four offered combinations of magnetic field intensity vectors on two sides of a magnetic–magnetic interface ($\mu_1 = 2\mu_2$ and $\mathbf{J}_s = 0$); for Conceptual Question 5.14.

CONCEPTUAL QUESTION 5.15 Boundary conditions for the flux density vector. The following cases shown in Figure 5.5 represent possible magnetic flux density vectors, $\mathbf{B}_1$ and $\mathbf{B}_2$, on the two sides of a magnetic–magnetic interface:

(A) Case (a) only.

(B) Case (b) only.

(C) Case (c) only.

(D) Case (d) only.

(E) More than one case.

(F) None of the cases.

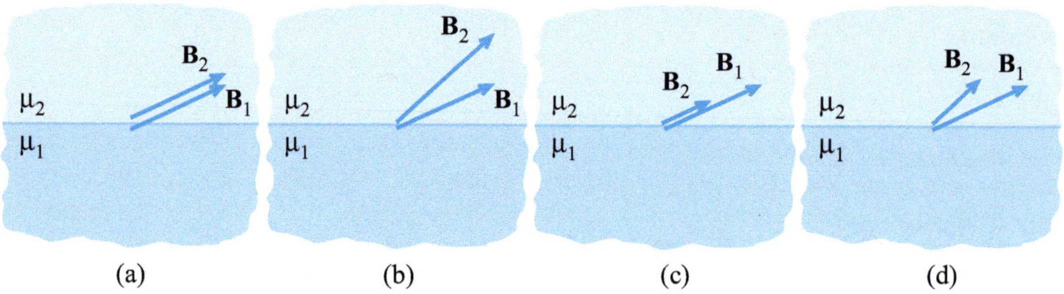

Figure 5.5 Four offered combinations of **B** vectors on two sides of a boundary surface between two magnetic media with $\mu_1 = 2\mu_2$ and $\mathbf{J}_s = 0$; for Conceptual Question 5.15.

CONCEPTUAL QUESTION 5.16 **Refraction of magnetic field lines.** Figure 5.6 shows magnetic field lines near a boundary between two linear magnetic media that is free of conduction currents ($\mathbf{J}_s = 0$). Which of the two media has higher permeability?

(A) Medium 1.

(B) Medium 2.

(C) Need more information.

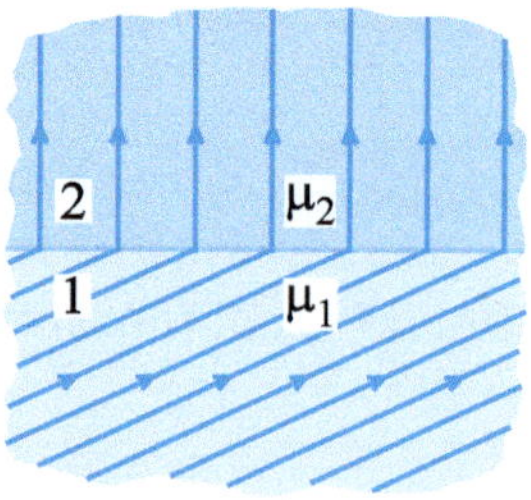

Figure 5.6 Refraction of magnetic field lines at an interface between two magnetic media, with $\mathbf{J}_s = 0$; for Conceptual Question 5.16.

CONCEPTUAL QUESTION 5.17 **Magnetic field vector near a PMC surface.** A current element $I\,\mathrm{dl}$ $(I > 0)$ is situated in air above a large block of a ferromagnetic material that can be considered to be a perfect magnetic conductor (PMC). The element is parallel to the surface of the block. The magnetic field intensity vector $\mathbf{H}$ in air immediately above the surface is

(A) as in Figure 5.7(a).

(B) as in Figure 5.7(b).

(C) as in Figure 5.7(c).

(D) as in Figure 5.7(d).

(E) zero.

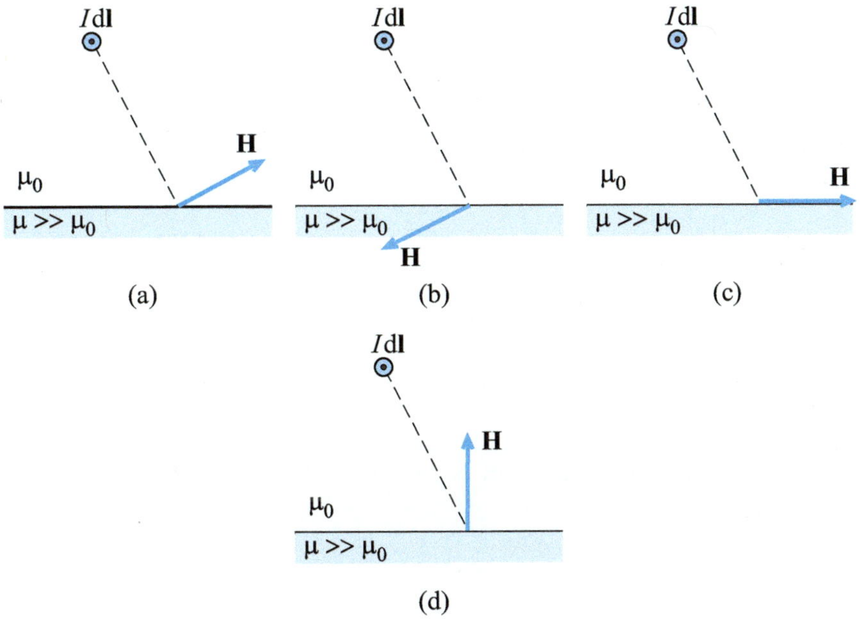

Figure 5.7 Current element $I\,\mathrm{dl}$ above a PMC block, in air; for Conceptual Question 5.17.

5.4 Image Theory for the Magnetic Field

Magnetostatic systems often include current conductors in the presence of large flat ferromagnetic bodies, which can be approximated by a PMC [see Eq. (5.6)] half-space. By utilizing image theory, similarly to the procedure depicted in Figure 1.32, we can remove the ferromagnetic body from the system and replace it by an image of the original current distribution, as shown in Figure 5.8 for the case of a straight current conductor parallel to the PMC plane, to obtain a free-space system.

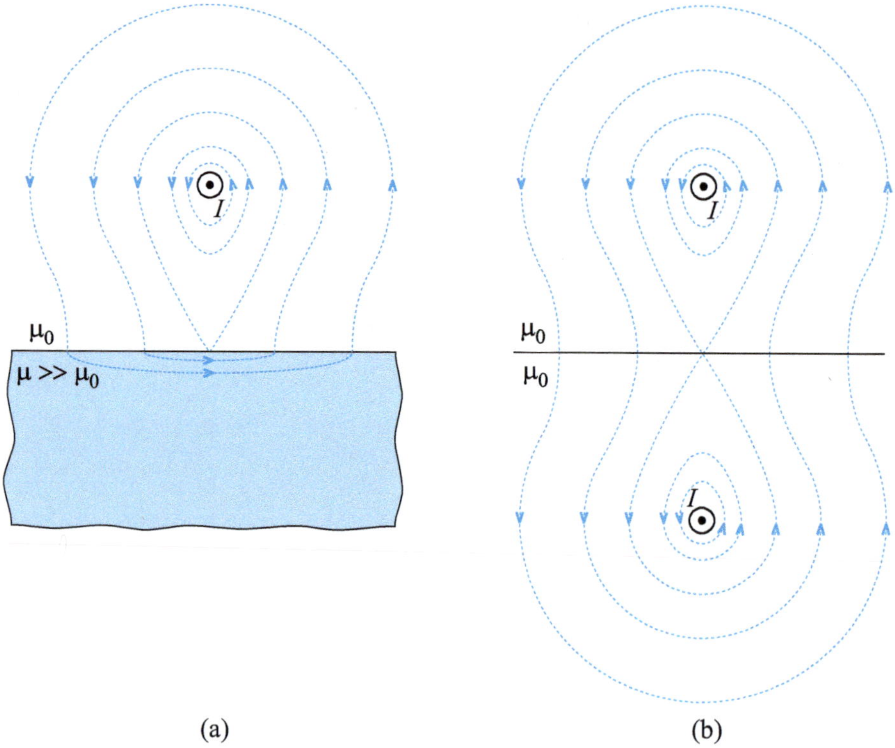

Figure 5.8 (a) Straight current conductor parallel to the interface of a ferromagnetic (or PMC) half-space. (b) By image theory, the influence of the ferromagnetic material on the magnetic field in the upper half-space can be represented by a positive image of the original current.

CONCEPTUAL QUESTION 5.18 **Force on a conductor above a ferromagnetic plane.** Magnetic force with which a large ferromagnetic block (namely, its magnetization current), considered to be a PMC plane, acts on a current conductor (wire with a steady current I) running parallel to it, at some distance in air, is always

(A) zero.

(B) attractive (plane attracts the conductor).

(C) repulsive (plane repels the conductor).

(D) nonzero, but neither attractive nor repulsive (it is parallel to the plane).

CONCEPTUAL QUESTION 5.19 Image theory for a line current parallel to a PMC corner. Figure 5.9 shows the application of image theory to a very long and thin metallic wire with a steady current of intensity I placed in air parallel to a PMC ($\mu_\mathrm{r} \to \infty$) 90° corner. The magnetic field in air is the same in the original system and the system with images (and PMC corner removed) if the intensities of the three image currents in the figure are

(A) $I_1 = I_2 = -I$ and $I_3 = I$.

(B) $I_1 = I_2 = I$ and $I_3 = -I$.

(C) $I_1 = I_2 = I_3 = I$.

(D) $I_1 = I_2 = I$ and $I_3 = 2I$.

(E) $I_1 = I_2 = I$ and $I_3 = 0$.

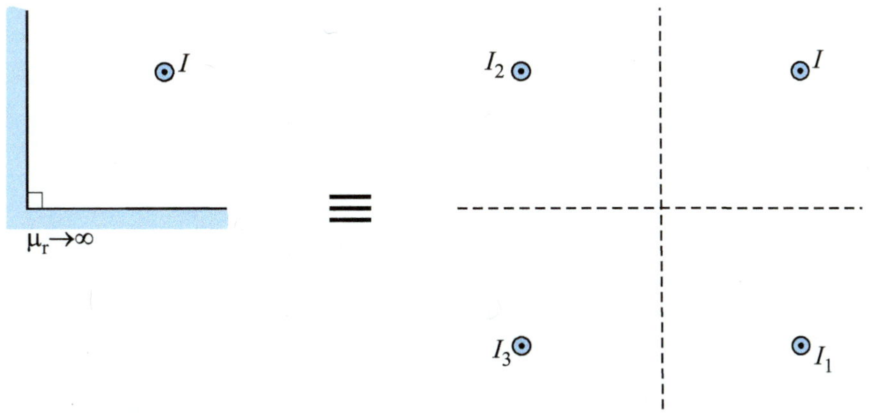

Figure 5.9 Image theory for a current-carrying metallic wire in the presence of a 90° PMC corner (cross section of the structure); for Conceptual Question 5.19.

5.5 Magnetization Curves and Hysteresis

In this section, we consider the B–H relationship for ferromagnetic materials. This relationship, being nonlinear in general, is usually given as a graph showing B

(ordinate) as a function of H (abscissa), and a curve representing the function $B(H)$ on such a diagram is called a magnetization curve. Shown in Figure 5.10(a) is a typical initial magnetization curve for a ferromagnetic sample, where the material is completely demagnetized and both B and H are zero before a field is applied. Very strong magnetic fields are usually required to reach the state of saturation, where the magnetization curve flattens off completely. As we next reduce H, the effects of hysteresis begin to show [hysteresis (derived from a Greek word meaning "to lag") implies that B lags behind H], and we do not retrace the initial-magnetization curve, as can be seen in Figure 5.10(b). Even after H becomes zero, B does not go to zero, but to a value $B = B_r$, termed the remanent (residual) magnetic flux density. Note that the existence of a remanent flux density in a ferromagnetic material makes permanent magnets possible. With a further decrease and then increase in H, a full cycle in the BH diagram is completed [Figure 5.10(b)]. The loop traced out by the magnetization curve during this cycle is referred to as the hysteresis loop.

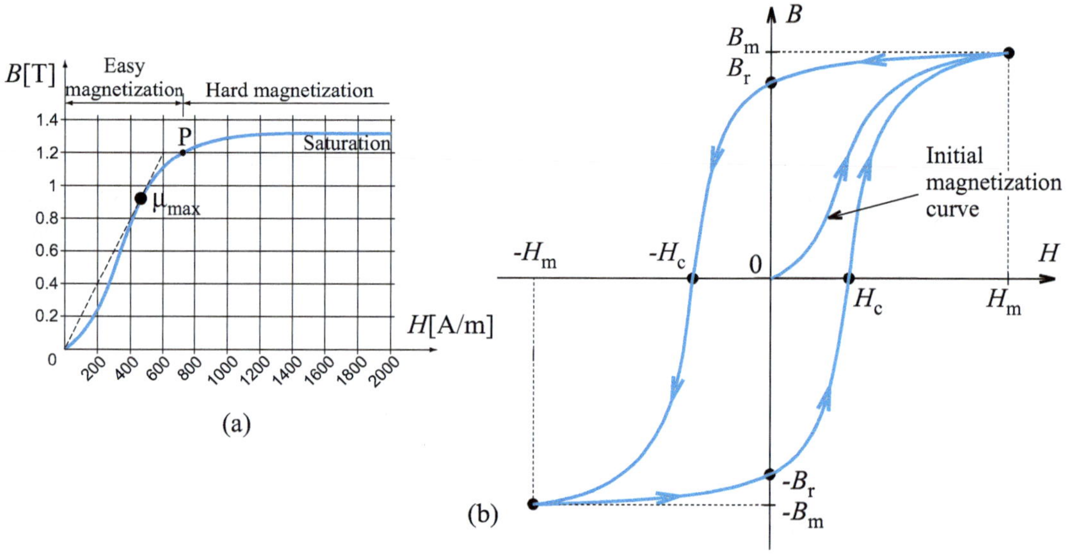

Figure 5.10 Typical initial magnetization curve (a) and hysteresis loop (b) for a ferromagnetic material.

CONCEPTUAL QUESTION 5.20 **Increasing current of a coil over a toroidal magnetic core.** In the structure in Figure 5.11, a coil of wire is wound uniformly and densely about a thin toroidal magnetic core. The current of the coil is I, and the magnetic field intensity and flux density in the core are H and B, respectively. If I is doubled,

(A) both H and B double as well.

(B) H doubles and B remains approximately the same.

(C) H remains approximately the same and B doubles.

(D) H doubles and B increases by about 50%.

(E) None of the above.

(F) Need more information.

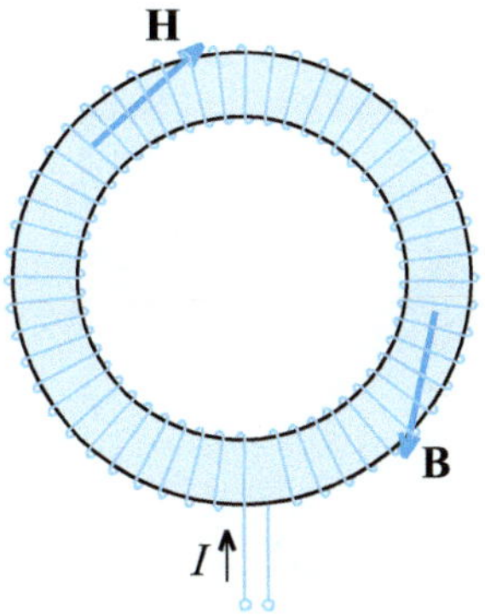

Figure 5.11 Thin toroidal magnetic core with a uniform and dense winding with a steady current; for Conceptual Question 5.20.

CONCEPTUAL QUESTION 5.21 Magnetization curves for increasing and decreasing the applied field. Magnetization curves for increasing and then decreasing the applied magnetic field in a nonlinear ferromagnetic material are

(A) the same.

(B) not the same.

CONCEPTUAL QUESTION 5.22 Rise/fall of the magnetic flux density in an iron core. A uniform and dense wire winding is placed over a thin toroidal core made of iron, as shown in Figure 5.12. If the current I of the coil is increased gradually and smoothly from zero to a very large value, the value of the magnetic flux density, B, in the core

(A) rises at approximately the same rate during the entire process.

(B) rises slowly first and more rapidly later.

(C) rises rapidly first and more slowly later.

(D) does not change.

(E) first rises and then falls.

(F) rises and falls in a cyclic fashion.

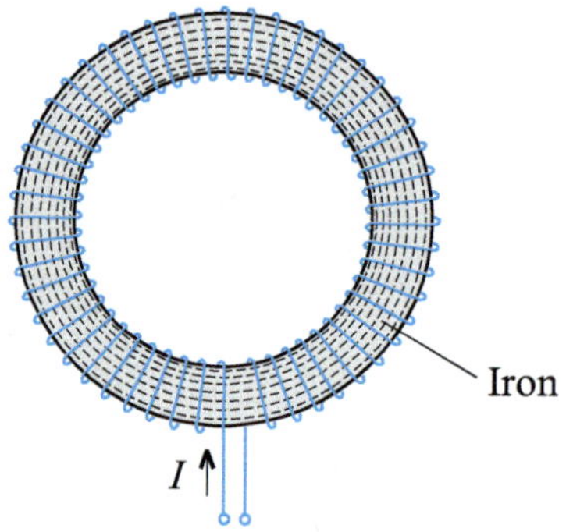

Figure 5.12 Thin toroidal iron core with an excitation coil; for Conceptual Question 5.22.

CONCEPTUAL QUESTION 5.23 **Decreasing current of a coil over an iron core.** Let the current I of the coil on the iron core in Figure 5.12 be established at a very large (positive) value. If I is then reduced to zero, the values of the magnetic field intensity (H) and flux density (B) in the core encounter the following changes:

(A) Both H and B drop to zero.

(B) H drops to zero and B retains its starting (saturation) value.

(C) H retains its saturation value and B drops to zero.

(D) H drops to zero and B drops to a nonzero value.

(E) H drops to a nonzero value and B drops to zero.

(F) Both H and B drop but not to zero.

CONCEPTUAL QUESTION 5.24 **Hysteresis loop.** The hysteresis loop is a property of a ferromagnetic material that does not depend on the range of variation of the applied magnetic field (H).

(A) True.

(B) False.

5.6 Magnetic Circuits

A magnetic circuit in general is a collection of bodies and media that form a way along which the magnetic field lines close upon themselves, i.e., it is a circuit of the magnetic flux flow. The name arises from the similarity to electric circuits. In practical applications, including transformers, generators, motors, relays, magnetic recording devices, etc., magnetic circuits are formed from ferromagnetic cores of various shapes, that may or may not have air gaps, with current-carrying windings wound about parts of the cores. Figure 5.13(a) shows a typical magnetic circuit. With assumptions that the field is restricted to the branches of the magnetic circuit including air gaps (flux leakage and fringing are negligible) and is uniform in every branch (air gap), we now apply the law of conservation of magnetic flux [Eq. (4.16)] to a closed surface S placed about a node (junction of branches) and the generalized Ampère's law [Eq. (5.3)] to a contour C placed along a closed path of flux lines in a magnetic circuit, as indicated in Figure 5.13(b), to obtain

$$\oint_S \mathbf{B} \cdot \mathrm{d}\mathbf{S} = 0 \quad \longrightarrow \quad \sum_{i=1}^{M} B_i S_i = 0\,, \qquad \oint_C \mathbf{H} \cdot \mathrm{d}\mathbf{l} = I_C \quad \longrightarrow \quad \sum_{j=1}^{P} H_j l_j = \sum_{k=1}^{Q} N_k I_k$$

$$\text{(Kirchhoff's laws for magnetic circuits)}\,, \qquad (5.9)$$

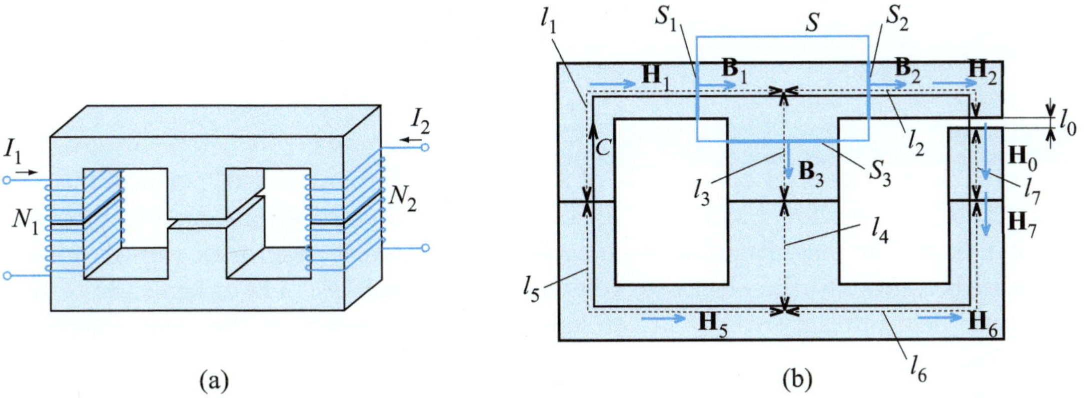

Figure 5.13 (a) Typical magnetic circuit. (b) A closed surface S about a node and a closed path C along the axes of branches of a circuit – for the formulation of Kirchhoff's laws for magnetic circuits.

where B_i ($i = 1, 2, \ldots, M$) and H_j ($j = 1, 2, \ldots, P$) are the magnetic flux densities and field intensities in the branches. Equations (5.9) are referred to as Kirchhoff's laws for magnetic circuits. In addition to these circuital laws, we need, for a solution of the circuit, the "element laws," namely, the relationships $B = B(H)$ (magnetization curves) for the branches of the circuit, which are most frequently nonlinear [Figure 5.10(a)].

CONCEPTUAL QUESTION 5.25 **Approximations in analysis of magnetic circuits.** In analysis of magnetic circuits, a set of approximations is introduced to simplify the computation. However, consider the following possible assumptions: (a) The magnetic flux is concentrated exclusively inside the branches of the ferromagnetic core and air gaps. (b) Magnetic materials of the core can be considered to be linear. (c) Magnetic materials of the branches are never in the state of saturation. (d) There is no magnetic field in air gaps. (e) Lengths of air gaps can be considered to be zero. (f) The fringing magnetic flux near the edges of air gaps can be neglected. (g) The magnetic field intensity (H) is the same in all branches of the circuit. (h) The magnetic field is uniform throughout the volume of each branch of the circuit. (i) Magnetic fluxes are the same in all branches of the circuit. Which of the assumptions constitute the set of approximations used in the analysis?

(A) All assumptions, (a)–(i).

(B) Assumptions (a)–(c) and (g)–(i).

(C) Assumptions (b)–(e) and (i).

(D) Assumptions (a), (e), and (g).

(E) Assumptions (a), (f), and (h).

CONCEPTUAL QUESTION 5.26 **Air gap in a simple linear magnetic circuit.** A thin toroidal core, made of a ferromagnetic material of permeability μ, has an air gap, as shown in Figure 5.14. There is a time-invariant current through the winding. The magnitude of the magnetic field intensity vector in the ferromagnetic with respect to the clockwise reference direction is H. The magnitude of the magnetic field intensity vector in the gap (H_0) with respect to the same reference direction is

(A) $H_0 = H$.

(B) $H_0 = 0$.

(C) $H_0 = \mu_0 H$.

(D) $H_0 = \mu_0 H/\mu$.

(E) $H_0 = \mu H/\mu_0$.

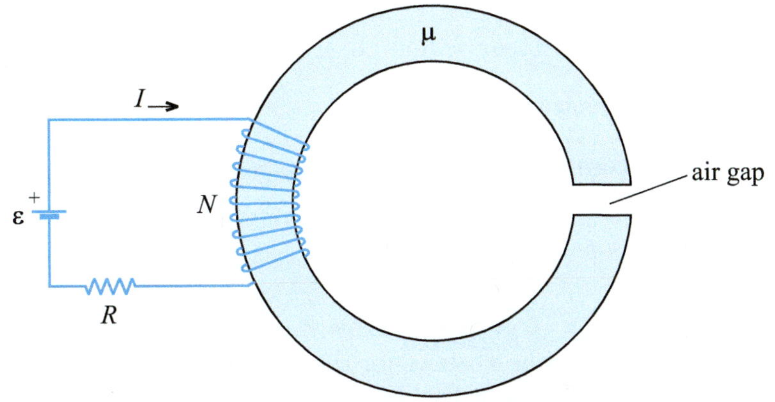

Figure 5.14 Simple linear magnetic circuit with an air gap; for Conceptual Question 5.26.

CONCEPTUAL QUESTION 5.27 Nonlinear magnetic circuit with three branches. Consider the nonlinear magnetic circuit in Figure 5.15(a) and the idealized (linearized) initial magnetization curve of its ferromagnetic core in Figure 5.15(b). For the numeration and orientation (reference directions of magnetic

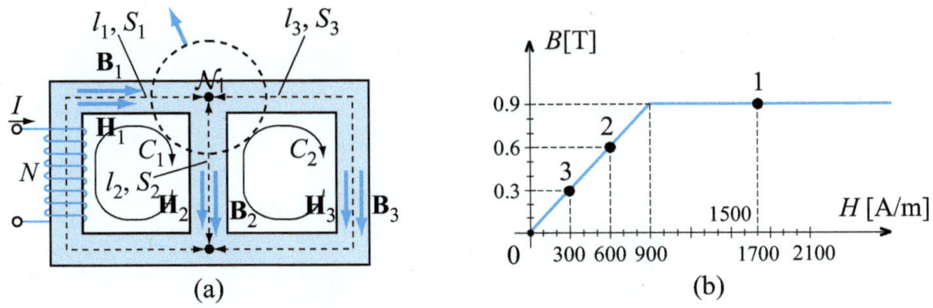

Figure 5.15 Analysis of a nonlinear magnetic circuit with three branches: (a) circuit geometry, with the adopted reference directions of B_i and H_i, $i = 1, 2, 3$, in the branches, and (b) idealized initial magnetization curve of the core material, with indicated operating points (1, 2, and 3) for the branches (the combination of positions of points shown does not necessarily represent the true or possible solution of the circuit); for Conceptual Question 5.27.

flux densities and field intensities) of the circuit branches shown in Figure 5.15(a), the following combination of positions of the operating points for the branches on the magnetization curve represents a possible solution for the circuit:

(A) branch 1 in saturation, branches 2 and 3 in the linear regime, as in Fig. 5.15(b).

(B) branches 1 and 2 in the linear regime, branch 3 in saturation.

(C) branches 1 and 3 in the linear regime, branch 2 in saturation.

(D) all three branches in saturation.

(E) none of the above combinations.

CONCEPTUAL QUESTION 5.28 **Kirchhoff's laws for a complex magnetic circuit.** Consider Kirchhoff's laws for the magnetic circuit with several nodes and branches sketched in Figure 5.16. In particular, what is the minimum number of equations written in accordance with the law of conservation of magnetic flux (Kirchhoff's "current" law), $N_{\text{eqs for nodes}}$, and what is the minimum number of equations based on the generalized Ampère's law (Kirchhoff's "voltage" law), $N_{\text{eqs for closed paths}}$, that would combined lead to the solution for the magnetic flux densities and field intensities in the branches of this circuit?

(A) $N_{\text{eqs for nodes}} = 4$ and $N_{\text{eqs for closed paths}} = 6$.

(B) $N_{\text{eqs for nodes}} = 4$ and $N_{\text{eqs for closed paths}} = 8$.

(C) $N_{\text{eqs for nodes}} = 3$ and $N_{\text{eqs for closed paths}} = 3$.

(D) $N_{\text{eqs for nodes}} = 4$ and $N_{\text{eqs for closed paths}} = 4$.

(E) $N_{\text{eqs for nodes}} = 8$ and $N_{\text{eqs for closed paths}} = 4$.

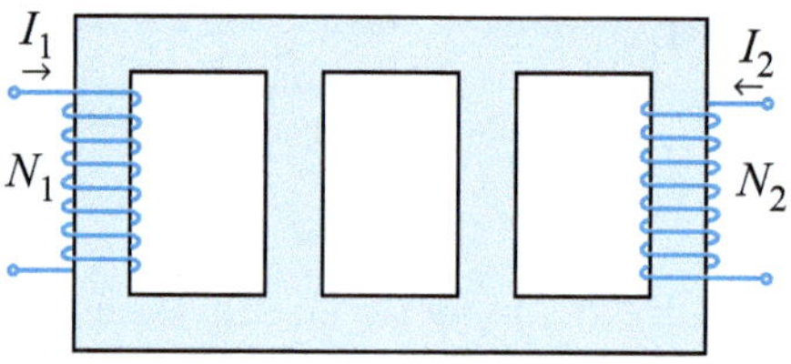

Figure 5.16 Complex magnetic circuit with the entire core made of the same ferromagnetic material; for Conceptual Question 5.28.

CONCEPTUAL QUESTION 5.29 Direct and reverse problems in magnetic circuit analysis. For a typical nonlinear magnetic circuit (with a nonlinear magnetization curve of the core material), such as the one in Figures 5.13(a) and 5.10(a), a direct problem in the analysis is generally formulated as follows: for a given excitation of the circuit, i.e., magnetomotive forces (NI products for coils in the branches), find the response in terms of one or more magnetic fluxes in branches of the circuit. On the other side, a reverse problem reads: for a given response (magnetic flux in one or more branches of the circuit), find the unknown excitation (one or more magnetomotive forces) that produces it. Which problem is simpler to solve?

(A) Direct problem.

(B) Reverse problem.

(C) The two problems are generally of the same complexity.

(D) Each problem is more complex to solve in about the same number of cases.

CONCEPTUAL QUESTION 5.30 Possibility of a branch with a zero magnetic flux. In a magnetic circuit with several nodes and branches, nonlinear magnetization curves of the core materials, and more than one magnetomotive force (winding with current), it is possible to have a branch with a zero magnetic flux.

(A) True.

(B) False.

5.7 Magnetic Energy

Every system of conducting loops with currents contains a certain amount of energy, called magnetic energy, in a manner analogous to a system of charged conducting bodies storing electric energy. To describe the localization and distribution of the magnetic energy of the system, we define and use the magnetic energy density (energy per unit volume), w_{m}. At a point in an arbitrary magnetic medium, w_{m} can be computed as

$$w_{\mathrm{m}} = \int_0^B H \, \mathrm{d}B \quad \text{(magnetic energy density; unit: J/m}^3\text{)} . \qquad (5.10)$$

Having in mind a typical initial magnetization curve of a nonlinear ferromagnetic material [e.g., that in Figure 5.10(a)], we note that $H\,\mathrm{d}B$ is proportional to the area of a thin strip of "length" H (length measured in A/m) and "width" $\mathrm{d}B$ (width measured in T) positioned between the curve and the B-axis at the "height" B with respect to the H-axis, as indicated in Figure 5.17(a). This means that w_m in Eq. (5.10) is proportional to the area of the curvilinear triangle OPQ in Figure 5.17(a). For a linear material, the area of the triangle comes out to be $w_\mathrm{m} = BH/2 = \mu H^2/2$ [note the duality with the corresponding expressions for the electric energy density in Eq. (2.20)], as depicted in Figure 5.17(b). The total magnetic energy in a volume v amounts to

$$W_\mathrm{m} = \int_v w_\mathrm{m}\,\mathrm{d}v \quad \text{(magnetic energy; unit: J)}\,. \tag{5.11}$$

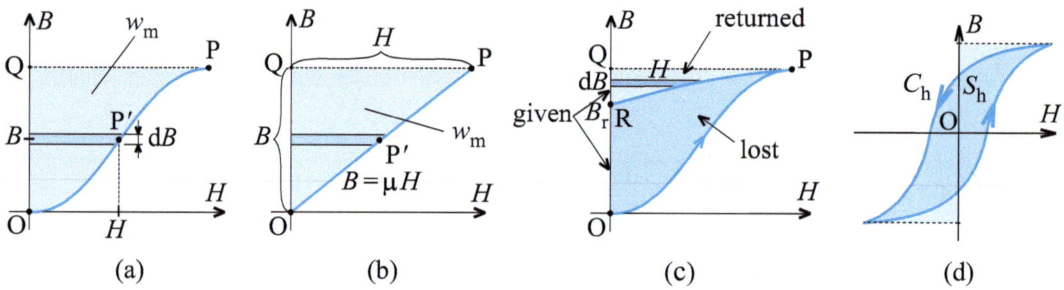

Figure 5.17 (a) Computation of the magnetic energy density in Eq. (5.10) for a nonlinear magnetic material. (b) Linear case. (c) Evaluation of hysteresis losses as the difference between the energy given to the field and the energy returned to the sources in the process of magnetization and demagnetization of a ferromagnetic material. (d) Hysteresis losses in a full hysteresis cycle in the material.

In ferromagnetic materials that exhibit hysteresis effects, function $B(H)$ has multiple branches [see Figure 5.10(b)]. Thus, if H is reduced from its value at the point P in Figure 5.17(a) to zero, the energy returned by the field is smaller than the energy previously spent by the sources and given to the field, as can be seen in Figure 5.17(c). The difference is lost to heat in the material in the process of its magnetization and demagnetization (hysteresis losses). The energy density of hysteresis losses is proportional to the area of the curvilinear triangle OPR in Figure 5.17(c). Finally, the density of the energy lost in one complete magnetization–demagnetization hysteresis cycle is determined by the area enclosed by the hysteresis loop, S_h, in Figure 5.17(d).

CONCEPTUAL QUESTION 5.31 **Change of magnetic field intensity/flux density and energy.** Two coils with linear magnetic cores contain the same amounts of magnetic energy. If the magnetic field intensity (H) at every point in the core of the first coil becomes twice larger, while the magnetic flux density (B) at every point in the core of the second coil is halved, the energy stored in the first coil in the new steady state is

(A) 1/4 of

(B) 1/16 of

(C) 4 times

(D) 16 times

(E) the same as

that stored in the second coil.

CONCEPTUAL QUESTION 5.32 **Energy spent for establishing the field in a nonlinear material.** Consider a magnetic circuit with a ferromagnetic core whose idealized initial magnetization curve is shown in Figure 5.18. Final (established) values of the magnetic flux density and field intensity in the core (operating point P of the circuit, Figure 5.18) are B and H, respectively. With $B_{\rm k}$ and $H_{\rm k}$ denoting the magnetic flux density and field intensity at the point K ("knee" point between the two segments of the curve) in Figure 5.18, the following holds true for the density $w_{\rm m}$ of energy spent for establishing the field in the core, namely, to change it from zero to (B, H):

(A) $0 < w_{\rm m} < B_{\rm k}H_{\rm k}/2$.

(B) $w_{\rm m} = B_{\rm k}H_{\rm k}/2$.

(C) $B_{\rm k}H_{\rm k}/2 < w_{\rm m} < BH/2$.

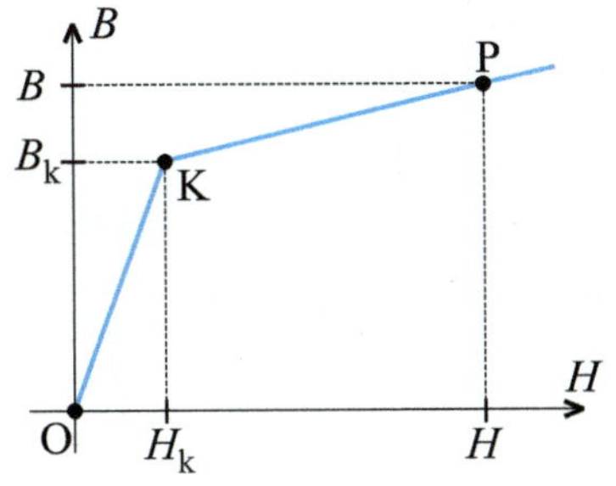

Figure 5.18 Idealized initial magnetization curve of a ferromagnetic core; for Conceptual Question 5.32.

(D) $w_m = (B - B_k)(H - H_k)/2$.

(E) $w_m = BH/2$.

(F) $w_m > BH/2$.

CONCEPTUAL QUESTION 5.33 **Energy spent in magnetization–demagnetization.** In two equally sized pieces of different ferromagnetic materials, a uniform magnetic field is first established, at the same intensity (H_m), and then reduced to zero ($H = 0$), during which process the operating point describes the respective paths shown in Figure 5.19. The net magnetic energy spent in the magnetization–demagnetization of the piece in case (a) is

(A) twice

(B) three times

(C) a half of

(D) a third of

(E) the same as

that in case (b).

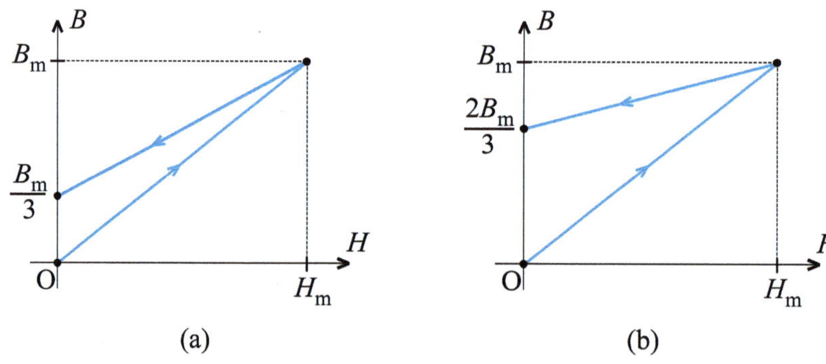

Figure 5.19 Magnetization–demagnetization of two different ferromagnetic materials; for Conceptual Question 5.33.

CONCEPTUAL QUESTION 5.34 **Energy density of hysteresis losses in the core of a solenoid.** A very long solenoidal coil is wound about a nonlinear ferromagnetic core whose idealized hysteresis loop is shown in Figure 5.20. There is a time-harmonic (steady-state sinusoidal) current of intensity $i(t) = I_0 \sin \omega t$ (I_0 is the amplitude and ω is the angular or radian frequency of the current) flowing through the winding. The magnetic field is uniform throughout the core, and saturation is not reached in the material. The energy density of hysteresis losses in the core in one full magnetization–demagnetization hysteresis cycle of the material equals

(A) $w_\mathrm{h} = 0$.

(B) $w_\mathrm{h} = B_\mathrm{m} H_\mathrm{m}/2$.

(C) $w_\mathrm{h} = B_\mathrm{m} H_\mathrm{m}$.

(D) $w_\mathrm{h} = 2B_\mathrm{m} H_\mathrm{m}$.

(E) $w_\mathrm{h} = 4B_\mathrm{m} H_\mathrm{m}$.

(F) None of the above.

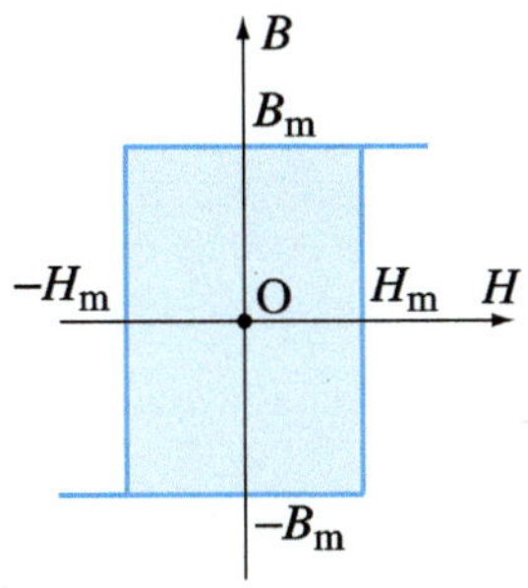

Figure 5.20 Idealized hysteresis loop of a ferromagnetic core filling a solenoidal coil; for Conceptual Question 5.34.

CONCEPTUAL QUESTION 5.35 Dependence of the power of hysteresis losses on frequency. There is a uniform time-harmonic magnetic field of intensity $H(t) = H_0 \cos(2\pi f t)$ throughout the volume of a ferromagnetic body. If the frequency (f) of the field is doubled, the new time-average power of hysteresis losses in the body equals c times its previous value, where

(A) $c = 2$.

(B) $c > 2$.

(C) $c = 1/2$.

(D) $c < 1/2$.

(E) $c = 1$.

---

6 TIME-VARYING ELECTROMAGNETIC FIELD

Introduction

We now introduce time variation of electric and magnetic fields into our electromagnetic model. The new field is the time-varying electromagnetic field, which is caused by time-varying charges and currents. As opposed to static fields, the electric and magnetic fields constituting the time-varying electromagnetic field are coupled to each other and cannot be analyzed separately. Moreover, the mutual induction (generation) of time-varying electric and magnetic fields is the basis of propagation of electromagnetic waves and of electromagnetic radiation. The first essentially new feature that is not present under the static assumption is electromagnetic induction, where we also study the related concept of inductance. The other crucial step is addition of a new type of current, so-called displacement current, to the static version of the generalized Ampère's law. The full set of general Maxwell's equations – in integral and differential notation, and in the form of boundary conditions – is studied and used in the time domain, as well as in the complex (frequency) domain, which is usually considerably more efficient. The time retardation (lagging in time of fields behind their sources) is quantified and Lorenz (retarded) electromagnetic potentials are introduced and evaluated. The chapter also discusses Poynting's theorem, as an expression of the principle of conservation of energy for electromagnetic phenomena.

6.1 Induced Electric Field Intensity Vector

We know from Chapter 1 that a point charge Q in free space is a source of an electric field, predicted by Coulomb's law and described by Eq. (1.4). On the other side, the Biot–Savart law (Chapter 4) tells us that there will also be a magnetic field, given by Eq. (4.4), if this charge moves with some velocity $\mathbf{v}$ in space. We now introduce a third field, which will exist in the space around the charge whenever the velocity $\mathbf{v}$ changes in time, i.e., whenever the acceleration (or deceleration) $\mathbf{a} = \mathrm{d}\mathbf{v}/\mathrm{d}t$ of the charge is not zero. This new field is an electric field in its nature. It is called the induced electric field and its intensity vector is given by

$$\mathbf{E}_{\mathrm{ind}}(t) = -\frac{\mu_0}{4\pi}\frac{Q\,\frac{\mathrm{d}\mathbf{v}}{\mathrm{d}t}}{R} \quad \text{(point charge in an accelerated motion)} . \tag{6.1}$$

A generalization, by the principle of superposition, of this expression leads to the following expression for $\mathbf{E}_{\mathrm{ind}}$ due to a time-varying current of intensity $i(t)$ flowing along a line (wire) l [see Figure 4.4(a)]:

$$\mathbf{E}_{\rm ind}(t) = -\frac{\mu_0}{4\pi}\int_l \frac{(\mathrm{d}i/\mathrm{d}t)\,\mathrm{d}\mathbf{l}}{R} \quad \text{(induced electric field intensity vector; unit: V/m)} \; . \tag{6.2}$$

More precisely, the above field expression holds only for a slowly time-varying (low-frequency) current, and the field $\mathbf{E}_{\rm ind}$ thus constitutes the slowly time-varying or quasistatic electromagnetic field, as will be discussed in Section 6.11. Comparing Eqs. (6.2) and (4.17), we conclude that $\mathbf{E}_{\rm ind} = -\partial\mathbf{A}/\partial t$, where $\mathbf{A}(t)$ is the magnetic vector potential due to the current $i(t)$.

CONCEPTUAL QUESTION 6.1 **Electric field due to a moving point charge.** A point charge Q ($Q > 0$) moves in free space with a velocity v ($v > 0$). The resultant electric field intensity vector, $\mathbf{E}$, at a point (M) on a line that is perpendicular to the vector $\mathbf{v}$ is as in[1]

(A) Figure 6.1(a).

(B) Figure 6.1(b).

(C) Figure 6.1(c).

(D) Figure 6.1(d).

(E) Need more information.

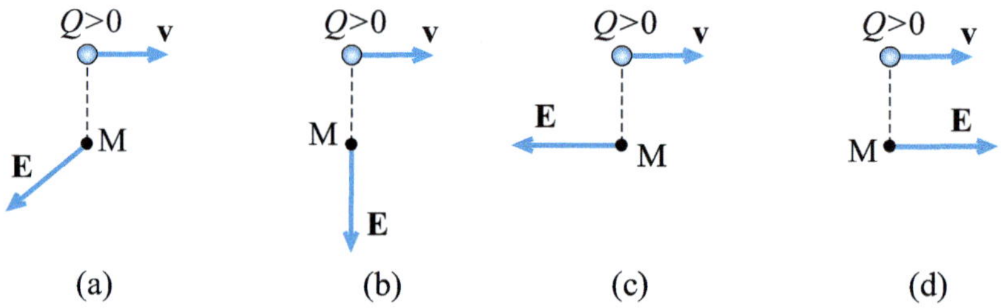

Figure 6.1 Point charge $Q > 0$ moving with a velocity v in free space – discussion of the resultant electric field intensity vector, $\mathbf{E}$; for Conceptual Question 6.1.

CONCEPTUAL QUESTION 6.2 **Finding charge velocity based on the electric field vector.** If the electric field vector due to a (moving) point charge $Q > 0$ is as in Figure 6.2, and both v_0 and t_0 are positive constants, which of the

[1] *For every conceptual question in this text, exactly one answer is correct.*

following expressions/values for the velocity v of the charge may result in such a situation?

(A) $v = v_0 t/t_0$.

(B) $v = v_0$.

(C) $v = 0$.

(D) More than one of the above cases.

(E) None of the above cases.

(F) Need more information.

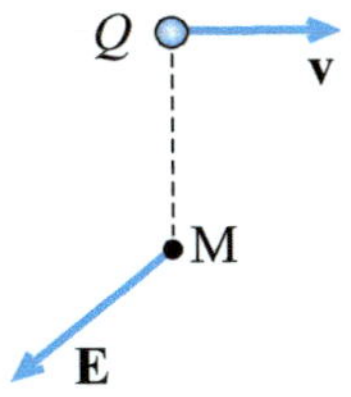

Figure 6.2 Discussion of the unknown charge velocity based on a known induced electric field intensity vector; for Conceptual Question 6.2.

CONCEPTUAL QUESTION 6.3 **Another situation involving a moving/stationary charge.** Assuming that the resultant electric field vector due to a (moving) point charge is as in Figure 6.3, the following is a possible velocity v of the charge (v_0 and t_0 are positive constants):

(A) $v = v_0 t/t_0$.

(B) $v = v_0$.

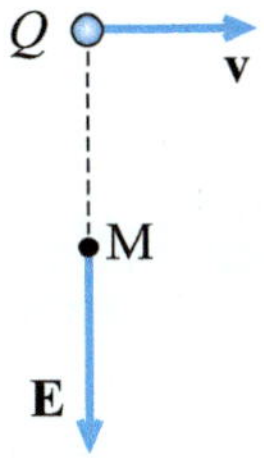

Figure 6.3 Another case of a given induced electric field vector due to a (moving) charge $Q > 0$; for Conceptual Question 6.3.

(C) $v = 0$.

(D) More than one of the above cases.

(E) None of the above cases.

(F) Need more information.

CONCEPTUAL QUESTION 6.4 **Induced electric field at two instants of time.** There is a slowly time-varying current of intensity $i(t)$ flowing along a metallic wire loop in air. At two instants of time when $i = i_1$ and $i = i_2$, the induced electric field intensity vector at a point in space equals $\mathbf{E}_{\text{ind1}}$ and $\mathbf{E}_{\text{ind2}}$, respectively. If i_2 happens to be twice i_1, then $\mathbf{E}_{\text{ind2}}$ must be twice $\mathbf{E}_{\text{ind1}}$.

(A) True.

(B) False.

CONCEPTUAL QUESTION 6.5 **Induced electric field at the center of a circular current loop.** A circular metallic wire loop situated in free space carries a slowly time-varying current. If the radius of the loop is doubled and the current intensity remains the same, the magnitude of the induced electric field intensity vector, $\mathbf{E}_{\text{ind}}$, at the loop center

(A) increases.

(B) decreases.

(C) remains the same.

(D) Need more information.

CONCEPTUAL QUESTION 6.6 **Induced electric field of a rectangular current loop.** A slowly time-varying current of intensity $i(t)$ is established in a rectangular wire loop of edge lengths a and $b = a/2$ in air, as shown in Figure 6.4. At an instant of time when $\mathrm{d}i/\mathrm{d}t < 0$, the induced electric field intensity vector $\mathbf{E}_{\text{ind}}$ at the point M in the figure has only

(A) a positive x-component.

(B) a negative x-component.

(C) a positive y-component.

(D) a negative y-component.

(E) a positive z-component.

(F) a negative z-component.

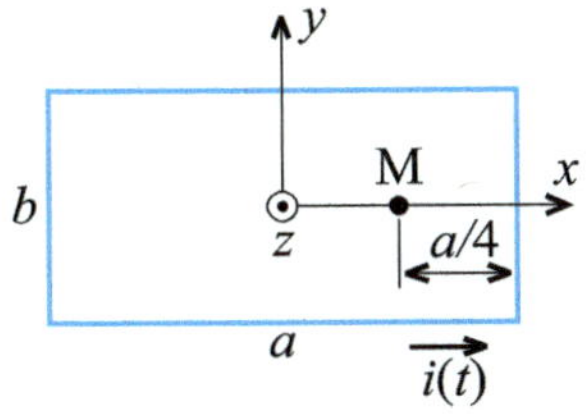

Figure 6.4 Rectangular current loop in air; for Conceptual Question 6.6.

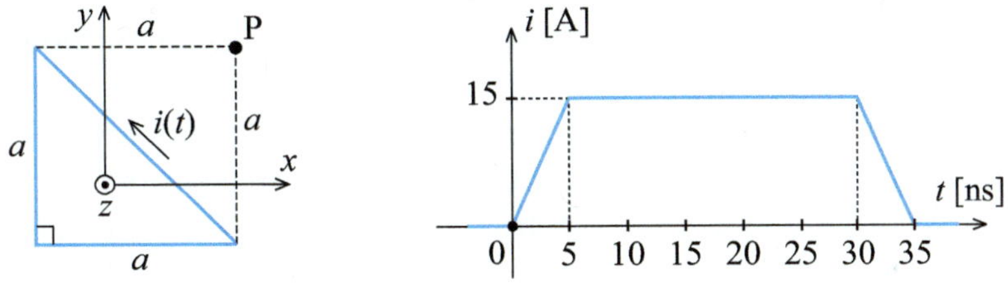

(a) (b)

Figure 6.5 Triangular current loop (a) carrying a pulse current (b); for Conceptual Question 6.7.

CONCEPTUAL QUESTION 6.7 **Induced electric field of a triangular pulse-current loop.** A loop in the form of a triangle representing a half of a square of side a, shown in Figure 6.5(a), is situated in free space and carries a pulse current of intensity $i(t)$, given in Figure 6.5(b). With $E_0 > 0$, the induced electric field intensity vector ($\mathbf{E}_{\text{ind}}$) at a point P located at the fourth vertex of the square [in Figure 6.5(a)] and time instant $t = 17.5$ ns [in Figure 6.5(b)] is given by

(A) $\mathbf{E}_{\text{ind}} = E_0(\hat{\mathbf{x}} - \hat{\mathbf{y}})$.

(B) $\mathbf{E}_{\text{ind}} = E_0(-\hat{\mathbf{x}} + \hat{\mathbf{y}})$.

(C) $\mathbf{E}_{\text{ind}} = E_0(\hat{\mathbf{x}} + \hat{\mathbf{y}})$.

(D) $\mathbf{E}_{\text{ind}} = E_0\,\hat{\mathbf{z}}$.

(E) $\mathbf{E}_{\text{ind}} = 0$.

6.2 Faraday's Law of Electromagnetic Induction

The line integral of the induced electric field intensity vector, $\mathbf{E}_{\rm ind}$ [Eq. (6.2)], along a line joining any two points M and N in space represents the electromotive force (emf) induced in the line:

$$e_{\rm ind} = \int_{\rm M}^{\rm N} \mathbf{E}_{\rm ind} \cdot {\rm d}\mathbf{l} \quad \text{(induced electromotive force – emf; unit: V)} \, , \qquad (6.3)$$

and the line can now be replaced by an equivalent ideal voltage generator whose emf is $e_{\rm ind}$, as shown in Figure 6.6(a). For a closed line (contour), Figure 6.6(b), the induced emf is given by

$$e_{\rm ind} = -\frac{{\rm d}\Phi}{{\rm d}t} \quad \text{or} \quad \oint_C \mathbf{E}_{\rm ind} \cdot {\rm d}\mathbf{l} = -\frac{\rm d}{{\rm d}t} \int_S \mathbf{B} \cdot {\rm d}\mathbf{S} \quad \text{(Faraday's law)} \, . \qquad (6.4)$$

Namely, it is equal to the negative of the time rate of change of the magnetic flux, Φ, through the contour, i.e., through a surface of arbitrary shape bounded by the contour [Eq. (4.15)] and oriented in accordance with the right-hand rule with respect to the orientation of the contour. This rule tells us that the flux is in the direction defined by the thumb of the right hand when the other fingers point in the direction of the emf, as indicated in Figure 6.6(b). Equation (6.4) is known as Faraday's law of electromagnetic induction. It is the most important governing law of the slowly time-varying (quasistatic) electromagnetic field and the explicit relation between the electric and magnetic fields that change in time. Faraday's law represents Maxwell's first equation for the time-varying electromagnetic field. It is essentially different from Maxwell's first equation for the time-invariant electromagnetic field in Eq. (1.13).

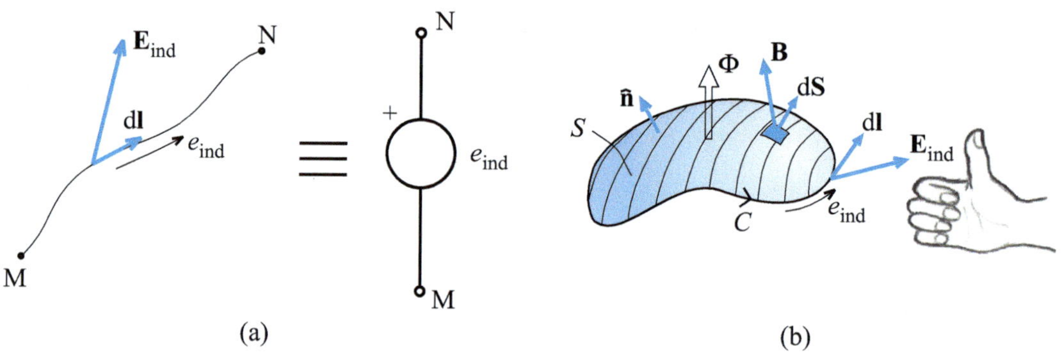

Figure 6.6 (a) Induced emf in a line joining two points in space. (b) Arbitrary contour in a time-varying magnetic field – for the statement of Faraday's law of electromagnetic induction.

The contour C in Figure 6.6(b) can be an imaginary (nonmaterial) contour, i.e., it does not need to be a conducting wire loop for Eq. (6.4) to be true. However, in the case when C does represent a conducting wire contour, there is a current in the wire, called the induced current, of intensity $i_{\text{ind}} = e_{\text{ind}}/R$ given for the same reference direction as e_{ind}, where R is the total resistance of the contour. Because of the minus sign in Eq. (6.4), the magnetic field due to the induced current opposes (tends to cancel) the change in the magnetic flux that caused the induced emf and current in the first place, and this is referred to as Lenz's law.[2]

This section is devoted to the application of Faraday's law of electromagnetic induction in evaluating the induced emf in stationary contours due to time variations of the magnetic field – this kind of electromagnetic induction is called transformer induction, because it is the basis of current and voltage transformation by magnetic coupling between circuits. The electromagnetic induction due to motion of conductors in magnetic fields will be introduced and studied in the next section.

CONCEPTUAL QUESTION 6.8 **Induced emf and current along a straight wire.** Consider a straight metallic wire of finite length situated in a quasistatic (low-frequency) electromagnetic field. Which of the two quantities, the induced electromotive force (emf), e_{ind}, and current intensity, i, must be zero along the wire?

(A) e_{ind} only.

(B) i only.

(C) Both quantities.

(D) Neither of the quantities.

CONCEPTUAL QUESTION 6.9 **Wire loop in a uniform time-harmonic magnetic field.** A rectangular wire loop is situated in a uniform low-frequency time-harmonic magnetic field of flux density $B(t) = B_0 \sin \omega t$ $(B_0 > 0)$. The vector

[2]Note that in the case of a PEC (perfect electric conductor) [Eq. (3.8)] wire contour, e.g., a contour made from a superconductor, which has a zero total resistance ($R = 0$), so that $e_{\text{ind}} = Ri = 0$, Eq. (6.4) gives $\text{d}\Phi/\text{d}t = 0$, that is, $\Phi = \text{const}$ through the contour. This means that the total existing magnetic flux through a superconducting contour cannot be changed. For example, if the magnetic field in which a superconducting contour resides (external or primary magnetic field) is changed and/or the contour is moved in the field, a current is induced in the contour whose magnetic field (secondary field) completely cancels the change of the magnetic flux through the contour (Lenz's law in its extreme form), such that the total flux Φ through the contour remains constant.

B is perpendicular to the plane of the loop, as shown in Figure 6.7. The magnetic field due to the induced current can be neglected. The induced emf in the loop is of the following form ($\mathcal{E}_0$ is a positive constant):

(A) $e_{\text{ind}}(t) = \mathcal{E}_0 \sin \omega t$.

(B) $e_{\text{ind}}(t) = -\mathcal{E}_0 \cos 2\omega t$.

(C) $e_{\text{ind}}(t) = -\mathcal{E}_0$.

(D) $e_{\text{ind}}(t) = -\mathcal{E}_0 \cos \omega t$.

(E) $e_{\text{ind}}(t) = 0$.

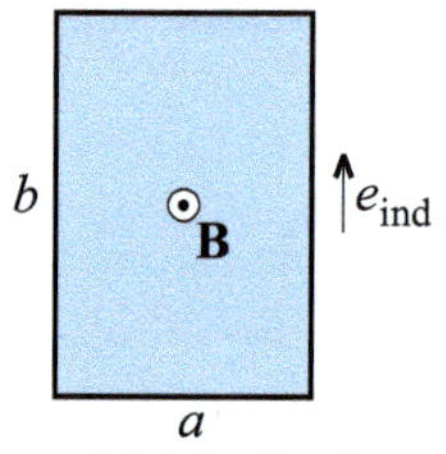

Figure 6.7 Wire loop in a uniform low-frequency time-harmonic magnetic field of flux density vector **B**; for Conceptual Question 6.9.

CONCEPTUAL QUESTION 6.10 Induced emf in a broken wire contour. Assume that the rectangular wire loop placed in a uniform low-frequency magnetic field with $B(t) = B_0 \sin \omega t$ ($B_0 > 0$) and **B** being normal to the loop plane (Figure 6.7) is broken, i.e., there is a small air gap at one of the corners of the loop making the loop discontinuous (open) at that point. Under these circumstances, the induced emf in the loop is of the following form ($\mathcal{E}_0 > 0$ is a constant):

(A) $e_{\text{ind}}(t) = \mathcal{E}_0 \sin \omega t$.

(B) $e_{\text{ind}}(t) = -\mathcal{E}_0 \cos 2\omega t$.

(C) $e_{\text{ind}}(t) = -\mathcal{E}_0$.

(D) $e_{\text{ind}}(t) = -\mathcal{E}_0 \cos \omega t$.

(E) $e_{\text{ind}}(t) = 0$.

CONCEPTUAL QUESTION 6.11 **Magnetic flux due to the induced current in a superconducting contour.** A rectangular superconducting contour (the resistance of the contour is $R = 0$) of area S is first situated outside any magnetic field, and there is no current in it. The contour is then brought in a uniform time-invariant magnetic field of flux density B and positioned so that the vector $\mathbf{B}$ is perpendicular to the plane of the contour, as shown in Figure 6.8. In the new steady state, the magnetic flux through the contour due to the current induced in it, computed with respect to the same orientation as that of the vector $\mathbf{B}$, equals

(A) $\Phi_{\text{ind}} = BS$,

(B) $\Phi_{\text{ind}} = -BS$,

(C) $\Phi_{\text{ind}} = kBS$,

(D) $\Phi_{\text{ind}} = -kBS$,

(E) $\Phi_{\text{ind}} = 0$,

where k is a dimensionless constant and $0 < k < 1$.

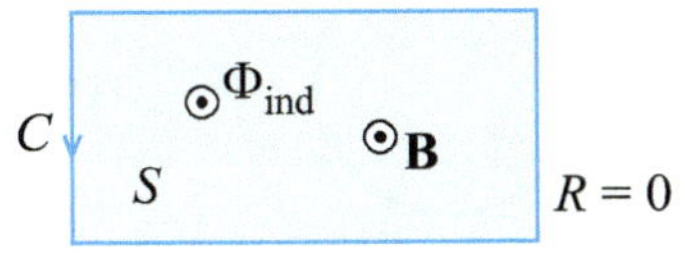

Figure 6.8 Superconducting contour brought in a uniform time-invariant magnetic field; for Conceptual Question 6.11.

CONCEPTUAL QUESTION 6.12 **Voltage vs. induced emf for an open-ended wire conductor.** Consider an open-ended wire conductor of arbitrary shape whose ends are denoted as points M and N. The conductor is placed in a low-frequency time-harmonic electromagnetic field, and the induced emf along the wire is $e_{\text{ind}}(t)$ given for the reference direction from M to N. The voltage between points M and N equals

(A) $v_{\text{MN}}(t) = e_{\text{ind}}(t)$.

(B) $v_{\text{MN}}(t) = -e_{\text{ind}}(t)$.

(C) $v_{\text{MN}}(t) = 0$.

(D) None of the above.

CONCEPTUAL QUESTION 6.13 **Open-circuited loop around a solenoid.** An air-filled infinitely long solenoidal coil with a circular cross section of radius a carries a slowly time-varying current. The magnetic flux through a surface spanned over one turn of the coil is $\Phi(t)$. The magnetic field outside the coil is zero. An open-circuited circular wire loop of radius b ($b > a$) is placed coaxially around the solenoid, as shown in Figure 6.9. The voltage between the terminals of the loop, for the reference orientation of the terminals in the figure, equals

(A) $v(t) = \mathrm{d}\Phi/\mathrm{d}t$.

(B) $v(t) = -\mathrm{d}\Phi/\mathrm{d}t$.

(C) $v(t) = (b/a)^2\,\mathrm{d}\Phi/\mathrm{d}t$.

(D) $v(t) = -(b/a)^2\,\mathrm{d}\Phi/\mathrm{d}t$.

(E) $v(t) = 0$.

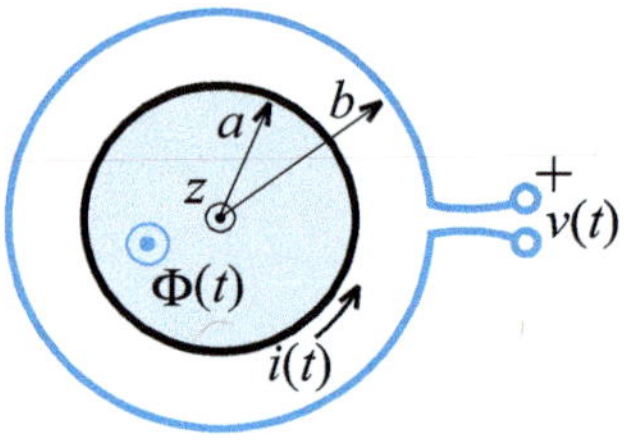

Figure 6.9 Open-circuited loop around an air-filled infinitely long solenoid with a slowly time-varying current (cross section of the structure); for Conceptual Question 6.13.

CONCEPTUAL QUESTION 6.14 **Induced emf along a circular loop.** Consider the induced electromotive force, e_{ind}, along the loop in Figure 6.10. If the radius of the loop is doubled (becomes $2b$), while both a and $\Phi(t)$ are not changed, we have that e_{ind}

(A) doubles.

(B) quadruples.

(C) remains the same.

(D) is halved.

(E) becomes a quarter of the previous value.

(F) None of the above.

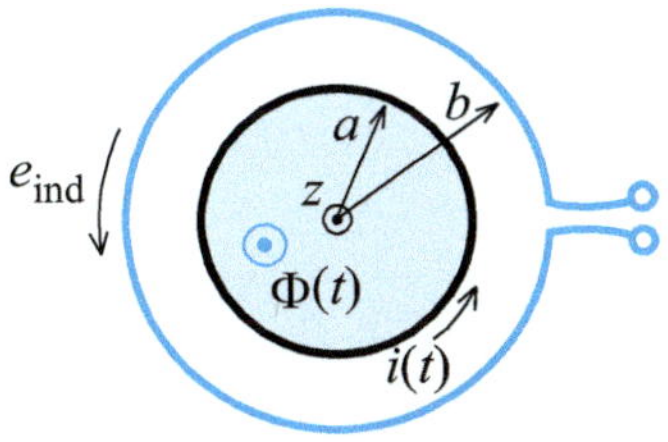

Figure 6.10 Induced emf along an open-circuited loop around a solenoid with a slowly time-varying magnetic flux ($\mathbf{B} = 0$ outside the solenoid); for Conceptual Question 6.14.

CONCEPTUAL QUESTION 6.15 **Induced electric field along a circular loop.** When the radius of the loop (b) in Figure 6.11 is doubled, the induced electric field intensity, $E_{\rm ind}$, along the loop

(A) doubles.

(B) quadruples.

(C) remains the same.

(D) is halved.

(E) becomes a quarter of the previous value.

(F) None of the above.

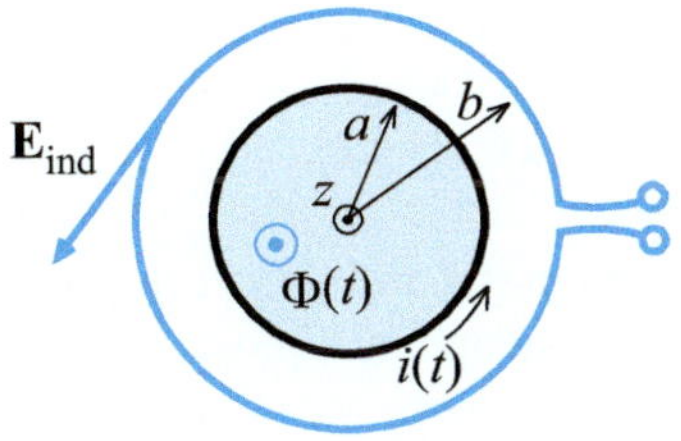

Figure 6.11 Induced electric field intensity vector along an open-circuited loop around an infinitely long solenoidal coil; for Conceptual Question 6.15.

CONCEPTUAL QUESTION 6.16 **Induced emf and electric field along an imaginary contour.** If the loop in Figure 6.9 is an imaginary (nonmaterial) contour, in place of a conducting wire loop, as shown in Figure 6.12, which of the two quantities, the induced electromotive force ($e_{\rm ind}$) and induced electric field intensity ($E_{\rm ind}$), along the contour remain the same as along the conducting wire?

(A) e_{ind} only.

(B) E_{ind} only.

(C) Both quantities.

(D) Neither of the quantities.

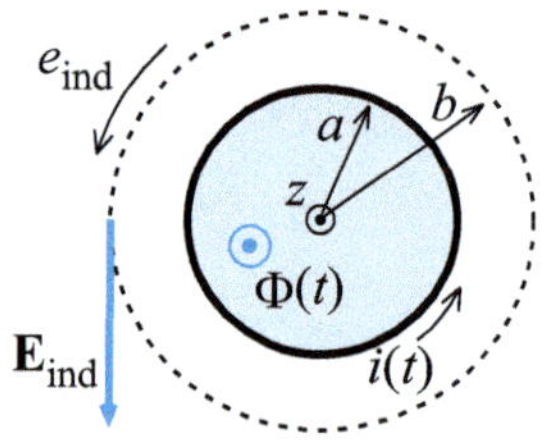

Figure 6.12 Imaginary (nonmaterial) contour (versus a conducting wire loop) around an infinitely long solenoid; for Conceptual Question 6.16.

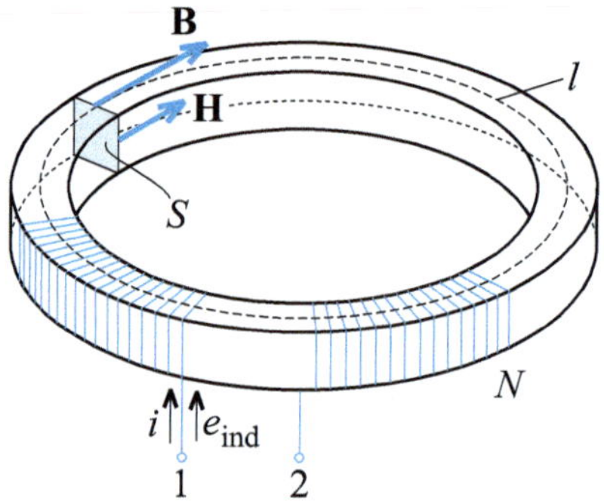

Figure 6.13 Analysis of electromagnetic induction in a coil with a core made from a nonlinear ferromagnetic material; for Conceptual Question 6.17.

CONCEPTUAL QUESTION 6.17 Electromagnetic induction in a nonlinear magnetic circuit. Figure 6.13 shows a coil of wire, with a low-frequency time-harmonic current $i(t) = I_0 \cos \omega t$, wound uniformly and densely about a thin toroidal core made of a nonlinear ferromagnetic material that exhibits hysteresis effects. Consider the magnetic flux density, $B(t)$, and magnetic field intensity, $H(t)$, in the core, as well as the induced emf, $e_{\mathrm{ind}}(t)$, in the coil. Which of these quantities are time-harmonic functions?

(A) $B(t)$ only.

(B) $H(t)$ only.

(C) $B(t)$ and $H(t)$ only.

(D) $B(t)$ and $e_{\rm ind}(t)$ only.

(E) All three quantities.

(F) None of the quantities.

6.3 Electromagnetic Induction Due to Motion and Total Induction

Consider a conductor moving with a velocity $\mathbf{v}$ in a static (time-invariant) magnetic field of the flux density $\mathbf{B}$. The field exerts the magnetic force, $\mathbf{F}_{\rm m}$, given by Eq. (4.1), on each of the charge carriers in the conductor. This force "pushes" the carriers to move, and separates positive and negative excess charges in the conductor. We can formally divide $\mathbf{F}_{\rm m}$ by the charge of a carrier (Q) and obtain $\mathbf{F}_{\rm m}/Q = \mathbf{v}\times\mathbf{B}$. This new quantity, expressed in V/m, is termed the induced electric field intensity vector due to motion,

$$\mathbf{E}_{\rm ind} = \mathbf{v}\times\mathbf{B} \quad \text{(induced electric field intensity vector due to motion; unit: V/m)}\,. \tag{6.5}$$

It generates an induced electromotive force, as given by Eq. (6.3). Hence, the emf along a line through a conductor between points M and N [Figure 6.6(a)] and along a contour (closed line) [Figure 6.6(b)] due to motion in a time-invariant magnetic field is

$$e_{\rm ind} = \int_{\rm M}^{\rm N} \mathbf{E}_{\rm ind}\cdot{\rm d}\mathbf{l} = \int_{\rm M}^{\rm N} (\mathbf{v}\times\mathbf{B})\cdot{\rm d}\mathbf{l}\,, \quad e_{\rm ind} = \oint_C (\mathbf{v}\times\mathbf{B})\cdot{\rm d}\mathbf{l} \quad \text{(motional emf)}\,. \tag{6.6}$$

This emf is referred to as the emf due to motional induction or simply motional emf. Note that the velocity of different parts of the contour need not be the same, including cases when some parts are stationary while others move in arbitrary directions. In other words, the motion of the contour may include translation, rotation, and deformation (changing shape and size) of the contour in an arbitrary manner.

When a contour moves and/or changes in a static magnetic field, the magnetic flux through the contour generally changes with time. In fact, it is possible to relate the emf induced in the contour to the rate of change of the flux, i.e., to express the motional emf in terms of Faraday's concept of changing flux through the contour exactly as in Eq. (6.4). So, the same form of Faraday's law of electromagnetic induction holds for both transformer and motional emf in a contour, as well as due to a combination of the two mechanisms – total induction,

$$e_{\text{ind}} = -\frac{\mathrm{d}\Phi}{\mathrm{d}t} \quad \text{(emf due to transformer or motional or total induction)} . \tag{6.7}$$

CONCEPTUAL QUESTION 6.18 **Moving bar in a uniform magnetostatic field.** A metallic bar (of finite length) moves uniformly with a velocity $\mathbf{v}$ in a steady uniform magnetic field of flux density $\mathbf{B}$. Figure 6.14 shows five cases with different positions of the bar and of its velocity vector with respect to the magnetic field lines. There is a nonzero emf induced in the bar ($e_{\text{ind}} \neq 0$) for

(A) case (a) only.

(B) case (b) only.

(C) case (d) only.

(D) cases (a) and (c) only.

(E) cases (b), (c), and (d) only.

(F) none of the cases shown.

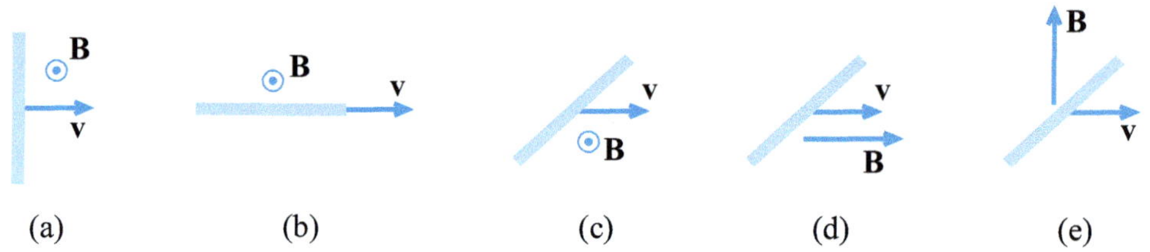

Figure 6.14 Metallic bar moving with a velocity $\mathbf{v}$ in a magnetic field of flux density $\mathbf{B}$ – five different mutual positions of the bar and vectors $\mathbf{v}$ and $\mathbf{B}$; for Conceptual Question 6.18.

CONCEPTUAL QUESTION 6.19 **Moving loop in a magnetic field.** A rectangular conducting wire loop moves with a constant velocity v ($\mathbf{v} = v\,\hat{\mathbf{x}}$) in a magnetic field of flux density vector $\mathbf{B}$. The ambient medium is air. Referring to Figure 6.15, and with B_0, ω, and a being positive constants, there is a nonzero emf induced in the loop ($e_{\text{ind}} \neq 0$) if

(A) $\mathbf{B} = B_0 \cos\omega t\,\hat{\mathbf{x}}$.

(B) $\mathbf{B} = B_0 \cos\omega t\,\hat{\mathbf{z}}$.

(C) $\mathbf{B} = B_0\,\hat{\mathbf{z}}$.

(D) $\mathbf{B} = B_0 x\, \hat{\mathbf{z}}/a$.

(E) $\mathbf{B} = B_0 y\, \hat{\mathbf{z}}/a$.

(F) more than one of the above cases.

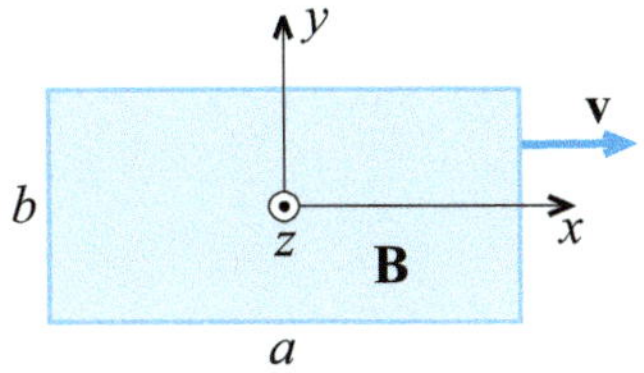

Figure 6.15 Conducting wire loop moving with a velocity $\mathbf{v}$ in a magnetic field; for Conceptual Question 6.19.

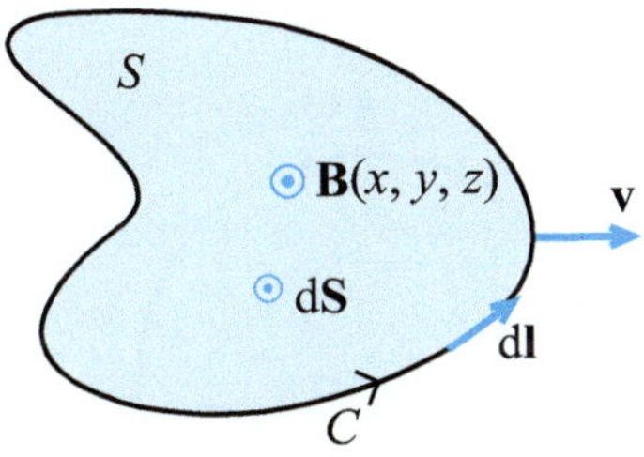

Figure 6.16 Planar metallic wire loop moving with a velocity $\mathbf{v}$ in a time-constant magnetic field of flux density $\mathbf{B}(x, y, z)$; for Conceptual Question 6.20.

CONCEPTUAL QUESTION 6.20 **Computing the motional emf in a loop.** A planar metallic wire loop moves with a velocity $\mathbf{v}$ in a nonuniform static magnetic field of flux density $\mathbf{B}$, as depicted in Figure 6.16. The magnetic field due to the induced current in the loop is negligible. Consider the following two expressions computed for this loop: $A_1 = -\frac{\mathrm{d}}{\mathrm{d}t}\int_S \mathbf{B} \cdot \mathrm{d}\mathbf{S}$ and $A_2 = \oint_C (\mathbf{v} \times \mathbf{B}) \cdot \mathrm{d}\mathbf{l}$, where the reference directions of $\mathrm{d}\mathbf{l}$ and $\mathrm{d}\mathbf{S}$ are interconnected by the right-hand rule. Which of the following is true for the induced emf, e_{ind}, in the loop?

(A) $e_{\text{ind}} = A_1 + A_2$.

(B) $e_{\text{ind}} = A_1 - A_2$.

(C) $e_{\text{ind}} = A_1 = A_2$.

(D) $e_{\text{ind}} = A_1$ and $e_{\text{ind}} \neq A_2$.

(E) $e_{\text{ind}} = A_2$ and $e_{\text{ind}} \neq A_1$.

CONCEPTUAL QUESTION 6.21 **Rotating loop in a static magnetic field.** Figure 6.17 shows a circular conducting wire loop that rotates with a constant angular velocity ω about its axis in a uniform time-invariant magnetic field of flux density B. The vector $\mathbf{B}$ is perpendicular to the plane of drawing. The magnetic field due to induced currents can be neglected. With $\mathcal{E}_0$ being a positive constant and $T = 2\pi/\omega$, the induced emf in the loop is of the following form:

(A) $e_{\text{ind}}(t) = \mathcal{E}_0 \cos \omega t$.

(B) $e_{\text{ind}}(t) = \mathcal{E}_0 \left(1 - \text{e}^{-t/T}\right)$.

(C) $e_{\text{ind}}(t) = \mathcal{E}_0 t/T$.

(D) $e_{\text{ind}}(t) = -\mathcal{E}_0$.

(E) $e_{\text{ind}}(t) = 0$.

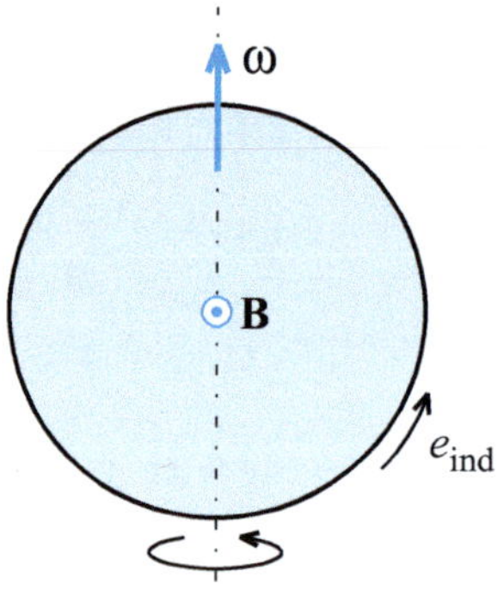

Figure 6.17 Circular conducting wire loop rotating in a uniform time-invariant magnetic field; for Conceptual Question 6.21.

CONCEPTUAL QUESTION 6.22 **Moving line – closed/open, imaginary/conducting.** Figure 6.18 portrays four cases of either a (closed) contour or open line (with two ends) that are either imaginary (nonmaterial) or made of a metallic wire – moving uniformly with a velocity $\mathbf{v}$ in a time-invariant (static) magnetic field of flux density $\mathbf{B}$, in a vacuum. For a specific orientation and variation in space of the vector $\mathbf{B}$, it is possible for an emf to be induced ($e_{\text{ind}} \neq 0$) along the contour/line in

(A) cases (a) and (c) only.

(B) case (c) only.

(C) cases (c) and (d) only.

(D) cases (a), (c), and (d) only.

(E) all four cases.

(F) none of the cases.

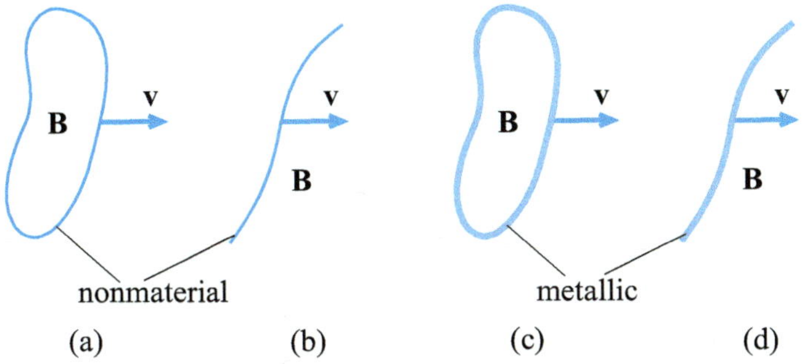

Figure 6.18 Imaginary or conducting contour or open line moving in a time-invariant magnetic field; for Conceptual Question 6.22.

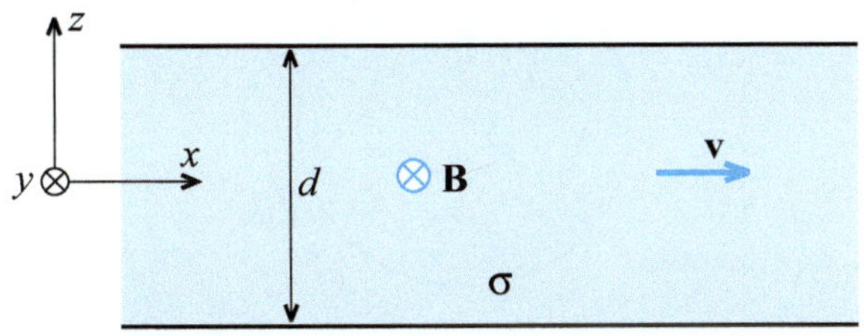

Figure 6.19 Conducting fluid flow through a tube with a time-constant magnetic field; for Conceptual Question 6.23.

CONCEPTUAL QUESTION 6.23 **Fluid flow through a tube with a static magnetic field.** A liquid of conductivity σ flows with a constant velocity $\mathbf{v} = v\,\hat{\mathbf{x}}$ through a tube of width d, in which a uniform time-invariant magnetic field of flux density $\mathbf{B} = B\,\hat{\mathbf{y}}$ is applied, as depicted in Figure 6.19. The generated voltage across the width of the tube equals

(A) $V = vB$.

(B) $V = \sigma vB$.

(C) $V = vBd$.

(D) $V = \sigma vBd$.

(E) $V = 0$.

(F) Need more information.

CONCEPTUAL QUESTION 6.24 **Eddy currents in a rotating cylinder.** A very long conducting cylinder of radius a uniformly rotates with angular velocity ω about its axis in a uniform time-invariant magnetic field of flux density $\mathbf{B}$, as shown in Figure 6.20. Streamlines of induced volume currents, so-called eddy currents, inside the cylinder away from its ends (bases) are

(A) circular (circles centered at the cylinder axis).

(B) radial (with respect to the cylinder axis).

(C) axial (parallel to the cylinder axis).

(D) nonexistent (eddy currents are not induced in this case).

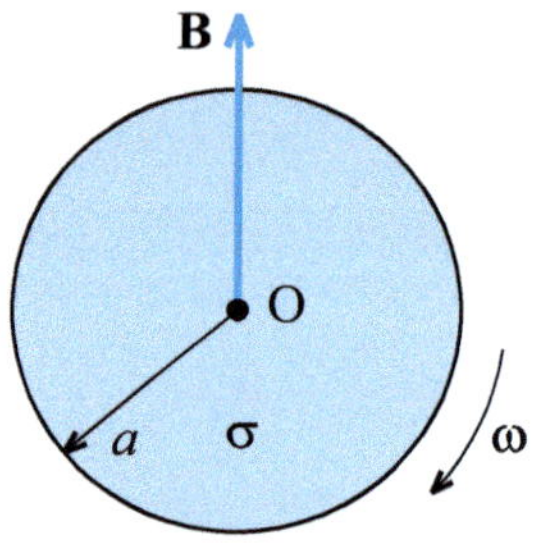

Figure 6.20 Cross section of a conducting cylinder rotating in a uniform magnetostatic field; for Conceptual Question 6.24.

CONCEPTUAL QUESTION 6.25 **Rotating loop in a dynamic magnetic field.** A circular conducting wire loop uniformly rotates with angular velocity ω about its axis in a low-frequency time-harmonic uniform magnetic field of flux density $B(t) = B_0 \cos \omega t$ ($B_0 > 0$), as shown in Figure 6.21, and the magnetic field due to induced currents is negligible. The induced emf in the loop, with $\mathcal{E}_0 > 0$, can be represented as

(A) $e_{\text{ind}}(t) = \mathcal{E}_0 \cos \omega t$.

(B) $e_{\text{ind}}(t) = \mathcal{E}_0 \sin 2\omega t$.

(C) $e_{\text{ind}}(t) = -\mathcal{E}_0$.

(D) $e_{\text{ind}}(t) = \mathcal{E}_0 \sin \omega t$.

(E) $e_{\text{ind}}(t) = 0$.

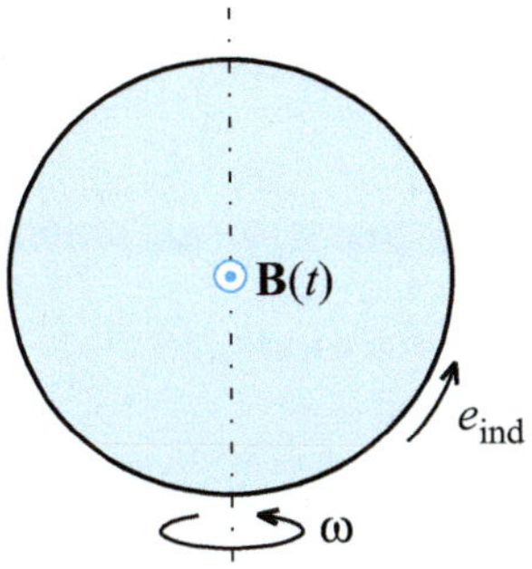

Figure 6.21 Circular wire loop rotating in a time-harmonic uniform magnetic field, with the angular frequency of the field being equal to the angular velocity of the loop rotation, ω; for Conceptual Question 6.25.

CONCEPTUAL QUESTION 6.26 Rotating loop in a rotating magnetic field. Assume that the flux density vector $\mathbf{B}$ of a uniform time-constant magnetic field in Figure 6.22, where a circular conducting wire loop uniformly rotates with an angular velocity ω, also rotates, in the same direction as the loop, with an angular velocity ω_B, where $\omega_B > \omega$ (vectors $\boldsymbol{\omega}$ and $\boldsymbol{\omega}_B$ are collinear and in the same direction). With $\mathcal{E}_0$, $\mathcal{E}_{01}$, and $\mathcal{E}_{02}$ standing for respective constants and the magnetic field due to induced currents being negligible, the induced emf in the loop is given by

(A) $e_{\text{ind}}(t) = \mathcal{E}_0 \cos \omega t$.

(B) $e_{\text{ind}}(t) = \mathcal{E}_0 \cos \omega_B t$.

(C) $e_{\text{ind}}(t) = \mathcal{E}_{01} \cos \omega t + \mathcal{E}_{02} \cos \omega_B t$.

(D) $e_{\text{ind}}(t) = \mathcal{E}_0 \cos(\omega_B - \omega)t$.

(E) $e_{\text{ind}}(t) = \mathcal{E}_0 \cos(\omega_B + \omega)t$.

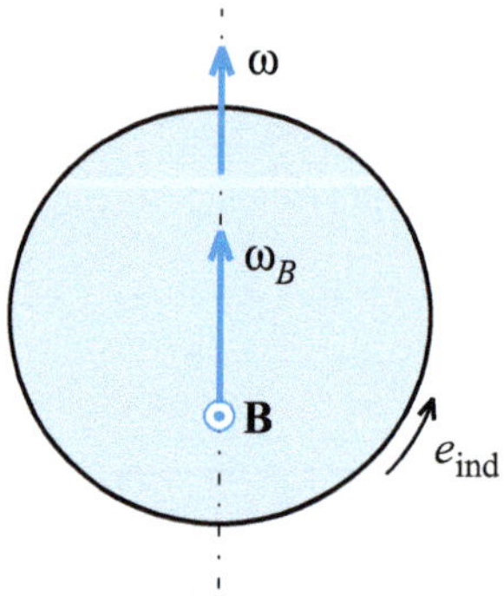

Figure 6.22 Circular wire loop rotating with an angular velocity ω in a rotating magnetic field of angular frequency ω_B, where $\omega_B > \omega$. This device represents an elementary asynchronous motor; for Conceptual Question 6.26.

6.4 Self-Inductance

In general, inductance can be interpreted as a measure of transformer electromagnetic induction in a system of conducting contours (circuits) with time-varying currents in a linear magnetic medium. Briefly, self-inductance is a measure of the magnetic flux and induced emf in a single isolated contour (or in one of the contours in a system) due to its own current. Similarly, a current in one contour causes magnetic flux through another contour and induced emf in it, and mutual inductance is used to characterize this coupling between the contours.

Consider a stationary conducting wire contour (loop), C, in a linear, homogeneous or inhomogeneous, magnetic medium, and assume that a slowly time-varying current of intensity i is established in the contour, as shown in Figure 6.23(a). This current produces a magnetic field whose flux density vector, $\mathbf{B}$, at any point of space and any instant of time is linearly proportional to i [see Eq. (4.6)], as well as an induced electric field of intensity $\mathbf{E}_{\text{ind}}$, given by Eq. (6.2), and an induced emf along C, e_{ind}, according to Eq. (6.4). From Eq. (4.15), the magnetic flux, Φ, through a surface S bounded by C [Figure 6.23(a)] is also linearly proportional to i, so we can write

$$L = \frac{\Phi}{i} , \quad e_{\text{ind}} = -\frac{d\Phi}{dt} = -L\frac{di}{dt} \quad (L - \text{self-inductance; unit: H}) , \tag{6.8}$$

where L is termed the self-inductance or just inductance of the contour, and its unit is henry (H). More precisely, this is the external inductance, since it takes into account only the flux Φ of the magnetic field that exists outside the conductor of the loop. Because the surrounding medium is magnetically linear, L depends only on the medium permeability and on the shape and dimensions of the contour, and not on the current intensity i. Note that, while the emf definition of self-inductance

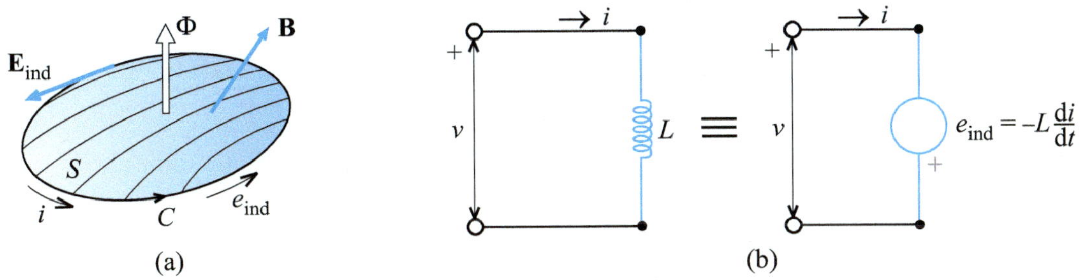

Figure 6.23 (a) Current contour in a linear magnetic medium – for the definition of self-inductance, L. (b) Circuit-theory representation of an inductor and equivalent controlled voltage generator.

in Eqs. (6.8) does not make any sense for steady currents, the flux definition can be used in practically the same way under both dynamic and static conditions.

For two-conductor transmission lines, the external inductance per unit length of the line, L', is defined analogously to the p.u.l. capacitance in Eq. (2.12): $L' = L_{\text{p.u.l.}} = L/i = \Phi'/i = \Phi/(il)$ (unit: F/m), with Φ' standing for the flux per unit length of the line, computed through a surface spanned between the line conductors.

Figure 6.23(b) shows the circuit-theory representation of an inductor, a device with its inductance L as its basic property. Having in mind Figure 6.6(a) and Eqs. (6.8), the inductor can now be replaced, with respect to its terminals, by an equivalent ideal controlled (by the time derivative of the current i) voltage generator, as indicated in Figure 6.23(b), from which the voltage v of the inductor is easily found.

Just as a capacitor stores electric energy, an inductor stores magnetic energy, which is given by [see Eqs. (2.19) and (6.8)] $W_{\text{m}} = Li^2/2 = \Phi i/2 = \Phi^2/(2L)$, and can also be computed using Eq. (5.11).

CONCEPTUAL QUESTION 6.27 **Sign of self-inductance.** The self-inductance, L, of an arbitrary metallic wire contour in a linear magnetic medium is always nonnegative.

(A) True.

(B) False.

CONCEPTUAL QUESTION 6.28 **Inductance of an imaginary contour.** Consider a conducting wire loop and an imaginary (nonmaterial) contour of exactly the same shape and size, in free space. Their inductances (external self-inductances) are the same.

(A) True.

(B) False.

CONCEPTUAL QUESTION 6.29 **Typical values of self-inductances in practice.** Given the typical values of self-inductances in practice, $L = 1$ mH is

(A) a small inductance.

(B) a medium-valued inductance.

(C) a large inductance.

CONCEPTUAL QUESTION 6.30 **Wire contour in a dc regime.** For a metallic wire contour in free space, which of the three quantities, the magnetic flux through the contour, Φ, the induced electromotive force (emf) along the contour, e_{ind}, and inductance of the contour, L, must be zero if the current in the contour is time-invariant?

(A) Φ and e_{ind} only.

(B) L only.

(C) e_{ind} and L only.

(D) e_{ind} only.

(E) All three quantities.

(F) None of the quantities.

CONCEPTUAL QUESTION 6.31 **Inductance of a contour with a steady current.** Letting L_{dc} and L_{ac} be the inductances of a contour obtained in its dc and ac (alternating current) regimes of operation, namely, as $L_{\text{dc}} = \Phi_1/I$ and $L_{\text{ac}} = \Phi_2/i$ assuming a dc current I and a slowly time-varying current i to flow in the contour, respectively, we have that

(A) $L_{\text{dc}} < L_{\text{ac}}$.

(B) $L_{\text{dc}} = L_{\text{ac}}$.

(C) $L_{\text{dc}} > L_{\text{ac}}$.

CONCEPTUAL QUESTION 6.32 **Inductance of a loop and the induced emf.** A metallic wire loop in air carries a slowly time-varying current of intensity $i(t)$. The inductance of the loop is L, its resistance is R, and the magnetic flux through the loop is $\Phi(t)$. For the notation in Figure 6.24, the induced emf in the loop is given by

(A) $e_{\text{ind}}(t) = Li(t)$.

(B) $e_{\text{ind}}(t) = L\,\mathrm{d}i/\,\mathrm{d}t$.

(C) $e_{\text{ind}}(t) = -L\,\mathrm{d}i/\,\mathrm{d}t$.

(D) $e_{\text{ind}}(t) = -\,\mathrm{d}\Phi/\,\mathrm{d}t + L\,\mathrm{d}i/\,\mathrm{d}t$.

(E) $e_{\text{ind}}(t) = L\,\mathrm{d}i/\,\mathrm{d}t + Ri(t)$.

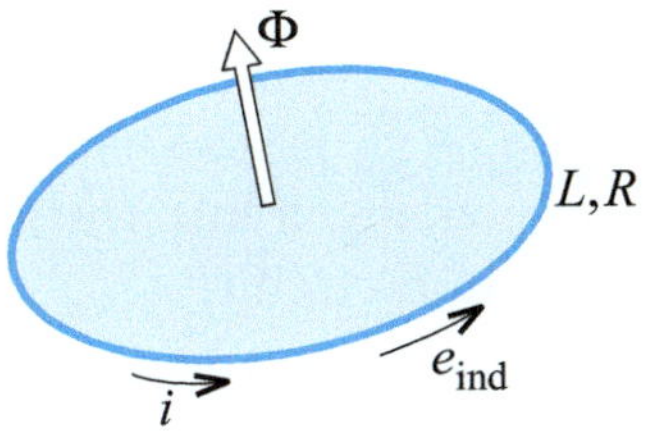

Figure 6.24 Wire loop with a slowly time-varying current in air; for Conceptual Question 6.32.

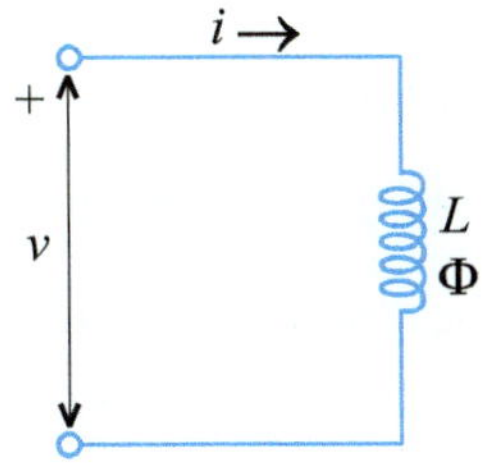

Figure 6.25 Inductor with a slowly time-varying current; for Conceptual Question 6.33.

CONCEPTUAL QUESTION 6.33 **Inductance of an inductor.** A linear inductor of inductance L carries a slowly time-varying current of intensity $i(t)$, and the magnetic flux of the inductor is $\Phi(t)$. For the notation in Figure 6.25, the voltage across the inductor terminals amounts to

(A) $v(t) = L\,\mathrm{d}i/\,\mathrm{d}t$.

(B) $v(t) = -L\,\mathrm{d}i/\,\mathrm{d}t$.

(C) $v(t) = -\,\mathrm{d}\Phi/\,\mathrm{d}t + L\,\mathrm{d}i/\,\mathrm{d}t$.

(D) $v(t) = \mathrm{d}\Phi/\,\mathrm{d}t - L\,\mathrm{d}i/\,\mathrm{d}t$.

(E) $v(t) = -(1/L)\int i(t)\,\mathrm{d}t$.

CONCEPTUAL QUESTION 6.34 **Circuit-theory model of an ideal inductor.** The only effect modeled by an inductor (ideal inductor) in circuit theory is the emf due to self-induction, e_{ind}.

(A) True.

(B) False.

CONCEPTUAL QUESTION 6.35 **Localization of the magnetic field in an electric circuit.** It is assumed in circuit theory that the magnetic field is concentrated only in the inductors in a circuit.

(A) True.

(B) False.

CONCEPTUAL QUESTION 6.36 **Emf due to connecting conductors in a circuit.** In the circuit-theory model, the emf due to the connecting conductors (lines between the elements in circuit layouts)

(A) depends on the shape and size of the conductors and on their current.

(B) depends on the current in the conductors and not on their shape and size.

(C) is always assumed to not exist.

CONCEPTUAL QUESTION 6.37 **Self-inductance of a loop in the presence of another loop.** The self-inductance of a loop in the presence of another loop is sometimes different from its self-inductance when isolated, in free space.

(A) True.

(B) False.

CONCEPTUAL QUESTION 6.38 Six changes to a circular loop with a steady current. Consider a circular copper loop carrying a steady current, in air, and the following changes to the loop, one at the time: (a) change the wire material from copper to aluminum, (b) double the loop radius, (c) extend the loop so it becomes an ellipse with the major to minor axis ratio of four but the same circumference as the original loop, (d) add a ferromagnetic core so that the loop encircles it, (e) double the current of the loop, and (f) reverse the direction of the loop current. Which of these changes would result in a change of the loop inductance?

(A) Changes (a)–(d) only.

(B) Changes (b)–(f) only.

(C) Changes (b)–(d) only.

(D) Changes (a) and (d) only.

(E) Changes (b)–(d) and (f) only.

(F) All changes, (a)–(f).

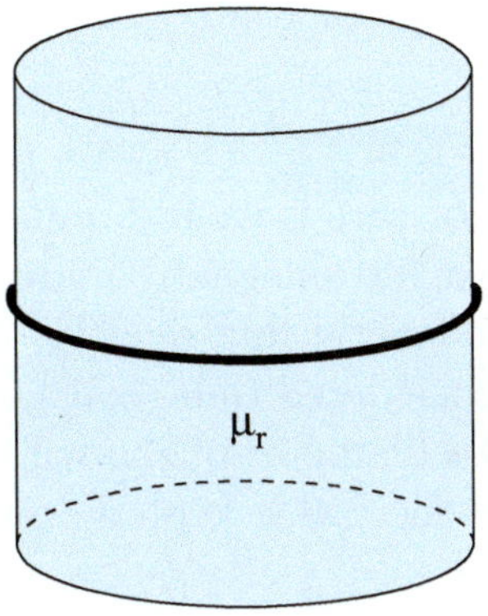

Figure 6.26 Wire loop with a linear ferromagnetic core; for Conceptual Question 6.39.

CONCEPTUAL QUESTION 6.39 Inductance of a loop with a linear magnetic core. If a piece of a ferromagnetic material of relative permeability μ_r is placed as a core of a wire loop, as indicated in Figure 6.26, the inductance of the loop, L, is related to that, L_0, of the same loop with no core as follows:

(A) $L = \mu_r L_0$.

(B) $L_0 < L < \mu_r L_0$.

(C) $L = L_0$.

(D) $L = L_0/\mu_r$.

(E) $L = \mu_r\mu_0 L_0$.

(F) $L = L_0/(\mu_r\mu_0)$.

(μ_0 is the permeability of a vacuum).

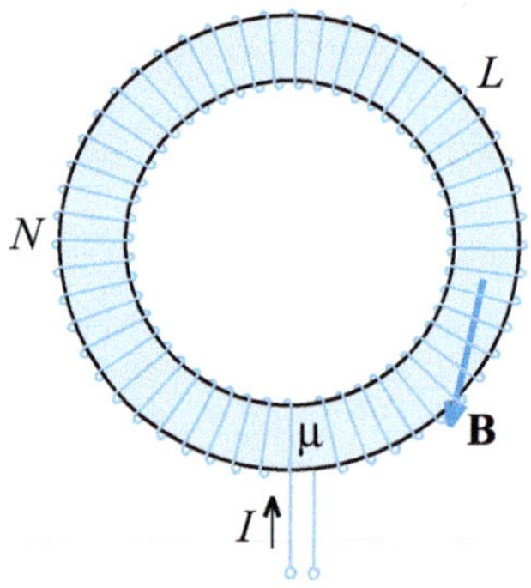

Figure 6.27 Thin toroidal linear magnetic core with a uniform and dense winding carrying a steady current; for Conceptual Question 6.40.

CONCEPTUAL QUESTION 6.40 **Doubling the number of wire turns in a coil.** A coil with N turns of wire is wound uniformly and densely about a thin toroidal core made from a linear ferromagnetic material of relative permeability μ_r (Figure 6.27). Consider the magnetic flux density, B, inside the core (note that, the core being thin, B can be considered to be constant throughout its volume) and inductance, L, of the coil. If the diameter of the wire in the coil is halved and N is doubled, while the current I in the coil is kept the same, we have that

(A) both B and L double.

(B) both B and L remain the same.

(C) both B and L quadruple.

(D) B doubles and L quadruples.

(E) B remains the same and L quadruples.

(F) B remains the same and L is halved.

CONCEPTUAL QUESTION 6.41 **Definition of external inductance for a two-wire line.** The external inductance per unit length of a thin symmetrical two-wire transmission line in air is not defined, because the line does not represent a current loop.

(A) True.

(B) False.

CONCEPTUAL QUESTION 6.42 **Definition of external inductance for a coaxial cable.** The external inductance per unit length of an air-filled coaxial cable is zero, because the cable is a closed (perfectly shielded) system, and its magnetic field is concentrated exclusively inside the cable.

(A) True.

(B) False.

CONCEPTUAL QUESTION 6.43 **Nonlinear inductor.** The inductance of a nonlinear inductor, namely, an inductor filled with a magnetically nonlinear material, depends on the intensity i of the inductor current, $L = L(i)$.

(A) True.

(B) False.

CONCEPTUAL QUESTION 6.44 **Energies of two inductors with the same magnetic flux.** Two linear inductors of inductances L and $2L$, respectively, have the same magnetic flux, Φ. The magnetic energy stored in the inductor with twice as large inductance is

(A) twice

(B) four times

(C) a half of

(D) a quarter of

(E) the same as

that stored in the other inductor.

6.5 Mutual Inductance

Consider now two stationary conducting wire contours, C_1 and C_2, in a linear (homogeneous or inhomogeneous) magnetic medium, as shown in Figure 6.28(a). Let the first contour (primary circuit) carry a slowly time-varying current of intensity i_1. As a result, a magnetic field, of flux density $\mathbf{B}_1$, is produced everywhere, and $\mathbf{B}_1$, which is a function of both the spatial coordinates and time, is linearly proportional to i_1. Some of the lines of $\mathbf{B}_1$ pass through the second contour (secondary circuit), i.e., through a surface S_2 bounded by C_2. These lines constitute the magnetic flux through the second contour due to the current i_1, which can be expressed as $\Phi_2 = \int_{S_2} \mathbf{B}_1 \cdot \mathrm{d}\mathbf{S}_2$. So, by computing or measuring the flux Φ_2 or the associated induced emf, e_{ind2}, in the second contour [Figure 6.28(a)], the mutual inductance between the two magnetically coupled contours (circuits) can be evaluated as

$$L_{21} = \frac{\Phi_2}{i_1}, \quad e_{\mathrm{ind2}} = -\frac{\mathrm{d}\Phi_2}{\mathrm{d}t} = -L_{21}\frac{\mathrm{d}i_1}{\mathrm{d}t} \quad (L_{21} \text{ – mutual inductance; unit: H}) . \tag{6.9}$$

Because of reciprocity (in a linear system, transfer functions remain the same if the source location and the location at which the response to the source is observed are interchanged), $L_{12} = L_{21}$. Note that the symbol M is also used to denote mutual inductance. Its magnitude depends on the shape, size, and mutual position of the contours, and on the magnetic properties (permeability) of the medium. The mutual inductance can be both positive and negative, depending on the adopted reference orientation of each of the contours for their given mutual position. Namely, if a positive current i_1 in the contour C_1 gives rise to a positive magnetic flux Φ_2 for the orientation of the surface S_2 that is in accordance to the right-hand rule with respect to the orientation of the contour C_2, the mutual inductance is positive. Otherwise, it is negative.[3]

Shown in Figure 6.28(b) is the circuit-theory representation of two magnetically coupled circuits. It consists of two coupled ideal inductors, where, in addition to modeling the emf due to self-induction in each inductor, the effect of the emf due to mutual induction between the inductors is also modeled. The mutual inductance between the inductors is customarily written as $L_{12} = \pm k\sqrt{L_1 L_2}$, where k is a

[3]In some texts, the information about the sign of the mutual inductance for specific reference orientations of contours is not included in its definition, i.e., mutual inductance is defined as always being nonnegative.

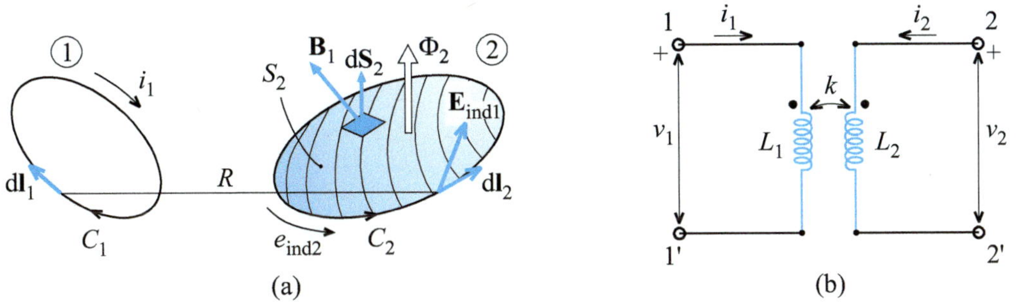

Figure 6.28 (a) Two magnetically coupled conducting contours – for the definition of mutual inductance. (b) Circuit-theory representation of two coupled inductors.

positive dimensionless constant called the coefficient of (magnetic) coupling of the inductors (circuits) and defined as

$$k = \frac{|L_{12}|}{\sqrt{L_1 L_2}}, \quad 0 \leq k \leq 1 \quad \text{(coefficient of magnetic coupling)} . \tag{6.10}$$

It cannot be greater than unity. As the sign of L_{12} depends on the adopted reference directions of currents i_1 and i_2, and therefore cannot be given as a single piece of information (positive or negative) along with k, independently from the current directions, we use a so-called two-dot notation to include this information in the representation in Figure 6.28(b), by placing two big dots near the particular ends of the two inductors. According to this notation (convention), if both currents (i_1 and i_2) enter the inductors at ends marked by a big dot [as in Figure 6.28(b)], the mutual inductance, for that particular combination of reference directions of currents, is positive. The same is true if both currents leave the inductors at marked (dotted) ends. Otherwise, if one current enters and the other leaves the inductor at marked ends, the mutual inductance is negative.

The magnetic energy of the system in Figure 6.28 is $W_m = (\Phi_1 i_1 + \Phi_2 i_2)/2$, where, by means of Eqs. (6.8) and (6.9) and the superposition principle, $\Phi_1 = L_1 i_1 + L_{12} i_2$ and similarly for Φ_2.

CONCEPTUAL QUESTION 6.45 Mutual inductance between two wire contours. Two conducting wire contours, C_1 and C_2, in air carry slowly time-varying currents of intensities $i_1(t)$ and $i_2(t)$, respectively, as shown in Figure 6.29. The mutual inductance L_{21} between the contours will not change if

(A) the size of C_1 changes and C_2 remains the same.

(B) the size of C_2 changes and C_1 remains the same.

(C) both contours change in size.

(D) both contours remain the same.

(E) More than one of the above cases.

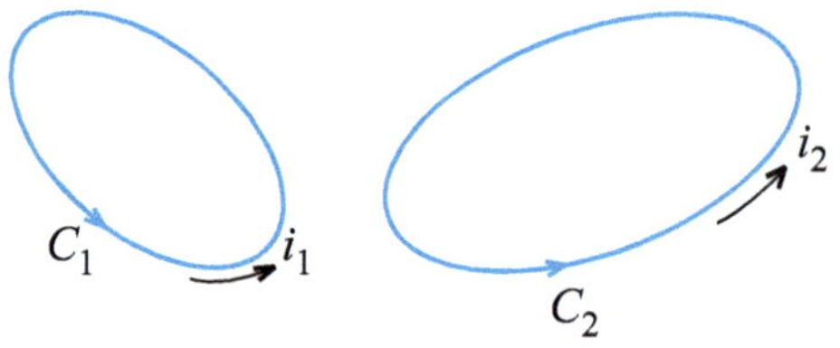

Figure 6.29 Two wire contours with slowly time-varying currents in air; for Conceptual Question 6.45.

CONCEPTUAL QUESTION 6.46 **Change of currents in magnetically coupled contours.** The mutual inductance L_{21} of the two contours in Figure 6.29 will change if

(A) the current i_1 is doubled and i_2 remains the same.

(B) the current i_2 is doubled and i_1 remains the same.

(C) both currents are doubled.

(D) both currents remain the same.

(E) More than one of the above cases.

(F) None of the above cases.

CONCEPTUAL QUESTION 6.47 **Change of orientation of one of the contours.** If in Figure 6.30(a) the orientation of the contour C_1 is reversed and that of C_2 remains the same, so the new situation is shown in Figure 6.30(b), which of the mutual inductances L_{12} and L_{21} of the contours will change?

(A) L_{12} only.

(B) L_{21} only.

(C) Both inductances.

(D) Neither of the inductances.

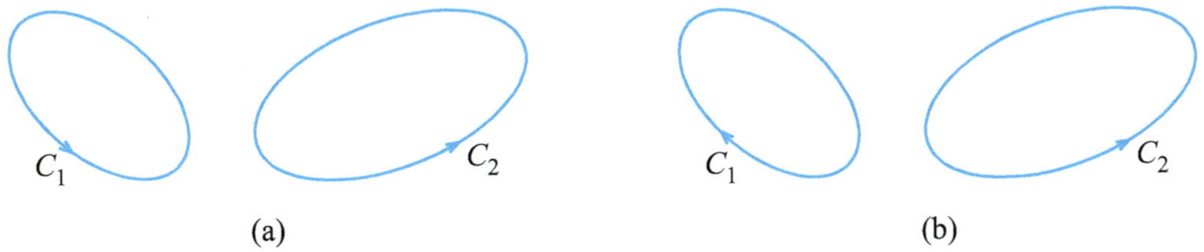

Figure 6.30 (a) Two wire contours in free space and (b) the same contours but with the orientation of one of them (C_1) reversed; for Conceptual Question 6.47.

CONCEPTUAL QUESTION 6.48 **Magnetic flux through one of the coupled contours.** The magnetic flux through the contour C_2 in Figure 6.31 is given by

(A) $\Phi_2 = L_{21}i_1$.

(B) $\Phi_2 = -L_{21}i_1$.

(C) $\Phi_2 = L_{12}i_1$.

(D) $\Phi_2 = -L_{12}i_1$.

(E) more than one of the expressions above.

(F) none of the expressions above.

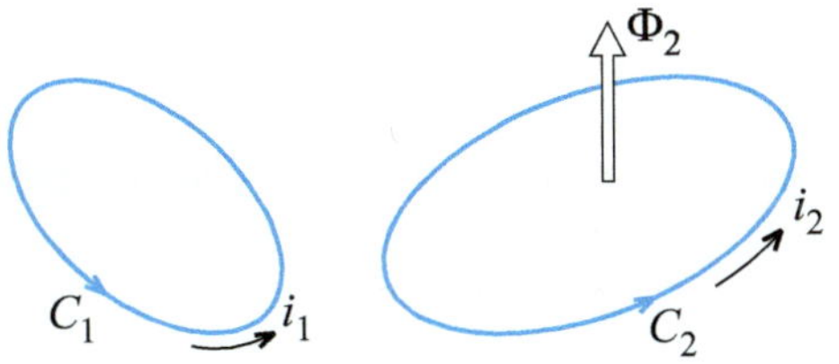

Figure 6.31 Magnetic flux through the second contour – for two conducting wire contours in air carrying slowly time-varying currents; for Conceptual Question 6.48.

CONCEPTUAL QUESTION 6.49 **Adding one or two ferromagnetic cores.** Consider placing a piece or two pieces of a linear ferromagnetic material of relative permeability μ_r, as in Figure 6.32, as a core of one or both magnetically coupled wire contours in free space (Figure 6.32 shows the new situation with a ferromagnetic piece inserted only in the first contour). The mutual inductance L_{21} of the contours will increase in magnitude if

(A) a piece is inserted only in the contour C_1.

(B) a piece is inserted only in C_2.

(C) pieces are inserted in each of the contours.

(D) All three above cases.

(E) None of the above cases.

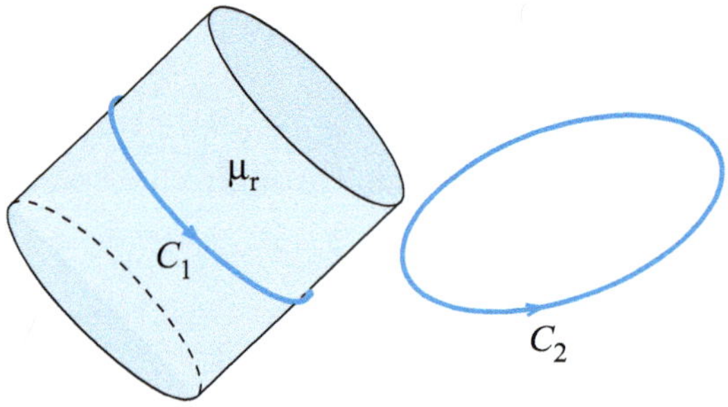

Figure 6.32 Adding linear ferromagnetic core(s) to one or both magnetically coupled wire contours in free space; for Conceptual Question 6.49.

CONCEPTUAL QUESTION 6.50 **Two mutual inductances of magnetically coupled contours.** Consider two magnetically coupled contours with a piece of a linear ferromagnetic material added as a core of the contour C_1 – in Figure 6.32. For this situation (and the particular orientation of the contours in Figure 6.32), mutual inductances L_{12} and L_{21} of the contours are related as

(A) $L_{12} = L_{21}$.

(B) $L_{12} = -L_{21}$.

(C) $|L_{12}| < |L_{21}|$.

(D) $|L_{12}| > |L_{21}|$.

CONCEPTUAL QUESTION 6.51 **Four mutual positions of two wire loops.** Out of the four mutual positions of two circular wire loops shown in Figure 6.33, the magnitude of the mutual inductance between the loops is the largest in

(A) case (a).

(B) case (b).

(C) case (c).

(D) case (d).

(E) No difference.

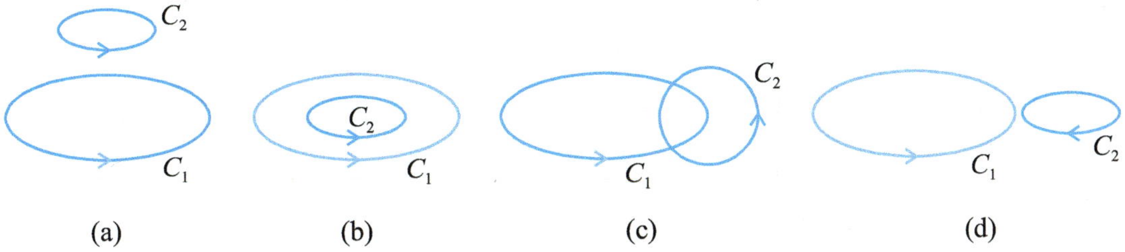

Figure 6.33 Two circular loops with four different mutual positions; for Conceptual Question 6.51.

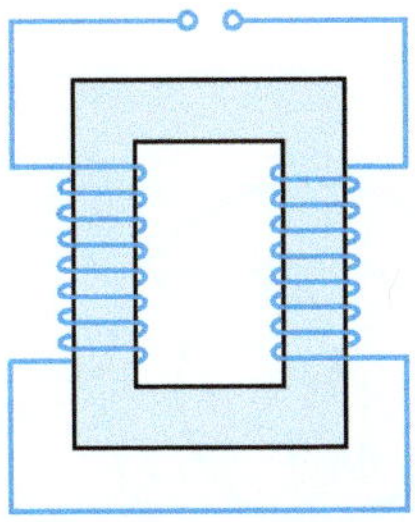

Figure 6.34 Series connection of two coils wound on a cardboard core; for Conceptual Question 6.52.

CONCEPTUAL QUESTION 6.52 **Two coils connected in series.** Figure 6.34 shows two coils wound on a cardboard core. The mutual inductance L_{12} of the coils is

(A) positive.

(B) negative.

(C) zero.

(D) Need more information.

CONCEPTUAL QUESTION 6.53 **Two coils connected in parallel.** For two coils shown in Figure 6.35, the mutual inductance L_{12} of the coils is

(A) positive.

(B) negative.

(C) zero.

(D) Need more information.

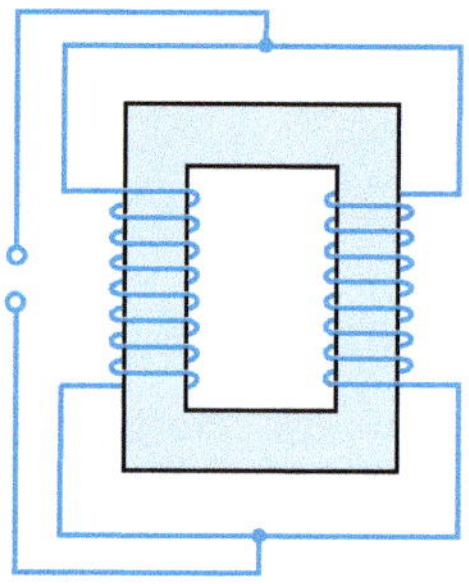

Figure 6.35 Parallel connection of two coils on a cardboard core; for Conceptual Question 6.53.

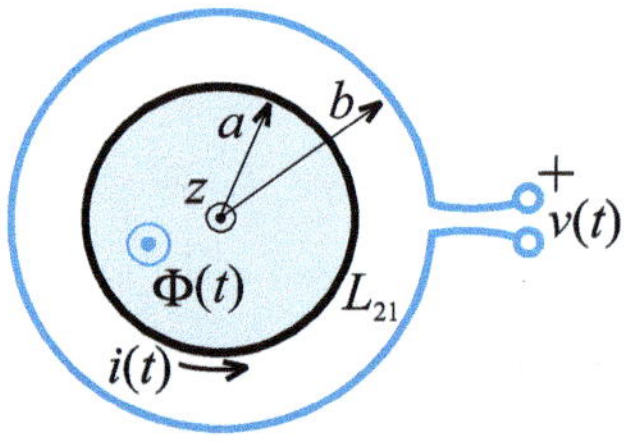

Figure 6.36 Mutual inductance between an air-filled infinitely long solenoid and an open-circuited coaxial loop; for Conceptual Question 6.54.

CONCEPTUAL QUESTION 6.54 **Magnetic coupling between a solenoid and a loop.** Consider the open-circuited loop around a solenoidal coil with a slowly time-varying current of intensity $i(t)$ in Figure 6.36, and let the mutual inductance between the solenoid and the loop be L_{21}. For the given (in the figure) reference orientations of the turns of the solenoid and the terminals of the loop, the voltage between the loop terminals amounts to

(A) $v(t) = |L_{21}|\,\mathrm{d}i/\,\mathrm{d}t$.

(B) $v(t) = -|L_{21}|\,\mathrm{d}i/\,\mathrm{d}t$.

(C) $v(t) = (a/b)^2 |L_{21}| \, \mathrm{d}i/\mathrm{d}t$.

(D) $v(t) = -(a/b)^2 |L_{21}| \, \mathrm{d}i/\mathrm{d}t$.

(E) $v(t) = 0$.

CONCEPTUAL QUESTION 6.55 **Computing the solenoid-loop mutual inductance.** For the solenoid-loop system in Figure 6.36, the mutual inductance L_{21} can be computed as $L_{21} = \Phi/i$, where Φ is the magnetic flux through a surface spanned over one turn of the coil and both the coil turns and the loop are oriented in the counterclockwise direction.

(A) True.

(B) False.

CONCEPTUAL QUESTION 6.56 **Short-circuited loop and open-circuited solenoid.** Consider first an open-circuited loop placed coaxially about an air-filled infinitely long solenoidal coil with a slowly time-varying current of intensity $i(t)$, shown in Figure 6.37(a). There is no magnetic field outside the coil. The resultant voltage between the terminals of the loop is $v(t)$. Assume then that the loop in Figure 6.37(a) is short-circuited and that a slowly time-varying current is established in it, of intensity $i(t)$ with respect to the counterclockwise reference direction, as depicted in Figure 6.37(b). In addition, let the solenoidal coil be very long but of finite length (the loop around the solenoid is at the middle of its length), with N turns of wire, and open-circuited. For this arrangement and with $v(t)$ denoting the open-circuited loop voltage in Figure 6.37(a), the induced emf in the open-circuited coil in Figure 6.37(b) with respect to the counterclockwise reference orientation of its turns is given by

(A) $e_{\mathrm{ind}}(t) \approx v(t)$.

(B) $e_{\mathrm{ind}}(t) \approx -v(t)$.

(C) $e_{\mathrm{ind}}(t) \approx Nv(t)$.

(D) $e_{\mathrm{ind}}(t) \approx -Nv(t)$.

(E) $e_{\mathrm{ind}}(t) = 0$.

(F) None of the above.

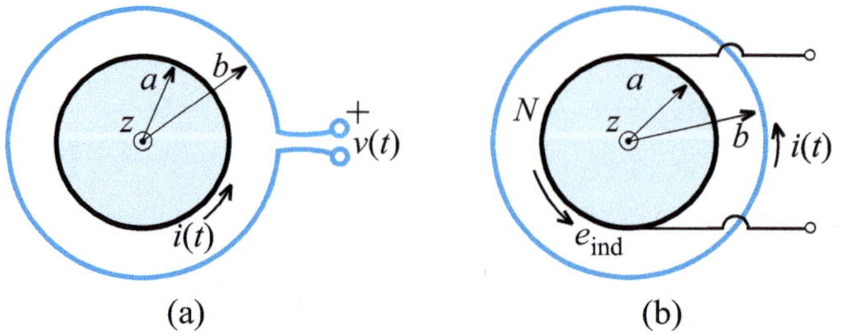

Figure 6.37 (a) Solenoid with current $i(t)$ and an open-circuited coaxial loop. (b) Short-circuited loop with current $i(t)$ and an open-circuited coaxial solenoidal coil. Shown are cross sections of the structures; for Conceptual Question 6.56.

CONCEPTUAL QUESTION 6.57 **Two-dot convention for the sign of mutual inductance.** The mutual inductance L_{12} of two magnetically coupled inductors shown in Figure 6.38 is

(A) positive.

(B) negative.

(C) Need more information.

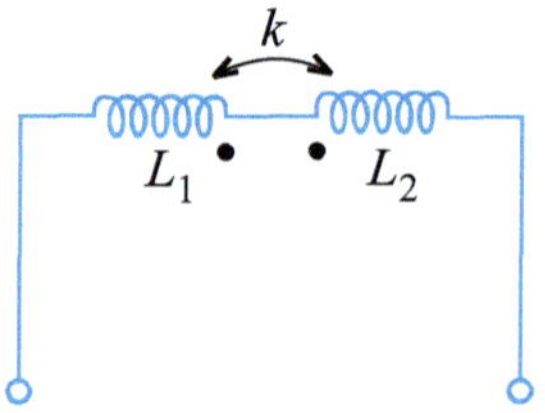

Figure 6.38 Two magnetically coupled inductors; for Conceptual Question 6.57.

CONCEPTUAL QUESTION 6.58 **Equivalent inductance of two inductors in series.** Compare the equivalent inductance of two inductors of inductances L_1 and L_2 connected in series when they are magnetically coupled and when they are not, respectively. This inductance is greater when the inductors are

(A) coupled.

(B) uncoupled.

(C) No difference.

(D) Need more information.

CONCEPTUAL QUESTION 6.59 **Coupling coefficient, two coils above each other.** Two identical solenoidal coils are wound on a cylindrical cardboard core, one above the other. If this core is replaced by one made of a linear ferromagnetic material, the coefficient of coupling (k) between the two coils

(A) increases.

(B) decreases.

(C) remains the same.

(D) Need more information.

CONCEPTUAL QUESTION 6.60 **Coupling coefficient, coils inside each other.** Consider two air-filled solenoidal coils of equal lengths and different radii positioned coaxially with respect to each other such that their centers coincide. How does the coupling coefficient of the coils, k, change if a linear ferromagnetic core is inserted to completely fill the inner coil?

(A) It increases.

(B) It decreases.

(C) It remains the same.

(D) Need more information.

CONCEPTUAL QUESTION 6.61 **Coupling coefficient, coils on a toroidal core.** Two coils are wound uniformly and densely in two layers, one on top of the other, about a thin toroidal cardboard core. How does the coupling coefficient of the coils change if this core is replaced by a linear ferromagnetic one?

(A) It increases.

(B) It decreases.

(C) It remains the same.

(D) Need more information.

CONCEPTUAL QUESTION 6.62 **Self-inductance of a loop on a PMC plane.** A planar metallic wire loop lies in air on a ferromagnetic plane with $\mu_r \to \infty$, as portrayed in Figure 6.39. There is the following relationship between the self-inductance of this loop (in the presence of the plane), L, and its self-inductance when isolated, in free space (air), L_0:

(A) $L = L_0$.

(B) $L = 2L_0$.

(C) $L \gg L_0$.

(D) $L = L_0/2$.

(E) $L \ll L_0$.

(F) None of the above.

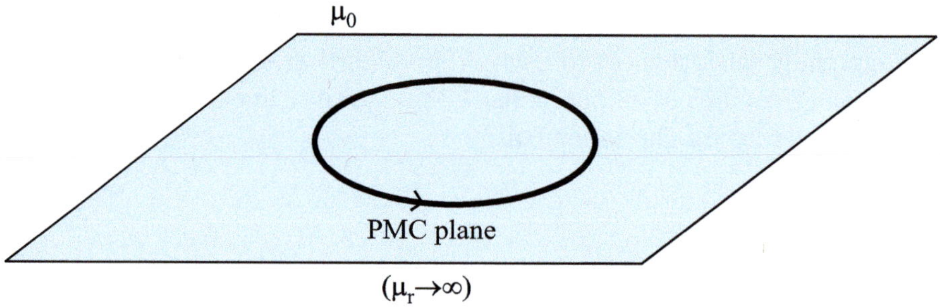

Figure 6.39 Planar metallic wire loop lying on a ferromagnetic (or PMC) plane; for Conceptual Question 6.62.

CONCEPTUAL QUESTION 6.63 **Energy of two coupled coils.** If the mutual inductance (L_{12}) of two magnetically coupled coils carrying steady currents I_1 and I_2 is made to be twice that value [$L_{12}^{(\text{new})} = 2L_{12}^{(\text{old})}$] while keeping their self-inductances and currents unchanged, the magnetic energy of the system

(A) doubles.

(B) increases but does not exactly double.

(C) decreases.

(D) remains the same.

(E) Need more information.

6.6 Displacement Current

The rest of this chapter is devoted to the rapidly time-varying (e.g., high-frequency time-harmonic) electromagnetic field, which cannot be analyzed without taking into account the electromagnetic retardation effect. We first correct the quasistatic (or static) version of the generalized Ampère's law (Maxwell's second equation), Eq. (5.3), by adding the displacement current, in parallel to the conduction current. The addition of this new term in Maxwell's equations corresponds to the inclusion of the time retardation in the field expressions, and thus enables modeling of electromagnetic wave propagation and radiation.

The most general version of Ampère's law, that for the rapidly time-varying (high-frequency) electromagnetic field, reads

$$\oint_C \mathbf{H} \cdot \mathrm{d}\mathbf{l} = \int_S \left(\mathbf{J} + \frac{\partial \mathbf{D}}{\partial t} \right) \cdot \mathrm{d}\mathbf{S} \quad \text{(corrected generalized Ampère's law)} , \quad (6.11)$$

where the expression $\partial \mathbf{D}/\partial t$ has the dimension of a current density (it is expressed in $\mathrm{A/m^2}$), and this is the density of a new type of current that may exist even in air or a vacuum. It is called the displacement current density and is denoted by $\mathbf{J}_\mathrm{d}$,

$$\mathbf{J}_\mathrm{d} = \frac{\partial \mathbf{D}}{\partial t} \quad \text{(displacement current density vector; unit: A/m}^2\text{)} . \quad (6.12)$$

In the slowly time-varying (low-frequency) field (see Section 6.11), the rate of the time variation in the electric flux density vector, $\partial \mathbf{D}/\partial t$, at a point is slow enough to be neglected with respect to the conduction current density vector given by Eq. (3.7), $\mathbf{J} = \sigma \mathbf{E}$, so that Ampère's law in Eq. (6.11) can be approximated by its quasistatic or static version in Eq. (5.3). Exceptions are slowly time-varying fields in nonconducting (e.g., air) or poorly conducting media, where $\mathbf{J}$ is zero or very small, so that $\mathbf{J}_\mathrm{d}$ cannot be neglected even in the slowly time-varying case. Except in such cases, we generally consider that Ampère's law for the slowly time-varying or quasistatic electromagnetic field does not include the term $\partial \mathbf{D}/\partial t$.

CONCEPTUAL QUESTION 6.64 **Closed path in a high-frequency electromagnetic field.** A high-frequency time-harmonic electromagnetic field exists in a lossy medium. The current density vector in the medium is $\mathbf{J}$, the electric flux density vector is $\mathbf{D}$, and the magnetic flux density vector is $\mathbf{B}$. The line integral of the magnetic field intensity vector, $\mathbf{H}$, along an arbitrary planar contour (closed path) in the medium equals the flux of the following vector through a flat surface spanned over the contour:

(A) $\mathbf{J}$,

(B) $\mathbf{B}$,

(C) $\partial \mathbf{D}/\partial t$,

(D) $\mathbf{J} + \partial \mathbf{D}/\partial t$,

(E) $-\partial \mathbf{B}/\partial t$,

where the orientation of the contour and the orientation of the surface are adopted in accordance with the right-hand rule.

CONCEPTUAL QUESTION 6.65 **Displacement current causing a magnetic field.** Displacement current causes a magnetic field in the same way conduction current does.

(A) True.

(B) False.

CONCEPTUAL QUESTION 6.66 **Corrected generalized differential Ampère's law.** The net curl of the time-varying magnetic field intensity vector, $\mathbf{H}$, exists at a point in space whenever a time-varying electric field is present at that point.

(A) True.

(B) False.

CONCEPTUAL QUESTION 6.67 **Relationship between E and H vectors at a point.** At a point in air and an instant of time, $\partial \mathbf{E}/\partial t \neq 0$ and $\mathbf{H} = 0$. Is this possible?

(A) Yes.

(B) No.

(C) It is impossible to tell.

CONCEPTUAL QUESTION 6.68 **More on the relationship between E and H vectors.** At a point in air and an instant of time, $\nabla \times \mathbf{H} \neq 0$ and $\mathbf{E} = 0$. Is this possible?

(A) Yes.

(B) No.

(C) It is impossible to tell.

CONCEPTUAL QUESTION 6.69 **Comparison of conduction and displacement current densities.** The amplitude of the time-harmonic displacement current density in a conducting medium is never larger than that of the conduction current density at the same point.

(A) True.

(B) False.

CONCEPTUAL QUESTION 6.70 **Displacement current density at high and low frequencies.** The time-harmonic displacement current density vector at a point in a conducting medium can be neglected

(A) at high frequencies.

(B) at low frequencies.

(C) always.

(D) never.

CONCEPTUAL QUESTION 6.71 **Displacement current at 60 Hz.** At 60 Hz, the displacement current can be neglected in all electromagnetic problems.

(A) True.

(B) False.

CONCEPTUAL QUESTION 6.72 **Current flow through leads of an ideal capacitor.** Which of the following types of current cannot flow through the terminals (leads) of an ideal capacitor in an electric circuit?

(A) Time-invariant current.

(B) Slowly time-varying current.

(C) Rapidly time-varying current.

(D) Any time-varying current.

(E) Any current.

(F) None of the above.

CONCEPTUAL QUESTION 6.73 **Current flow through a nonideal capacitor.** The following type(s) of current cannot flow through a nonideal capacitor in an electric circuit:

(A) Time-invariant current.

(B) Slowly time-varying current.

(C) Rapidly time-varying current.

(D) Any time-varying current.

(E) Any current.

(F) None of the above.

CONCEPTUAL QUESTION 6.74 **Relationship between conduction and displacement currents.** If the conduction current density at a point of a lossy medium of parameters ε, μ_0, and σ is given by $J(t) = J_0 \sin \omega t$, the displacement current density at that point is of the following form (J_0, ω, and J_{d0} are positive constants):

(A) $J_d(t) = J_{d0} \sin \omega t$.

(B) $J_d(t) = -J_{d0} \sin \omega t$.

(C) $J_d(t) = J_{d0} \cos \omega t$.

(D) $J_d(t) = J_{d0}$.

(E) $J_d(t) = 0$.

(F) None of the above.

CONCEPTUAL QUESTION 6.75 **Conduction to displacement current ratio.** For a sample of conducting material with conductivity σ and permittivity ε that is occupied by a time-varying electric field of intensity $E(t) = E_0 \cos \omega t$ (E_0 and ω are positive constants), the ratio of the amplitudes (peak values) of the conduction and displacement current densities is given by

(A) $|J|_{\text{max}}/|J_{\text{d}}|_{\text{max}} = \varepsilon E_0/\sigma$.

(B) $|J|_{\text{max}}/|J_{\text{d}}|_{\text{max}} = \sigma E_0/\varepsilon$.

(C) $|J|_{\text{max}}/|J_{\text{d}}|_{\text{max}} = \omega\varepsilon/\sigma$.

(D) $|J|_{\text{max}}/|J_{\text{d}}|_{\text{max}} = \sigma/(\omega\varepsilon)$.

(E) $|J|_{\text{max}}/|J_{\text{d}}|_{\text{max}} = 0$.

(F) $|J|_{\text{max}}/|J_{\text{d}}|_{\text{max}} \to \infty$.

CONCEPTUAL QUESTION 6.76 **Displacement current in air.** Displacement current can exist in air.

(A) True.

(B) False.

CONCEPTUAL QUESTION 6.77 **Displacement current in a vacuum.** Displacement current can exist in a vacuum.

(A) True.

(B) False.

6.7 Maxwell's Equations for the High-Frequency Electromagnetic Field

Having now in place the corrected version of the generalized Ampère's law, Eq. (6.11), we are ready [also see Eqs. (6.4), (5.3), (2.5), and (4.16)] to summarize the full set of Maxwell's equations for the most general field – a rapidly time-varying or high-frequency electromagnetic field. We list them here in differential form, together with the three constitutive equations describing material properties of electromagnetic media [see Eqs. (2.6), (5.5), and (3.7)]:

$$\begin{cases} \nabla \times \mathbf{E} = -\frac{\partial \mathbf{B}}{\partial t} & \text{(Maxwell's first equation, differential)} \\ \nabla \times \mathbf{H} = \mathbf{J} + \frac{\partial \mathbf{D}}{\partial t} & \text{(Maxwell's second equation, differential)} \\ \nabla \cdot \mathbf{D} = \rho & \text{(Maxwell's third equation, differential)} \\ \nabla \cdot \mathbf{B} = 0 & \text{(Maxwell's fourth equation, differential)} \\ \mathbf{D} = \mathbf{D}(\mathbf{E})\ [\mathbf{D} = \varepsilon \mathbf{E}] & \text{(constitutive equation for } \mathbf{D}) \\ \mathbf{B} = \mathbf{B}(\mathbf{H})\ [\mathbf{B} = \mu \mathbf{H}] & \text{(constitutive equation for } \mathbf{B}) \\ \mathbf{J} = \mathbf{J}(\mathbf{E})\ [\mathbf{J} = \sigma \mathbf{E}] & \text{(constitutive equation for } \mathbf{J}) \end{cases} . \tag{6.13}$$

In addition, the current and charge densities, $\mathbf{J}$ and ρ, are related to each other by the continuity equation in Eq. (3.5), or by its integral form in Eq. (3.4).

Note that the above differential equations are valid always and everywhere, except at points where material properties of electromagnetic media change abruptly from one value to another, that is, at boundary surfaces between electromagnetically different media. At such points, basic field vector functions change abruptly as well across the boundary surface, which means that their spatial derivatives in the direction normal to the surface are not defined.

Note also that Maxwell's equations are not completely independent from one another. For instance, taking the divergence of both sides of the differential form of Maxwell's second equation and combining the obtained equation with the corresponding version of the continuity equation, Eq. (3.5), and the identity $\nabla \cdot (\nabla \times \mathbf{a}) = 0$, we can derive Maxwell's third equation, in Eqs. (6.13).

CONCEPTUAL QUESTION 6.78 **Differential Maxwell's equations at high and low frequencies.** How many of the four differential Maxwell's equations for the rapidly time-varying electromagnetic field are the same as in the slowly time-varying case?

(A) None.

(B) One.

(C) Two.

(D) Three.

(E) All.

CONCEPTUAL QUESTION 6.79 Integral Maxwell's equations at high and low frequencies. Considering the four Maxwell's equations in integral form, how many of them are the same for both rapidly and slowly time-varying electromagnetic fields?

(A) None.

(B) One.

(C) Two.

(D) Three.

(E) All.

CONCEPTUAL QUESTION 6.80 Maxwell's equations for high-frequency and static fields. How many of the four differential Maxwell's equations for the rapidly time-varying electromagnetic field have the same form as the corresponding equations governing the time-invariant field?

(A) None.

(B) One.

(C) Two.

(D) Three.

(E) All.

CONCEPTUAL QUESTION 6.81 Statements of Maxwell's equations. Consider the following four statements: (a) A magnetic field that changes in time produces an electric field. (b) A time-varying electric field generates a magnetic field. (c) Electric charges are sources of an electric field. (d) Magnetic charges are sources of a magnetic field. How many of the statements are true?

(A) None.

(B) One.

(C) Two.

(D) Three.

(E) All.

CONCEPTUAL QUESTION 6.82 **Limitations to circulation and flux Maxwell's equations.** Can contours C for applying the two circulation Maxwell's equations in integral form in an electromagnetic system with rapidly time-varying fields be adopted completely independently from closed surfaces S adopted for applying the two flux equations?

(A) Yes.

(B) No.

CONCEPTUAL QUESTION 6.83 **Validity of Maxwell's equations in differential form.** Maxwell's equations in differential form are valid for all possible electromagnetic fields and all points in an arbitrary electromagnetic system.

(A) True.

(B) False.

CONCEPTUAL QUESTION 6.84 **Deriving Maxwell's equations from one another.** Consider the four Maxwell's equations in differential form for the rapidly time-varying electromagnetic field, and the following two statements: (a) Two divergence equations can be derived from two curl equations and the associated continuity equation. (b) Two curl equations can be derived from two divergence equations and the continuity equation. Which of the statements is true?

(A) Statement (a) only.

(B) Statement (b) only.

(C) Both statements.

(D) Neither of the statements.

CONCEPTUAL QUESTION 6.85 **Zero and nonzero dynamic-field quantities in a PEC.** Consider a time-varying electromagnetic field in a homogeneous linear medium and the following quantities describing it or being associated with it: electric field intensity vector ($\mathbf{E}$), magnetic field intensity vector ($\mathbf{H}$), electric flux density vector ($\mathbf{D}$), magnetic flux density vector ($\mathbf{B}$), volume current density vector ($\mathbf{J}$), and volume charge density (ρ). Which of these quantities must be zero if the medium is a perfect electric conductor (PEC), with $\sigma \to \infty$?

(A) **E** and **D** only.

(B) **E**, **D**, and ρ only.

(C) **H**, **B**, and **J** only.

(D) **E**, **H**, **D**, and **B** only.

(E) All quantities.

(F) None of the quantities.

CONCEPTUAL QUESTION 6.86 **Zero and nonzero static-field quantities in a PEC.** Which of the field and source volume quantities involved in Maxwell's equations, **E**, **H**, **D**, **B**, **J**, and ρ, must be zero for a time-constant electromagnetic field in a perfect electric conductor ($\sigma \to \infty$)?

(A) **E** and **D** only.

(B) **E**, **D**, and ρ only.

(C) **H**, **B**, and **J** only.

(D) **E**, **H**, **D**, and **B** only.

(E) All quantities.

(F) None of the quantities.

6.8 Boundary Conditions for the High-Frequency Electromagnetic Field

General electromagnetic boundary conditions – for the rapidly time-varying or high-frequency electromagnetic field, namely, for the tangential components of vectors **E** and **H** and normal components of vectors **D** and **B**, respectively, at a boundary surface between two electromagnetic media (regions 1 and 2) have the same form as in Eqs. (2.7) and (5.7); in vector notation, they are given by

$$\begin{cases} \hat{\mathbf{n}} \times \mathbf{E}_1 - \hat{\mathbf{n}} \times \mathbf{E}_2 = 0 & \text{(boundary condition for } \mathbf{E}_\mathrm{t}) \\ \hat{\mathbf{n}} \times \mathbf{H}_1 - \hat{\mathbf{n}} \times \mathbf{H}_2 = \mathbf{J}_\mathrm{s} & \text{(boundary condition for } \mathbf{H}_\mathrm{t}) \\ \hat{\mathbf{n}} \cdot \mathbf{D}_1 - \hat{\mathbf{n}} \cdot \mathbf{D}_2 = \rho_\mathrm{s} & \text{(boundary condition for } \mathbf{D}_\mathrm{n}) \\ \hat{\mathbf{n}} \cdot \mathbf{B}_1 - \hat{\mathbf{n}} \cdot \mathbf{B}_2 = 0 & \text{(boundary condition for } \mathbf{B}_\mathrm{n}) \end{cases} , \tag{6.14}$$

where $\hat{\mathbf{n}}$ is the normal unit vector on the surface, directed from region 2 to region 1 (Figures 2.4 and 5.3). In addition, as discussed in Section 3.3, the boundary condition for normal components of the vector $\mathbf{J}$ is of the same form as that for $\mathbf{D}$, with ρ_s replaced by $-\partial\rho_\mathrm{s}/\partial t$, so $\hat{\mathbf{n}} \cdot \mathbf{J}_1 - \hat{\mathbf{n}} \cdot \mathbf{J}_2 = -\partial\rho_\mathrm{s}/\partial t$.

As an important special case, if region 2 is a perfect electric conductor (PEC), with $\sigma \to \infty$, in which there can be no time-varying electromagnetic field, Eqs. (6.14) become

$$\hat{\mathbf{n}} \times \mathbf{E} = 0 , \quad \hat{\mathbf{n}} \times \mathbf{H} = \mathbf{J}_\mathrm{s} , \quad \hat{\mathbf{n}} \cdot \mathbf{D} = \rho_\mathrm{s} , \quad \hat{\mathbf{n}} \cdot \mathbf{B} = 0 \quad \text{(on a PEC surface)} , \tag{6.15}$$

with $\hat{\mathbf{n}}$ directed from the conductor outward. The surface current and charge densities, $\mathbf{J}_\mathrm{s}$ and ρ_s, are mutually related by the corresponding version of the continuity equation, $\nabla_\mathrm{s} \cdot \mathbf{J}_\mathrm{s} = -\partial\rho_\mathrm{s}/\partial t$, which invokes the surface version of the (volume) divergence operator in Eq. (3.5) and is referred to as the continuity equation for (PEC) plates.

CONCEPTUAL QUESTION 6.87 Dynamic and static boundary conditions. Which of the following boundary conditions for the respective components of rapidly time-varying (high-frequency) electric and magnetic field intensity vectors, $\mathbf{E}$ and $\mathbf{H}$, and flux density vectors, $\mathbf{D}$ and $\mathbf{B}$, at an interface between two different media has a different form when compared to the same condition for time-constant electric and magnetic fields?

(A) Boundary condition for tangential components of $\mathbf{E}$.

(B) Boundary condition for tangential components of $\mathbf{H}$.

(C) Boundary condition for normal components of $\mathbf{D}$.

(D) Boundary condition for normal components of $\mathbf{B}$.

(E) More than one of the above boundary conditions.

(F) None of the above boundary conditions.

CONCEPTUAL QUESTION 6.88 **Continuity of H and D vector components across a boundary.** Which of the components of rapidly time-varying magnetic field intensity vectors, **H**, and electric flux density vector, **D**, must be the same on the two sides of a boundary surface between two arbitrary electromagnetic media?

(A) Tangential components of **H**.

(B) Normal components of **H**.

(C) Tangential components of **D**.

(D) Normal components of **D**.

(E) More than one of the above selections of components.

(F) None of the above selections of components.

CONCEPTUAL QUESTION 6.89 **Reversing the direction of the normal on a boundary surface.** How many of the four general (high-frequency) electromagnetic boundary conditions, for the tangential components of vectors **E** and **H** and normal components of vectors **D** and **B**, respectively, at a boundary surface between two arbitrary electromagnetic media will retain the same form if they are rewritten adopting the normal unit vector on the surface, $\hat{\mathbf{n}}$, to be directed from region 1 to region 2?

(A) None.

(B) One.

(C) Two.

(D) Three.

(E) All.

CONCEPTUAL QUESTION 6.90 **Zero and nonzero field components near a PEC surface.** Which of the following components of rapidly time-varying electric and magnetic field intensity vectors, **E** and **H**, must be zero at a point in air immediately above the surface of a horizontal PEC (perfect electric conductor) screen?

(A) Tangential component of **E** and normal component of **H**.

(B) Normal component of **E** and tangential component of **H**.

(C) Tangential components of both **E** and **H**.

(D) Normal components of both **E** and **H**.

(E) More than one of the above combinations.

(F) None of the above combinations.

CONCEPTUAL QUESTION 6.91 **Boundary condition for the current density vector.** Does the boundary condition for the normal components of the rapidly time-varying or high-frequency current density vector, **J**, at a conductor–conductor boundary, corresponding to the continuity equation for rapid time variations of currents and charges, have the same form as its static version?

(A) Yes.

(B) No.

CONCEPTUAL QUESTION 6.92 **Relationship between surface currents and charges on a PEC surface.** On a flat PEC surface lying in the xy-plane of the Cartesian coordinate system, there are high-frequency surface currents and charges. The surface current density vector, $\mathbf{J}_\mathrm{s}$, given by its x- and y-components, and charge density, ρ_s, at an arbitrary point of the surface are interrelated as

(A) $J_{\mathrm{s}x} + J_{\mathrm{s}y} = -\partial\rho_\mathrm{s}/\partial t$.

(B) $\partial J_{\mathrm{s}x}/\partial x + \partial J_{\mathrm{s}y}/\partial y = \rho_\mathrm{s}$.

(C) $\partial J_{\mathrm{s}x}/\partial x + \partial J_{\mathrm{s}y}/\partial y = \partial\rho_\mathrm{s}/\partial t$.

(D) $\partial J_{\mathrm{s}x}/\partial x + \partial J_{\mathrm{s}y}/\partial y = -\partial\rho_\mathrm{s}/\partial t$.

(E) $\partial(J_{\mathrm{s}x} + J_{\mathrm{s}y})/\partial t = -\partial\rho_\mathrm{s}/\partial x - \partial\rho_\mathrm{s}/\partial y$.

(F) They are not related to each other.

6.9 Time-Harmonic Electromagnetics

Consider a time-harmonic (steady-state sinusoidal) voltage of frequency (repetition rate) f and amplitude (peak value) V_0. Its instantaneous value, that is, the value at an instant t, can be written as

$$v(t) = V_0 \cos(\omega t + \theta) = V\sqrt{2}\cos(\omega t + \theta) \quad (V = V_{\mathrm{rms}}) \, , \quad \omega = 2\pi f \, , \quad T = \frac{2\pi}{\omega} = \frac{1}{f}$$
$$\text{(time-harmonic voltage)} \, , \quad (6.16)$$

where ω is the angular frequency (or radian frequency) and T is time period of time-harmonic oscillation [units are hertz (Hz) for f (Hz = 1/s) and radian per second (rad/s) for ω], while θ is the initial phase (phase at an instant $t = 0$) of the voltage. The root-mean-square (rms) value of $v(t)$, by definition, is found as

$$V_{\mathrm{rms}} = \sqrt{\frac{1}{T}\int_0^T v^2(t)\,\mathrm{d}t} \quad \text{(rms value)} \, , \qquad (6.17)$$

so it amounts to[4] $V = V_{\mathrm{rms}} = V_0/\sqrt{2}$. In fact, it is more convenient to use rms values of time-harmonic quantities than their maximum values (amplitudes). Most instruments are calibrated to read rms values of measured quantities. In addition, all expressions for time-average powers, energies, and power and energy densities in the time-harmonic operation in circuit theory and electromagnetics can be computed just as for the time-invariant operation, if rms values of currents, voltages, field intensities, and other quantities are used. For example, the power of Joule's (ohmic) losses in a resistor of resistance R equals $P_{\mathrm{J}} = RI^2$ (dc), $P_{\mathrm{J}}(t) = Ri^2(t)$ (instantaneous), and

$$(P_{\mathrm{J}})_{\mathrm{ave}} = \frac{1}{T}\int_0^T P_{\mathrm{J}}(t)\,\mathrm{d}t = R\frac{1}{T}\int_0^T i^2(t)\,\mathrm{d}t = \frac{1}{2}\,RI_0^2 = RI^2 \quad \text{(time-average power)} \, . \qquad (6.18)$$

Time-harmonic expressions for electromagnetic quantities that vary also in space are written in a completely analogous way to that in Eq. (6.16), while keeping in mind that both the rms value and initial phase are, in general, functions of spatial coordinates. In addition, for a vector, separate expressions are written for each of its components. For instance, the Cartesian x-component of the time-harmonic electric field intensity vector, $\mathbf{E}$, can be expressed as $E_x(x,y,z,t) = E_x(x,y,z)\sqrt{2}\cos[\omega t + \theta_x(x,y,z)]$, and similarly for E_y and E_z.

[4]As rms quantities will be used regularly throughout the rest of this text, we drop the subscripts ('rms') identifying them. With such a convention, V_{rms} will be denoted simply as V, H_{rms} as H, and so on.

CONCEPTUAL QUESTION 6.93 Electric circuits with time-harmonic voltages and currents. In an electric circuit with all time-harmonic excitations (voltage and current generators) of the same frequency, all responses (voltages across circuit elements and currents in circuit branches) in the steady state are also time-harmonic quantities with the same frequency or are zero.

(A) True for all circuits.

(B) True for some circuits, not all.

(C) False for all circuits.

CONCEPTUAL QUESTION 6.94 Linear electromagnetic system with a time-harmonic field sample. Consider a linear electromagnetic system, in which all electromagnetic materials are linear, but may be inhomogeneous. If an x-component of the electric field intensity vector at a point in this system is found to be a time-harmonic quantity of frequency f, then all nonzero components (x-, y-, and z-components) of the electric and magnetic field intensity vectors at all points of the system are also time-harmonic quantities of frequency f.

(A) True.

(B) False.

(C) Need more information.

CONCEPTUAL QUESTION 6.95 Shift of time reference for a time-harmonic quantity. Change of the initial phase (θ) of a time-harmonic field or circuit quantity essentially amounts to a shift of time reference ($t = 0$).

(A) True.

(B) False.

CONCEPTUAL QUESTION 6.96 rms and peak-value voltages at a household electric outlet. If an ac voltmeter plugged into a household electric outlet reads 110 V, which is an rms voltage, the maximum value of the voltage at the outlet is (approximately)

(A) 55 V.

(B) 77.8 V.

(C) 110 V.

(D) 155.6 V.

(E) 220 V.

(F) Depends on the frequency.

CONCEPTUAL QUESTION 6.97 **Time-average value of time-harmonic quantities.** The time-average value of an arbitrary time-harmonic circuit or field quantity equals

(A) a half of the amplitude of the quantity.

(B) the amplitude of the quantity divided by $\sqrt{2}$.

(C) $\sqrt{2}$ times the rms value of the quantity.

(D) zero.

(E) Need more information.

CONCEPTUAL QUESTION 6.98 **rms value of a periodic rectangular-pulse voltage.** Consider the voltage $v(t)$ in the form of a periodic alternating rectangular-pulse time function of amplitude V_0 and period T, shown in Figure 6.40. The time-average value of this voltage, V_{ave}, and its root-mean-square (rms) value, V_{rms}, amount to

(A) $V_{\text{ave}} = V_0/2$ and $V_{\text{rms}} = V_0/\sqrt{2}$.

(B) $V_{\text{ave}} = V_{\text{rms}} = V_0$.

(C) $V_{\text{ave}} = 0$ and $V_{\text{rms}} = V_0/2$.

(D) $V_{\text{ave}} = 0$ and $V_{\text{rms}} = V_0/\sqrt{2}$.

(E) $V_{\text{ave}} = 0$ and $V_{\text{rms}} = V_0$.

(F) $V_{\text{ave}} = V_{\text{rms}} = 0$.

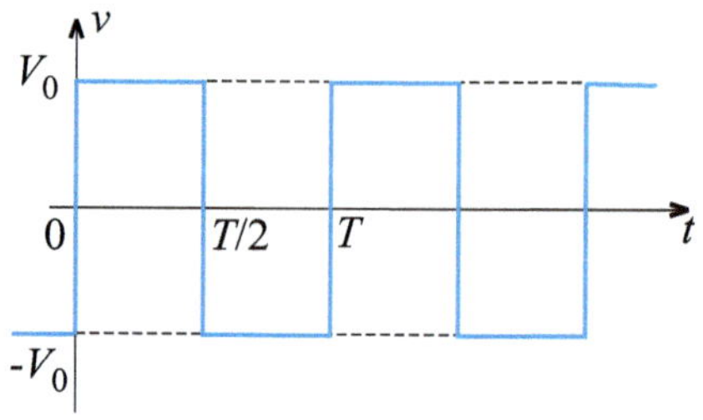

Figure 6.40 Periodic alternating rectangular-pulse voltage; for Conceptual Question 6.98.

CONCEPTUAL QUESTION 6.99 **Time-average energy density in terms of the rms field intensity.** The instantaneous electric energy density at a point in a dielectric medium of permittivity ε is given by $w_e(t) = \varepsilon E^2(t)/2$, where $E(t)$ is the instantaneous electric field intensity at that point. If $E(t)$ is a time-harmonic field of rms intensity E_{rms}, the associated time-average electric energy density is

(A) $(w_e)_{ave} = \varepsilon E_{rms}^2/2$.

(B) $(w_e)_{ave} = \varepsilon E_{rms}^2/4$.

(C) $(w_e)_{ave} = \varepsilon E_{rms}^2$.

(D) $(w_e)_{ave} = \varepsilon E_{rms}^2/\sqrt{2}$.

(E) $(w_e)_{ave} = 0$.

(F) Need more information.

CONCEPTUAL QUESTION 6.100 **Peak-value energy density in terms of the rms field intensity.** Consider a point in a dielectric medium of permittivity ε with a time-harmonic electric field of rms intensity E_{rms}. The maximum value of the electric energy density [which in time is computed as $w_e(t) = \varepsilon E^2(t)/2$] at this point equals

(A) $(w_e)_{max} = \varepsilon E_{rms}^2/2$.

(B) $(w_e)_{max} = \varepsilon E_{rms}^2/\sqrt{2}$.

(C) $(w_e)_{max} = \varepsilon E_{rms}^2$.

(D) $(w_e)_{max} = \sqrt{2}\varepsilon E_{rms}^2$.

(E) $(w_{\rm e})_{\rm max} = 2\varepsilon E_{\rm rms}^2$.

(F) Need more information.

6.10 Complex Representatives of Time-Harmonic Field and Circuit Quantities

Time-harmonic quantities can be graphically represented as uniformly rotating vectors, called phasors, in the Cartesian xy-plane, as shown in Figure 6.41(a), where the projection on the x-axis of a vector of magnitude V_0 rotating with a constant angular velocity ω equals $v(t)$ in Eq. (6.16). In addition, we can formally proclaim the x- and y-axes of Figure 6.41(a) to be the real and imaginary axes of the complex plane, as indicated in Figure 6.41(b), and use complex numbers to represent time-harmonic quantities. A complex number[5] $\underline{c}$ is a number composed of two real numbers, a and b, and it corresponds to a point, (a, b), or to a vector [position vector of the point (a, b) with respect to the coordinate origin] in the complex plane, as illustrated in Figure 6.41(c). Its rectangular (algebraic) and polar (exponential) forms read

$$\underline{c} = a + {\rm j}b = c\,{\rm e}^{{\rm j}\phi} , \quad a = {\rm Re}\{\underline{c}\} , \quad b = {\rm Im}\{\underline{c}\} , \quad {\rm j} = \sqrt{-1} , \quad c = |\underline{c}| = \sqrt{a^2 + b^2} , \tag{6.19}$$

so a and b represent the real and imaginary parts, respectively, of $\underline{c}$, and j stands for the imaginary unit, while c is the magnitude (or modulus) and ϕ the phase angle (argument) of $\underline{c}$. From the right-angled triangle with arms $|a|$ and $|b|$ in Figure 6.41(c) and Eqs. (6.19),

$${\rm e}^{{\rm j}\phi} = \cos\phi + {\rm j}\sin\phi \quad \text{(Euler's identity)} , \tag{6.20}$$

and this relation is known as Euler's identity. It is now clear that the projection of the rotating vector in Figure 6.41(a) on the real axis in Figure 6.41(b) equals the real part of the complex number.

As all the phasors representing time-harmonic quantities in a system rotate with the same angular velocity, the picture with all the vectors frozen at instant $t = 0$ contains all relevant data. We can, therefore, disregard rotation of phasors, or drop the time factor ${\rm e}^{{\rm j}\omega t}$ from the associated equations, and use the complex

[5]In this text, letters denoting complex numbers and complex variables are underlined, which is in compliance with the recommendation of the International Electrotechnical Commission (IEC).

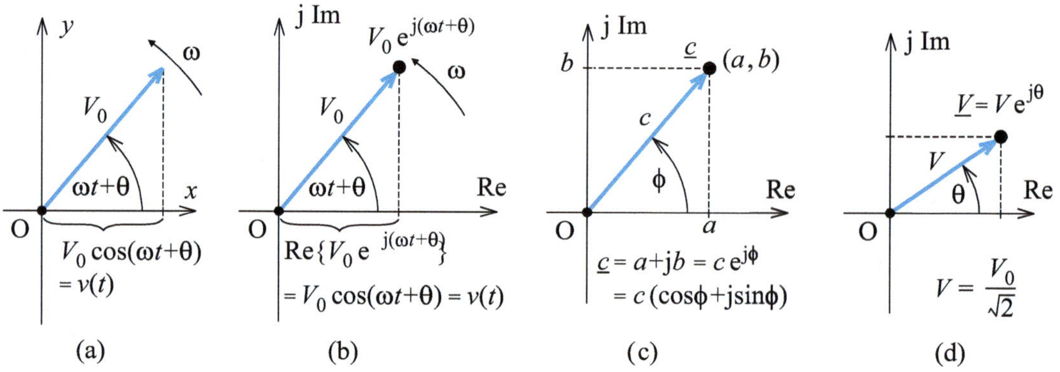

Figure 6.41 Representing time-harmonic quantities by phasors and complex numbers: (a) a phasor (rotating vector) whose magnitude and angular velocity equal the amplitude (peak value) and angular frequency, respectively, of the instantaneous quantity [see Eq. (6.16)], (b) a complex number with magnitude (modulus) and phase angle (argument) equal to the amplitude and instantaneous phase of the instantaneous quantity, (c) different forms of a complex number, in general, and (d) the final adopted complex root-mean-square (rms) representative of a time-harmonic quantity with the time factor $e^{j\omega t}$ suppressed.

representation shown in Figure 6.41(d), where the complex rms representative of a time-harmonic quantity is defined as

$$v(t) = V\sqrt{2}\cos(\omega t + \theta) \quad \longleftrightarrow \quad \underline{V} = V\,e^{j\theta}\,, \quad v(t) = \text{Re}\left\{\underline{V}\sqrt{2}\,e^{j\omega t}\right\}$$

$$\text{(time-complex conversion; } \underline{V} - \text{complex rms value)}\,. \quad (6.21)$$

The magnitude of the complex quantity is represented with the rms value, $|\underline{V}| = V = V_{\text{rms}}$, rather than with the amplitude of the corresponding instantaneous quantity because, as already mentioned and illustrated in two examples (reading of instruments and computation of time-average power) in the previous section, it is more convenient to deal with rms quantities. Note, however, that many electromagnetics texts do use the latter representation, namely, complex amplitude representatives, in the form $\underline{V} = V_0\,e^{j\theta}$ ($V_0 = V\sqrt{2}$). Note also that alternative notations to $\underline{V}$ for complex (phasor) quantities include $\tilde{V}$ (tilde over the letter), V_s (subscript s), etc.

A notable feature of the time-complex conversion in Eq. (6.21), given by $dv/dt \leftrightarrow j\omega\underline{V}$, allows us to replace all time derivatives in field/circuit equations by the factor $j\omega$, which enormously simplifies the analysis. For instance, the complex-domain equivalent of Maxwell's first equation in differential form, Eqs. (6.13), reads $\nabla\times\underline{\mathbf{E}} = -j\omega\underline{\mathbf{B}}$.

CONCEPTUAL QUESTION 6.101 Rotation of phasors in the complex plane. As the phasors representing multiple time-harmonic quantities in an electromagnetic system or in an electric circuit rotate in the complex plane, the following parameters for the quantities read from the diagram change in time, i.e., are not always the same:

(A) amplitudes of all time-harmonic quantities.

(B) instantaneous phases of all quantities.

(C) phase differences between individual quantities.

(D) angular frequencies of all quantities.

(E) more than one set of parameters above.

(F) none of the parameters above.

CONCEPTUAL QUESTION 6.102 **Reversing the direction of phasor rotation.** If the rotation of phasors representing field and circuit time-harmonic quantities is reversed, namely, if the phasors are set to rotate in the clockwise (mathematically negative) direction, that is equivalent to all the quantities in an electromagnetic system or electric circuit containing the same factor $e^{-j\omega t}$ (in place of $e^{j\omega t}$), which then appears on both sides of all of the field/circuit equations governing the system/circuit.

(A) True.

(B) False.

CONCEPTUAL QUESTION 6.103 **Phasor rotation in both directions for a system/circuit.** In the representation and analysis of a time-harmonic electromagnetic system or electric circuit using phasors, it is possible to have some phasors rotate in the clockwise and the other ones in the counter-clockwise directions.

(A) True.

(B) False.

CONCEPTUAL QUESTION 6.104 **Absolute and relative phases of time-harmonic quantities.** Which of the following phase parameters are important for the proper representation of time-harmonic quantities in the analysis of an electromagnetic system or electric circuit?

(A) Instantaneous (at an arbitrary time t) phases of all quantities.

(B) Only initial (at $t = 0$) phases of all quantities.

(C) Only phase differences between individual quantities.

(D) More than one of the above sets of phase parameters.

(E) None of the above sets of parameters.

CONCEPTUAL QUESTION 6.105 **Sine reference in describing time-harmonic quantities.** If the sine reference, instead of the cosine one, is used to describe time-harmonic field and circuit quantities, e.g., $v(t) = V_0 \sin(\omega t + \theta)$, then the representation of these quantities by rotating vectors (phasors) in the complex plane becomes irrelevant and cannot be used.

(A) True.

(B) False.

CONCEPTUAL QUESTION 6.106 **Time-to-complex transformation.** The complex rms equivalent of a time-harmonic voltage given by $v(t) = 10\sqrt{2}\sin(10^9 t)$ V (t in s) is

(A) $\underline{V} = -10$ V.

(B) $\underline{V} = -10\sqrt{2}$ V.

(C) $\underline{V} = -\mathrm{j}10$ V.

(D) $\underline{V} = -\mathrm{j}10\sqrt{2}$ V.

(E) $\underline{V} = \mathrm{j}10\,\mathrm{e}^{-\mathrm{j}\pi/2}$ V.

(F) $\underline{V} = 5\sqrt{2}(1-\mathrm{j})$ V.

CONCEPTUAL QUESTION 6.107 **Complex-to-time transformation.** If the complex rms current is $\underline{I} = (-1+\mathrm{j})$ A, the corresponding instantaneous current is given by

(A) $i(t) = \cos(\omega t - \pi/4)$ A,

(B) $i(t) = \sqrt{2}\cos(\omega t - \pi/4)$ A,

(C) $i(t) = \sqrt{2}\cos(\omega t + 3\pi/4)$ A,

(D) $i(t) = 2\cos(\omega t - \pi/4)$ A,

(E) $i(t) = 2\cos(\omega t + \pi/4)$ A,

(F) $i(t) = 2\cos(\omega t + 3\pi/4)$ A,

where ω is the angular frequency.

CONCEPTUAL QUESTION 6.108 **Transferring a time-harmonic field vector to complex domain.** If the magnetic field intensity vector is expressed as $\mathbf{H} = \sqrt{2}\,\mathrm{e}^{-0.02z}\left[\cos(10^8 t - z)\,\hat{\mathbf{x}} + \cos(10^8 t - z + 90^\circ)\,\hat{\mathbf{y}}\right]$ A/m (t in s; z in m), the associated complex rms field vector equals

(A) $\underline{\mathbf{H}} = -0.02z\,\mathrm{e}^{-\mathrm{j}10^{-8}z}\,(\hat{\mathbf{x}} + 0.5\,\hat{\mathbf{y}})$ A/m

(B) $\underline{\mathbf{H}} = -0.02z\,\mathrm{e}^{-\mathrm{j}10^{-8}z}\,(1 + \pi/2)\,\hat{\mathbf{z}}$ A/m

(C) $\underline{\mathbf{H}} = -z\,\mathrm{e}^{-0.02z}\,(\hat{\mathbf{x}} + \mathrm{j}\,\hat{\mathbf{y}})$ A/m

(D) $\underline{\mathbf{H}} = \mathrm{e}^{-0.02z}\,\mathrm{e}^{-\mathrm{j}z}\,(\hat{\mathbf{x}} + \mathrm{j}\,\hat{\mathbf{y}})$ A/m

(E) $\underline{\mathbf{H}} = \mathrm{e}^{-0.02z}\,(\hat{\mathbf{x}} + \mathrm{j}\,\hat{\mathbf{y}})$ A/m

(F) $\underline{\mathbf{H}} = \mathrm{e}^{-\mathrm{j}1.02z}\,(\hat{\mathbf{x}} + \mathrm{j}\,\hat{\mathbf{y}})$ A/m

(z in m).

CONCEPTUAL QUESTION 6.109 **Converting a complex vector to the instantaneous expression.** The instantaneous counterpart of the complex rms electric field intensity vector given by $\underline{\mathbf{E}} = E_0\,\mathrm{e}^{\mathrm{j}\theta}\sin\beta z\,\mathrm{e}^{-\mathrm{j}\beta x}\,\hat{\mathbf{x}} + E_0\cos\beta z\,\mathrm{e}^{-\mathrm{j}\beta x}\,\hat{\mathbf{z}}$, assuming that the operating angular frequency is ω, is as follows:

(A) $\mathbf{E}(t) = E_0\sqrt{2}\cos\theta\sin\beta z\cos(\omega t - \beta x)\,\hat{\mathbf{x}} + E_0\sqrt{2}\cos\beta z\cos(\omega t - \beta x)\,\hat{\mathbf{z}}$.

(B) $\mathbf{E}(t) = E_0\sqrt{2}\sin\beta z\cos(\omega t - \beta x + \theta)\,\hat{\mathbf{x}} + E_0\sqrt{2}\cos\beta z\cos(\omega t - \beta x)\,\hat{\mathbf{z}}$.

(C) $\mathbf{E}(t) = E_0\sqrt{2}\cos(\omega t - \beta z - \beta x + \theta)\,\hat{\mathbf{x}} + E_0\sqrt{2}\cos(\omega t + \beta z - \beta x)\,\hat{\mathbf{z}}$.

(D) $\mathbf{E}(t) = E_0\sqrt{2}\cos\theta\cos\beta x\cos(\omega t - \beta z)\,\hat{\mathbf{x}} + E_0\sqrt{2}\cos\beta x\cos(\omega t + \beta z)\,\hat{\mathbf{z}}$.

(E) $\mathbf{E}(t) = -E_0\sqrt{2}\sin\beta z\cos\beta z\cos(\omega t - \beta x + \theta)\,\hat{\mathbf{y}}$.

(F) $\mathbf{E}(t) = -E_0\sqrt{2}\sin\beta z\cos\beta z\cos\beta x\cos(\omega t + \theta)\,\hat{\mathbf{y}}$.

CONCEPTUAL QUESTION 6.110 **Complex equivalent of a partial differential equation.** The complex-domain equivalent of the partial differential equation $\partial E_x/\partial z = -\mu\partial H_y/\partial t$ reads

(A) $\partial\underline{E}_x/\partial z = -\mathrm{j}\omega\mu\underline{H}_y$.

(B) $\mathrm{j}\omega\underline{E}_x = -\mu\partial\underline{H}_y/\partial t$.

(C) $\mathrm{j}\omega\underline{E}_x = -\mathrm{j}\omega\mu\underline{H}_y$.

(D) $\partial\underline{E}_x/\partial z = \mathrm{j}\omega\mu\underline{H}_y$.

(E) $\partial\underline{E}_x/\partial z = \mathrm{j}\mu\underline{H}_y/\omega$.

(F) $\partial\underline{E}_x/\partial z = -\mu\partial\underline{H}_y/\partial t$.

CONCEPTUAL QUESTION 6.111 **Transferring a second-order PDE to the complex domain.** The second-order partial differential equation $\nabla^2 V - \varepsilon\mu\partial^2 V/\partial t^2 = -\rho/\varepsilon$ is written in complex notation as follows:

(A) $\nabla^2\underline{V} - \varepsilon\mu\partial^2\underline{V}/\partial t^2 = -\underline{\rho}/\varepsilon$.

(B) $\nabla^2\underline{V} = -\underline{\rho}/\varepsilon$.

(C) $\nabla^2\underline{V} - \mathrm{j}\omega\varepsilon\mu\underline{V} = -\underline{\rho}/\varepsilon$.

(D) $\nabla^2\underline{V} + \mathrm{j}\omega\varepsilon\mu\underline{V} = \mathrm{j}\omega\underline{\rho}/\varepsilon$.

(E) $\nabla^2\underline{V} + \omega^2\varepsilon\mu\underline{V} = -\underline{\rho}/\varepsilon$.

(F) $\nabla^2\underline{V} - \omega^2\varepsilon\mu\underline{V} = -\underline{\rho}/\varepsilon$.

CONCEPTUAL QUESTION 6.112 **Drawing a complex vector.** A complex vector, in general, is a set of six numbers, three real and three imaginary parts of its components. This is why a complex vector, unlike its instantaneous counterpart, cannot be drawn as an arrow in space, except in some special cases.

(A) True.

(B) False.

6.11 Lorenz Electromagnetic Potentials

The time-retardation concept is one of the most important phenomena in electromagnetics. It basically tells us that there is a time lag between a change of the field sources, i.e., of time-varying charges and currents, and the associated change of the fields, so that the values of field intensities at a distance from the sources depend on the values of charge and current densities at an earlier time. In other words, it takes some time for the effect of a change of charges and currents to be "felt" at distant field points. The time lag equals the time needed for electromagnetic disturbances to propagate over the corresponding distance.

Namely, we consider an arbitrary distribution of volume currents and charges in a source domain of volume v, as shown in Figure 6.42(a). Let the current and charge densities, $\mathbf{J}$ and ρ, in v be known functions of spatial coordinates and time. If their variation in time is rapid (high frequency), the electric scalar potential, V, and magnetic vector potential, $\mathbf{A}$, at an arbitrary point, P, in space (field or observation point), that in the static case have the forms given in Eqs. (1.12) and (4.17), are evaluated as

$$V(t) = \frac{1}{4\pi\varepsilon}\int_v \frac{\rho(t - R/c)\,\mathrm{d}v}{R}\,, \quad \mathbf{A}(t) = \frac{\mu}{4\pi}\int_v \frac{\mathbf{J}(t - R/c)\,\mathrm{d}v}{R}\,, \quad c = \frac{1}{\sqrt{\varepsilon\mu}}$$

(Lorenz electric scalar and magnetic vector potentials) (6.22)

(we have $Q'\,\mathrm{d}l$ and $i\,\mathrm{d}\mathbf{l}$ in place of $\rho\,\mathrm{d}v$ and $\mathbf{J}\,\mathrm{d}v$, respectively, for line charges and currents). These potentials are called the Lorenz or retarded electromagnetic potentials. The constant c is the velocity of propagation of electromagnetic disturbances in the medium (of parameters ε and μ). If the medium is air (vacuum), substituting the values for ε_0 and μ_0 from Eqs. (1.2) and (4.3) yields $c_0 = 1/\sqrt{\varepsilon_0\mu_0} \approx 3\times10^8$ m/s, which equals the speed of light (and other electromagnetic waves) in free space. Since the time t in Eqs. (6.22) is the time at the point P, while $t' = t - R/c$ is the time at the source point (P′) in Figure 6.42(a), we conclude that there is a time delay (retardation) between the sources and the potentials equal to

$$\tau = \frac{R}{c} \quad \text{(time lag between source and observation points)}\,. \qquad (6.23)$$

In other words, the electromagnetic disturbances caused by a time variation of elementary sources $\rho\,\mathrm{d}v$ and $\mathbf{J}\,\mathrm{d}v$ at the point P′ propagate over the distance R in the form of spherical electromagnetic waves [Figure 6.42(a)] of velocity c and are

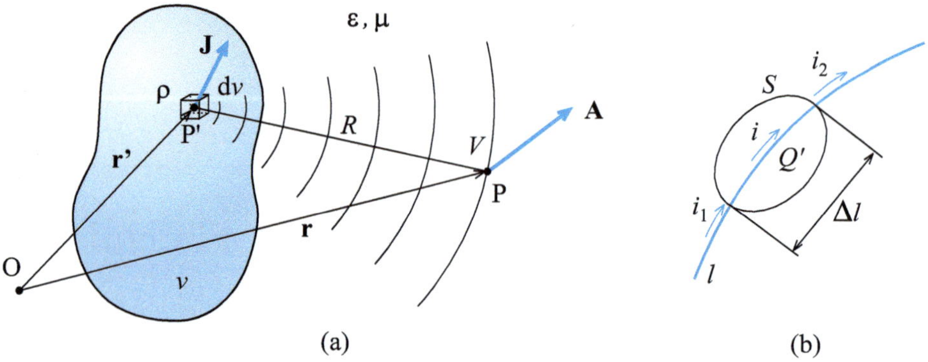

Figure 6.42 (a) Evaluation of electromagnetic potentials due to rapidly time-varying volume currents and charges in a linear, homogeneous, and lossless medium. (b) Illustration of the continuity equation for rapidly time-varying line currents.

conveyed to the potentials (and fields) at the point P after the propagation time τ in Eq. (6.23).

Of course, the high-frequency source distributions $\mathbf{J}$ and ρ in Figure 6.42(a) cannot be specified independently, but must be related to one another through the continuity equation, Eq. (3.5), and the analogous relationship exists between i and Q'. Note that the latter relationship, known as the continuity equation for wires, tells us that the intensity of a rapidly time-varying current, as opposed to a time-constant and a slowly time-varying current, in a wire conductor can change along the conductor, as illustrated in Figure 6.42(b), where, from Eq. (3.4), $i_2 - i_1 = -\partial(Q'\Delta l)/\partial t$. The continuity relation between the source distributions (e.g., between $\mathbf{J}$ and ρ) results in an analogous differential relation between potentials $\mathbf{A}$ and V, i.e., the continuity equation for potentials, known as the Lorenz condition (or Lorenz gauge).

However, if the time τ for all combinations of source and field points in a domain of interest is much shorter than the time of change of the sources [e.g., the period of change of time-harmonic (steady-state sinusoidal) charges and currents, $T = 1/f$, where f is the frequency of the sources – see Eqs. (6.16)], the retardation effect in the system under consideration can be neglected. This means that the system size and the rate of change of charges and currents are such that electromagnetic disturbances (waves) propagate over the entire system (or the useful part of the system) before the sources have changed significantly. We refer to such charges and currents as slowly time-varying or low-frequency sources and the corresponding electromagnetic fields as slowly time-varying (low-frequency) or quasistatic fields. More specifically, if the electromagnetic system under consideration satisfies the low-frequency or quasistatic condition, we have $\tau \approx 0$ in Eq. (6.23), and the potentials V and $\mathbf{A}$ in Eqs. (6.22) become non-retarded, quasistatic electric and magnetic potentials.

CONCEPTUAL QUESTION 6.113 **Time lag between source and observation points.** Consider an electromagnetic system with rapidly time-varying sources (currents and charges) in a linear, homogeneous, and lossless medium. If the time variation of sources becomes even more rapid, the time lag between source and field (observation) points

(A) increases.

(B) decreases.

(C) remains the same.

(D) Need more information.

CONCEPTUAL QUESTION 6.114 **Lorenz potentials due to volume currents and charges.** Consider an arbitrary distribution of high-frequency time-harmonic volume currents, of density $\mathbf{J}$, and charges, of density ρ, in a domain v in air, and the associated magnetic vector potential, $\mathbf{A}$, and electric scalar potential, V, at an arbitrary point in space (where none of the potentials is zero). If the magnitude of $\mathbf{J}$ is doubled everywhere in v, so the new current density vector is $\mathbf{J}_{\text{new}} = 2\mathbf{J}$, the new potentials are given by

(A) $\mathbf{A}_{\text{new}} = 2\mathbf{A}$ and $V_{\text{new}} = V$.

(B) $\mathbf{A}_{\text{new}} = \mathbf{A}$ and $V_{\text{new}} = 2V$.

(C) $\mathbf{A}_{\text{new}} = 2\mathbf{A}$ and $V_{\text{new}} = 2V$.

(D) $\mathbf{A}_{\text{new}} = 4\mathbf{A}$ and $V_{\text{new}} = V$.

(E) None of the above.

CONCEPTUAL QUESTION 6.115 **Relationship between Lorenz potentials due to volume sources.** The differential relationship between potentials $\mathbf{A}$ and V due to a volume distribution of high-frequency currents and charges in free space is given by

(A) $\nabla \times \mathbf{A} = -\varepsilon_0\mu_0 \nabla V$.

(B) $\nabla \times \mathbf{A} = -\varepsilon_0\mu_0 V$.

(C) $\nabla \cdot \mathbf{A} = -\varepsilon_0\mu_0\, \partial V/\partial t$.

(D) $\nabla \cdot \mathbf{A} = -\varepsilon_0\mu_0 V$.

(E) $\mathbf{A}$ and V are not related to each other.

CONCEPTUAL QUESTION 6.116 **Relationship between Lorenz potentials due to line sources.** For potentials $\mathbf{A}$ and V due to a high-frequency current of intensity i and charge of density Q' along a metallic wire in free space, we have that

(A) $\nabla \times \mathbf{A} = -\varepsilon_0\mu_0 \nabla V$.

(B) $\nabla \times \mathbf{A} = -\varepsilon_0\mu_0 V$.

(C) $\nabla \cdot \mathbf{A} = -\varepsilon_0\mu_0\, \partial V/\partial t$.

(D) $\nabla \cdot \mathbf{A} = -\varepsilon_0\mu_0 V$.

(E) $\mathbf{A}$ and V are not related to each other.

CONCEPTUAL QUESTION 6.117 **Relationship between Lorenz potentials due to surface sources.** Consider potentials due to a distribution of high-frequency surface current and charge of densities $\mathbf{J}_\mathrm{s}$ and ρ_s, respectively, in free space. The differential relationship between them is as follows:

(A) $\nabla \times \mathbf{A} = -\varepsilon_0\mu_0 \nabla V$.

(B) $\nabla \times \mathbf{A} = -\varepsilon_0\mu_0 V$.

(C) $\nabla \cdot \mathbf{A} = -\varepsilon_0\mu_0\, \partial V/\partial t$.

(D) $\nabla \cdot \mathbf{A} = -\varepsilon_0\mu_0 V$.

(E) $\mathbf{A}$ and V are not related to each other.

CONCEPTUAL QUESTION 6.118 **Relationship between static electric and magnetic potentials.** Potentials $\mathbf{A}$ and V due to time-constant volume currents and charges in free space are related as

(A) $\nabla \times \mathbf{A} = -\varepsilon_0\mu_0 \nabla V$.

(B) $\nabla \times \mathbf{A} = -\varepsilon_0\mu_0 V$.

(C) $\nabla \cdot \mathbf{A} = -\varepsilon_0\mu_0\, \partial V/\partial t$.

(D) $\nabla \cdot \mathbf{A} = -\varepsilon_0\mu_0 V$.

(E) $\mathbf{A}$ and V are not related to each other.

CONCEPTUAL QUESTION 6.119 **Current in a wire with open ends.** Which of the following types of current can flow through a straight metallic wire of finite length (wire with open ends)?

(A) Time-invariant current.

(B) Slowly time-varying current.

(C) Rapidly time-varying current.

(D) Any time-varying current.

(E) Any current.

(F) None of the above.

CONCEPTUAL QUESTION 6.120 **Change of current amplitude along a wire loop.** A circular wire loop of radius $a = 30$ cm in free space carries a time-harmonic current whose amplitude significantly varies along the loop. Which of the following are possible frequencies of this current?

(A) 0 Hz (dc), 60 Hz, 100 kHz, and 1 GHz.

(B) 0 Hz (dc), 60 Hz, and 100 kHz.

(C) 60 Hz and 100 kHz.

(D) 60 Hz, 100 kHz, and 1 GHz.

(E) 100 kHz and 1 GHz.

(F) 1 GHz.

CONCEPTUAL QUESTION 6.121 **Slowly time-varying electromagnetic field.** With D designating the maximum dimension of the domain of interest (containing all sources and all field points of interest for the analysis) in a time-varying electromagnetic field in free space and Δt the minimum time of change of the sources, the criterion by which we determine whether or not the field can be considered as slowly time varying is the following:

(A) $D = c_0 \Delta t$ ($c_0 = 3 \times 10^8$ m/s).

(B) $D > c_0 \Delta t$.

(C) $D < c_0 \Delta t$.

(D) $D \gg c_0 \Delta t$.

(E) $D \ll c_0 \Delta t$.

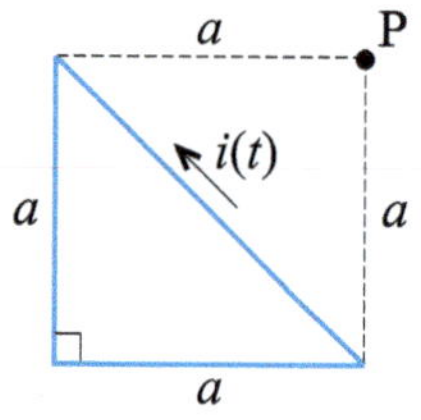

Figure 6.43 Triangular current loop with a time-harmonic current in air; for Conceptual Question 6.122.

CONCEPTUAL QUESTION 6.122 **Low-frequency time-harmonic current in a triangular loop.** A loop in the form of a triangle representing a half of a square of side a carries a time-harmonic current of intensity $i(t) = I_0 \cos(2\pi f t)$, as shown in Figure 6.43, and the medium is air. The electromagnetic field due to this current is to be evaluated at a point P located at the fourth vertex of the square. In this evaluation, $i(t)$ can be considered as a low-frequency current for the following frequencies f:

(A) 1 kHz.

(B) 1 kHz and 1 MHz.

(C) 1 kHz, 1 MHz, and 1 GHz.

(D) None of the above frequencies.

(E) Need more information.

CONCEPTUAL QUESTION 6.123 Current loop as a quasistatic system. Consider the triangular loop with current $i(t) = I_0 \cos(2\pi f t)$ and the observation (field) point P in Figure 6.43. With c_0 denoting the speed of light and other electromagnetic waves in free space ($c_0 = 3 \times 10^8$ m/s), this system can be considered as quasistatic if

(A) $f \ll c_0/a$.

(B) $f \ll c_0 a$.

(C) $f < 2\pi c_0/a$.

(D) $f \gg c_0/a$.

(E) $f \ll 1$ MHz.

(F) $f \gg 1$ MHz.

CONCEPTUAL QUESTION 6.124 Triangular loop with a pulse current as an EMI source. A source of electromagnetic interference (EMI) can be approximated by a triangular current loop – in Figure 6.44(a), where $a = 5$ cm – situated in free space. The contour carries a current whose intensity, $i(t)$, is a pulse function of time, shown in Figure 6.44(b). The EMI is evaluated at the point P in Figure 6.44(a). Can this current be considered as slowly time varying and the system as quasistatic?

(A) Yes.

(B) No.

(C) It is impossible to tell.

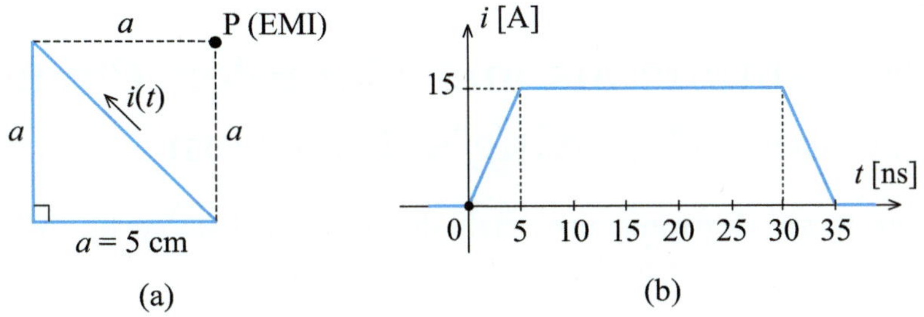

Figure 6.44 (a) Triangular current loop approximating a source of EMI. (b) Pulse current intensity (with nonzero rise and fall times) in the loop; for Conceptual Question 6.124.

CONCEPTUAL QUESTION 6.125 **Pulse current in a larger triangular loop.** Assuming that $a = 1$ m for the triangular current loop and the position of the point P at which the EMI is evaluated in Figure 6.44(a), can the pulse current in Figure 6.44(b) be considered as slowly time-varying and the system as quasistatic?

(A) Yes.

(B) No.

(C) It is impossible to tell.

CONCEPTUAL QUESTION 6.126 **Travel of electromagnetic disturbances in quasistatic systems.** In a quasistatic system, the velocity with which electromagnetic disturbances (waves) propagate from one point in the system to another can effectively be considered to be

(A) zero.

(B) infinite.

CONCEPTUAL QUESTION 6.127 **Using high-frequency potentials at low frequencies.** High-frequency expressions for electric and magnetic potentials due to various current and charge distributions can as well be used at low frequencies.

(A) True.

(B) False.

6.12 Instantaneous and Complex Poynting Vector, Poynting's Theorem

Consider an arbitrary electromagnetic field described by field vectors **E** and **H**. The associated instantaneous (at any instant of time) Poynting vector, defined as

$$\boldsymbol{\mathcal{P}} = \mathbf{E} \times \mathbf{H} \quad \text{(Poynting vector; unit: W/m}^2\text{)} \tag{6.24}$$

(note that **S** is also widely used to denote the Poynting vector), has the dimension of a surface power density (power per unit area), and is expressed in W/m^2 (the unit for E, V/m, times the unit for H, A/m). Hence, the instantaneous power transferred through any (open or closed) surface S (power flow) can be obtained as

$$P_{\mathrm{f}} = \int_S \boldsymbol{\mathcal{P}} \cdot \mathrm{d}\mathbf{S} \quad \text{(instantaneous power flow through a surface } S\text{; unit: W)} . \tag{6.25}$$

In the case of a time-harmonic field, the complex Poynting vector is given by

$$\underline{\boldsymbol{\mathcal{P}}} = \underline{\mathbf{E}} \times \underline{\mathbf{H}}^* \quad \text{(complex Poynting vector)} , \tag{6.26}$$

where $\underline{\mathbf{E}}$ and $\underline{\mathbf{H}}$ are complex rms electric and magnetic field intensity vectors [Eq. (6.21)], so that the time average of the instantaneous Poynting vector, in Eq. (6.24), equals [see Eq. (6.18)]

$$\boldsymbol{\mathcal{P}}_{\mathrm{ave}} = \frac{1}{T}\int_0^T \boldsymbol{\mathcal{P}}(t)\,\mathrm{d}t = \mathrm{Re}\{\underline{\boldsymbol{\mathcal{P}}}\} \quad \text{(time-average Poynting vector)} , \tag{6.27}$$

and represents the time-average (active) power flow per unit area. The complex and time-average net power flows through a surface S are computed as

$$\underline{S} = \int_S \underline{\boldsymbol{\mathcal{P}}} \cdot \mathrm{d}\mathbf{S} \quad \text{and} \quad P_{\mathrm{ave}} = \int_S \boldsymbol{\mathcal{P}}_{\mathrm{ave}} \cdot \mathrm{d}\mathbf{S}$$
$$\text{(complex and time-average power flows)} , \tag{6.28}$$

respectively.

Poynting's theorem represents the mathematical expression of the principle of conservation of energy as applied to electromagnetic fields. It is derived directly from Maxwell's equations in either the time or the complex domain, and can be used also in static situations. For an arbitrary spatial domain of volume v filled with a linear, generally inhomogeneous, and lossy material, it can be written as

$$P_{\mathrm{g}}(t) = P_{\mathrm{J}}(t) + \frac{\mathrm{d}W_{\mathrm{em}}}{\mathrm{d}t} + P_{\mathrm{f}}(t) \quad \text{(Poynting's theorem)} . \tag{6.29}$$

On the left-hand side of this equation, P_{g} is the total instantaneous generated power of external electric energy volume sources (generators), analogous to ideal voltage and current generators in circuit theory, in the domain v. On the right-hand side of the equation, P_{J} is the total instantaneous power of Joule's (ohmic) losses, the density of which is given in Eq. (3.9), in v, $W_{\mathrm{em}} = W_{\mathrm{e}} + W_{\mathrm{m}}$ is the total stored instantaneous electromagnetic (electric plus magnetic) energy, with the energy densities w_{e} and w_{m} being those in Eq. (2.20) and Figure 5.17(b), respectively, and P_{f} is the total instantaneous net power leaving the domain through the surface S enclosing it, Eq. (6.25).

CONCEPTUAL QUESTION 6.128 **Flux of the Poynting vector.** In a dynamic electromagnetic system, the electric field vector, $\mathbf{E}$, is perpendicular to a surface S at its every point and is directed in the same way across S. The flux of the Poynting vector, $\boldsymbol{\mathcal{P}}$, through this surface computed with respect to the same orientation of S as the direction of $\mathbf{E}$ is

(A) positive.

(B) negative.

(C) zero.

(D) Need more information.

CONCEPTUAL QUESTION 6.129 **Time-average from complex Poynting vector.** If the complex Poynting vector at a point equals $\underline{\boldsymbol{\mathcal{P}}} = (2+\mathrm{j}4)\,\hat{\mathbf{x}}\ \mathrm{W/m^2}$, the time-average Poynting vector at that point amounts to

(A) $\boldsymbol{\mathcal{P}}_{\mathrm{ave}} = 1\,\hat{\mathbf{x}}\ \mathrm{W/m^2}$.

(B) $\boldsymbol{\mathcal{P}}_{\mathrm{ave}} = \sqrt{2}\,\hat{\mathbf{x}}\ \mathrm{W/m^2}$.

(C) $\boldsymbol{\mathcal{P}}_{\mathrm{ave}} = 2\,\hat{\mathbf{x}}\ \mathrm{W/m^2}$.

(D) $\boldsymbol{\mathcal{P}}_{\mathrm{ave}} = 4\,\hat{\mathbf{x}}\ \mathrm{W/m^2}$.

(E) $\boldsymbol{\mathcal{P}}_{\mathrm{ave}} = 6\,\hat{\mathbf{x}}\ \mathrm{W/m^2}$.

CONCEPTUAL QUESTION 6.130 **Time-average power flow.** At all points of a surface whose area vector is $\mathbf{S} = 1\,\hat{\mathbf{z}}\ \mathrm{m^2}$, the complex rms electric and magnetic field intensity vectors are $\underline{\mathbf{E}} = 3\,\mathrm{e}^{\mathrm{j}4}\,\hat{\mathbf{x}}\ \mathrm{V/m}$ and $\underline{\mathbf{H}} = 2\,\mathrm{e}^{\mathrm{j}4}\,\hat{\mathbf{y}}\ \mathrm{A/m}$, respectively. The time-average power transferred through S in the direction of $\mathbf{S}$ equals

(A) $P_{\mathrm{ave}} = 3\ \mathrm{W}$.

(B) $P_{\mathrm{ave}} = 6\ \mathrm{W}$.

(C) $P_{\mathrm{ave}} = 0$.

CONCEPTUAL QUESTION 6.131 **Reactive power flow.** For a surface with $\mathbf{S} = 1\,\hat{\mathbf{z}}\ \mathrm{m}^2$ and complex electric and magnetic field vectors $\underline{\mathbf{E}} = 3\,\mathrm{e}^{\mathrm{j}4}\,\hat{\mathbf{x}}$ V/m and $\underline{\mathbf{H}} = 2\,\mathrm{e}^{\mathrm{j}4}\,\hat{\mathbf{y}}$ A/m at the surface, the reactive power delivered through S in the direction of $\mathbf{S}$ amounts to

(A) $Q = 6$ W.

(B) $Q = -3$ W.

(C) $Q = 0$.

CONCEPTUAL QUESTION 6.132 **Positive and negative terms in Poynting's theorem.** Consider all terms in Poynting's theorem for a domain v: the power of generators (in v), P_{g}, the power of Joule's losses, P_{J}, the time-rate of change of the stored electromagnetic energy, $\mathrm{d}W_{\mathrm{em}}/\mathrm{d}t$, and the power flow (through the boundary surface S), P_{f}. How many of the four terms can be both positive and negative?

(A) None.

(B) One.

(C) Two.

(D) Three.

(E) All.

CONCEPTUAL QUESTION 6.133 **Power flow from a domain with no generators.** A domain v, occupied by an electromagnetic field, does not contain any generators. The outward net instantaneous power flow from this domain to the space exterior to it, $P_{\mathrm{f}}(t)$,

(A) must be positive.

(B) must be zero.

(C) must be nonnegative (positive or zero).

(D) must be negative.

(E) must be negative or zero.

(F) can be positive, negative, and zero.

CONCEPTUAL QUESTION 6.134 **Electromagnetic energy inside an air-filled PEC cavity.** There is a time-harmonic electromagnetic field inside an air-filled perfectly closed rectangular box (cavity) whose walls are made of a perfect electric conductor (PEC). The box interior is free of any generators. The total stored instantaneous electromagnetic energy, W_{em}, inside the box is

(A) zero.

(B) constant in time, but not zero.

(C) time harmonic.

(D) time varying, but not time harmonic.

CONCEPTUAL QUESTION 6.135 **Poynting's theorem for static fields.** If the region of concern (v) is occupied by a time-invariant (static) electromagnetic field, how many of the four terms in the general Poynting's theorem, the power of generators (P_{g}), the loss power (P_{J}), the stored electromagnetic energy change rate ($\text{d}W_{\text{em}}/\text{d}t$), and the power flow ($P_{\text{f}}$), vanish?

(A) None.

(B) One.

(C) Two.

(D) Three.

(E) The theorem does not apply to static fields.

CONCEPTUAL QUESTION 6.136 **Power and energy assumptions in the circuit-theory model.** In the circuit-theory model of an arbitrary linear *RLC* circuit based on Kirchhoff's current and voltage laws, and element laws for a resistor, inductor, capacitor, and ideal voltage and current generators, some assumptions are made when applying Poynting's theorem to the circuit or its various parts. In this regard, consider the following five statements: (a) Power of external electric energy volume sources (P_{g}) in the circuit is exclusively that of ideal voltage and current generators. (b) Power of Joule's losses (P_{J}) is dissipated only in resistors. (c) Electric energy (W_{e}) is concentrated only in capacitors. (d) Magnetic energy (W_{m}) is concentrated only in inductors. (e) Power flow (P_{f}) in the circuit occurs only through connecting conductors (between elements). Which of these statements are true – for the circuit-theory model?

(A) Statement (a) only.

(B) Statements (a) and (b) only.

(C) Statements (a)–(d) only.

(D) Statements (a) and (c)–(e) only.

(E) All statements, (a)–(e).

(F) None of the statements.

CONCEPTUAL QUESTION 6.137 **Power flow from an RLC load.** Consider a series RLC circuit with an ideal voltage generator of slowly time-varying emf $e_{\mathrm{g}}(t)$, as in Figure 6.45. The net outward flux of the Poynting vector through a surface enclosing the RLC load (Figure 6.45) equals

(A) $\oint_S \boldsymbol{\mathcal{P}} \cdot \mathrm{d}\mathbf{S} = e_{\mathrm{g}} i$.

(B) $\oint_S \boldsymbol{\mathcal{P}} \cdot \mathrm{d}\mathbf{S} = -e_{\mathrm{g}} i$.

(C) $\oint_S \boldsymbol{\mathcal{P}} \cdot \mathrm{d}\mathbf{S} = R i^2$.

(D) $\oint_S \boldsymbol{\mathcal{P}} \cdot \mathrm{d}\mathbf{S} = -R i^2$.

(E) $\oint_S \boldsymbol{\mathcal{P}} \cdot \mathrm{d}\mathbf{S} = 0$.

(F) Need more information.

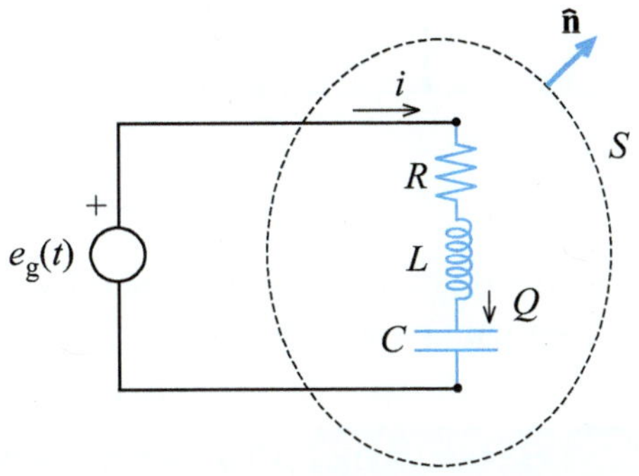

Figure 6.45 Application of Poynting's theorem to a series RLC circuit with a slowly time-varying current; for Conceptual Question 6.137.

CONCEPTUAL QUESTION 6.138 **Another application of Poynting's theorem to an RLC circuit.** An application of Poynting's theorem to the closed surface S in Figure 6.45 gives

(A) $Ri^2 + \mathrm{d}\,[Q^2/(2C) + Li^2/2]/\mathrm{d}t = 0$.

(B) $e_\mathrm{g}i = Ri^2 + \mathrm{d}\,[Q^2/(2C) + Li^2/2]/\mathrm{d}t$.

(C) $e_\mathrm{g}i = Ri^2$.

(D) $-Ri^2 + Q^2/(2C) + Li^2/2 = 0$.

(E) $Ri^2 + Q^2/(2C) - Li^2/2 = 0$.

(F) Poynting's theorem does not apply to this case.

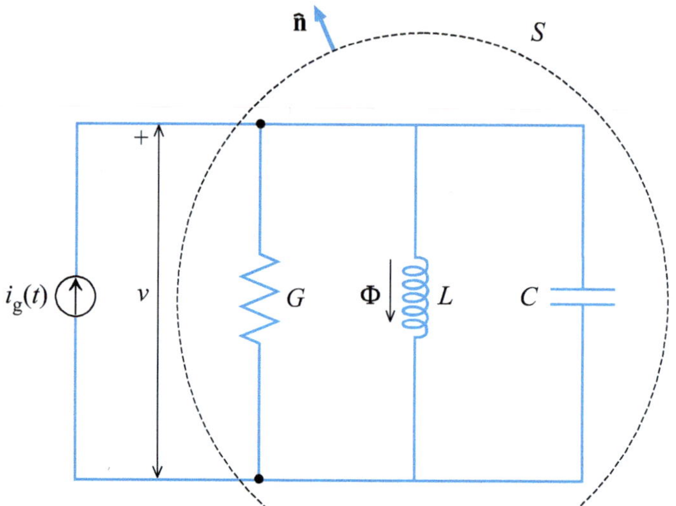

Figure 6.46 Application of Poynting's theorem to a parallel GLC circuit with a slowly time-varying current; for Conceptual Question 6.139.

CONCEPTUAL QUESTION 6.139 **Poynting's theorem for a parallel RLC circuit.** Consider a simple parallel RLC or GLC circuit driven by an ideal current generator of slowly time-varying current intensity $i_\mathrm{g}(t)$, as shown in Figure 6.46. An application of Poynting's theorem to a closed surface S placed about a GLC load (Figure 6.46) results in

(A) $Gv^2 + \mathrm{d}\,[\Phi^2/(2L) + Cv^2/2]/\mathrm{d}t = 0$.

(B) $vi_{\rm g} = Gv^2 + {\rm d}\,[\Phi^2/(2L) + Cv^2/2]/{\rm d}t$.

(C) $vi_{\rm g} = Gv^2$.

(D) None of the above.

(E) Poynting's theorem does not apply to this case.

CONCEPTUAL QUESTION 6.140 **Power flow along a lossless coaxial cable, time domain.** A lossless coaxial cable with conductor radii a and b (a is the radius of the inner conductor and b is the inner radius of the outer conductor, where $a < b$) and dielectric parameters ε and μ_0 is driven, at one end, by a low-frequency time-harmonic generator, and is terminated at the other end in an unknown complex impedance load. The instantaneous voltage and current of the cable are $v(t)$ and $i(t)$, respectively, in every cross section (v is measured from the inner to the outer conductor, and the orientation of i in the inner conductor is toward the load). The flux of the instantaneous Poynting vector through a cross section of the cable computed with respect to the reference direction toward the load equals

(A) $\int_S \mathcal{P} \cdot {\rm d}\mathbf{S} = vi$.

(B) $\int_S \mathcal{P} \cdot {\rm d}\mathbf{S} = -vi$.

(C) $\int_S \mathcal{P} \cdot {\rm d}\mathbf{S} = \pi(b^2 - a^2)|v||i|/(b-a)^2$.

(D) $\int_S \mathcal{P} \cdot {\rm d}\mathbf{S} = \pi\varepsilon v^2/\ln(b/a)$.

(E) Need to know the load impedance.

(F) Depends on which cross section is considered.

CONCEPTUAL QUESTION 6.141 **Power flow along a coaxial cable, complex domain.** Consider a lossless coaxial cable with conductor radii a and b ($b > a$), fed by a low-frequency time-harmonic generator and terminated in an unknown impedance load, and assume that we know the complex rms voltage and current of the cable: they are $\underline{V}$ and $\underline{I}$, respectively, in every cross section. The flux of the complex Poynting vector through a cross section of the cable given for the reference direction toward the load amounts to

(A) $\int_S \underline{\mathcal{P}} \cdot {\rm d}\mathbf{S} = \underline{V}\,\underline{I}^*$.

(B) $\int_S \underline{\mathcal{P}} \cdot {\rm d}\mathbf{S} = \underline{V}\,\underline{I}$.

(C) $\int_S \underline{\mathbf{\mathcal{P}}} \cdot \mathrm{d}\mathbf{S} = \pi(b^2 - a^2)\underline{V}\,\underline{I}^*/(b-a)^2$.

(D) $\int_S \underline{\mathbf{\mathcal{P}}} \cdot \mathrm{d}\mathbf{S} = 0$.

(E) Need to know the load impedance.

(F) Depends on which cross section is considered.

7 UNIFORM PLANE ELECTROMAGNETIC WAVES

Introduction

Electromagnetic waves, i.e., traveling electric and magnetic fields, are the most important consequence of general Maxwell's equations, discussed in the preceding chapter. We now proceed with analysis of electromagnetic wave propagation, to describe the properties of waves as they propagate away from their sources – rapidly time-varying currents and charges in a source region, such as the one in Figure 6.42(a), which is simply a transmitting antenna. Far away from it, the elementary spherical waves originated by the sources form a unified global spherical wavefront, which – if considered only over a receiving aperture (e.g., at the receiving end of a wireless link) – can be treated as if it were a part of a uniform plane wave. Such a wave has planar wavefronts and uniform (constant) distributions of fields over every plane perpendicular to the direction of wave propagation. Most importantly, we can completely remove the spherical wave from the analysis and assume that a uniform plane wave illuminating the aperture exists in the entire space. Once this model is established, we then deal with uniform plane waves only, and study their propagation not only in unbounded media with and without losses (this chapter) but also in the presence of planar interfaces between material regions with different electromagnetic properties (next chapter).

7.1 Wave Equations

We consider an electromagnetic wave whose electric and magnetic field intensity vectors are **E** and **H**, respectively, in an unbounded region filled with a linear, homogeneous, and lossless ($\sigma = 0$) material of permittivity ε and permeability μ, which is assumed to be source-free ($\mathbf{J} = 0$ and $\rho = 0$), and use general Maxwell's equations in differential form, Eqs. (6.13). These are first-order partial differential equations with spatial coordinates and time as independent variables and **E** and **H** as unknowns (unknown functions, to be determined, of space and time). They can be combined, taking the curl of Maxwell's first equation and substituting the expression for $\nabla \times \mathbf{H}$ from the second equation and $\nabla \cdot \mathbf{E} = 0$ from the third one, to give a second-order partial differential equation in terms of **E** alone. In an entirely analogous fashion, starting with the curl of Maxwell's second equation, we can obtain a second-order partial differential equation in **H**, so we have

$$\nabla^2\mathbf{E} - \varepsilon\mu\frac{\partial^2\mathbf{E}}{\partial t^2} = 0\,, \quad \nabla^2\mathbf{H} - \varepsilon\mu\frac{\partial^2\mathbf{H}}{\partial t^2} = 0 \quad \text{(wave equations)}\,. \tag{7.1}$$

These are three-dimensional source-free wave equations for **E** and **H**, respectively, commonly referred to simply as wave equations, and their complex form is known as the Helmholtz equations. However, Eqs. (7.1) are not independent from each other, because they are both obtained from the same two curl Maxwell's equations. Therefore, they are not sufficient for obtaining both **E** and **H**, and Eqs. (6.13) and "back substitution" must be employed as well.

CONCEPTUAL QUESTION 7.1 **Electromagnetic and other waves.** How many of the following types of waves are not electromagnetic waves: (1) X-rays, (2) infrared waves, (3) microwaves, (4) string waves, (5) laser beams, (6) radio waves, (7) ocean waves, (8) sound waves, (9) membrane waves, (10) light waves, (11) ultrasound waves, (12) γ-rays, (13) ultraviolet waves, and (14) seismic waves?[1]

(A) Five.

(B) Six.

(C) Seven.

(D) Eight.

(E) Nine.

(F) Ten.

CONCEPTUAL QUESTION 7.2 **Solving Maxwell's equations for field vectors.** Any solution for the electric and magnetic field intensity vectors, **E** and **H**, to the full set of source-free Maxwell's equations automatically satisfies both wave equations, namely, the electric- and magnetic-field wave equations.

(A) True.

(B) False.

CONCEPTUAL QUESTION 7.3 **Solving wave equations for field vectors.** Any solution to the set of two wave equations, for fields **E** and **H**, automatically satisfies all source-free Maxwell's equations.

[1] *For every conceptual question in this text, exactly one answer is correct.*

(A) True.

(B) False.

CONCEPTUAL QUESTION 7.4 **Obtaining wave equations from Maxwell's equations and vice versa.** Consider the following statements: (a) Wave equations for field vectors **E** and **H** are a consequence of source-free Maxwell's equations; (b) Source-free Maxwell's equations can be obtained from the two wave equations. Which of the statements is true?

(A) Statement (a) only.

(B) Statement (b) only.

(C) Both statements.

(D) Neither statement.

CONCEPTUAL QUESTION 7.5 **Scalar wave equations for Cartesian field components.** The wave equation for the vector **E** expressed as a function of Cartesian coordinates and free space as the propagation medium reduces to three scalar partial differential equations (scalar wave equations) with E_x, E_y, and E_z, respectively, as unknowns, and similarly for the H-field vector wave equation.

(A) True.

(B) False.

CONCEPTUAL QUESTION 7.6 **Electric-field Helmholtz equation.** The E-field Helmholtz equation (wave equation in the complex domain) is given by

(A) $\nabla^2 \underline{\mathbf{E}} - \mathrm{j}\omega\varepsilon\mu\underline{\mathbf{E}} = 0$.

(B) $\nabla^2 \underline{\mathbf{E}} + \mathrm{j}\omega\varepsilon\mu\underline{\mathbf{E}} = 0$.

(C) $\nabla^2 \underline{\mathbf{E}} + \omega^2\varepsilon\mu\underline{\mathbf{E}} = 0$.

(D) $\nabla^2 \underline{\mathbf{E}} - \omega^2\varepsilon\mu\underline{\mathbf{E}} = 0$.

(E) $\nabla^2 \underline{\mathbf{E}} + \varepsilon\mu\underline{\mathbf{E}}/\omega^2 = 0$.

(F) $\nabla^2 \underline{\mathbf{E}} - \varepsilon\mu\underline{\mathbf{E}}/\omega^2 = 0$.

CONCEPTUAL QUESTION 7.7 **Solving Helmholtz equations for field vectors.** Being a consequence of complex Maxwell's equations, the two Helmholtz equations for the electric and magnetic field intensity vectors, $\underline{\mathbf{E}}$ and $\underline{\mathbf{H}}$, respectively, are alone sufficient for obtaining both $\underline{\mathbf{E}}$ and $\underline{\mathbf{H}}$.

(A) True.

(B) False.

7.2 Time-Domain Analysis of Uniform Plane Waves

We adopt the rectangular (Cartesian) coordinate system such that the direction of wave propagation is along the z-axis, and that the electric field vector of the wave is of the form $\mathbf{E} = E_x(z,t)\,\hat{\mathbf{x}}$, as indicated in Figure 7.1. With this, the E-field wave equation in Eqs. (7.1) simplifies to a one-dimensional scalar wave equation, whose solution is any twice-differentiable function f of the variable $t' = t - z/c$,

$$\frac{\partial^2 E_x}{\partial z^2} - \varepsilon\mu\frac{\partial^2 E_x}{\partial t^2} = 0 \quad \longrightarrow \quad E_x = f\left(t - \frac{z}{c}\right) \quad (E\text{-field of a uniform plane wave}), \tag{7.2}$$

which can be verified by direct substitution. Note the analogy with spherical wave functions in Eqs. (6.22). The wave in Eq. (7.2) travels with the velocity c in the positive z direction, where

$$c = \frac{1}{\sqrt{\varepsilon\mu}} \quad \text{(velocity of uniform plane waves; unit: m/s)}. \tag{7.3}$$

For a vacuum or air (free space), Eqs. (1.2) and (4.3) give

$$c_0 = \frac{1}{\sqrt{\varepsilon_0\mu_0}} = 299{,}792{,}458 \text{ m/s} \approx 3 \times 10^8 \text{ m/s} \quad \text{(velocity of waves in free space)}. \tag{7.4}$$

This constant is commonly referred to as the speed of light.

To find the solution for the magnetic field intensity vector of the wave, from the solution for $\mathbf{E}$ in Eq. (7.2), we invoke Maxwell's first equation, Eqs. (6.13), which results in $\mathbf{H} = H_y(z,t)\,\hat{\mathbf{y}}$ (Figure 7.1) and

$$\frac{\partial E_x}{\partial z} = -\mu\frac{\partial H_y}{\partial t} \quad \longrightarrow \quad H_y = \frac{1}{\eta} f\left(t - \frac{z}{c}\right) \quad (H\text{-field of a uniform plane wave}), \tag{7.5}$$

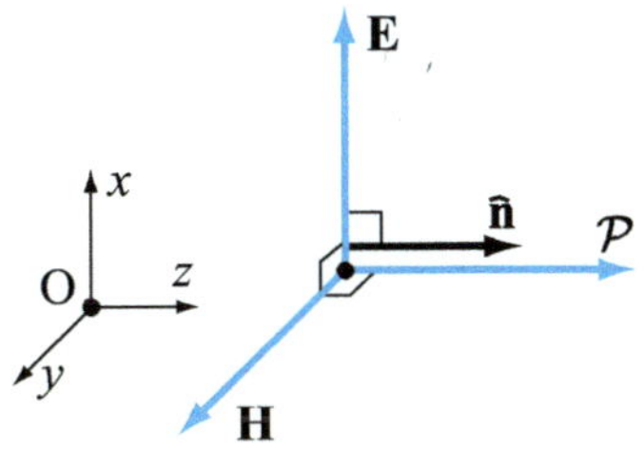

Figure 7.1 Electric field intensity vector (**E**), magnetic field intensity vector (**H**), propagation unit vector ($\hat{\mathbf{n}}$), and Poynting vector ($\boldsymbol{\mathcal{P}}$) of a uniform plane electromagnetic wave propagating in an unbounded medium.

where η, having the unit of impedance, Ω [(V/m)/(A/m) = V/A = Ω], stands for the so-called intrinsic impedance of the medium, given by

$$\eta = \frac{E}{H} = \sqrt{\frac{\mu}{\varepsilon}} \quad \text{(intrinsic impedance of a medium; unit: } \Omega) \,. \tag{7.6}$$

If the medium is air (vacuum),

$$\eta_0 = \sqrt{\frac{\mu_0}{\varepsilon_0}} \approx 120\pi \ \Omega \approx 377 \ \Omega \quad \text{(intrinsic impedance of free space)} \,. \tag{7.7}$$

From Eq. (6.24), the Poynting vector of the wave is $\boldsymbol{\mathcal{P}} = \mathbf{E} \times \mathbf{H} = EH\hat{\mathbf{n}}$ ($\hat{\mathbf{n}} = \hat{\mathbf{z}}$), Figure 7.1. Of course, the direction of $\boldsymbol{\mathcal{P}}$ coincides with the direction of the wave propagation ($\hat{\mathbf{n}}$).

CONCEPTUAL QUESTION 7.8 Dot products of E, H, and $\hat{\mathbf{n}}$ vectors for a plane wave. If the electric and magnetic field vectors of a uniform plane electromagnetic wave are **E** and **H**, and the propagation unit vector (defining the direction of wave travel) is $\hat{\mathbf{n}}$, we have the following:

(A) $\hat{\mathbf{n}} \cdot \mathbf{E} \neq 0$, $\hat{\mathbf{n}} \cdot \mathbf{H} \neq 0$, and $\mathbf{E} \cdot \mathbf{H} \neq 0$.

(B) $\hat{\mathbf{n}} \cdot \mathbf{E} \neq 0$, $\hat{\mathbf{n}} \cdot \mathbf{H} \neq 0$, and $\mathbf{E} \cdot \mathbf{H} = 0$.

(C) $\hat{\mathbf{n}} \cdot \mathbf{E} = 0$, $\hat{\mathbf{n}} \cdot \mathbf{H} = 0$, and $\mathbf{E} \cdot \mathbf{H} \neq 0$.

(D) $\hat{\mathbf{n}} \cdot \mathbf{E} = 0$, $\hat{\mathbf{n}} \cdot \mathbf{H} = 0$, and $\mathbf{E} \cdot \mathbf{H} = 0$.

(E) More than one of the above sets of relations are possible.

(F) None of the above.

CONCEPTUAL QUESTION 7.9 **Vector relations between E and H vectors of a plane wave.** A uniform plane electromagnetic wave propagates in a lossless dielectric medium, in the direction defined by a unit vector $\hat{\mathbf{n}}$ (propagation unit vector). Using the intrinsic impedance of the medium, η, the following vector relations between the electric and magnetic field vectors of the wave can be written independently of any given coordinate system:

(A) $\mathbf{H} = \hat{\mathbf{n}} \times \mathbf{E}/\eta$ and $\mathbf{E} = \eta\mathbf{H} \times \hat{\mathbf{n}}$.

(B) $\mathbf{H} = \hat{\mathbf{n}} \times \mathbf{E}/\eta$ and $\mathbf{E} = -\eta\mathbf{H} \times \hat{\mathbf{n}}$.

(C) $\mathbf{H} = -\hat{\mathbf{n}} \times \mathbf{E}/\eta$ and $\mathbf{E} = \eta\mathbf{H} \times \hat{\mathbf{n}}$.

(D) $\mathbf{H} = -\hat{\mathbf{n}} \times \mathbf{E}/\eta$ and $\mathbf{E} = -\eta\mathbf{H} \times \hat{\mathbf{n}}$.

(E) More than one of the above pairs of relations.

(F) It is impossible to relate **E** and **H** independently of a coordinate system.

CONCEPTUAL QUESTION 7.10 **Cross product of E and H vectors.** For a uniform plane wave traveling, with a propagation unit vector $\hat{\mathbf{n}}$, through a lossless dielectric medium whose intrinsic impedance is η, the following relation holds true:

(A) $\mathbf{E} \times \mathbf{H} = \hat{\mathbf{n}}$.

(B) $\mathbf{E} \times \mathbf{H} = \eta\hat{\mathbf{n}}$.

(C) $\mathbf{E} \times \mathbf{H} = \hat{\mathbf{n}}/\eta$.

(D) $\mathbf{E} \times \mathbf{H} = 0$.

(E) More than one of the above relations are possible.

(F) None of the above.

CONCEPTUAL QUESTION 7.11 **Electric field vector of a backward propagating wave.** The electric field intensity vector of a uniform plane electromagnetic wave traveling in the negative z direction (the propagation unit vector of the wave is $\hat{\mathbf{n}} = -\hat{\mathbf{z}}$) through a lossless electromagnetic medium of permittivity ε and permeability μ can be expressed as [$f(\cdot)$ is an arbitrary twice-differentiable function and $c = 1/\sqrt{\varepsilon\mu}$]

(A) $\mathbf{E} = f(t + z/c)\,\hat{\mathbf{x}}$.

(B) $\mathbf{E} = f(t + z/c)\,\hat{\mathbf{y}}$.

(C) $\mathbf{E} = f(t + z/c)\,(-\,\hat{\mathbf{z}})$.

(D) $\mathbf{E} = f(t - z/c)\,(-\,\hat{\mathbf{z}})$.

(E) More than one of the above expressions are possible.

(F) None of the above.

CONCEPTUAL QUESTION 7.12 **Magnetic field vector of a backward propagating wave.** Consider a uniform plane electromagnetic wave traveling in the negative z direction through a dielectric medium whose intrinsic impedance is η. If the electric field vector of the wave has an x-component only, $\mathbf{E} = E\,\hat{\mathbf{x}}$, the associated magnetic field vector is given by

(A) $\mathbf{H} = E\,\hat{\mathbf{x}}/\eta$.

(B) $\mathbf{H} = -E\,\hat{\mathbf{x}}/\eta$.

(C) $\mathbf{H} = E\,\hat{\mathbf{y}}/\eta$.

(D) $\mathbf{H} = -E\,\hat{\mathbf{y}}/\eta$.

(E) $\mathbf{H} = E\,\hat{\mathbf{z}}/\eta$.

(F) $\mathbf{H} = -E\,\hat{\mathbf{z}}/\eta$.

CONCEPTUAL QUESTION 7.13 **Plane wave propagation along a different axis.** The electric field vector of an electromagnetic wave propagating in a nonmagnetic ($\mu = \mu_0$) dielectric of permittivity ε is given by $\mathbf{E} = f(t + y/c)\,\hat{\mathbf{z}}$. The magnetic field vector of the wave is

(A) $\mathbf{H} = \sqrt{\varepsilon/\mu_0}\,f(t + y/c)\,\hat{\mathbf{z}}$.

(B) $\mathbf{H} = \sqrt{\varepsilon/\mu_0}\,f(t - y/c)\,\hat{\mathbf{z}}$.

(C) $\mathbf{H} = \sqrt{\varepsilon/\mu_0}\,f(t + y/c)\,\hat{\mathbf{x}}$.

(D) $\mathbf{H} = \sqrt{\varepsilon/\mu_0}\,f(t + y/c)\,(-\,\hat{\mathbf{x}})$.

(E) $\mathbf{H} = \sqrt{\mu_0/\varepsilon}\, f(t + y/c)\, \hat{\mathbf{x}}$.

(F) $\mathbf{H} = \sqrt{\mu_0/\varepsilon}\, f(t + y/c)\, (-\hat{\mathbf{x}})$.

CONCEPTUAL QUESTION 7.14 **Validity and uniqueness of a solution to a wave equation.** The expression $\mathbf{E} = f_1(t - z/c)\,\hat{\mathbf{x}} + f_2(t + z/c)\,\hat{\mathbf{x}}$, where $f_1(\cdot)$ and $f_2(\cdot)$ are arbitrary twice-differentiable functions and $c = 1/\sqrt{\varepsilon\mu}$, is

(A) the only solution

(B) one of the many solutions

(C) not a solution

of the wave equation for the electric field vector in a lossless electromagnetic medium of permittivity ε and permeability μ.

CONCEPTUAL QUESTION 7.15 **Solutions to the magnetic field wave equation.** If $f_1(\cdot)$ and $f_2(\cdot)$ are arbitrary twice-differentiable functions, $c = 1/\sqrt{\varepsilon\mu}$, and $\eta = \sqrt{\mu/\varepsilon}$, which of the following two expressions are solutions of the wave equation for the magnetic field vector in a lossless medium of parameters ε and μ?

(A) $\mathbf{H} = f_1(t - z/c)\,\hat{\mathbf{y}}/\eta + f_2(t + z/c)\,\hat{\mathbf{y}}/\eta$.

(B) $\mathbf{H} = f_1(t - z/c)\,\hat{\mathbf{y}}/\eta - f_2(t + z/c)\,\hat{\mathbf{y}}/\eta$.

(C) Both expressions in (A) and (B).

(D) Neither of the expressions.

CONCEPTUAL QUESTION 7.16 **Wave solutions to Maxwell's equations.** If the expression $\mathbf{E} = f_1(t - z/c)\,\hat{\mathbf{x}} + f_2(t + z/c)\,\hat{\mathbf{x}}$, where $f_1(\cdot)$ and $f_2(\cdot)$ are arbitrary twice-differentiable functions and $c = 1/\sqrt{\varepsilon\mu}$, is a solution for the vector $\mathbf{E}$ of Maxwell's equations for the rapidly time-varying electromagnetic field in a medium of parameters ε, μ, and $\sigma = 0$, which of the following two expressions, (A) and (B), is the associated solution for the vector $\mathbf{H}$ of Maxwell's equations?

(A) $\mathbf{H} = f_1(t - z/c)\,\hat{\mathbf{y}}/\eta + f_2(t + z/c)\,\hat{\mathbf{y}}/\eta$.

(B) $\mathbf{H} = f_1(t - z/c)\,\hat{\mathbf{y}}/\eta - f_2(t + z/c)\,\hat{\mathbf{y}}/\eta$.

(C) Both expressions in (A) and (B).

(D) Neither of the expressions.

7.3 Time-Harmonic Uniform Plane Waves and Complex-Domain Analysis

In the case of harmonic (steady-state sinusoidal) time variations of uniform plane electromagnetic waves, the function $f(t')$ in Eqs. (7.2) and (7.5), where $t' = t - z/c$, acquires the form given by Eq. (6.16), so that the expressions for the electric and magnetic fields become

$$E_x = E_0\sqrt{2}\cos(\omega t - \beta z + \theta_0)\,, \quad H_y = \frac{E_0}{\eta}\sqrt{2}\cos(\omega t - \beta z + \theta_0)$$

$$\text{(time-harmonic uniform plane wave)}\,, \quad (7.8)$$

where E_0 is the rms value and θ_0 the initial (for $t = 0$) phase in the plane $z = 0$ of the electric field intensity of the wave, ω is its angular frequency, and $H_y = E_x/\eta$. The constant β, in units of rad/m, is the phase coefficient or wavenumber (note that the symbol k is also used to denote the wavenumber), given by

$$\beta = \frac{\omega}{c} = \omega\sqrt{\varepsilon\mu} = \frac{2\pi f}{c} = \frac{2\pi}{\lambda} \quad \text{(phase coefficient or wavenumber; unit: rad/m)}\,, \quad (7.9)$$

with λ being the wavelength of the wave, measured in meters and defined as the distance traveled by a wave during one time period T [Eqs. (6.16)],

$$\lambda = cT = \frac{c}{f} \quad \text{(wavelength; unit: m)}\,. \quad (7.10)$$

The phase velocity of the wave, that is, the velocity with which any constant-phase point in Eqs. (7.8) moves in the positive z direction (in Figure 7.1), $v_p = \omega/\beta$, comes out to be $v_p = c = 1/\sqrt{\varepsilon\mu}$, Eq. (7.3).

Applying the time-complex conversion in Eq. (6.21) to the expressions for the instantaneous field intensities in Eqs. (7.8), we obtain the following expressions for complex rms field intensities of the wave in Figure 7.1:

$$\underline{E}_x = \underline{E}_0\,\mathrm{e}^{-\mathrm{j}\beta z}\,, \quad \underline{H}_y = \frac{\underline{E}_0}{\eta}\,\mathrm{e}^{-\mathrm{j}\beta z}\,, \quad \underline{E}_0 = E_0\,\mathrm{e}^{\mathrm{j}\theta_0} \quad \text{(complex rms fields)}\,. \quad (7.11)$$

CONCEPTUAL QUESTION 7.17 **Space period of a time-harmonic plane wave.** The field intensities of a time-harmonic uniform plane wave traveling in a lossless medium with phase coefficient β are periodic in space (e.g., along the z-axis), and the space period is

(A) π.

(B) 2π.

(C) β.

(D) 2β.

(E) $2\pi/\beta$.

(F) not defined – the functions are not periodic in space.

CONCEPTUAL QUESTION 7.18 **Magnetic field from electric field in time domain.** The electric field of an electromagnetic wave propagating through free space is given by $\mathbf{E} = 100\cos(3\times10^8 t + x)\,\hat{\mathbf{z}}$ V/m (t in s; x in m). The magnetic field intensity vector of the wave is

(A) $\mathbf{H} = 37.7\cos(3\times10^8 t - x)\,\hat{\mathbf{y}}$ kA/m

(B) $\mathbf{H} = 0.265\cos(3\times10^8 t - x)\,\hat{\mathbf{z}}$ A/m

(C) $\mathbf{H} = 0.265\cos(3\times10^8 t + x)\,\hat{\mathbf{y}}$ A/m

(D) $\mathbf{H} = 0.265\cos(3\times10^8 t + x)\,(-\hat{\mathbf{y}})$ A/m

(E) $\mathbf{H} = 37.7\cos(3\times10^8 t - y)\,\hat{\mathbf{x}}$ kA/m

(F) $\mathbf{H} = 0.265\cos(3\times10^8 t + y)\,(-\hat{\mathbf{x}})$ A/m

(t in s; x and y in m).

CONCEPTUAL QUESTION 7.19 **Electric field from magnetic field in complex domain.** The complex rms magnetic field intensity vector of an electromagnetic wave traveling through air can be expressed as $\underline{\mathbf{H}} = 0.1\,\mathrm{e}^{\mathrm{j}10y}\,\hat{\mathbf{x}}$ A/m (y in m). The complex electric field vector of the wave is given by

(A) $\underline{\mathbf{E}} = 265\,\mathrm{e}^{-\mathrm{j}10y}\,\hat{\mathbf{z}}$ μV/m

(B) $\underline{\mathbf{E}} = 37.7\,\mathrm{e}^{-\mathrm{j}10y}\,\hat{\mathbf{x}}$ V/m

(C) $\underline{\mathbf{E}} = 37.7\,\mathrm{e}^{\mathrm{j}10y}\,\hat{\mathbf{z}}$ V/m

(D) $\underline{\mathbf{E}} = 37.7\,\mathrm{e}^{\mathrm{j}10y}\,(-\hat{\mathbf{z}})$ V/m

(E) $\underline{\mathbf{E}} = 265\,\mathrm{e}^{\mathrm{j}10z}\,(-\hat{\mathbf{y}})$ μV/m

(F) $\underline{\mathbf{E}} = 37.7\,\mathrm{e}^{-\mathrm{j}10z}\,\hat{\mathbf{y}}$ V/m

(y and z in m).

CONCEPTUAL QUESTION 7.20 **Finding angular frequency of a wave.** The angular (radian) frequency (ω) of a wave whose magnetic field vector is given by $\underline{\mathbf{H}} = 0.1\,\mathrm{e}^{\mathrm{j}10y}\,\hat{\mathbf{x}}$ A/m (y in m) as it propagates in air amounts to

(A) 0.477×10^7 rad/s.

(B) 3×10^7 rad/s.

(C) 3×10^8 rad/s.

(D) 0.477×10^9 rad/s.

(E) 3×10^9 rad/s.

(F) Need more information.

CONCEPTUAL QUESTION 7.21 **Induced emf in a contour due to a plane wave.** A rectangular contour C of side lengths a and b is placed in the field of a uniform plane time-harmonic electromagnetic wave of angular frequency ω propagating in free space. The magnetic field vector of the wave is perpendicular to the plane of the contour, and the electric field vector is parallel to the pair of contour edges that are b long. For this situation, consider the following two integrals: $\underline{I}_1 = \oint_C \underline{\mathbf{E}} \cdot \mathrm{d}\mathbf{l}$ and $\underline{I}_2 = \int_S \mathrm{j}\omega\mu_0 \underline{\mathbf{H}} \cdot \mathrm{d}\mathbf{S}$, where S is the surface spanned over the contour, oriented in accordance with the right-hand rule with respect to the orientation of the contour. These integrals are related as

(A) $\underline{I}_1 = \underline{I}_2$.

(B) $\underline{I}_1 = -\underline{I}_2$.

(C) $\underline{I}_1 = \eta_0 \underline{I}_2$ ($\eta_0 = \sqrt{\mu_0/\varepsilon_0}$).

(D) $\underline{I}_1 = -\eta_0 \underline{I}_2$.

(E) The two integrals are not related to each other.

(F) Need more information.

CONCEPTUAL QUESTION 7.22 **Time-harmonic quantities for a time-harmonic wave.** Consider the following instantaneous quantities for a uniform plane time-harmonic electromagnetic wave propagating in a lossless dielectric: the electric field intensity vector ($\mathbf{E}$), the magnetic field intensity vector ($\mathbf{H}$), the Poynting vector ($\boldsymbol{\mathcal{P}}$), the electric energy density ($w_{\rm e}$), and the magnetic energy density ($w_{\rm m}$). Which of these quantities are time harmonic?

(A) $\mathbf{E}$ only.

(B) $\mathbf{E}$ and $\mathbf{H}$ only.

(C) $\mathbf{E}$, $\mathbf{H}$, and $\boldsymbol{\mathcal{P}}$ only.

(D) $\mathbf{E}$, $\mathbf{H}$, $w_{\rm e}$, and $w_{\rm m}$ only.

(E) All quantities.

(F) None of the quantities.

CONCEPTUAL QUESTION 7.23 **Instantaneous and time-average electric energy densities.** The instantaneous electric field intensity of a time-harmonic uniform plane electromagnetic wave traveling in a dielectric of permittivity ε is E, while the rms value of this field is E_0. The instantaneous and time-average electric energy densities of the wave are, respectively, given by

(A) $w_{\rm e} = \varepsilon E^2$ and $(w_{\rm e})_{\rm ave} = \varepsilon E_0^2$.

(B) $w_{\rm e} = \varepsilon E^2/2$ and $(w_{\rm e})_{\rm ave} = \varepsilon E_0^2/2$.

(C) $w_{\rm e} = \varepsilon E^2/4$ and $(w_{\rm e})_{\rm ave} = \varepsilon E_0^2/4$.

(D) $w_{\rm e} = \varepsilon E^2/2$ and $(w_{\rm e})_{\rm ave} = \varepsilon E_0^2/4$.

(E) $w_{\rm e} = \varepsilon E^2/4$ and $(w_{\rm e})_{\rm ave} = \varepsilon E_0^2/2$.

(F) $w_{\rm e} = \varepsilon E^2$ and $(w_{\rm e})_{\rm ave} = \varepsilon E_0^2/2$.

CONCEPTUAL QUESTION 7.24 **Equality of electric and magnetic energy densities.** The instantaneous electric and magnetic energy densities, w_{e} and w_{m}, of a uniform plane time-harmonic electromagnetic wave traveling through air are the same at

(A) all points of space and all instants of time.

(B) all points of space and some (not all) characteristic instants of time.

(C) all instants of time and some (not all) characteristic points of space.

(D) no points of space and no instants of time.

CONCEPTUAL QUESTION 7.25 **Complex Poynting vector of a plane wave.** The complex rms electric field intensity vector of a time-harmonic electromagnetic wave traveling through a dielectric medium whose intrinsic impedance is η is given by $\underline{\mathbf{E}} = E_0\,\mathrm{e}^{-\mathrm{j}\beta z}\,\hat{\mathbf{x}}$. The complex Poynting vector of the wave equals

(A) $\underline{\mathcal{P}} = 0$.

(B) $\underline{\mathcal{P}} = E_0^2\,\hat{\mathbf{z}}/\eta$.

(C) $\underline{\mathcal{P}} = E_0^2\,\mathrm{e}^{-\mathrm{j}\beta z}\,\hat{\mathbf{z}}/\eta$.

(D) $\underline{\mathcal{P}} = E_0^2\,\mathrm{e}^{-\mathrm{j}2\beta z}\,\hat{\mathbf{z}}/\eta$.

(E) $\underline{\mathcal{P}} = E_0^2\,\mathrm{e}^{-2\beta z}\,\hat{\mathbf{z}}/\eta$.

(F) Need more information.

CONCEPTUAL QUESTION 7.26 **Instantaneous and time-average Poynting vector.** Consider the instantaneous Poynting vector, $\boldsymbol{\mathcal{P}}$, of a uniform plane time-harmonic electromagnetic wave propagating through a lossless medium, as well as its time average, $\boldsymbol{\mathcal{P}}_{\mathrm{ave}}$. Which of these two vectors does not depend on spatial coordinates?

(A) $\boldsymbol{\mathcal{P}}$ only.

(B) $\boldsymbol{\mathcal{P}}_{\mathrm{ave}}$ only.

(C) Both vectors.

(D) Neither of the vectors.

CONCEPTUAL QUESTION 7.27 Oscillation in time of the Poynting vector. The electric and magnetic field vectors of a time-harmonic uniform traveling plane wave oscillate in time at an angular frequency ω. At what angular frequency does the Poynting vector of the wave oscillate?

(A) ω.

(B) ω^2.

(C) 2ω.

(D) $\omega/2$.

(E) zero.

(F) None of the above.

7.4 Arbitrarily Directed Uniform Plane Waves

Next, we consider a uniform plane wave whose propagation direction is completely arbitrary with respect to a given coordinate system, i.e., it does not coincide with an axis (x, y, or z) of the system and is defined by the unit vector $\hat{\mathbf{n}} = n_x\,\hat{\mathbf{x}} + n_y\,\hat{\mathbf{y}} + n_z\,\hat{\mathbf{z}}$, as shown in Figure 7.2. To find the expressions for the field vectors of this wave at an arbitrary point P in space, the position vector of which (with respect to the

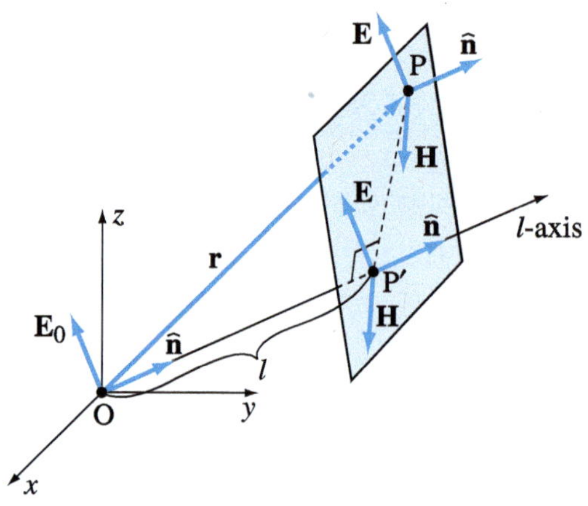

Figure 7.2 Uniform plane wave whose propagation does not coincide with any of the axes of the adopted global Cartesian coordinate system.

coordinate origin O) is $\mathbf{r} = x\,\hat{\mathbf{x}} + y\,\hat{\mathbf{y}} + z\,\hat{\mathbf{z}}$ (the Cartesian coordinates of P are x, y, and z), we realize that the distance $l = \mathbf{r} \cdot \hat{\mathbf{n}}$ in Figure 7.2 plays the role of z in Eqs. (7.11). Therefore, the complex electric and magnetic rms field intensity vectors at the point P (and in the entire plane $l = \text{const}$) are given by

$$\underline{\mathbf{E}} = \underline{\mathbf{E}}_0\, \mathrm{e}^{-\mathrm{j}\beta l} = \underline{\mathbf{E}}_0\, \mathrm{e}^{-\mathrm{j}\beta \mathbf{r}\cdot\hat{\mathbf{n}}} = \underline{\mathbf{E}}_0\, \mathrm{e}^{-\mathrm{j}\beta(x n_x + y n_y + z n_z)}\,, \quad \underline{\mathbf{H}} = \frac{1}{\eta}\,\hat{\mathbf{n}} \times \underline{\mathbf{E}}$$

$$\text{(arbitrarily directed uniform plane wave)}\,. \quad (7.12)$$

Analogous expressions can be written in the time domain, as in Eqs. (7.8). Of course, Eqs. (7.12) with $\hat{\mathbf{n}} = \hat{\mathbf{z}}$ ($n_x = n_y = 0$ and $n_z = 1$) and $\underline{\mathbf{E}}_0 = \underline{E}_0\,\hat{\mathbf{x}}$ simplify to Eqs. (7.11) for the wave in Figure 7.1 (traveling in the positive z direction).

CONCEPTUAL QUESTION 7.28 **Dot products of E, H, and n̂, arbitrary propagation direction.** Consider a uniform plane time-harmonic electromagnetic wave propagating through a lossless dielectric medium in the direction defined by the vector $\hat{\mathbf{x}} + \hat{\mathbf{y}}$ in a rectangular coordinate system, so that the propagation unit vector of the wave is $\hat{\mathbf{n}} = \sqrt{2}(\hat{\mathbf{x}} + \hat{\mathbf{y}})/2$ ($|\hat{\mathbf{n}}| = 1$). Which of the following statements about dot products $\hat{\mathbf{n}} \cdot \mathbf{E}$, $\hat{\mathbf{n}} \cdot \mathbf{H}$, and $\mathbf{E} \cdot \mathbf{H}$, where $\mathbf{E}$ and $\mathbf{H}$ are, respectively, the electric and magnetic field vectors of the wave computed at an arbitrary point in space, $\mathrm{P}(x, y, z)$, hold true?

(A) $\hat{\mathbf{n}} \cdot \mathbf{E} \neq 0$, $\hat{\mathbf{n}} \cdot \mathbf{H} \neq 0$, and $\mathbf{E} \cdot \mathbf{H} \neq 0$.

(B) $\hat{\mathbf{n}} \cdot \mathbf{E} \neq 0$, $\hat{\mathbf{n}} \cdot \mathbf{H} \neq 0$, and $\mathbf{E} \cdot \mathbf{H} = 0$.

(C) $\hat{\mathbf{n}} \cdot \mathbf{E} = 0$, $\hat{\mathbf{n}} \cdot \mathbf{H} = 0$, and $\mathbf{E} \cdot \mathbf{H} \neq 0$.

(D) $\hat{\mathbf{n}} \cdot \mathbf{E} = 0$, $\hat{\mathbf{n}} \cdot \mathbf{H} = 0$, and $\mathbf{E} \cdot \mathbf{H} = 0$.

(E) More than one of the above sets of relations are possible.

(F) None of the above.

CONCEPTUAL QUESTION 7.29 **Complex Poynting vector, arbitrary propagation direction.** The propagation unit vector of a time-harmonic uniform plane wave is given by $\hat{\mathbf{n}} = \sqrt{2}(\hat{\mathbf{x}} - \hat{\mathbf{z}})/2$, in a rectangular coordinate system. The complex rms electric field intensity of the wave at the coordinate origin is $\underline{E}_0$, the operating frequency is f, and the medium is air. The complex Poynting vector of the wave at an arbitrary point in space, $\mathrm{P}(x, y, z)$, comes out to be

(A) $\underline{\boldsymbol{\mathcal{P}}} = 0$.

(B) $\underline{\boldsymbol{\mathcal{P}}} = |\underline{E}_0|^2 \, \hat{\mathbf{n}}/\eta_0 \; (\eta_0 = \sqrt{\mu_0/\varepsilon_0})$.

(C) $\underline{\boldsymbol{\mathcal{P}}} = |\underline{E}_0|^2 \, \mathrm{e}^{-\mathrm{j}\beta\sqrt{2}(x-z)/2} \, \hat{\mathbf{n}}/\eta_0 \; (\beta = 2\pi f \sqrt{\varepsilon_0 \mu_0})$.

(D) $\underline{\boldsymbol{\mathcal{P}}} = |\underline{E}_0|^2 \, \mathrm{e}^{-\mathrm{j}\beta\sqrt{2}(x-z)} \, \hat{\mathbf{n}}/\eta_0$.

(E) $\underline{\boldsymbol{\mathcal{P}}} = \underline{E}_0 \underline{E}_0^* \, \mathrm{e}^{-\mathrm{j}\beta\sqrt{2}(x-z)} \, (\hat{\mathbf{x}} - \hat{\mathbf{z}})/\eta_0$.

(F) None of the above.

CONCEPTUAL QUESTION 7.30 **Electric field of an arbitrarily directed plane wave.** Consider a uniform plane time-harmonic electromagnetic wave whose propagation direction is completely arbitrary with respect to a given Cartesian coordinate system. The propagation unit vector of the wave is $\hat{\mathbf{n}}$ ($|\hat{\mathbf{n}}| = 1$) and its complex electric field rms intensity vector at the coordinate origin O is $\underline{\mathbf{E}}_0$. The phase coefficient of the wave is β. The expression for the electric field vector of the wave at an arbitrary point P in space, whose Cartesian coordinates are x, y, and z, and the position vector of which (with respect to O) is $\mathbf{r}$, is given by

(A) $\underline{\mathbf{E}} = \underline{\mathbf{E}}_0 \, \mathrm{e}^{-\mathrm{j}\beta(x+y+z)}$.

(B) $\underline{\mathbf{E}} = \underline{\mathbf{E}}_0 \, \mathrm{e}^{-\mathrm{j}\beta(\hat{\mathbf{x}}+\hat{\mathbf{y}}+\hat{\mathbf{z}}) \cdot \hat{\mathbf{n}}}$.

(C) $\underline{\mathbf{E}} = \underline{\mathbf{E}}_0 \, \mathrm{e}^{-\mathrm{j}\beta\mathbf{r} \cdot \hat{\mathbf{n}}}$.

(D) $\underline{\mathbf{E}} = \underline{\mathbf{E}}_0 \, \mathrm{e}^{-\mathrm{j}\beta\mathbf{r} \cdot (\hat{\mathbf{x}}+\hat{\mathbf{y}}+\hat{\mathbf{z}})}$.

(E) More than one of the above expressions.

(F) None of the above expressions.

CONCEPTUAL QUESTION 7.31 **Equiphase plane of an arbitrarily directed plane wave.** Cartesian components of the propagation unit vector of a uniform plane time-harmonic electromagnetic wave traveling in air are n_x, n_y, and n_z ($\hat{\mathbf{n}}$ does not coincide with any of the coordinate unit vectors or their opposites). The equation of an arbitrary equiphase plane of the wave, namely, a plane at every point of which the initial phase (phase at an instant $t = 0$) of the wave is the same, can be written as

(A) $n_x^2 + n_y^2 + n_z^2 = 1$.

(B) $x - n_x = y - n_y = z - n_z = \text{const}$.

(C) $x\,\hat{\mathbf{x}} + y\,\hat{\mathbf{y}} + z\,\hat{\mathbf{z}} = \text{const.}$

(D) $xn_x + yn_y + zn_z = \text{const.}$

(E) $xn_x + yn_y + zn_z = 0.$

(F) None of the above.

CONCEPTUAL QUESTION 7.32 **Propagation along the diagonal of the first Cartesian octant.** With E_0 and k being the appropriate constants, which of the following expressions represents the complex electric field vector of a uniform plane time-harmonic electromagnetic wave propagating in free space along the main diagonal of the first octant of the Cartesian coordinate system, so that its direction of propagation makes equal angles with all three coordinate axes?

(A) $\underline{\mathbf{E}} = E_0(\,\hat{\mathbf{x}} + \,\hat{\mathbf{y}} + \,\hat{\mathbf{z}})\,\mathrm{e}^{-\mathrm{j}k(x+y+z)}.$

(B) $\underline{\mathbf{E}} = E_0(\,\hat{\mathbf{x}} + \,\hat{\mathbf{y}} + \,\hat{\mathbf{z}})\,\mathrm{e}^{-\mathrm{j}kxyz}.$

(C) $\underline{\mathbf{E}} = E_0(\,\hat{\mathbf{x}} + \,\hat{\mathbf{y}} + \,\hat{\mathbf{z}})\,\mathrm{e}^{-\mathrm{j}k(x-2y+z)}.$

(D) $\underline{\mathbf{E}} = E_0(\,\hat{\mathbf{x}} - 2\,\hat{\mathbf{y}} + \,\hat{\mathbf{z}})\,\mathrm{e}^{-\mathrm{j}k(x+y+z)}.$

(E) $\underline{\mathbf{E}} = E_0(\,\hat{\mathbf{x}} + \,\hat{\mathbf{y}} + \,\hat{\mathbf{z}})\,\mathrm{e}^{-\mathrm{j}kz}.$

(F) More than one of the above expressions.

7.5 Theory of Time-Harmonic Waves in Lossy Media

For a uniform plane time-harmonic electromagnetic wave in a linear and homogeneous medium that exhibits losses ($\sigma \neq 0$), Eqs. (7.11) become

$$\underline{E}_x = \underline{E}_0\,\mathrm{e}^{-\underline{\gamma} z} = \underline{E}_0\,\mathrm{e}^{-\alpha z}\,\mathrm{e}^{-\mathrm{j}\beta z}\,, \quad \underline{\gamma} = \alpha + \mathrm{j}\beta\,, \quad \underline{H}_y = \frac{\underline{E}_0}{\underline{\eta}}\,\mathrm{e}^{-\underline{\gamma} z}\,, \quad \underline{\eta} = |\underline{\eta}|\,\mathrm{e}^{\mathrm{j}\phi}$$

$$\text{(plane wave in a lossy medium)}\,, \tag{7.13}$$

where the attenuation coefficient, α [unit is Np/m (neper per meter)], and phase coefficient, β, which combined give the complex propagation coefficient, $\underline{\gamma}$, of the wave, and the magnitude of the complex intrinsic impedance of the medium, $|\underline{\eta}|$, and its phase angle (argument), ϕ, are given by

$$\alpha = \omega\sqrt{\frac{\varepsilon\mu}{2}}\left[\sqrt{1+\left(\frac{\sigma}{\omega\varepsilon}\right)^2}-1\right]^{1/2} \quad (\text{Np/m}) , \quad \beta = \omega\sqrt{\frac{\varepsilon\mu}{2}}\left[\sqrt{1+\left(\frac{\sigma}{\omega\varepsilon}\right)^2}+1\right]^{1/2} ,$$

$$|\underline{\eta}| = \frac{\sqrt{\frac{\mu}{\varepsilon}}}{\left[1+\left(\frac{\sigma}{\omega\varepsilon}\right)^2\right]^{1/4}} , \quad \phi = \frac{1}{2}\arctan\frac{\sigma}{\omega\varepsilon} \quad (\alpha,\ \beta,\ |\underline{\eta}|,\ \phi - \text{arbitrary medium}) . \tag{7.14}$$

CONCEPTUAL QUESTION 7.33 **Phase coefficient for an arbitrary lossy medium.** The real part of the complex propagation coefficient, $\underline{\gamma}$, of a uniform plane time-harmonic electromagnetic wave in an arbitrary lossy medium of parameters ε, μ, and σ being the attenuation coefficient, α, the imaginary part is the phase coefficient of the wave, β, which has the same value as β of the same wave propagating in a lossless medium of parameters ε, μ, and $\sigma = 0$.

(A) True.

(B) False.

CONCEPTUAL QUESTION 7.34 **Rate of attenuation of the time-average Poynting vector.** If α is the attenuation coefficient of a uniform plane time-harmonic wave propagating in a lossy medium, the rate of attenuation in the direction of wave propagation of the time-average Poynting vector of the wave is determined by

(A) this same α.

(B) twice α.

(C) α squared.

(D) a half of α.

(E) None of the above.

CONCEPTUAL QUESTION 7.35 **Power flow dependence on the impedance phase angle.** A time-harmonic plane wave propagates through a lossy medium whose complex intrinsic impedance is $\underline{\eta} = |\underline{\eta}|\,e^{j\phi}$. The time-average surface power density of the wave in the direction of its propagation is proportional to

(A) $e^{j\phi}$.

(B) $e^{-j\phi}$.

(C) $\cos\phi$.

(D) $\sin\phi$.

(E) None of the above.

CONCEPTUAL QUESTION 7.36 **Attenuation of the magnetic field.** The attenuation of the magnetic field of a time-harmonic electromagnetic wave traveling in a conducting medium is

(A) always larger than

(B) always smaller than

(C) always equal to

(D) sometimes larger and sometimes smaller than

(E) sometimes equal to and sometimes smaller than

the attenuation of the electric field of the wave.

CONCEPTUAL QUESTION 7.37 **Instantaneous electric field of an attenuated wave.** A time-harmonic uniform plane wave of angular frequency ω propagates through a lossy medium in the positive z direction. The rms intensity and initial (for $t = 0$) phase in the plane $z = 0$ of the electric field of the wave are E_0 and θ_0, respectively. The attenuation and phase coefficients for the medium at the given frequency are α and β. The instantaneous electric field intensity of the wave is given by

(A) $E(z,t) = E_0\sqrt{2}\,e^{-(\alpha+\beta)z}\cos(\omega t + \theta_0)$.

(B) $E(z,t) = E_0\sqrt{2}\cos(\omega t - \alpha z - \beta z + \theta_0)$.

(C) $E(z,t) = E_0\sqrt{2}\cos(\omega t - \beta z + \theta_0 - \alpha)$.

(D) $E(z,t) = E_0\sqrt{2}\,\mathrm{e}^{-\alpha z}\cos(\omega t - \beta z + \theta_0)$.

(E) $E(z,t) = E_0\sqrt{2}\,\alpha\cos(\omega t - \beta z + \theta_0)$.

CONCEPTUAL QUESTION 7.38 Instantaneous magnetic field of an attenuated wave. Consider a uniform plane wave of angular frequency ω traveling in the positive z direction in a lossy medium whose attenuation and phase coefficients are α and β, and the complex intrinsic impedance is $\underline{\eta} = |\underline{\eta}|\,\mathrm{e}^{\mathrm{j}\phi}$. If the rms intensity and initial phase for $z = 0$ of the electric field of the wave are E_0 and θ_0, respectively, the instantaneous magnetic field intensity of the wave is

(A) $H(z,t) = E_0\sqrt{2}\,\mathrm{e}^{-\alpha z-\phi}\cos(\omega t - \beta z + \theta_0)/|\underline{\eta}|$.

(B) $H(z,t) = E_0\sqrt{2}\,\mathrm{e}^{-\alpha z}\cos(\omega t - \beta z + \theta_0 - \phi)/|\underline{\eta}|$.

(C) $H(z,t) = E_0\sqrt{2}\,\mathrm{e}^{-(\alpha+\beta)z-\phi}\cos(\omega t + \theta_0)/|\underline{\eta}|$.

(D) $H(z,t) = E_0\sqrt{2}\cos(\omega t - \alpha z - \beta z + \theta_0 - \phi)/|\underline{\eta}|$.

(E) $H(z,t) = E_0\sqrt{2}\,\alpha\,\mathrm{e}^{-\phi}\cos(\omega t - \beta z + \theta_0)/|\underline{\eta}|$.

CONCEPTUAL QUESTION 7.39 Phase velocity for an arbitrary lossy medium. Consider a uniform plane time-harmonic electromagnetic wave of frequency f traveling through an arbitrary lossy medium of parameters ε, μ, and σ. The phase velocity of this wave, v_p ($v_\mathrm{p} = \omega/\beta$), is

(A) a constant, equal to 3×10^8 m/s.

(B) a constant, different from 3×10^8 m/s.

(C) a function of frequency (f).

CONCEPTUAL QUESTION 7.40 Phase difference between electric and magnetic fields. Denoting by E and H the electric and magnetic field intensities, respectively, of a uniform plane time-harmonic wave in an arbitrary lossy medium, we have that

(A) E and H are in phase.

(B) E always lags H.

(C) H always lags E.

(D) H lags E in some cases, E lags H in others.

7.6 Good Dielectrics and Good Conductors

For wave propagation in good dielectrics, namely, materials whose permittivity and conductivity at a given frequency satisfy condition $\sigma \ll \omega\varepsilon$, Eqs. (7.14) can be approximated to

$$\sigma \ll \omega\varepsilon \quad \longrightarrow \quad \alpha \approx \frac{\sigma}{2}\sqrt{\frac{\mu}{\varepsilon}}\,, \quad \beta \approx \omega\sqrt{\varepsilon\mu}\,, \quad \underline{\eta} \approx \sqrt{\frac{\mu}{\varepsilon}} \quad \text{(good dielectrics)}\,. \tag{7.15}$$

On the other side, in the case of good conductors, for which $\sigma \gg \omega\varepsilon$, Eqs. (7.14) are simplified to

$$\sigma \gg \omega\varepsilon \quad \longrightarrow \quad \alpha \approx \beta \approx \sqrt{\frac{\omega\mu\sigma}{2}}\,, \quad |\underline{\eta}| \approx \sqrt{\frac{\omega\mu}{\sigma}}\,, \quad \phi \approx 45^\circ \quad \text{(good conductors)}\,. \tag{7.16}$$

CONCEPTUAL QUESTION 7.41 Basic propagation parameters for a good dielectric. Which of the basic propagation parameters, the attenuation coefficient (α), the phase coefficient (β), the magnitude of the complex intrinsic impedance ($|\underline{\eta}|$), and the phase angle of the impedance (ϕ), for a good dielectric of material parameters ε, μ, and σ are practically the same as those for a perfect dielectric of parameters ε, μ, and $\sigma = 0$, at the same frequency?

(A) β only.

(B) β, $|\underline{\eta}|$, and ϕ only.

(C) $|\underline{\eta}|$ and ϕ only.

(D) α, β, and $|\underline{\eta}|$ only.

(E) All parameters.

(F) None of the parameters.

CONCEPTUAL QUESTION 7.42 **Is fresh water a good dielectric or conductor?** Fresh water with relative permittivity $\varepsilon_r = 80$ and conductivity $\sigma = 10^{-3}$ S/m acts as

(A) a good dielectric.

(B) a good conductor.

(C) a conductive medium midway between a good conductor and a good dielectric.

(D) None of the above.

(E) Need more information.

CONCEPTUAL QUESTION 7.43 **Magnetic field strength in good conductors.** Consider a uniform plane time-harmonic electromagnetic wave propagating inside a conductor. The better the conductor (larger σ)

(A) the stronger

(B) the weaker

(C) no difference

the magnetic field in it, for the same electric field strength of the wave in the material.

CONCEPTUAL QUESTION 7.44 **Magnitude of the intrinsic impedance of copper.** At radio and microwave frequencies, the magnitude of the complex intrinsic impedance ($|\underline{\eta}|$) of copper is

(A) much larger than

(B) approximately the same as

(C) much smaller than

the intrinsic impedance of free space ($\eta_0 = 377\ \Omega$).

CONCEPTUAL QUESTION 7.45 **Phase of the intrinsic impedance of a good conductor.** In a good conductor, the magnetic field of a uniform plane time-harmonic electromagnetic wave lags the electric field approximately by

(A) $-90°$.

(B) $-45°$.

(C) 0.

(D) $45°$.

(E) $90°$.

(F) $180°$.

CONCEPTUAL QUESTION 7.46 **Intrinsic impedance of a PEC.** The magnitude of the complex intrinsic impedance of a perfect electric conductor is

(A) $|\underline{\eta}| = 0$.

(B) $|\underline{\eta}| \to \infty$.

(C) $|\underline{\eta}| \approx 377\ \Omega$.

(D) Depends on the frequency.

7.7 Skin Effect

An electromagnetic wave incident on the surface of a conductor attenuates rapidly with distance from the surface. The skin depth, δ, is defined as the depth into the conductor (distance from the conductor surface) at which the amplitude of the electric field of the wave is attenuated to 1/e (or about 36.8%) of its initial value, i.e., value at the surface. Using Eqs. (7.13) and (7.16), it equals

$$\delta = \frac{1}{\alpha} \quad \text{(skin depth; unit: m)}\,, \qquad \delta \approx \frac{1}{\sqrt{\pi \mu f \sigma}} \quad \text{(for good conductors)}\,, \tag{7.17}$$

where the first expression holds true for any conducting material, while the second one is obtained in the case of a good conductor. Note that, from Eqs. (7.13) and (7.17), at locations more than about 5δ away from the surface of a conductor, the penetrating wave retains less than one percent of its intensity at the surface. The phenomenon of predominant localization of fields, currents, and power in the skin of a conducting body is referred to as the skin effect.

CONCEPTUAL QUESTION 7.47 **Skin effect vs. Joule's losses in a good conductor.** The skin effect, i.e., the rapid spatial decrease of the wave amplitude, in a good conductor is caused by local Joule's losses (electric power is lost to heat) throughout the volume of the material.

(A) True.

(B) False.

CONCEPTUAL QUESTION 7.48 **Shielding effectiveness of an aluminum foil vs. frequency.** In order to prevent the electric and magnetic fields from entering or leaving a room, the walls of the room are shielded with a 1-mm-thick aluminum foil. The best protection is achieved at a frequency of

(A) 1 kHz.

(B) 10 kHz.

(C) 100 kHz.

(D) 1 MHz.

(E) No difference.

CONCEPTUAL QUESTION 7.49 **Shielding against the electric vs. magnetic field.** A 1-mm-thick aluminum shield at 100 kHz provides

(A) better isolation for the electric field.

(B) better isolation for the magnetic field.

(C) the same isolation for the electric and magnetic fields.

CONCEPTUAL QUESTION 7.50 **Hollow vs. solid metallic conductors at microwave frequencies.** There is practically no difference in the (electrical) performance between hollow and solid metallic (e.g., copper or aluminum) conductors in various types of antennas and transmission lines at microwave frequencies.

(A) True.

(B) False.

CONCEPTUAL QUESTION 7.51 Hollow vs. solid conductors at the power frequency. Considering copper or aluminum conductors in various types of power-frequency (60 Hz) devices and systems, there is practically no difference in the performance between hollow and solid conductors.

(A) True.

(B) False.

CONCEPTUAL QUESTION 7.52 Radio communication in freshwater and salty lakes. Due to the salt content, sea (salty) water is much more electrically conductive than fresh water. If a radio transmitter and a radio receiver are submerged in a lake, the communication is better for

(A) a freshwater lake.

(B) a salty lake.

(C) No difference.

CONCEPTUAL QUESTION 7.53 Choice of frequency for radio communication with a submarine. Out of the following, what is the most suitable frequency for radio communication with a submerged submarine?

(A) 1 kHz.

(B) 10 kHz.

(C) 100 kHz.

(D) 1 MHz.

(E) 10 MHz.

CONCEPTUAL QUESTION 7.54 Skin depth of iron vs. copper. Given that iron has about six times lower conductivity than copper, the skin depth of iron turns out to be

(A) larger

(B) smaller

than that of copper at the same frequency.

CONCEPTUAL QUESTION 7.55 **Skin depth of a PEC.** The skin depth of a perfect electric conductor amounts to

(A) $\delta = 0$.

(B) $\delta \to \infty$.

(C) $\delta \approx 1$ nm.

(D) Depends on the frequency (only).

(E) Depends on the frequency and the permeability of the material.

7.8 Wave Propagation in Plasmas

Plasmas are ionized gases which, in addition to neutral atoms and molecules, include a large enough number of ionized atoms and molecules and free electrons that macroscopic electromagnetic effects caused by Coulomb forces between charged particles are notable. The phase coefficient of a uniform plane electromagnetic wave of frequency f propagating through a plasma medium is given by

$$\beta = \frac{2\pi f}{c_0}\sqrt{1-\frac{f_{\mathrm{p}}^2}{f^2}}\,, \quad f_{\mathrm{p}} = 9\sqrt{N} \quad (N \text{ in m}^{-3};\ f_{\mathrm{p}} \text{ in Hz}) \quad (\text{plasma medium})\,, \tag{7.18}$$

where c_0 is the free-space wave velocity, Eq. (7.4), f_{p} is the so-called plasma frequency, and N is the concentration of free electrons in the gas. For $f < f_{\mathrm{p}}$, β becomes purely imaginary and thus effectively acts like a large attenuation coefficient α in Eqs. (7.13), so that the wave does not propagate.

An important example of a plasma medium is the upper region of the earth's atmosphere, from about 50 to 500 km altitude above the earth's surface, called the ionosphere. It consists of a highly rarefied gas that is ionized by the sun's radiation and plays an essential role in a number of radio-wave applications.

CONCEPTUAL QUESTION 7.56 **Frequency response of a plasma medium.** Given the frequencies at which electromagnetic waves can and cannot propagate through it, plasma (ionized gas) behaves like

(A) a low-pass filter.

(B) a high-pass filter.

(C) a band-pass filter.

(D) no filter at all.

CONCEPTUAL QUESTION 7.57 **Propagation at the plasma frequency.** Consider an ionized gas whose plasma frequency is f_p. A plane wave of frequency $f = f_\mathrm{p}$

(A) can propagate through the gas.

(B) cannot propagate through the gas.

(C) Need more information.

CONCEPTUAL QUESTION 7.58 **Wave frequency much higher than the plasma frequency.** A plane wave of frequency f propagates in a plasma medium of plasma frequency f_p. For $f \gg f_\mathrm{p}$, the plasma behaves like

(A) an open circuit in circuit theory.

(B) a short circuit in circuit theory.

(C) free space.

CONCEPTUAL QUESTION 7.59 **Concentration of free electrons in a plasma halved.** A time-harmonic plane wave propagates through an ionized gas (plasma) whose concentration of free electrons equals $N = 6 \times 10^{11}\ \mathrm{m}^{-3}$. Would this same wave (with the same frequency) be able to propagate through a plasma medium with $N = 3 \times 10^{11}\ \mathrm{m}^{-3}$?

(A) Yes.

(B) No.

(C) Need more information.

CONCEPTUAL QUESTION 7.60 **Concentration of free electrons doubled.** Would a wave of a given frequency traveling in a plasma medium with the concentration of free electrons of $N = 6 \times 10^{11}\ \text{m}^{-3}$ be able to propagate through a plasma with $N = 12 \times 10^{11}\ \text{m}^{-3}$?

(A) Yes.

(B) No.

(C) Need more information.

CONCEPTUAL QUESTION 7.61 **Wave incidence from the earth on the ionosphere.** A uniform plane time-harmonic electromagnetic wave incident from the earth's surface normally on the ionosphere

(A) passes through it.

(B) bounces off it.

(C) Need more information.

CONCEPTUAL QUESTION 7.62 **Wave incidence from space on the ionosphere.** A plane time-harmonic wave vertically incident from space onto the upper boundary of the ionosphere

(A) passes through it.

(B) bounces off it.

(C) Need more information.

7.9 Dispersion and Group Velocity

The phase velocity in some lossy and/or complex media is frequency dependent, since the phase coefficient, β, of a time-harmonic electromagnetic wave propagating through the medium is a nonlinear function of the angular (radian) frequency, ω, of the wave. If an electromagnetic signal that can be decomposed onto multiple time-harmonic waves of different frequencies (and generally different amplitudes and phases) – Fourier components of the signal – is transmitted through such a medium, the different frequency components propagate at different phase velocities, $v_p(\omega)$, and thus arrive with different phase delays at the receiving point. This means that the relative phases of Fourier components are changed, so that the signal shape is changed as well (signal is distorted). In other words, the medium causes the distortion of the signal by dispersing its frequency components. This phenomenon is generally known as dispersion, and media (or wave-guiding structures) with

$$v_p = \frac{\omega}{\beta(\omega)} = v_p(\omega) \quad \text{(phase velocity in dispersive media)} \tag{7.19}$$

are referred to as dispersive media (or structures). The graphical representation of a nonlinear β–ω relationship for the medium is called the dispersion diagram, Figure 7.3.

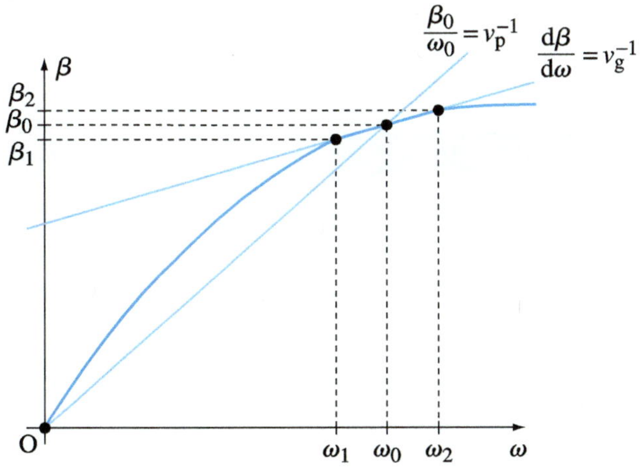

Figure 7.3 Sketch of the β–ω relationship of a dispersive medium, so-called dispersion diagram, with graphical interpretations of the definitions of phase and group velocities for the medium, Eqs. (7.19) and (7.20).

A measure of the speed of propagation of a group of frequencies that constitute a wave packet is called the group velocity and is denoted as v_g. Its reciprocal geometrically represents the slope of the β–ω curve in Figure 7.3, and hence

$$v_g(\omega) = \frac{1}{d\beta(\omega)/d\omega} \quad \text{(group velocity)} . \tag{7.20}$$

In general, this is the velocity of travel of electromagnetic energy and information carried by an electromagnetic wave through a given medium, and is also often called the energy velocity or signal velocity.

CONCEPTUAL QUESTION 7.63 **Definition of a dispersive propagation medium.** Which of the three quantities, the phase coefficient, β, the phase velocity, $v_{\rm p}$, and the group velocity, $v_{\rm g}$, are independent of frequency for nondispersive propagation media and are functions of frequency for dispersive media?

(A) β only.

(B) $v_{\rm p}$ only.

(C) β and $v_{\rm p}$ only.

(D) $v_{\rm p}$ and $v_{\rm g}$ only.

(E) All three quantities.

(F) None of the quantities.

CONCEPTUAL QUESTION 7.64 **Broadening of time pulses as they travel in space.** A signal in the form of a rectangular pulse in time loses its sharp edges and broadens as it propagates through an electromagnetic medium. This medium is

(A) dispersive.

(B) nondispersive.

(C) It is impossible to tell.

CONCEPTUAL QUESTION 7.65 **Signal shape at reception for air medium.** A signal consisting of multiple frequency (Fourier) components travels through air (free space). At the receiving point, the superposition of the individual components is always an exact replica of the original signal.

(A) True.

(B) False.

(C) It is impossible to tell.

CONCEPTUAL QUESTION 7.66 Examples of dispersive propagation media. Consider the following media: perfect (lossless) dielectrics, low-loss dielectrics (good dielectrics), good conductors, and plasmas. Which of them are dispersive?

(A) Perfect and low-loss dielectrics only.

(B) Low-loss dielectrics, good conductors, and plasmas only.

(C) Good conductors and plasmas only.

(D) Plasmas only.

(E) All four types of media.

(F) None of the media.

CONCEPTUAL QUESTION 7.67 VLF ship-to-submarine communication. In a VLF (very low frequency) undersea ship-to-submarine communication system, plane waves are launched into the ocean, as a ship is trying to communicate a message to a submerged submarine. Assume that a signal with two different frequency components, at $f_1 = 7.77$ kHz and $f_2 = 10$ kHz, respectively, is launched at the ocean surface and is received at different depths in the ocean. Also assume that each of the two signal components is strong enough to be received at all depths considered, despite a large attenuation of seawater. Consider the following statements: (a) At certain depths, the respective electric and magnetic fields, at the two frequencies, may be in counter-phase and actually cancel each other at signal reception. (b) At certain depths, the fields at the two frequencies may add in phase at signal reception. Which of the statements is true?

(A) Statement (a) only.

(B) Statement (b) only.

(C) Both statements.

(D) Neither of the statements.

CONCEPTUAL QUESTION 7.68 Group velocity in a dispersive medium. In a dispersive propagation medium, the group velocity ($v_{\rm g}$) is

(A) equal to the phase velocity ($v_{\rm p}$).

(B) always lower than v_p.

(C) always higher than v_p.

(D) lower than v_p for some media, higher than v_p for others.

(E) zero.

CONCEPTUAL QUESTION 7.69 **Group velocity in a nondispersive medium.** Considering group and phase velocities in a nondispersive propagation medium, we have that v_g is

(A) equal to v_p.

(B) always lower than v_p.

(C) always higher than v_p.

(D) lower than v_p for some media, higher than v_p for others.

(E) zero.

CONCEPTUAL QUESTION 7.70 **Energy velocity.** In a dispersive medium or wave-guiding structure, transport of electromagnetic energy by a propagating electromagnetic wave occurs at

(A) the phase velocity, v_p, of the wave.

(B) the group velocity, v_g, of the wave.

(C) the velocity $c_0 = 3 \times 10^8$ m/s.

(D) None of the above.

CONCEPTUAL QUESTION 7.71 **Possible violation of the theory of special relativity.** With v_p and v_g being the phase and group velocities, respectively, of an electromagnetic wave in a certain dispersive medium and $c_0 = 3 \times 10^8$ m/s standing for the speed of light in free space, which of the following assumptions is impossible (from the standpoint of the theory of special relativity, which states that energy and matter cannot travel faster than c_0)?

(A) $v_{\rm p} > c_0$.

(B) $v_{\rm g} > c_0$.

(C) Both assumptions.

(D) Neither assumption.

7.10 Polarization of Electromagnetic Waves

All time-harmonic electromagnetic waves considered so far in this chapter exhibit so-called linear polarization (LP), with the tip of the electric field intensity vector of the wave (and the same is true for the magnetic one) at a given point in space tracing a straight line in the course of time, as shown in Figure 7.4(a). However, if two waves with mutually orthogonal linear polarizations at the same frequency co-propagate in the same direction, the polarization of the resultant wave depends on

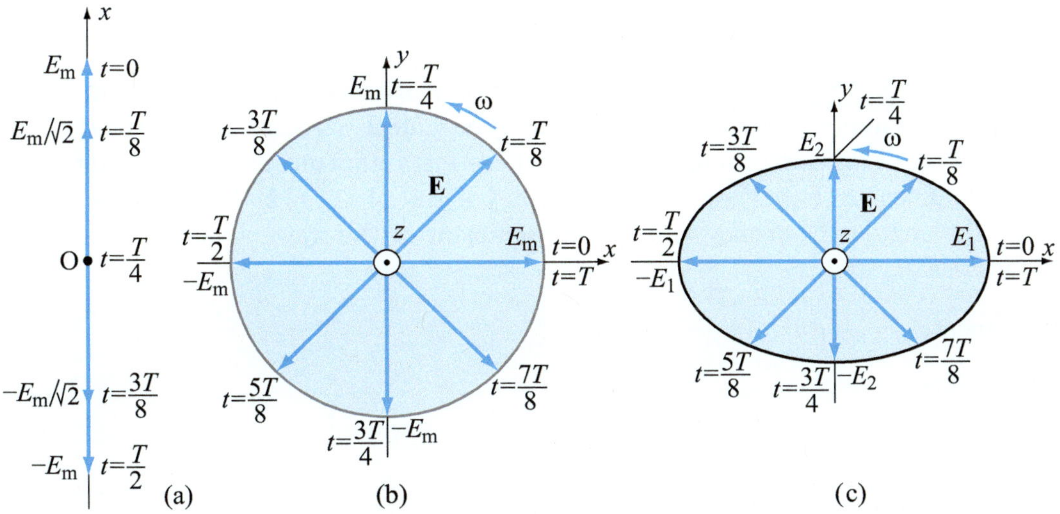

Figure 7.4 (a) Linearly polarized time-harmonic uniform plane wave, Eqs. (7.8) with $z = 0$, $\theta_0 = 0$, and $E_{\rm m} = E_0\sqrt{2}$ (amplitude). (b) Circularly polarized plane wave, Eqs. (7.21). (c) Elliptical polarization, Eqs. (7.22). Rotations in (b) and (c) correspond to the plus sign (+ in place of ±) at the beginning of expressions for E_y, and the waves are referred to as right-hand (RH) CP and EP waves, respectively (when the thumb of the right hand points into the direction of the wave travel, the other fingers curl in the direction of rotation of $\mathbf{E}$); a minus sign would result in a change of the rotation direction and give rise to left-hand (LH) CP and EP waves.

the relative amplitudes (or rms values) and phases of its individual LP components. For instance, if the two transverse components have the same amplitudes but are out of phase by $\mp 90°$ (i.e., they are in time-phase quadrature),

$$E_x = E_{\rm m} \cos(\omega t - \beta z) , \quad E_y = \pm E_{\rm m} \sin(\omega t - \beta z) \quad \text{(circular polarization)} , \tag{7.21}$$

the wave is circularly polarized (CP). Namely, the resultant vector **E** for $z = \text{const}$ rotates with an angular velocity equal to ω and its tip describes a circle, of radius $E_{\rm m}$, as a function of time, as depicted in Figure 7.4(b). If we then change one of the amplitudes of the transverse components of the wave in Eqs. (7.21) so that they are no longer the same,

$$E_x = E_1 \cos(\omega t - \beta z) , \quad E_y = \pm E_2 \sin(\omega t - \beta z) \quad \text{(elliptical polarization)} \tag{7.22}$$

($E_1 \neq E_2$), the tip of the resultant vector **E** traces an ellipse (polarization ellipse) in the plane $z = \text{const}$, Figure 7.4(c), and the wave is said to be elliptically polarized (EP). Finally, in the most general case, that with an arbitrary phase difference between the two components of **E**, in addition to an arbitrary ratio between their amplitudes, the wave is also EP, but the polarization ellipse is tilted with respect to the x-axis; so, elliptical polarization is the most general polarization of time-harmonic vectors.

CONCEPTUAL QUESTION 7.72 **Determination of polarization state of a plane wave – case 1.** Determine the type (linear, circular, or elliptical) of the polarization of an electromagnetic wave whose instantaneous electric field intensity vector is given by $\mathbf{E}(x, t) = [2\cos(\omega t + \beta x)\,\hat{\mathbf{y}} - \sin(\omega t + \beta x)\,\hat{\mathbf{z}}]$ V/m, where ω and β are the angular frequency and phase coefficient of the wave. The polarization of the wave is

(A) linear.

(B) circular.

(C) elliptical.

(D) Need more information.

CONCEPTUAL QUESTION 7.73 **Determination of polarization state – case 2.** If the electric field of a wave can be expressed as $\mathbf{E}(x, t) = [2\cos(\omega t + \beta x)\,\hat{\mathbf{y}} - \cos(\omega t + \beta x)\,\hat{\mathbf{z}}]$ V/m, its polarization is

(A) linear.

(B) circular.

(C) elliptical.

(D) Need more information.

CONCEPTUAL QUESTION 7.74 **Determination of polarization state – case 3.** The polarization of a wave whose electric field vector is given by $\mathbf{E}(x,t) = [\cos(\omega t + \beta x)\,\hat{\mathbf{y}} - \sin(\omega t + \beta x)\,\hat{\mathbf{z}}]$ V/m turns out to be

(A) linear.

(B) circular.

(C) elliptical.

(D) Need more information.

CONCEPTUAL QUESTION 7.75 **Determination of polarization state – case 4.** The polarization of a wave with $\mathbf{E}(x,t) = [2\cos(\omega t + \beta x + \pi/3)\,\hat{\mathbf{y}} - \sin(\omega t + \beta x)\,\hat{\mathbf{z}}]$ V/m is

(A) linear.

(B) circular.

(C) elliptical.

(D) Need more information.

CONCEPTUAL QUESTION 7.76 **Determination of polarization state – case 5.** If $\mathbf{E}(x,t) = [\cos(\omega t + \beta x + \pi/3)\,\hat{\mathbf{y}} - \sin(\omega t + \beta x)\,\hat{\mathbf{z}}]$ V/m, the wave polarization is

(A) linear.

(B) circular.

(C) elliptical.

(D) Need more information.

CONCEPTUAL QUESTION 7.77 **Determination of polarization state – case 6.** What is the polarization of a wave with $\mathbf{E}(x, t) = [\sin(\omega t + \beta x + \pi/4)\,\hat{\mathbf{y}} - \sin(\omega t + \beta x - \pi/4)\,\hat{\mathbf{z}}]$ V/m?

(A) Linear.

(B) Circular.

(C) Elliptical.

(D) Need more information.

CONCEPTUAL QUESTION 7.78 **Determination of polarization state – case 7.** The wave polarization if $\mathbf{E}(x, t) = [\sin(\omega t + \beta x + \pi/4)\,\hat{\mathbf{y}} - 2\sin(\omega t + \beta x - 3\pi/4)\,\hat{\mathbf{z}}]$ V/m is

(A) linear.

(B) circular.

(C) elliptical.

(D) Need more information.

CONCEPTUAL QUESTION 7.79 **Changing a right-hand circularly polarized wave.** If the amplitude (or rms value) of one of the two transverse components constituting a right-hand circularly polarized (RHCP) wave is doubled, while the amplitude of the other component and phases of both components are kept unchanged, the new wave is

(A) also RHCP.

(B) left-hand circularly polarized (LHCP).

(C) right-hand elliptically polarized (RHEP).

(D) left-hand elliptically polarized (LHEP).

(E) linearly polarized (LP).

(F) Need more information.

CONCEPTUAL QUESTION 7.80 **Polarization handedness for the magnetic field vector.** Polarization handedness (right- or left-handed) of an elliptically polarized wave, namely, whether the wave is RHEP or LHEP, determined by considering the magnetic field vector of the wave is opposite to that obtained by viewing the electric field vector.

(A) True.

(B) False.

CONCEPTUAL QUESTION 7.81 **Polarization ellipse for the magnetic field vector.** The polarization ellipse for the magnetic field vector of an elliptically polarized wave is orthogonal to that for the electric field vector of the wave.

(A) True.

(B) False.

---

8 REFLECTION AND TRANSMISSION OF PLANE WAVES

Introduction

Capitalizing on the concepts and techniques of the analysis of wave propagation in homogeneous and unbounded media of various electromagnetic properties from the previous chapter, we now proceed to develop the concepts and techniques for the analysis of wave interaction with planar boundaries between material regions. In general, as a wave encounters an interface separating two different media, it is partly reflected back to the incident medium (wave reflection) and partly transmitted to the medium on the other side of the interface (wave transmission), and hence the title of this chapter. The material will be presented as several separate cases of reflection and transmission (also referred to as refraction) of plane waves, in order of increasing complexity, from normal incidence (wave propagation direction is normal to the interface) on a perfectly conducting plane and normal incidence on a penetrable interface (between two arbitrary media), to oblique incidence (at an arbitrary angle) on these two types of interfaces. The material will include discussions of distributions of total fields and waves in the structure, as well as numerous concepts associated with these fields and waves. In all problems, however, the core of the solution will be the use of appropriate general electromagnetic boundary conditions, as a "connection" between the fields on different sides of the interfaces.

8.1 Normal Incidence on a Perfectly Conducting Plane

Consider a uniform plane linearly polarized time-harmonic electromagnetic wave of frequency f and rms electric field intensity $\underline{E}_{i0}$ propagating through a lossless ($\sigma = 0$) medium of permittivity ε and permeability μ. Let the wave be incident normally on an infinite flat surface (the direction of wave propagation is normal to the surface) of a perfect electric conductor (PEC), with $\sigma \to \infty$, as shown in Figure 8.1(a). Complex electric and magnetic field intensity vectors of the wave, which we refer to as the incident (or forward) wave, can be written as [see Eqs. (7.11)]

$$\underline{\mathbf{E}}_{i} = \underline{E}_{i0}\,\mathrm{e}^{-\mathrm{j}\beta z}\,\hat{\mathbf{x}}\,, \quad \underline{\mathbf{H}}_{i} = \frac{\underline{E}_{i0}}{\eta}\,\mathrm{e}^{-\mathrm{j}\beta z}\,\hat{\mathbf{y}}\,, \quad \beta = \omega\sqrt{\varepsilon\mu}\,, \quad \eta = \sqrt{\frac{\mu}{\varepsilon}} \quad \text{(incident wave)} \tag{8.1}$$

($\omega = 2\pi f$). This wave excites currents to flow on the PEC surface, which, in turn, are sources of a reflected (or backward) wave, propagating in the negative z

direction. Its field vectors are given by

$$\underline{\mathbf{E}}_{\rm r} = \underline{E}_{\rm r0}\, {\rm e}^{{\rm j}\beta z}\,\hat{\mathbf{x}}\,, \quad \underline{\mathbf{H}}_{\rm r} = \frac{\underline{E}_{\rm r0}}{\eta}\, {\rm e}^{{\rm j}\beta z}(-\hat{\mathbf{y}}) \quad \text{(reflected wave)}\,. \tag{8.2}$$

From the boundary condition for the vector $\mathbf{E}$ (more precisely, for its tangential component) in Eqs. (6.15) applied in the plane $z = 0$, we obtain

$$\underline{E}_{\rm i0} + \underline{E}_{\rm r0} = 0 \quad \longrightarrow \quad \underline{E}_{\rm r0} = -\underline{E}_{\rm i0} \quad \text{(boundary condition)}\,. \tag{8.3}$$

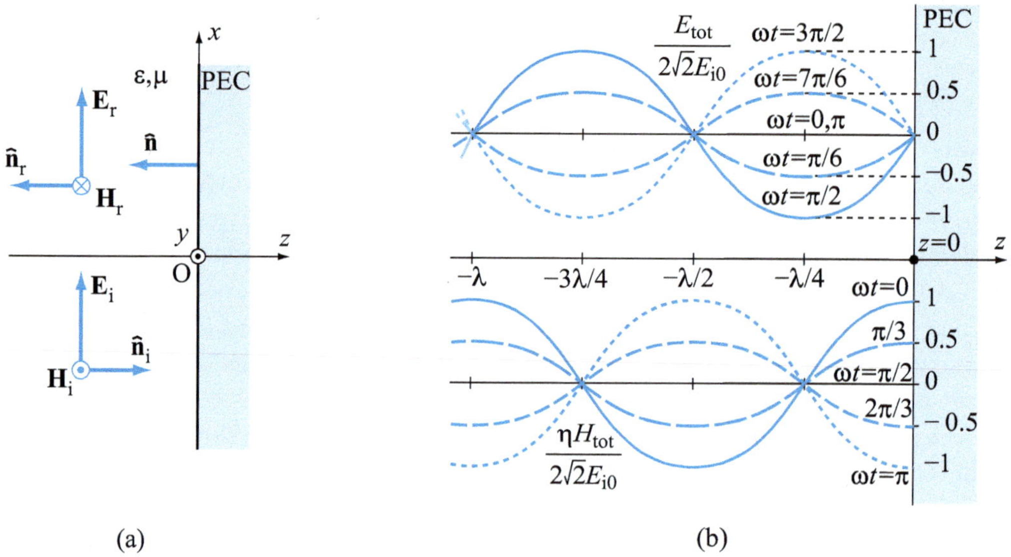

Figure 8.1 (a) Normal incidence of a uniform plane time-harmonic electromagnetic wave on a planar interface between a perfect dielectric and a perfect conductor. (b) Plots of normalized total electric and magnetic field intensities in Eqs. (8.5) against z at different instants of time.

Using Eqs. (8.1)–(8.3) and (6.20), the total complex electric and magnetic fields in the incident medium (for $z \leq 0$), i.e., the field vectors of the resultant wave, are

$$\underline{\mathbf{E}}_{\rm tot} = -2{\rm j}\underline{E}_{\rm i0}\sin\beta z\,\hat{\mathbf{x}}\,, \quad \underline{\mathbf{H}}_{\rm tot} = 2\frac{\underline{E}_{\rm i0}}{\eta}\cos\beta z\,\hat{\mathbf{y}} \quad \text{(resultant wave)}\,, \tag{8.4}$$

and, by means of Eq. (6.21), their time-domain counterparts come out to be

$$\mathbf{E}_{\rm tot}(t) = 2\sqrt{2}E_{\rm i0}\sin\beta z\sin\omega t\,\hat{\mathbf{x}}\,, \quad \mathbf{H}_{\rm tot}(t) = 2\sqrt{2}\,\frac{E_{\rm i0}}{\eta}\cos\beta z\cos\omega t\,\hat{\mathbf{y}} \quad (\underline{E}_{\rm i0} = E_{\rm i0})$$
$$\text{(instantaneous fields of a standing wave)}\,. \tag{8.5}$$

Figure 8.1(b) shows snapshots at different time instants of the resultant field intensities as a function of z. We see that the fields do not travel as the time advances, but stay where they are, only oscillating in time between the stationary zeros. So, they do not represent a traveling wave in either direction. The resultant wave, which is a superposition of two traveling waves, is thus termed a standing wave.

CONCEPTUAL QUESTION 8.1 **Normal incidence on a PEC – properties of a resultant wave.** A time-harmonic uniform plane electromagnetic wave with the rms intensity of the electric field E_{i0} and frequency f travels in air and is incident normally on an infinitely large screen made of a perfect electric conductor (PEC), as shown in Figure 8.2. Consider the resultant (incident plus reflected) wave in air and the following statements: (a) the resultant wave is a time-harmonic wave; (b) the resultant wave is a uniform wave; (c) the resultant wave is a plane wave; (d) the phase velocity of the resultant wave equals $c_0 = 3 \times 10^8$ m/s; (e) the group velocity of the resultant wave equals $c_0 = 3 \times 10^8$ m/s. Which of these statements are true?[1]

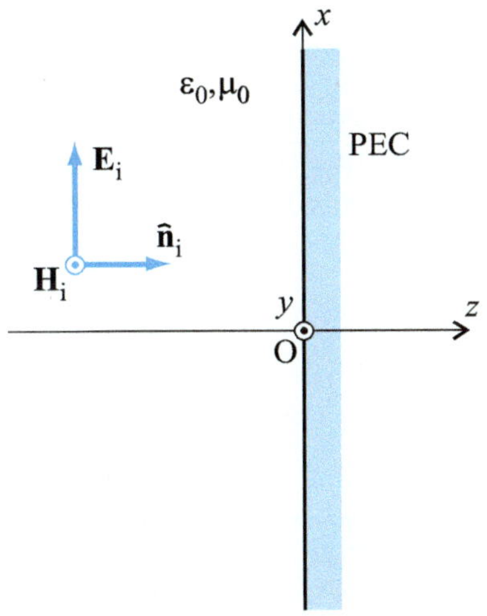

Figure 8.2 Normal incidence of a uniform plane time-harmonic electromagnetic wave from air on a PEC screen; for Conceptual Question 8.1.

(A) Statements (a) and (c) only.

(B) Statements (b) and (c) only.

(C) Statements (a)–(c) only.

(D) Statements (a)–(d) only.

(E) All statements, (a)–(e).

---

[1] *For every conceptual question in this text, exactly one answer is correct.*

CONCEPTUAL QUESTION 8.2 **Wavelength of a standing wave.** The wavelength of the resultant (incident plus reflected) wave in air in front of a PEC plane illuminated at normal incidence by a time-harmonic uniform plane wave of frequency f is

(A) $\lambda = c_0/f$ ($c_0 = 3 \times 10^8$ m/s).

(B) $\lambda = 2c_0/f$.

(C) $\lambda = 0.5c_0/f$.

(D) not defined.

(E) none of the above.

CONCEPTUAL QUESTION 8.3 **General form of standing time-harmonic waves.** In general, standing time-harmonic waves are recognized by the presence of the argument of the form $\omega t - \beta l$ in the time domain or the corresponding factor $e^{-j\beta l}$ in the complex domain, where l is an arbitrary length coordinate.

(A) True.

(B) False.

CONCEPTUAL QUESTION 8.4 **Electric to magnetic field ratio for a standing wave.** Consider two more statements regarding the resultant (incident plus reflected) electromagnetic wave in the incident region for normal incidence of a time-harmonic uniform plane wave from air upon a PEC screen: (1) the electric and magnetic field vectors of the resultant wave are perpendicular to each other at any instant of time and any point of space in front of the screen; (2) the ratio of the electric and magnetic field intensities of the resultant wave equals a constant, $\eta_0 = \sqrt{\mu_0/\varepsilon_0}$, at any time and any location in air. Which of these statements is true?

(A) Statement (1) only.

(B) Statement (2) only.

(C) Both statements.

(D) Neither of the statements.

CONCEPTUAL QUESTION 8.5 **Phase difference between electric and magnetic fields.** At every point of space, the electric and magnetic fields of the resultant (standing) electromagnetic wave in air in front of a PEC screen illuminated at normal incidence by a time-harmonic uniform plane wave are

(A) in phase.

(B) in counter-phase (180° out of phase with respect to each other).

(C) in time-phase quadrature (±90° out of phase with respect to each other).

(D) out of phase by a constant angle different from 0, 180°, and ±90°.

(E) out of phase by an angle that depends on the distance from the PEC screen.

CONCEPTUAL QUESTION 8.6 **Spatial dependence of phases of electric and magnetic fields.** Consider spatial dependence of the phases of both electric and magnetic fields of the resultant wave in front of a PEC screen for normal incidence of a time-harmonic uniform plane wave from air. For the z-axis adopted to be perpendicular to the screen with the origin ($z = 0$) in the PEC surface, these phases are

(A) linear functions of the spatial coordinate z.

(B) exponential functions of z.

(C) sinusoidal functions of z.

(D) constant.

CONCEPTUAL QUESTION 8.7 **Model with an infinite current sheet radiating in air.** In the structure consisting of a PEC screen in air illuminated at normal incidence by a time-harmonic uniform plane electromagnetic wave, surface currents (of density $\underline{\mathbf{J}}_{\mathrm{s}}$) are induced on the PEC surface by the incident wave, and the structure can, in turn, be replaced by an equivalent model with the infinite sheet of these currents situated in air. Namely, we can substitute the PEC by the material (air) occupying the other half-space, and consider the surface currents flowing in the plane previously representing the PEC surface to exist in an unbounded homogeneous medium (air-filled). In this model, the total electric field in the incident half-space (filled with air in the original structure, as well as in the equivalent model) equals

(A) the electric field of the incident wave.

(B) the electric field radiated by the current sheet.

(C) the sum of fields in (A) and (B).

(D) the difference of fields in (A) and (B).

(E) zero.

(F) More than one of the above answers are correct.

CONCEPTUAL QUESTION 8.8 **Total field in the region previously occupied by the PEC.** Consider an equivalent model with the infinite sheet of surface currents – induced on a PEC interface illuminated from air by a normally incident plane wave – being situated in air, with the PEC removed. In this model, the total electric field in the half-space previously occupied by the PEC is equal to

(A) the electric field of the incident wave.

(B) the electric field radiated by the current sheet.

(C) the sum of fields in (A) and (B).

(D) the difference of fields in (A) and (B).

(E) zero.

(F) More than one of the above answers are correct.

CONCEPTUAL QUESTION 8.9 **Insertion of a PEC sheet transversally to a standing wave.** Consider the standing electromagnetic wave resulting from a normally incident time-harmonic uniform plane wave with wavelength λ_0 reflecting in air off a perfectly conducting plane at $z = 0$, and assume that a perfectly conducting sheet is inserted in the plane $z = -h$ ($h > 0$), as shown in Figure 8.3. The structure of the electromagnetic field for $-h \leq z \leq 0$ will not change

(A) if $h = \lambda_0/8$.

(B) if $h = \lambda_0/4$.

(C) if $h = \lambda_0/2$.

(D) for arbitrary h.

(E) There is no such h.

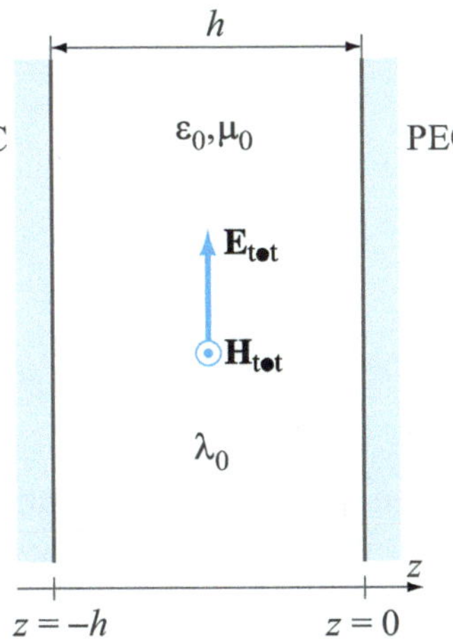

Figure 8.3 Insertion of a PEC sheet at $z = -h$ in front of the original reflecting PEC screen at $z = 0$, to obtain a self-contained structure with a standing electromagnetic plane wave trapped between the two parallel PEC planes (like two mirrors). This structure behaves like an electromagnetic resonator, and is known as the Fabry–Perot resonator; for Conceptual Question 8.9.

CONCEPTUAL QUESTION 8.10 **Current distribution over an illuminated PEC plane.** A uniform plane time-harmonic electromagnetic wave is normally incident on a PEC plane from a lossless medium, as in Figure 8.4. With reference to the Cartesian coordinate system in Figure 8.4, in which the electric field vector of the incident wave has an x-component only, the surface current density vector describing the distribution of surface currents over the PEC plane is of the following form:

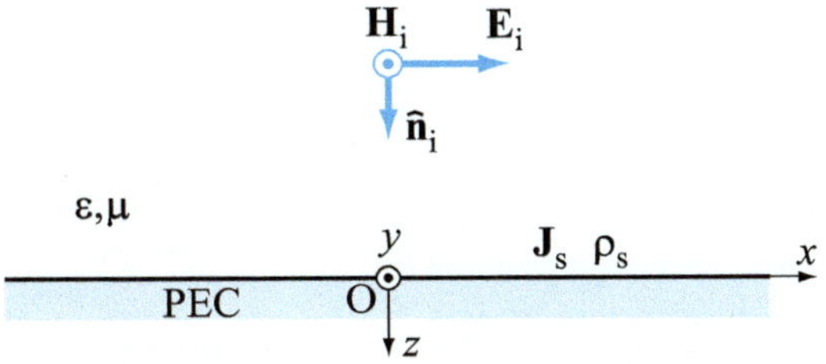

Figure 8.4 Surface currents and charges induced on a PEC plane by an incident uniform plane time-harmonic electromagnetic wave; for Conceptual Question 8.10.

(A) $\underline{\mathbf{J}}_s = \underline{J}_s(x)\,\hat{\mathbf{x}}$.

(B) $\underline{\mathbf{J}}_s = \underline{J}_s(y)\,\hat{\mathbf{x}}$.

(C) $\underline{\mathbf{J}}_s = \underline{J}_s(x)\,\hat{\mathbf{y}}$.

(D) $\underline{\mathbf{J}}_s = \underline{J}_s(y)\,\hat{\mathbf{y}}$.

(E) $\underline{\mathbf{J}}_s = \underline{J}_{s0}\,\hat{\mathbf{x}}$ ($\underline{J}_{s0} = \text{const}$).

(F) $\underline{\mathbf{J}}_s = \underline{J}_{s0}\,\hat{\mathbf{y}}$.

CONCEPTUAL QUESTION 8.11 **Charge distribution over an illuminated PEC plane.** The surface charge density representing the distribution of surface charges over a PEC plane impinged from air by a uniform plane time-harmonic wave at normal incidence (Figure 8.4) has the following form or value:

(A) $\underline{\rho}_s = \underline{\rho}_s(x)$.

(B) $\underline{\rho}_s = \underline{\rho}_s(y)$.

(C) $\underline{\rho}_s = \underline{\rho}_s(x, y)$.

(D) $\underline{\rho}_s = \underline{\rho}_{s0}$, where $\underline{\rho}_{s0} = \text{const}$ and $\underline{\rho}_{s0} \neq 0$.

(E) $\underline{\rho}_s = 0$.

CONCEPTUAL QUESTION 8.12 **Instants of time with no electric energy of a standing wave.** In a standing wave resulting from the normal incidence of a uniform plane time-harmonic electromagnetic wave from air upon a PEC plane, electric energy is never (at no instant of time) zero everywhere in air, in the entire half-space next to (e.g., above or in front of) the PEC plane.

(A) True.

(B) False.

CONCEPTUAL QUESTION 8.13 **Instants of time with no magnetic energy.** In the resultant wave in air for the normal incidence of a uniform plane time-harmonic electromagnetic wave on a PEC plane, there are instants of time when there is no magnetic energy in any portion of the half-space beside the PEC plane.

(A) True.

(B) False.

CONCEPTUAL QUESTION 8.14 **Locations with zero electric or magnetic energy density.** Considering instantaneous electric and magnetic energy densities, w_e and w_m, of the the resultant (standing) wave at different locations in air in front of (or above) the PEC plane illuminated at normal incidence by a time-harmonic uniform plane electromagnetic wave, there are

(A) locations where both w_e and w_m are zero

(B) locations where either w_e or w_m (but not both) is zero

(C) no locations where either w_e or w_m is zero

at all times.

CONCEPTUAL QUESTION 8.15 **Locations of electric and magnetic field maxima and minima.** For normal incidence of a uniform plane time-harmonic wave from air upon a PEC screen, consider the maxima and minima of the rms electric field intensity and those of the rms magnetic field intensity of the resultant wave in front of (above) the screen. The magnetic field maxima occur

(A) at distinct locations of electric field maxima.

(B) at distinct locations of electric field minima.

(C) at distinct locations coinciding with neither electric field maxima nor minima.

(D) everywhere (rms field intensity is constant).

CONCEPTUAL QUESTION 8.16 **Standing wave reception by a short wire dipole antenna.** A time-harmonic uniform plane electromagnetic wave with a wavelength λ_0 is incident normally from air on the earth's surface, which can be assumed to be perfectly flat, of infinite extent, and perfectly conducting, so a PEC ground plane, as shown in Figure 8.5. We wish to receive the wave, that is, the signal it carries, by a short wire dipole antenna (two straight wire arms of total length much smaller than λ_0 with a small gap between the antenna terminals), usually referred to as an electric probe, also shown in Figure 8.5. The rms electromotive force (emf) induced in the antenna is maximum if the dipole is positioned

(A) parallel to the x-axis and in the plane $z = -\lambda_0/4$ ($h = \lambda_0/4$).

(B) parallel to the y-axis and in the plane $z = -\lambda_0/4$ ($h = \lambda_0/4$).

(C) parallel to the z-axis and in the plane $z = -\lambda_0/4$ ($h = \lambda_0/4$).

(D) parallel to the x-axis and in the plane $z = -\lambda_0/2$ ($h = \lambda_0/2$).

(E) parallel to the y-axis and in the plane $z = -\lambda_0/2$ ($h = \lambda_0/2$).

(F) parallel to the z-axis and in the plane $z = -\lambda_0/2$ ($h = \lambda_0/2$).

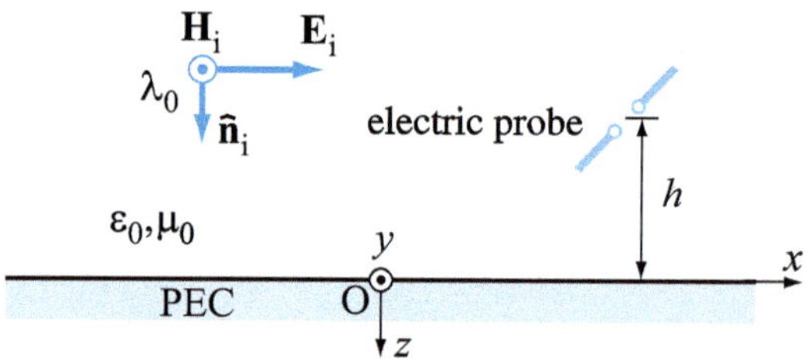

Figure 8.5 Reception of a uniform plane wave normally incident on the earth's surface (or a PEC ground plane) by a short wire dipole antenna; for Conceptual Question 8.16.

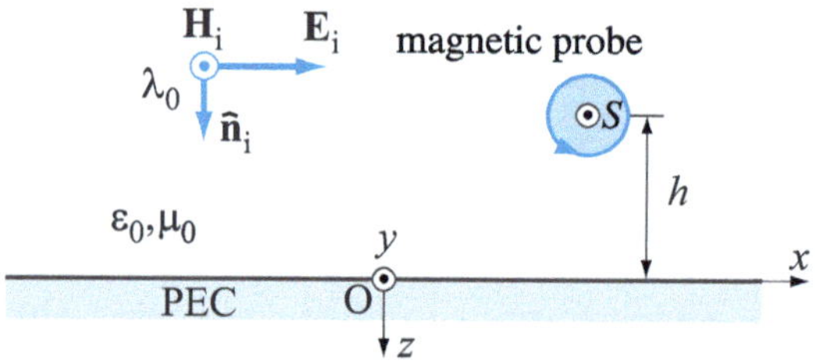

Figure 8.6 Reception of a uniform plane wave normally incident on a PEC ground plane by a small loop antenna; for Conceptual Question 8.17.

CONCEPTUAL QUESTION 8.17 **Magnetic probe in the field of a standing wave.** Consider a uniform plane wave of wavelength λ_0 incident from air normally on the earth's surface or a PEC ground plane, and assume that a small wire loop (the loop diameter, or other dimensions if not of circular shape, is small relative to λ_0), also called a magnetic probe, is used as a receiving antenna, as shown in Figure 8.6. The rms emf induced in the antenna is maximum if the loop is placed

(A) parallel to the xy-plane, with its center at $z = -\lambda_0/4$ ($h = \lambda_0/4$).

(B) parallel to the xz-plane, with its center at $z = -\lambda_0/4$ ($h = \lambda_0/4$).

(C) parallel to the yz-plane, with its center at $z = -\lambda_0/4$ ($h = \lambda_0/4$).

(D) parallel to the xy-plane, with its center at $z = -\lambda_0/2$ ($h = \lambda_0/2$).

(E) parallel to the xz-plane, with its center at $z = -\lambda_0/2$ ($h = \lambda_0/2$).

(F) parallel to the yz-plane, with its center at $z = -\lambda_0/2$ ($h = \lambda_0/2$).

CONCEPTUAL QUESTION 8.18 **Complex Poynting vector of a standing wave.** The complex Poynting vector ($\underline{\mathcal{P}}$) of the resultant wave in Figure 8.7 is

(A) purely real

(B) purely imaginary

(C) zero

(D) with equal nonzero real and imaginary parts

at every point in front of the PEC plane.

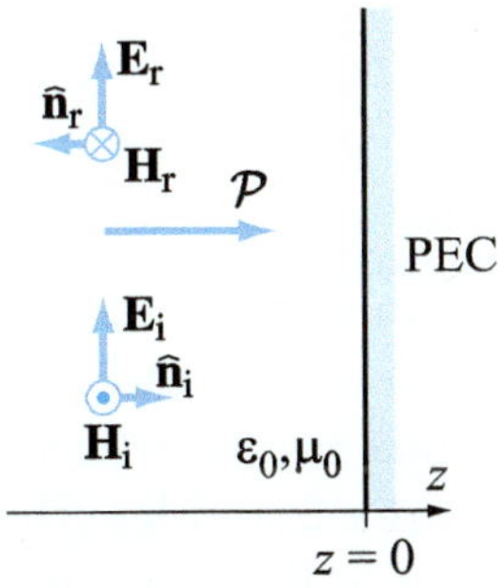

Figure 8.7 Poynting vector of a standing wave in front of a PEC screen illuminated by a uniform plane time-harmonic electromagnetic wave; for Conceptual Question 8.18.

CONCEPTUAL QUESTION 8.19 **Instantaneous Poynting vector of a standing wave.** All zeros of the instantaneous Poynting vector ($\mathcal{P}$) of the resultant wave for $-\infty < z \leq 0$ in Figure 8.7 are given by

(A) $z = -m\lambda/2$,

(B) $z = -m\lambda/4$,

(C) $z = -m\lambda/8$,

(D) $z = -(2m+1)\lambda/4$,

(E) $z = -(4m+1)\lambda/8$,

(F) no such z, namely, $\boldsymbol{\mathcal{P}}$ does not have any zeros for $-\infty < z \leq 0$,

where $m = 0, 1, 2, \ldots$, and λ is the free-space wavelength of the incident wave.

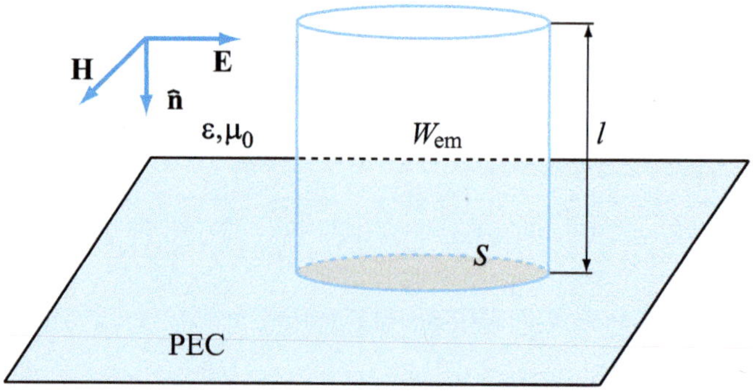

Figure 8.8 Analysis of the electromagnetic energy of the resultant wave in an imaginary cylinder placed on a reflecting PEC plane; for Conceptual Question 8.20.

CONCEPTUAL QUESTION 8.20 Electromagnetic energy in an imaginary cylinder. A time-harmonic uniform plane wave with angular frequency ω and wavelength λ travels in a lossless dielectric and is incident normally on a perfectly conducting plane. The instantaneous resultant electromagnetic energy stored in an imaginary cylinder with the basis area S and length $l = \lambda/4$, placed in the dielectric along the wave travel such that one of its bases lies in the PEC plane, as shown in in Figure 8.8, can be expressed as

(A) $W_{\text{em}}(t) = W_1 + W_2 \cos \omega t$,

(B) $W_{\text{em}}(t) = W_1 + W_2 \cos 2\omega t$,

(C) $W_{\text{em}}(t) = W_1 \cos \omega t$,

(D) $W_{\text{em}}(t) = W_1 \cos^2 \omega t$,

(E) $W_{\text{em}}(t) = W_1$,

(F) $W_{\text{em}}(t) = 0$,

where W_1 and W_2 are nonzero constants (given in joules).

CONCEPTUAL QUESTION 8.21 **Normal incidence on a PEC of an LHCP wave.** A left-hand circularly polarized time-harmonic uniform plane electromagnetic wave travels in a lossless dielectric medium and is incident on a PEC plane at normal incidence. The reflected wave is

(A) linearly polarized.

(B) left-hand circularly polarized.

(C) right-hand circularly polarized.

(D) left-hand elliptically polarized.

(E) right-hand elliptically polarized.

(F) unpolarized.

CONCEPTUAL QUESTION 8.22 **Normal incidence on a PEC of an RHEP wave.** For a right-hand elliptically polarized uniform plane wave normally incident on a PEC plane, the reflected wave is

(A) linearly polarized.

(B) left-hand circularly polarized.

(C) right-hand circularly polarized.

(D) left-hand elliptically polarized.

(E) right-hand elliptically polarized.

(F) unpolarized.

8.2 Normal Incidence on a Penetrable Planar Interface

We now consider a more general case with the medium on the right-hand side of the interface in Figure 8.1(a) being penetrable for the (normally) incident wave. Moreover, let both media be lossy, as indicated in Figure 8.9(a). Having in mind Eqs. (8.1), (8.2), and (7.13), we can write – for the field vectors in Figure 8.9(a):

$$\underline{\mathbf{E}}_{\mathrm{i}} = \underline{E}_{\mathrm{i}0}\,\mathrm{e}^{-\underline{\gamma}_1 z}\,\hat{\mathbf{x}}\,, \quad \underline{\mathbf{H}}_{\mathrm{i}} = \frac{\underline{E}_{\mathrm{i}0}}{\underline{\eta}_1}\,\mathrm{e}^{-\underline{\gamma}_1 z}\,\hat{\mathbf{y}} \quad \text{(incident wave)}\,, \tag{8.6}$$

$$\underline{\mathbf{E}}_{\mathrm{r}} = \underline{E}_{\mathrm{r}0}\,\mathrm{e}^{\underline{\gamma}_1 z}\,\hat{\mathbf{x}}\,, \quad \underline{\mathbf{H}}_{\mathrm{r}} = \frac{\underline{E}_{\mathrm{r}0}}{\underline{\eta}_1}\,\mathrm{e}^{\underline{\gamma}_1 z}(-\hat{\mathbf{y}}) \quad \text{(reflected wave)}\,, \tag{8.7}$$

$$\underline{\mathbf{E}}_{\mathrm{t}} = \underline{E}_{\mathrm{t}0}\,\mathrm{e}^{-\underline{\gamma}_2 z}\,\hat{\mathbf{x}}\,, \quad \underline{\mathbf{H}}_{\mathrm{t}} = \frac{\underline{E}_{\mathrm{t}0}}{\underline{\eta}_2}\,\mathrm{e}^{-\underline{\gamma}_2 z}\,\hat{\mathbf{y}} \quad \text{(transmitted wave)}\,. \tag{8.8}$$

In Eqs. (8.7) and (8.8), we have two unknown field intensities at $z = 0$, $\underline{E}_{\mathrm{r}0}$ and $\underline{E}_{\mathrm{t}0}$, so we invoke two boundary conditions – those for tangential components of vectors **E** and **H** in Eqs. (6.14) – to solve for $\underline{E}_{\mathrm{r}0}$ and $\underline{E}_{\mathrm{t}0}$, for a given $\underline{E}_{\mathrm{i}0}$,

$$\underline{E}_{\mathrm{i}0} + \underline{E}_{\mathrm{r}0} = \underline{E}_{\mathrm{t}0}\,, \quad \frac{\underline{E}_{\mathrm{i}0}}{\underline{\eta}_1} - \frac{\underline{E}_{\mathrm{r}0}}{\underline{\eta}_1} = \frac{\underline{E}_{\mathrm{t}0}}{\underline{\eta}_2} \quad (\underline{\mathbf{J}}_{\mathrm{s}} = 0) \quad \text{(boundary conditions)}\,, \tag{8.9}$$

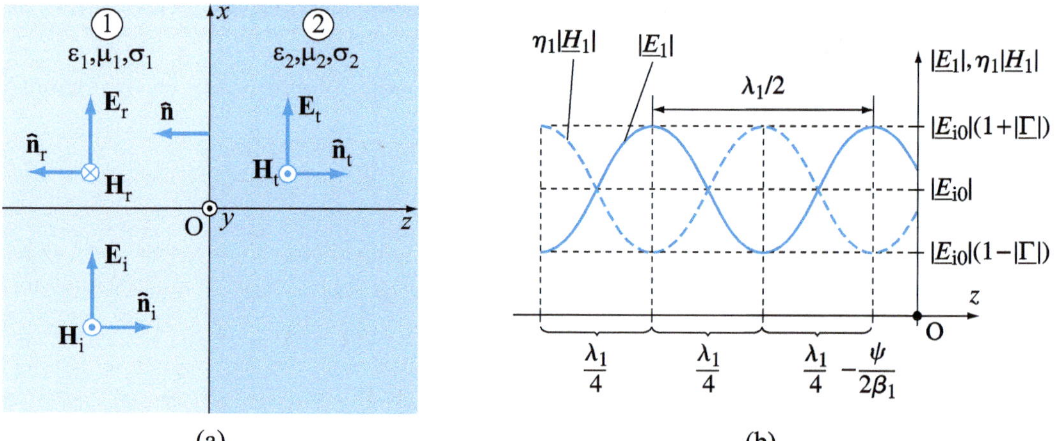

(a) (b)

Figure 8.9 (a) Normal incidence of a uniform plane time-harmonic electromagnetic wave on a planar interface between two media with arbitrary electromagnetic parameters. (b) Sketch of the magnitudes of the total electric and magnetic field intensity vectors in the incident medium (assumed to be lossless) as functions of z (standing wave patterns), for an arbitrary phase (ψ) of the reflection coefficient ($\underline{\Gamma} = |\underline{\Gamma}|\,\mathrm{e}^{\mathrm{j}\psi}$) in Eq. (8.10) ($\lambda_1 = 2\pi/\beta_1$ is the wavelength in the incident medium).

where the vector $\underline{\mathbf{J}}_s$ is taken to be zero since surface currents in the plane $z = 0$ can only exist if one of the two media is a perfect conductor, as in Figure 8.1(a). The solution of Eqs. (8.9) is expressed in terms of the so-called reflection and transmission coefficients, $\underline{\Gamma}$ and $\underline{\tau}$, as follows ($\underline{\eta}_1$ and $\underline{\eta}_2$ are intrinsic impedances of media 1 and 2, respectively):

$$\underline{\Gamma} = \frac{\underline{E}_{r0}}{\underline{E}_{i0}} = \frac{\underline{\eta}_2 - \underline{\eta}_1}{\underline{\eta}_1 + \underline{\eta}_2} \quad \text{(reflection coefficient; dimensionless)}\,, \tag{8.10}$$

$$\underline{\tau} = \frac{\underline{E}_{t0}}{\underline{E}_{i0}} = \frac{2\underline{\eta}_2}{\underline{\eta}_1 + \underline{\eta}_2} \quad \text{(transmission coefficient; dimensionless)}\,, \tag{8.11}$$

which completes the computation of the reflected and transmitted fields in Figure 8.9(a). Figure 8.9(b) shows the plots, so-called electric- and magnetic-field standing wave patterns, of the magnitudes of the total electric and magnetic field intensity vectors in the incident medium, $\underline{\mathbf{E}}_1 = \underline{\mathbf{E}}_i + \underline{\mathbf{E}}_r$ and $\underline{\mathbf{H}}_1 = \underline{\mathbf{H}}_i + \underline{\mathbf{H}}_r$, respectively, as functions of the coordinate z (plots are given for a lossless incident medium).

CONCEPTUAL QUESTION 8.23 **Reflection coefficient, interface between two perfect dielectrics.** A time-harmonic uniform plane wave is incident normally on a planar interface between two lossless dielectric media, as shown in Figure 8.10. The reflection coefficient, $\underline{\Gamma}$, for this case is

(A) purely real and positive.

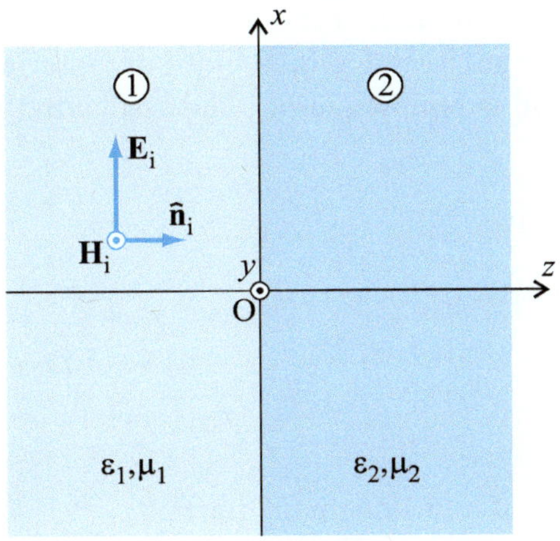

Figure 8.10 Normal incidence of a uniform plane time-harmonic electromagnetic wave from a lossless medium 1 on the planar surface of a penetrable lossless medium 2; for Conceptual Question 8.23.

(B) purely real and negative.

(C) zero.

(D) None of the above.

(E) Need more information.

CONCEPTUAL QUESTION 8.24 **Transmission coefficient, interface between perfect dielectrics.** The transmission coefficient, $\underline{\tau}$, for the normal incidence of a time-harmonic uniform plane wave at an interface between two perfect-dielectric media (Figure 8.10) is

(A) purely real and positive.

(B) purely real and negative.

(C) zero.

(D) None of the above.

(E) Need more information.

CONCEPTUAL QUESTION 8.25 **Reflection coefficient for air–good dielectric interface.** A uniform plane time-harmonic wave propagates in air and impinges normally the planar surface of a large block of material that can be considered to be a good dielectric and is nonmagnetic. The associated reflection coefficient ($\underline{\Gamma}$) is

(A) purely real and positive.

(B) purely real and negative.

(C) purely imaginary.

(D) zero.

(E) with equal nonzero real and imaginary parts.

(F) Need more information.

CONCEPTUAL QUESTION 8.26 **Transmission coefficient for air–good conductor interface.** The transmission coefficient ($\underline{\tau}$) for a uniform plane time-harmonic wave normally incident on an interface between air and a good conductor, at frequencies up to the visible-light region, turns out to be

(A) purely real and positive.

(B) purely real and negative.

(C) purely imaginary.

(D) zero.

(E) with equal nonzero real and imaginary parts.

(F) Need more information.

CONCEPTUAL QUESTION 8.27 **Transmission coefficient for air–PEC interface.** The coefficient $\underline{\tau}$ for a perfect electric conductor as the reflecting medium, and air as the incident medium, is

(A) purely real and positive.

(B) purely real and negative.

(C) purely imaginary.

(D) zero.

(E) with equal nonzero real and imaginary parts.

(F) Need more information.

CONCEPTUAL QUESTION 8.28 **Reflection coefficient for a PEC boundary.** The magnitude, $|\underline{\Gamma}|$, and phase angle, ψ, of the reflection coefficient ($\underline{\Gamma} = |\underline{\Gamma}|\, e^{j\psi}$) for an interface between a dielectric and a perfect conductor amount to

(A) $|\Gamma| = 1$ and $\psi = 0$.

(B) $|\Gamma| = -1$ and $\psi = \pi$.

(C) $|\Gamma| = 1$ and $\psi = 180°$.

(D) $|\Gamma| = 0$ and $\psi = 0$.

(E) $|\Gamma| \to \infty$ and ψ is not defined.

CONCEPTUAL QUESTION 8.29 **Air–glass and glass–air transmission.** If the reflection and transmission coefficients for the normal incidence of a plane wave from air (denoted as medium 1) into glass (medium 2), as shown in Figure 8.11(a), are $\underline{\Gamma}_{12}$ and $\underline{\tau}_{12}$, respectively, we have the following for the corresponding coefficients for the normal incidence from glass into air [Figure 8.11(b)], $\underline{\Gamma}_{21}$ and $\underline{\tau}_{21}$:

(A) $\underline{\Gamma}_{21} = -\underline{\Gamma}_{12}$ and $\underline{\tau}_{21} = \underline{\tau}_{12}$.

(B) $\underline{\Gamma}_{21} = \underline{\Gamma}_{12}$ and $\underline{\tau}_{21} = -\underline{\tau}_{12}$.

(C) $\underline{\Gamma}_{21} = -\underline{\Gamma}_{12}$ and $|\underline{\tau}_{21}| \neq |\underline{\tau}_{12}|$.

(D) $\underline{\Gamma}_{21} = \underline{\Gamma}_{12}$ and $|\underline{\tau}_{21}| \neq |\underline{\tau}_{12}|$.

(E) None of the above combinations.

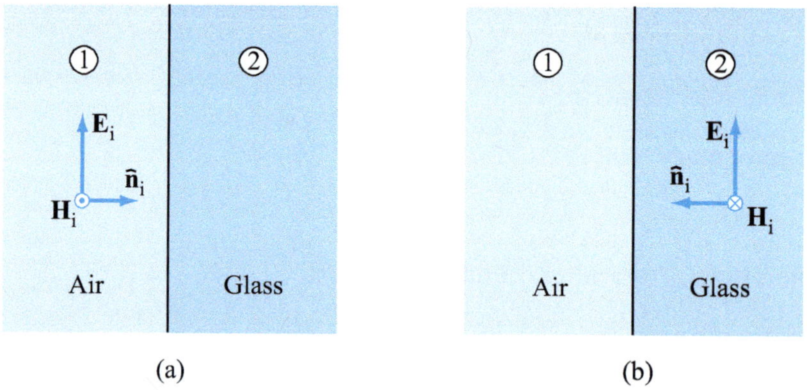

Figure 8.11 Normal incidence of a uniform plane time-harmonic electromagnetic wave (a) from air into glass and (b) from glass into air; for Conceptual Question 8.29.

CONCEPTUAL QUESTION 8.30 **Air–PEC and PEC–air transmission.** For air as medium 1 and a perfect electric conductor as medium 2, the reflection and transmission coefficients $\underline{\Gamma}_{21}$ and $\underline{\tau}_{21}$ (for the normal incidence from the PEC into air) can be related to the corresponding coefficients $\underline{\Gamma}_{12}$ and $\underline{\tau}_{12}$ as

(A) $\underline{\Gamma}_{21} = -\underline{\Gamma}_{12}$ and $\underline{\tau}_{21} = \underline{\tau}_{12}$.

(B) $\underline{\Gamma}_{21} = \underline{\Gamma}_{12}$ and $\underline{\tau}_{21} = -\underline{\tau}_{12}$.

(C) $\underline{\Gamma}_{21} = -\underline{\Gamma}_{12}$ and $|\underline{\tau}_{21}| \neq |\underline{\tau}_{12}|$.

(D) $\underline{\Gamma}_{21} = \underline{\Gamma}_{12}$ and $|\underline{\tau}_{21}| \neq |\underline{\tau}_{12}|$.

(E) None of the above combinations.

CONCEPTUAL QUESTION 8.31 **Dielectric interface with two combinations of permittivities.** A time-harmonic uniform plane wave is incident normally on a planar interface between two media that are nonmagnetic ($\mu_1 = \mu_2 = \mu_0$) in addition to being lossless ($\sigma_1 = \sigma_2 = 0$). Shown in Figure 8.12 is a time snapshot (at $t = 0$) of the instantaneous electric field intensities of incident, reflected, and

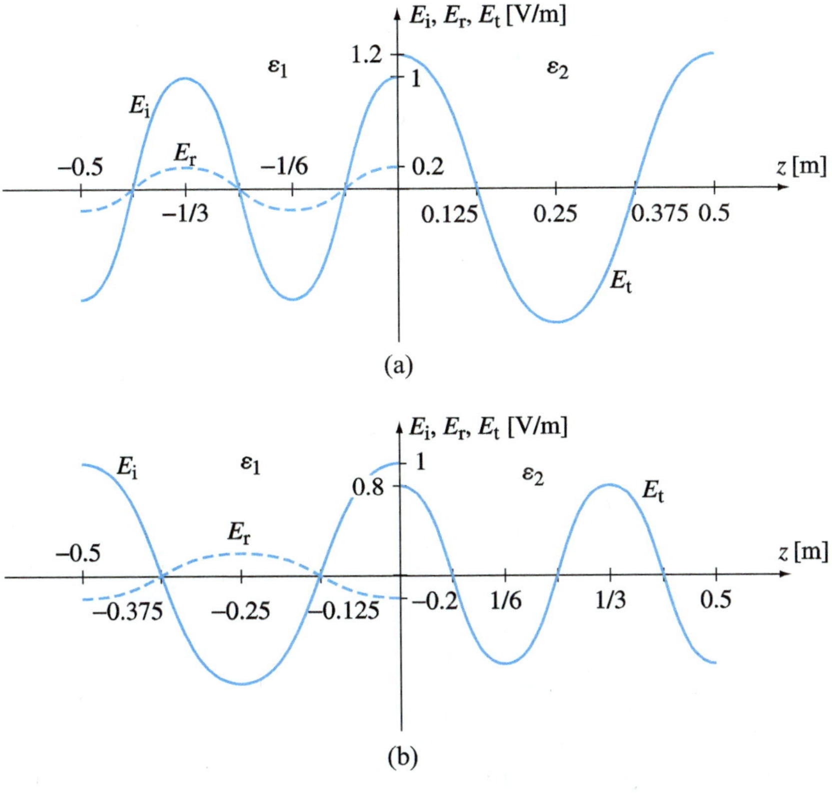

Figure 8.12 Plots of instantaneous electric field intensities of incident, reflected, and transmitted waves in a two-media structure with normal incidence of a uniform plane time-harmonic electromagnetic wave from medium 1 into medium 2 against z at $t = 0$ for $\mu_1 = \mu_2 = \mu_0$, $\sigma_1 = \sigma_2 = 0$, and two combinations of values of permittivities of the two media; for Conceptual Question 8.31.

transmitted waves (for the same reference direction of vectors $\mathbf{E}_\mathrm{i}$, $\mathbf{E}_\mathrm{r}$, and $\mathbf{E}_\mathrm{t}$) as a function of z, for two combinations of values of permittivities of the two media. The permittivities ε_1 and ε_2 for the situations in Figures 8.12(a) and (b), respectively, are related as follows:

(A) $\varepsilon_1 > \varepsilon_2$ in Figure 8.12(a) and $\varepsilon_1 < \varepsilon_2$ in Figure 8.12(b).

(B) $\varepsilon_1 < \varepsilon_2$ in Figure 8.12(a) and $\varepsilon_1 > \varepsilon_2$ in Figure 8.12(b).

(C) $\varepsilon_1 > \varepsilon_2$ in both Figures 8.12(a) and (b).

(D) $\varepsilon_1 < \varepsilon_2$ in both Figures 8.12(a) and (b).

(E) $\varepsilon_1 = \varepsilon_2$ in both Figures 8.12(a) and (b).

CONCEPTUAL QUESTION 8.32 **Frequencies of waves in two media.** For the two-media situation in Figure 8.12(a), the ratio of the frequency of the incident and reflected waves in medium 1, f_1, to the frequency of the transmitted wave in medium 2, f_2, is given by

(A) $f_1/f_2 = 1/2$.

(B) $f_1/f_2 = 2/3$.

(C) $f_1/f_2 = 1$.

(D) $f_1/f_2 = 3/2$.

(E) $f_1/f_2 = 2$.

(F) None of the above.

CONCEPTUAL QUESTION 8.33 **Reflection and transmission coefficients from a field picture.** For the field picture and combination of permittivities of the two media in Figure 8.13, the reflection and transmission coefficients, $\underline{\Gamma}$ and $\underline{\tau}$, are purely real and amount to

(A) $\Gamma = 0.2$ and $\tau = 0.8$.

(B) $\Gamma = 0.8$ and $\tau = 0.2$.

(C) $\Gamma = 0.2$ and $\tau = 1.2$.

(D) $\Gamma = -0.2$ and $\tau = 1.2$.

(E) $\Gamma = -0.2$ and $\tau = 0.8$.

(F) None of the above.

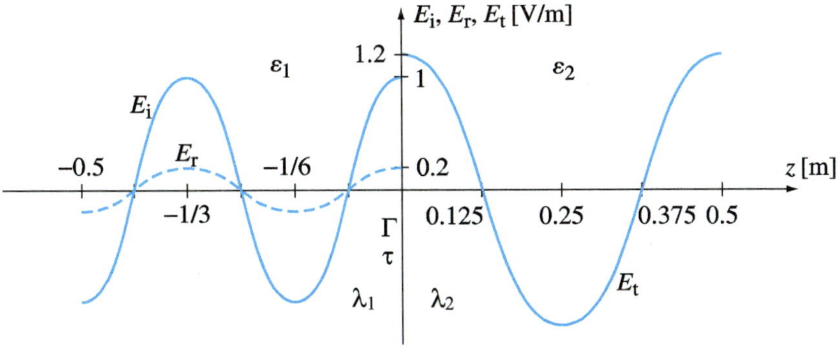

Figure 8.13 Time snapshot of the spatial distribution of instantaneous electric field intensities of incident, reflected, and transmitted waves for a combination of permittivities ε_1 and ε_2 of two nonmagnetic lossless media; for Conceptual Question 8.33.

CONCEPTUAL QUESTION 8.34 Wavelengths from waveforms in two media. For the waveforms in Figure 8.13, the ratio of the wavelength in medium 1, λ_1, to that in medium 2, λ_2, equals

(A) $\lambda_1/\lambda_2 = 1/2$.

(B) $\lambda_1/\lambda_2 = 2/3$.

(C) $\lambda_1/\lambda_2 = 1$.

(D) $\lambda_1/\lambda_2 = 3/2$.

(E) $\lambda_1/\lambda_2 = 2$.

(F) None of the above.

CONCEPTUAL QUESTION 8.35 Γ and τ from a different field picture. For the field picture and combination of media permittivities in Figure 8.14, we have

(A) $\Gamma = 0.2$ and $\tau = 0.8$.

(B) $\Gamma = 0.8$ and $\tau = 0.2$.

(C) $\Gamma = 0.2$ and $\tau = 1.2$.

(D) $\Gamma = -0.2$ and $\tau = 1.2$.

(E) $\Gamma = -0.2$ and $\tau = 0.8$.

(F) None of the above.

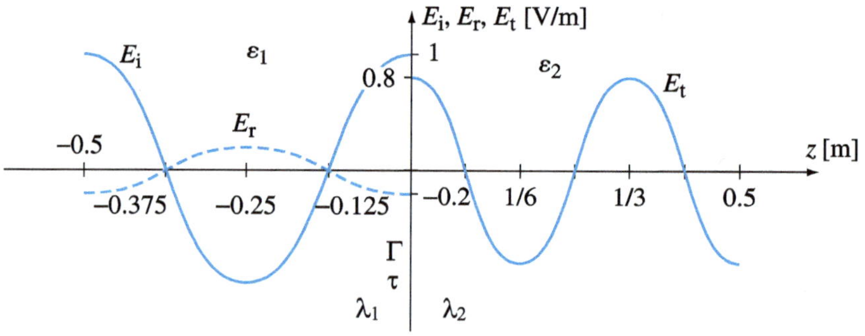

Figure 8.14 Electric field plots (at $t = 0$) for incident, reflected, and transmitted waves in a two-media structure for another combination of ε_1 and ε_2 ($\mu_1 = \mu_2 = \mu_0$, $\sigma_1 = \sigma_2 = 0$); for Conceptual Question 8.35.

CONCEPTUAL QUESTION 8.36 **Another evaluation of wavelengths.** From the waveforms in Figure 8.14, one can obtain that

(A) $\lambda_1/\lambda_2 = 1/2$.

(B) $\lambda_1/\lambda_2 = 2/3$.

(C) $\lambda_1/\lambda_2 = 1$.

(D) $\lambda_1/\lambda_2 = 3/2$.

(E) $\lambda_1/\lambda_2 = 2$.

(F) None of the above.

CONCEPTUAL QUESTION 8.37 **Relationship between reflection and transmission coefficients.** For an interface between two arbitrary lossless or lossy material media, the following relationship between the reflection and transmission coefficients, $\underline{\Gamma}$ and $\underline{\tau}$, holds true:

(A) $1 + \underline{\Gamma} = \underline{\tau}$.

(B) $\underline{\Gamma} + \underline{\tau} = 1$.

(C) $\underline{\Gamma} + \underline{\tau} = 2$.

(D) $\underline{\tau} = -\underline{\Gamma}$.

(E) $\underline{\Gamma} = 1 + \underline{\tau}$.

CONCEPTUAL QUESTION 8.38 **Reflected and transmitted powers.** A uniform plane time-harmonic electromagnetic wave propagates in a lossless medium and impinges normally the planar interface with another lossless medium, with different material parameters. If the associated reflection and transmission coefficients are $\underline{\Gamma}$ and $\underline{\tau}$, respectively, consider the following statements: (a) the percentage of the time-average incident power that is reflected from the interface equals $|\underline{\Gamma}|^2 \times 100$ (%); (b) the percentage of the incident power that is transmitted into the second medium amounts to $|\underline{\tau}|^2 \times 100$ (%). Which of the statements is true?

(A) Statement (a) only.

(B) Statement (b) only.

(C) Both statements.

(D) Neither of the statements.

CONCEPTUAL QUESTION 8.39 **Limits for the magnitude of the reflection coefficient.** Consider a boundary surface between two arbitrary material media. The range of possible values of the magnitude of the reflection coefficient, $|\underline{\Gamma}|$, is as follows:

(A) $0 \le |\underline{\Gamma}| \le 1$.

(B) $-1 \le |\underline{\Gamma}| \le 1$.

(C) $0 \le |\underline{\Gamma}| \le 2$.

(D) $1 \le |\underline{\Gamma}| < \infty$.

(E) $0 \le |\underline{\Gamma}| < \infty$.

CONCEPTUAL QUESTION 8.40 **Limits for the reflection coefficient in decibels.** The range of possible values of the reflection coefficient in decibels, $\Gamma_{\rm dB} = 20\log|\underline{\Gamma}|$ [$\log x \equiv \log_{10} x$ (common or decadic logarithm)], is given by

(A) $0\ \text{dB} \le \Gamma_{\rm dB} \le 1\ \text{dB}$.

(B) $1\ \text{dB} \le \Gamma_{\rm dB} < \infty$.

(C) $0\ \text{dB} \le \Gamma_{\rm dB} < \infty$.

(D) $-\infty < \Gamma_{\rm dB} \le 0\ \text{dB}$.

(E) $-\infty < \Gamma_{\rm dB} < \infty$.

CONCEPTUAL QUESTION 8.41 **Field maxima and minima in front of a dielectric interface.** Consider the normal incidence of a uniform plane time-harmonic electromagnetic wave at a planar interface between two different lossless dielectric media, as well as the maxima and minima of the rms electric and magnetic field intensities of the resultant electromagnetic wave in the incident medium. The magnetic field maxima occur

(A) at distinct locations of electric field maxima.

(B) at distinct locations of electric field minima.

(C) at distinct locations coinciding with neither electric field maxima nor minima.

(D) everywhere (rms field intensity is constant).

CONCEPTUAL QUESTION 8.42 **Finding the wave frequency from locations of field maxima.** A time-harmonic plane wave is launched to propagate through air and impinge normally upon and partially reflects from the surface of a dielectric material. The relative rms electric field intensities of the resultant wave are measured by an electric probe in the region in front of the material, along the incident wave propagation. By such measurement, it is found that the distance between successive field maxima is 0.5 m. What is the frequency of the wave?

(A) $f = 75$ MHz.

(B) $f = 150$ MHz.

(C) $f = 300$ MHz.

(D) $f = 600$ MHz.

(E) Need more information.

CONCEPTUAL QUESTION 8.43 Electric-field maximum at a boundary surface. A time-harmonic uniform plane wave in a perfect dielectric with intrinsic impedance of 100 Ω partially reflects, at normal incidence, from the surface of a material whose properties are unknown. It was found that the resultant wave in the incident region exhibits standing wave properties and that an electric-field maximum is located at the boundary surface. What can be concluded about the unknown material?

(A) The intrinsic impedance of the material is purely real and greater than 100 Ω.

(B) The unknown intrinsic impedance is purely real, nonzero, and less than 100 Ω.

(C) The intrinsic impedance of the material equals 100 Ω.

(D) The material is lossy.

(E) The material has the same parameters as a vacuum.

(F) The intrinsic impedance of the material is zero.

CONCEPTUAL QUESTION 8.44 Electric-field minimum at a boundary surface. A planar interface of a material with unknown properties is impinged by a normally incident time-harmonic uniform plane wave. Observing the resultant standing wave in the incident region, filled with a perfect dielectric with $\eta_1 = 100\ \Omega$, an electric-field minimum, which is not zero, is identified at the interface. From this, we can conclude – about the unknown material – that

(A) the intrinsic impedance of the material is purely real and greater than 100 Ω.

(B) the unknown intrinsic impedance is purely real, nonzero, and less than 100 Ω.

(C) the intrinsic impedance of the material equals 100 Ω.

(D) the material is lossy.

(E) the material has the same parameters as a vacuum.

(F) the intrinsic impedance of the material is zero.

CONCEPTUAL QUESTION 8.45 **Electric-field zero at a boundary surface.** If the resultant standing wave has a zero of the electric field at a boundary surface between a perfect dielectric with $\eta_1 = 100\ \Omega$ (incident medium) and a material with unknown parameters, illuminated by a time-harmonic uniform plane wave at normal incidence, what can be concluded about the unknown material?

(A) The intrinsic impedance of the material is purely real and greater than 100 Ω.

(B) The unknown intrinsic impedance is purely real, nonzero, and less than 100 Ω.

(C) The intrinsic impedance of the material equals 100 Ω.

(D) The material is lossy.

(E) The material has the same parameters as a vacuum.

(F) The intrinsic impedance of the material is zero.

CONCEPTUAL QUESTION 8.46 **RHCP incidence, polarization state of the reflected wave.** A right-hand circularly polarized (RHCP) uniform plane wave is incident normally from air on a dielectric half-space with parameters $\varepsilon_r = 4$, $\mu_r = 1$, and $\sigma = 0$. The reflected wave is

(A) linearly polarized (LP).

(B) RHCP.

(C) left-hand circularly polarized (LHCP).

(D) right-hand elliptically polarized (RHEP).

(E) left-hand elliptically polarized (LHEP).

(F) unpolarized.

CONCEPTUAL QUESTION 8.47 **Polarization state of the transmitted wave.** For the normal incidence of an RHCP uniform plane wave from air on an interface of a dielectric with $\varepsilon_r = 4$, $\mu_r = 1$, and $\sigma = 0$, the transmitted wave is

(A) LP.

(B) RHCP.

(C) LHCP.

(D) RHEP.

(E) LHEP.

(F) unpolarized.

CONCEPTUAL QUESTION 8.48 **Switching places of media, polarization of the reflected wave.** Assuming that an RHCP uniform plane wave is normally incident from a perfect-dielectric ($\varepsilon_{\rm r} = 4$, $\mu_{\rm r} = 1$, $\sigma = 0$) half-space into air, determine the polarization state of the reflected wave. It is

(A) LP.

(B) RHCP.

(C) LHCP.

(D) RHEP.

(E) LHEP.

(F) unpolarized.

CONCEPTUAL QUESTION 8.49 **Polarization of the transmitted wave with media switched.** For the normal incidence of an RHCP uniform plane wave from a perfect dielectric ($\varepsilon_{\rm r} = 4$, $\mu_{\rm r} = 1$, $\sigma = 0$) on a planar boundary surface with air, the transmitted wave is

(A) LP.

(B) RHCP.

(C) LHCP.

(D) RHEP.

(E) LHEP.

(F) unpolarized.

8.3 Oblique Incidence on a Perfect Conductor

In this section, we generalize the analysis of plane-wave reflections upon perfectly conducting surfaces for the normal incidence on the surface [Figure 8.1(a)] to the case of an arbitrary, oblique, incidence. Namely, we now let an incident uniform plane time-harmonic electromagnetic wave approach the PEC boundary at an arbitrary angle, a so-called incident angle, θ_i ($0 \le \theta_i \le 90°$) with respect to the normal on the boundary (for the normal incidence, $\theta_i = 0$). Furthermore, let the incident electric field intensity vector, $\underline{\mathbf{E}}_i$, be normal to the plane of incidence, defined by the direction of incident-wave propagation (incident ray) and normal on the boundary, as shown in Figure 8.15(a). An obliquely incident wave with such an orientation of $\underline{\mathbf{E}}_i$ is said to be normally (or perpendicularly) polarized. The other characteristic case, with $\underline{\mathbf{E}}_i$ lying in the plane of incidence (and, of course, being perpendicular to the direction of wave travel), is depicted in Figure 8.15(b). It is referred to as the parallel polarization of the incident wave. Using the field expression for an arbitrarily directed plane wave in Eqs. (7.12), for both the incident and reflected waves in Figure 8.15, and applying, in the plane $z = 0$, the boundary condition for the vector $\underline{\mathbf{E}}$ in Eqs. (6.15), we obtain

$$\theta_r = \theta_i \,, \quad \underline{E}_{r0} = -\underline{E}_{i0} \quad \text{(oblique incidence on a PEC)} \,, \tag{8.12}$$

for both wave polarizations, where the first relationship is known as Snell's law of reflection. With these facts, it is then straightforward to complete the solutions for

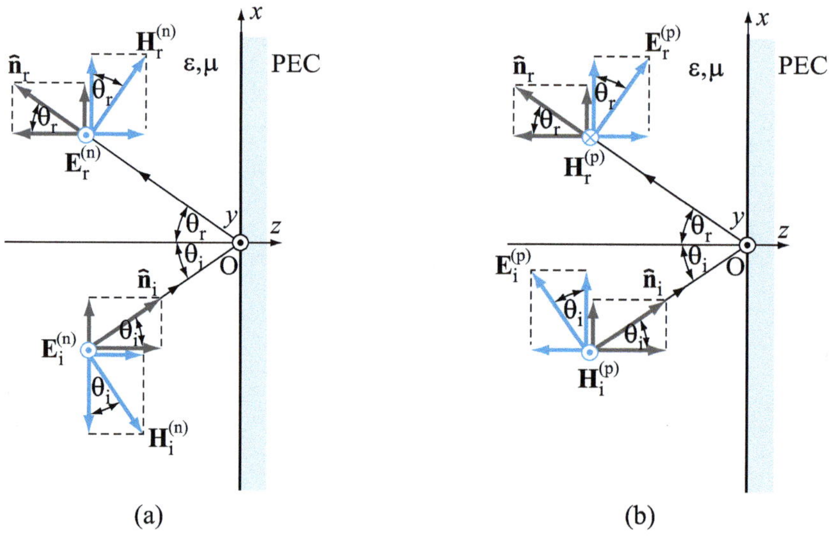

Figure 8.15 Oblique incidence, on a planar perfect electric conductor, of a uniform plane time-harmonic electromagnetic wave with (a) normal (perpendicular) polarization and (b) parallel polarization. The incident medium is a perfect dielectric, and the plane of drawing is also the plane of incidence.

fields $\underline{\mathbf{E}}_{\rm r}^{(\rm n)}$ and $\underline{\mathbf{H}}_{\rm r}^{(\rm n)}$ in Figure 8.15(a) and $\underline{\mathbf{E}}_{\rm r}^{(\rm p)}$ and $\underline{\mathbf{H}}_{\rm r}^{(\rm p)}$ in Figure 8.15(b), as well as for the total fields in the incident medium, $\underline{\mathbf{E}}_{\rm tot}^{(\rm n)} = \underline{\mathbf{E}}_{\rm i}^{(\rm n)} + \underline{\mathbf{E}}_{\rm r}^{(\rm n)}$, etc.

CONCEPTUAL QUESTION 8.50 **Obliquely incident wave with a normal-parallel polarization.** A uniform plane time-harmonic electromagnetic wave of frequency f and rms electric field intensity $E_{\rm i0}$ propagates in air and is incident obliquely, at an angle $\theta_{\rm i}$ ($0 < \theta_{\rm i} < 90°$), on a PEC screen, as shown in Figure 8.16. The electric field vector of the wave ($\mathbf{E}_{\rm i}$) makes an angle α ($0 < \alpha < 90°$) with the plane of incidence. The incident wave can be represented as a superposition of two uniform plane waves with normal (n) and parallel (p) polarizations, whose rms electric field intensities are given by

(A) $E_{\rm i0}^{(\rm n)} = E_{\rm i0} \cos\theta_{\rm i}$ and $E_{\rm i0}^{(\rm p)} = E_{\rm i0} \sin\theta_{\rm i}$,

(B) $E_{\rm i0}^{(\rm n)} = E_{\rm i0} \sin\theta_{\rm i}$ and $E_{\rm i0}^{(\rm p)} = E_{\rm i0} \cos\theta_{\rm i}$,

(C) $E_{\rm i0}^{(\rm n)} = E_{\rm i0} \cos\alpha$ and $E_{\rm i0}^{(\rm p)} = E_{\rm i0} \sin\alpha$,

(D) $E_{\rm i0}^{(\rm n)} = E_{\rm i0} \sin\alpha$ and $E_{\rm i0}^{(\rm p)} = E_{\rm i0} \cos\alpha$,

(E) $E_{\rm i0}^{(\rm n)} = E_{\rm i0} \cos\alpha$ and $E_{\rm i0}^{(\rm p)} = E_{\rm i0}$,

(F) $E_{\rm i0}^{(\rm n)} = E_{\rm i0}$ and $E_{\rm i0}^{(\rm p)} = E_{\rm i0} \sin\alpha$,

respectively.

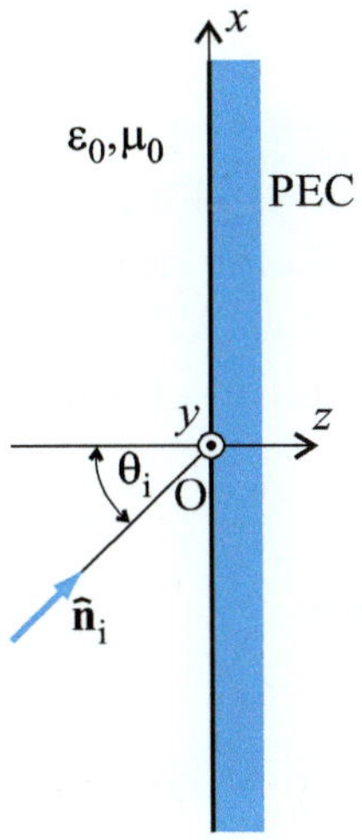

Figure 8.16 Oblique incidence of a uniform plane time-harmonic electromagnetic wave from air on a PEC screen; for Conceptual Question 8.50.

CONCEPTUAL QUESTION 8.51 Resultant wave with a combined polarization. Consider an obliquely incident – on a PEC screen from air – plane wave (Figure 8.16) with a combined normal-parallel polarization, which can be represented as a superposition of two plane waves with normal and parallel polarizations, whose rms electric field intensities are denoted as $E_{i0}^{(n)}$ and $E_{i0}^{(p)}$, respectively. The total electric field vector in the incident region can be obtained by adding together the corresponding total-field expression for the normally polarized incident wave, with complex rms electric field intensity $\underline{E}_{i0}$ in the expression substituted by $E_{i0}^{(n)}$, and that for the parallel polarization case, with $E_{i0}^{(p)}$ in place of $\underline{E}_{i0}$.

(A) True.

(B) False.

CONCEPTUAL QUESTION 8.52 Normal incidence, normal and parallel polarizations. The classification of incident waves into those with normal and parallel polarizations, respectively, does apply to normally incident waves as well.

(A) True.

(B) False.

CONCEPTUAL QUESTION 8.53 Incident electric field vector for normal polarization. A normally polarized uniform plane time-harmonic electromagnetic wave with phase coefficient β and rms electric field intensity E_{i0} is incident from air at an angle θ_i on a PEC plane. With reference to the Cartesian coordinate system in Figure 8.17, the electric field vector of the incident wave, $\underline{\mathbf{E}}_i = \underline{\mathbf{E}}_i^{(n)}$, at an

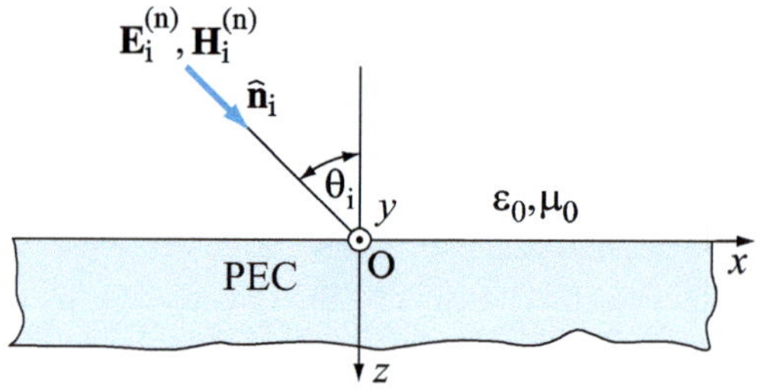

Figure 8.17 Oblique incidence of a normally polarized uniform plane time-harmonic electromagnetic wave from air on a PEC interface; for Conceptual Question 8.53.

arbitrary point in the incident medium, defined by coordinates x, y, and z ($z \leq 0$), is given by

(A) $\underline{\mathbf{E}}_{\rm i} = E_{\rm i0}\, {\rm e}^{-{\rm j}\beta(x\sin\theta_{\rm i}+z\cos\theta_{\rm i})}\,\hat{\mathbf{y}}$.

(B) $\underline{\mathbf{E}}_{\rm i} = E_{\rm i0}\, {\rm e}^{-{\rm j}\beta(x\sin\theta_{\rm i}-z\cos\theta_{\rm i})}\,\hat{\mathbf{y}}$.

(C) $\underline{\mathbf{E}}_{\rm i} = E_{\rm i0}\, {\rm e}^{-{\rm j}\beta y}(\sin\theta_{\rm i}\,\hat{\mathbf{x}} + \cos\theta_{\rm i}\,\hat{\mathbf{z}})$.

(D) $\underline{\mathbf{E}}_{\rm i} = E_{\rm i0}(\cos\theta_{\rm i}\,\hat{\mathbf{x}} + \sin\theta_{\rm i}\,\hat{\mathbf{z}})\,{\rm e}^{-{\rm j}\beta y}$.

(E) $\underline{\mathbf{E}}_{\rm i} = E_{\rm i0}(\cos\theta_{\rm i}\,\hat{\mathbf{x}} - \sin\theta_{\rm i}\,\hat{\mathbf{z}})\,{\rm e}^{-{\rm j}\beta(x\sin\theta_{\rm i}+z\cos\theta_{\rm i})}$.

(F) $\underline{\mathbf{E}}_{\rm i} = E_{\rm i0}(\cos\theta_{\rm i}\,\hat{\mathbf{x}} + \sin\theta_{\rm i}\,\hat{\mathbf{z}})\,{\rm e}^{-{\rm j}\beta(x\sin\theta_{\rm i}-z\cos\theta_{\rm i})}$.

CONCEPTUAL QUESTION 8.54 Incident electric field vector for parallel polarization. For parallel polarization of a plane wave incident from air at an angle $\theta_{\rm i}$ on a PEC plane (Figure 8.18), so for $\underline{\mathbf{E}}_{\rm i} = \underline{\mathbf{E}}_{\rm i}^{(\rm p)}$, we have that

(A) $\underline{\mathbf{E}}_{\rm i} = E_{\rm i0}\, {\rm e}^{-{\rm j}\beta(x\sin\theta_{\rm i}+z\cos\theta_{\rm i})}\,\hat{\mathbf{y}}$.

(B) $\underline{\mathbf{E}}_{\rm i} = E_{\rm i0}\, {\rm e}^{-{\rm j}\beta(x\sin\theta_{\rm i}-z\cos\theta_{\rm i})}\,\hat{\mathbf{y}}$.

(C) $\underline{\mathbf{E}}_{\rm i} = E_{\rm i0}\, {\rm e}^{-{\rm j}\beta y}(\sin\theta_{\rm i}\,\hat{\mathbf{x}} + \cos\theta_{\rm i}\,\hat{\mathbf{z}})$.

(D) $\underline{\mathbf{E}}_{\rm i} = E_{\rm i0}(\cos\theta_{\rm i}\,\hat{\mathbf{x}} + \sin\theta_{\rm i}\,\hat{\mathbf{z}})\,{\rm e}^{-{\rm j}\beta y}$.

(E) $\underline{\mathbf{E}}_{\rm i} = E_{\rm i0}(\cos\theta_{\rm i}\,\hat{\mathbf{x}} - \sin\theta_{\rm i}\,\hat{\mathbf{z}})\,{\rm e}^{-{\rm j}\beta(x\sin\theta_{\rm i}+z\cos\theta_{\rm i})}$.

(F) $\underline{\mathbf{E}}_{\rm i} = E_{\rm i0}(\cos\theta_{\rm i}\,\hat{\mathbf{x}} + \sin\theta_{\rm i}\,\hat{\mathbf{z}})\,{\rm e}^{-{\rm j}\beta(x\sin\theta_{\rm i}-z\cos\theta_{\rm i})}$.

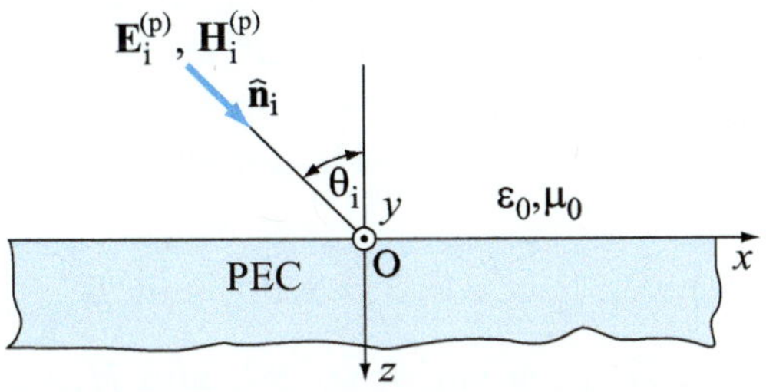

Figure 8.18 Uniform plane time-harmonic electromagnetic wave impinging a PEC plane at an oblique incidence with parallel polarization; for Conceptual Question 8.54.

CONCEPTUAL QUESTION 8.55 Field components of a wave with parallel polarization. For the obliquely incident plane wave with parallel polarization in Figure 8.18, the following is a complete list of nonzero electric and magnetic field components of the resultant wave in the incident medium:

(A) $\underline{E}_{\text{tot}\,x}^{(\text{p})}$ and $\underline{H}_{\text{tot}\,y}^{(\text{p})}$.

(B) $\underline{E}_{\text{tot}\,y}^{(\text{p})}$ and $\underline{H}_{\text{tot}\,x}^{(\text{p})}$.

(C) $\underline{E}_{\text{tot}\,x}^{(\text{p})}$, $\underline{E}_{\text{tot}\,z}^{(\text{p})}$, $\underline{H}_{\text{tot}\,y}^{(\text{p})}$, and $H_{\text{tot}\,z}^{(\text{p})}$.

(D) $\underline{E}_{\text{tot}\,y}^{(\text{p})}$, $\underline{H}_{\text{tot}\,x}^{(\text{p})}$, and $\underline{H}_{\text{tot}\,z}^{(\text{p})}$.

(E) $\underline{E}_{\text{tot}\,x}^{(\text{p})}$, $\underline{E}_{\text{tot}\,z}^{(\text{p})}$, and $\underline{H}_{\text{tot}\,y}^{(\text{p})}$.

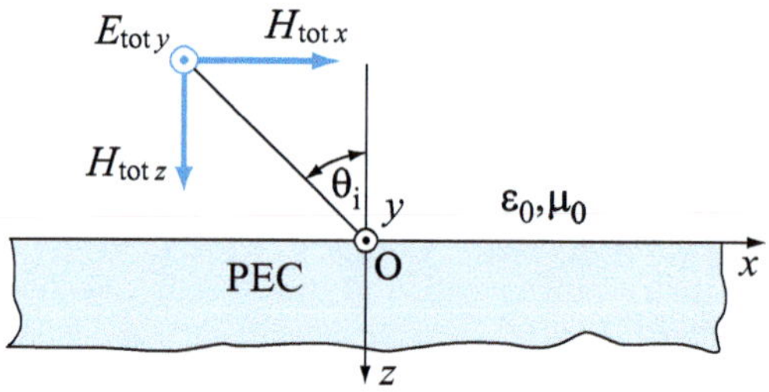

Figure 8.19 Oblique incidence of a uniform plane time-harmonic electromagnetic wave upon a PEC plane: y-component of the total electric field vector and x- and z-components of the total magnetic field vector in the incident medium; for Conceptual Question 8.56.

CONCEPTUAL QUESTION 8.56 Spatial dependence of total field components. Consider a wave incidence at an angle θ_{i} onto a PEC plane, Figure 8.19. The y-component of the total electric field vector and x- and z-components of the total magnetic field vector in the incident medium (for $z \leq 0$) for a certain polarization (normal or parallel) of the wave are proportional to the following functions of the coordinate z (note that the fields are also dependent on the coordinate x):

(A) $\underline{E}_{\text{tot}\,y} \propto \cos(\beta z \cos\theta_{\text{i}})$, $\underline{H}_{\text{tot}\,x} \propto \cos(\beta z \cos\theta_{\text{i}})$, and $\underline{H}_{\text{tot}\,z} \propto \cos(\beta z \cos\theta_{\text{i}})$.

(B) $\underline{E}_{\text{tot}\,y} \propto \cos(\beta z \cos\theta_{\text{i}})$, $\underline{H}_{\text{tot}\,x} \propto \sin(\beta z \cos\theta_{\text{i}})$, and $\underline{H}_{\text{tot}\,z} \propto \sin(\beta z \cos\theta_{\text{i}})$.

(C) $\underline{E}_{\text{tot}\,y} \propto \sin(\beta z \cos\theta_{\text{i}})$, $\underline{H}_{\text{tot}\,x} \propto \cos(\beta z \cos\theta_{\text{i}})$, and $\underline{H}_{\text{tot}\,z} \propto \sin(\beta z \cos\theta_{\text{i}})$.

(D) $\underline{E}_{\text{tot}\,y} \propto \sin(\beta z \cos\theta_{\text{i}})$, $\underline{H}_{\text{tot}\,x} \propto \cos(\beta z \cos\theta_{\text{i}})$, and $\underline{H}_{\text{tot}\,z} \propto \cos(\beta z \cos\theta_{\text{i}})$.

(E) $\underline{E}_{\text{tot}\,y} \propto \cos(\beta z \cos\theta_{\text{i}})$, $\underline{H}_{\text{tot}\,x} \propto \sin(\beta z \cos\theta_{\text{i}})$, and $\underline{H}_{\text{tot}\,z} \propto \cos(\beta z \cos\theta_{\text{i}})$.

CONCEPTUAL QUESTION 8.57 **Induced surface currents in a PEC screen, normal polarization.** A normally polarized uniform plane time-harmonic electromagnetic wave is incident obliquely from a lossless medium of permittivity ε and permeability μ on a flat horizontal screen made from a perfect electric conductor, as shown in Figure 8.20. Using the respective components of the total magnetic field vector in the incident medium, $\underline{\mathbf{H}}_{\text{tot}}$, given in a Cartesian coordinate system adopted as in Figure 8.20, the complex rms surface current density vector, $\underline{\mathbf{J}}_{\text{s}}$, describing the distribution of surface currents induced in the screen can be found as follows:

(A) $\underline{\mathbf{J}}_{\text{s}} = \underline{H}_{\text{tot}\,x}\big|_{z=0}\,\hat{\mathbf{x}}$.

(B) $\underline{\mathbf{J}}_{\text{s}} = \underline{H}_{\text{tot}\,y}\Big|_{z=0}\,\hat{\mathbf{y}}$.

(C) $\underline{\mathbf{J}}_{\text{s}} = \underline{H}_{\text{tot}\,y}\Big|_{z=0}\,\hat{\mathbf{x}}$.

(D) $\underline{\mathbf{J}}_{\text{s}} = \underline{H}_{\text{tot}\,x}\big|_{z=0}\,\hat{\mathbf{y}}$.

(E) $\underline{\mathbf{J}}_{\text{s}} = \underline{H}_{\text{tot}\,x}\big|_{z=0}\,(-\,\hat{\mathbf{y}})$.

(F) $\underline{\mathbf{J}}_{\text{s}} = \underline{H}_{\text{tot}\,z}\big|_{z=0}\,(-\,\hat{\mathbf{y}})$.

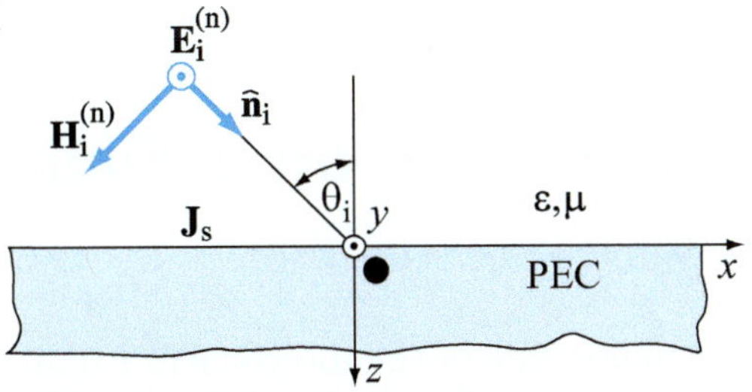

Figure 8.20 Surface currents induced on a PEC plane by a normally polarized obliquely incident uniform plane time-harmonic electromagnetic wave; for Conceptual Question 8.57.

CONCEPTUAL QUESTION 8.58 **Induced surface currents, parallel polarization.** For parallel polarization of a uniform plane wave incident obliquely from a lossless medium on a PEC plane (Figure 8.21), the density of induced surface currents on the plane, $\underline{\mathbf{J}}_{\text{s}}$, can be found – using the total magnetic field vector in the incident medium, $\underline{\mathbf{H}}_{\text{tot}}$ – as

(A) $\underline{\mathbf{J}}_{\text{s}} = \underline{H}_{\text{tot}\,x}\big|_{z=0}\,\hat{\mathbf{x}}$.

(B) $\underline{\mathbf{J}}_{\text{s}} = \underline{H}_{\text{tot}\,y}\Big|_{z=0}\,\hat{\mathbf{y}}$.

(C) $\underline{\mathbf{J}}_{\mathrm{s}} = \underline{H}_{\mathrm{tot}\,y}\big|_{z=0}\ \hat{\mathbf{x}}$.

(D) $\underline{\mathbf{J}}_{\mathrm{s}} = \underline{H}_{\mathrm{tot}\,x}|_{z=0}\ \hat{\mathbf{y}}$.

(E) $\underline{\mathbf{J}}_{\mathrm{s}} = \underline{H}_{\mathrm{tot}\,x}|_{z=0}\,(-\,\hat{\mathbf{y}})$.

(F) $\underline{\mathbf{J}}_{\mathrm{s}} = \underline{H}_{\mathrm{tot}\,z}|_{z=0}\,(-\,\hat{\mathbf{y}})$.

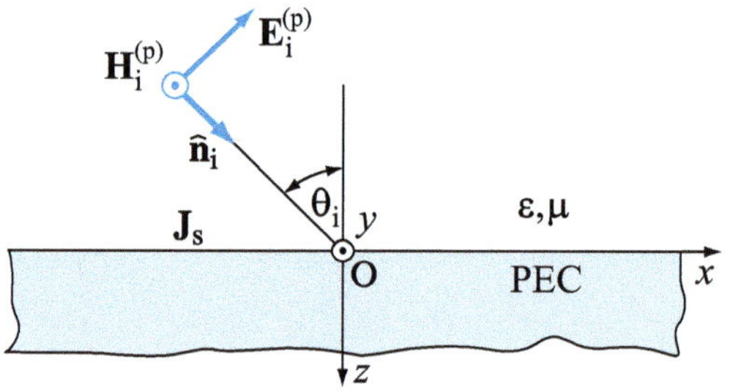

Figure 8.21 Surface currents induced on a PEC plane by an obliquely incident uniform plane wave with parallel polarization; for Conceptual Question 8.58.

8.4 Oblique Incidence on a Dielectric Boundary

If the medium on the right-hand side of the interface in Figure 8.15 is another lossless dielectric, a part of the incident energy will be transmitted through the interface, as in Figure 8.9 for the normal incidence case. In Figure 8.22, $\theta_{\mathrm{r}} = \theta_{\mathrm{i}}$, as in Eqs. (8.12), while the transmitted (refracted) angle, θ_{t}, is determined by Snell's law of refraction:

$$\frac{\sin\theta_{\mathrm{i}}}{\sin\theta_{\mathrm{t}}} = \frac{c_1}{c_2} = \frac{\beta_2}{\beta_1} = \sqrt{\frac{\varepsilon_2\mu_2}{\varepsilon_1\mu_1}} = \frac{n_2}{n_1} \quad \text{(Snell's law of refraction)}\,, \tag{8.13}$$

with c_1 and c_2 being the velocities [Eq. (7.3)] and β_1 and β_2 the phase coefficients [Eq. (7.9)] of waves in media 1 and 2, respectively, and $n_1 = c_0/c_1 = \sqrt{\varepsilon_{1\mathrm{r}}\mu_{1\mathrm{r}}}$ and $n_2 = c_0/c_2 = \sqrt{\varepsilon_{2\mathrm{r}}\mu_{2\mathrm{r}}}$ standing for the indices of refraction of the two media.

To find the unknown complex rms electric field intensities of the reflected and transmitted waves in the plane $z = 0$, $\underline{E}_{\mathrm{r0}}$ and $\underline{E}_{\mathrm{t0}}$, for a given intensity $\underline{E}_{\mathrm{i0}}$ at $z = 0$ of the incident wave, we apply the boundary conditions for tangential components of both electric and magnetic field vectors, as in Eqs. (8.9). Here, however, we need to distinguish between the normal and parallel polarizations of the incident

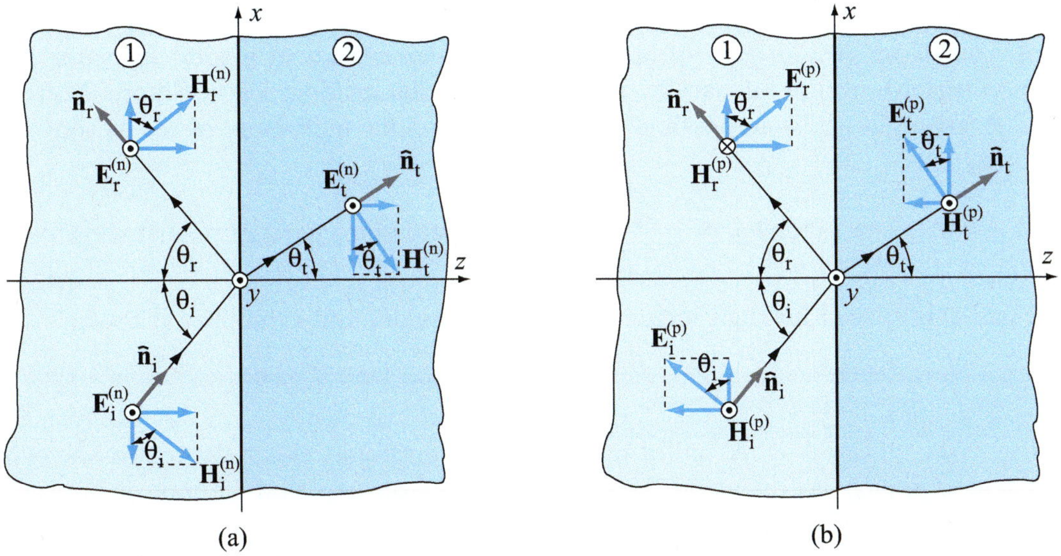

Figure 8.22 Reflection and refraction of an obliquely incident uniform plane time-harmonic electromagnetic wave at a planar dielectric boundary: electric and magnetic field vectors of the incident, reflected, and transmitted waves for normal (a) and parallel (b) polarizations of waves.

wave, which are depicted, respectively, in Figure 8.22(a) and (b), and in place of Eqs. (8.10) and (8.11) we obtain

$$\Gamma_{\rm n} = \left(\frac{\underline{E}_{\rm r0}}{\underline{E}_{\rm i0}}\right)_{\rm n} = \frac{\eta_2\cos\theta_{\rm i} - \eta_1\cos\theta_{\rm t}}{\eta_1\cos\theta_{\rm t} + \eta_2\cos\theta_{\rm i}}\,, \quad \tau_{\rm n} = \left(\frac{\underline{E}_{\rm t0}}{\underline{E}_{\rm i0}}\right)_{\rm n} = \frac{2\eta_2\cos\theta_{\rm i}}{\eta_1\cos\theta_{\rm t} + \eta_2\cos\theta_{\rm i}}$$

(Fresnel's coefficients for normal polarization) , (8.14)

$$\Gamma_{\rm p} = \left(\frac{\underline{E}_{\rm r0}}{\underline{E}_{\rm i0}}\right)_{\rm p} = \frac{\eta_2\cos\theta_{\rm t} - \eta_1\cos\theta_{\rm i}}{\eta_1\cos\theta_{\rm i} + \eta_2\cos\theta_{\rm t}}\,, \quad \tau_{\rm p} = \left(\frac{\underline{E}_{\rm t0}}{\underline{E}_{\rm i0}}\right)_{\rm p} = \frac{2\eta_2\cos\theta_{\rm i}}{\eta_1\cos\theta_{\rm i} + \eta_2\cos\theta_{\rm t}}$$

(Fresnel's coefficients for parallel polarization) . (8.15)

These coefficients are known as Fresnel's (reflection and transmission) coefficients.

If $\mu_1 = \mu_2$ and $\varepsilon_1 > \varepsilon_2$ in Figure 8.22 (for any polarization), there is an incident angle, $\theta_{\rm i}$, termed the critical angle and denoted as $\theta_{\rm ic}$, for which the transmitted angle, $\theta_{\rm t}$, acquires its maximum value, 90°. This means that there is no transmission into the second medium or that the incident wave is totally reflected. The total reflection condition remains satisfied for any incident angle exceeding the critical value, which is determined, from Eq. (8.13), as

$$\theta_{\rm t} = 90^\circ \quad \longrightarrow \quad \sin\theta_{\rm ic} = \sqrt{\frac{\varepsilon_2}{\varepsilon_1}} \quad (\varepsilon_1 > \varepsilon_2) \quad \longrightarrow \quad \theta_{\rm i} \geq \theta_{\rm ic} \quad \text{(total reflection)}\,. \tag{8.16}$$

On the other side, equating to zero the expression for Γ_p with $\mu_1 = \mu_2$ in Eqs. (8.15), we obtain that an opposite phenomenon, that of a total transmission (no reflection), occurs for parallel polarization of the incident wave, Figure 8.22(b), and θ_i given by (it is impossible to have $\Gamma_\text{n} = 0$ for any value of θ_i, as long as $\mu_1 = \mu_2$)

$$\Gamma_\text{p} = 0 \quad \longrightarrow \quad \tan\theta_\text{iB} = \sqrt{\frac{\varepsilon_2}{\varepsilon_1}} \quad \text{(total transmission, parallel polarization)} \, . \tag{8.17}$$

We call this special incident angle the Brewster angle, and symbolize it by θ_iB.

CONCEPTUAL QUESTION 8.59 **Oblique incidence on a dielectric–dielectric interface.** A uniform plane time-harmonic electromagnetic wave is incident obliquely at an angle θ_i ($0 < \theta_\text{i} < 90°$) on a planar boundary between two dielectric media, as depicted in Figure 8.23. Consider the reflected angle, θ_r, and transmitted (refracted) angle, θ_t, for this case. Which of them depends on the material parameters of the two media?

(A) θ_r only.

(B) θ_t only.

(C) Both angles.

(D) Neither of the angles.

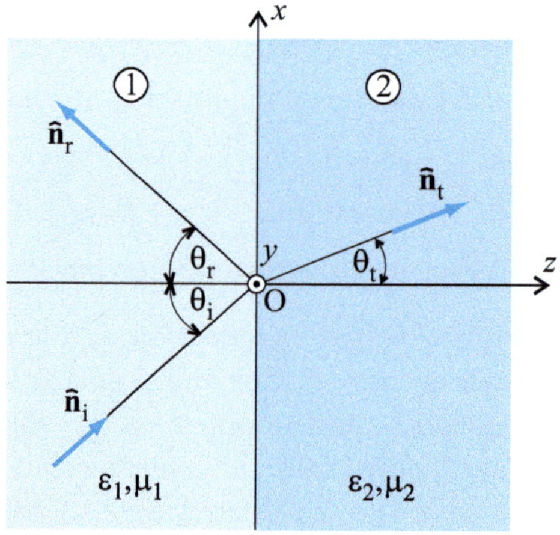

Figure 8.23 Oblique incidence of a uniform plane time-harmonic electromagnetic wave on a penetrable interface; for Conceptual Question 8.59.

CONCEPTUAL QUESTION 8.60 Boundary condition for the total electric field vector. Consider an oblique incidence of a plane wave on an interface between two lossless dielectric media, Figure 8.23, and the following two equations: (a) $\underline{E}_{i0}\cos\theta_i + \underline{E}_{r0}\cos\theta_r = \underline{E}_{t0}\cos\theta_t$ and (b) $\underline{E}_{i0} + \underline{E}_{r0} = \underline{E}_{t0}$. Which of the equations represents the boundary condition for tangential components of the total electric field vector for the normal polarization case and which one for the parallel polarization case?

(A) Equation (a) for normal polarization and equation (b) for parallel polarization.

(B) Equation (a) for parallel polarization and equation (b) for normal polarization.

(C) Equation (a) for both normal and parallel polarizations.

(D) Equation (b) for both normal and parallel polarizations.

CONCEPTUAL QUESTION 8.61 Boundary condition for the total magnetic field vector. For a boundary surface between two perfect-dielectric media illuminated by a time-harmonic uniform plane wave at oblique incidence (Figure 8.23), consider the boundary condition for tangential components of the total magnetic field vector and the following two equations: (a) $-\underline{E}_{i0}\cos\theta_i/\eta_1 + \underline{E}_{r0}\cos\theta_r/\eta_1 = -\underline{E}_{t0}\cos\theta_t/\eta_2$ and (b) $\underline{E}_{i0}/\eta_1 - \underline{E}_{r0}/\eta_1 = \underline{E}_{t0}/\eta_2$, where $\eta_1 = \sqrt{\mu_1/\varepsilon_1}$ and $\eta_2 = \sqrt{\mu_2/\varepsilon_2}$ are the intrinsic impedances of the two media. Which equation is for normal and which for parallel polarization?

(A) Equation (a) for normal polarization and equation (b) for parallel polarization.

(B) Equation (a) for parallel polarization and equation (b) for normal polarization.

(C) Equation (a) for both normal and parallel polarizations.

(D) Equation (b) for both normal and parallel polarizations.

CONCEPTUAL QUESTION 8.62 Fresnel's coefficients for parallel polarization. A uniform plane time-harmonic wave is incident obliquely from air on a flat surface of a lossless nonmagnetic dielectric, and the polarization of the wave is parallel. The incident angle is θ_i and the index of refraction of the dielectric is n. The associated reflection and transmission coefficients, Γ_p and τ_p, can be expressed in terms of

(A) $\sin\theta_i$ only.

(B) n only.

(C) $\sin\theta_\mathrm{i}$ and n only.

(D) None of the above.

CONCEPTUAL QUESTION 8.63 **Apparent shape of an oar immersed in water.** An oar immersed in the water appears to be broken at the water surface when viewed from a boat. The submerged part of the oar looks as if it is shorter than it actually is. This phenomenon can be explained using Snell's law of

(A) reflection for a light beam bouncing off the water surface.

(B) refraction for a light beam incident from air into the water.

(C) refraction for a light beam incident from the water into air.

(D) None of the above.

CONCEPTUAL QUESTION 8.64 **Why a submerged oar cannot be seen from a distant boat.** The submerged part of an oar, immersed in the water from a boat, cannot be seen from another, distant, boat (there are no obstacles in air for viewing the oar) because of

(A) the total reflection of light beams incident from air onto the water surface.

(B) the total reflection of light beams incident from the water onto its surface.

(C) the zero reflection of light beams incident from air onto the water surface.

(D) the zero reflection of light beams incident from the water onto its surface.

(E) None of the above.

CONCEPTUAL QUESTION 8.65 **Multiple total reflections in an optical fiber.** In a typical optical fiber, light is confined to traveling inside a cylindrical dielectric (glass or transparent-plastic) rod by means of multiple total reflections from the interface with a surrounding coaxial layer made of a different dielectric material, as shown in Figure 8.24. With n_core and n_cladding respectively denoting the refraction indices of the fiber core and outer layer (called cladding), it is necessary that

(A) $n_{\text{core}} < n_{\text{cladding}}$

(B) $n_{\text{core}} > n_{\text{cladding}}$

(C) $n_{\text{core}} \ll n_{\text{cladding}}$

(D) $n_{\text{core}} \gg n_{\text{cladding}}$

(E) $n_{\text{core}} = 1$ and $n_{\text{cladding}} \neq 1$

for total reflection to be possible.

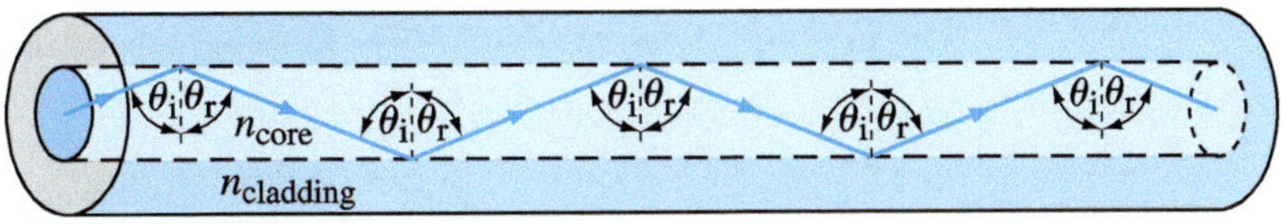

Figure 8.24 Sketch of wave propagation inside the core of an optical fiber, by means of multiple total reflections from the outer layer (cladding); for Conceptual Question 8.65.

CONCEPTUAL QUESTION 8.66 Reflection of an EP wave incident at Brewster angle. An elliptically polarized uniform plane electromagnetic wave in air is incident on a dielectric surface at the Brewster angle. The reflected wave is

(A) linearly polarized.

(B) circularly polarized.

(C) elliptically polarized.

(D) unpolarized.

(E) Need more information.

CONCEPTUAL QUESTION 8.67 Transmission of a CP wave incident at Brewster angle. A circularly polarized uniform plane wave is incident from air on a dielectric half-space at the Brewster angle. The transmitted wave is

(A) linearly polarized.

(B) circularly polarized.

(C) elliptically polarized.

(D) unpolarized.

(E) Need more information.

9 FIELD ANALYSIS OF TRANSMISSION LINES

Introduction

In addition to wireless links, which use free (unbounded) electromagnetic waves propagating in free space or material media (previous two chapters), electromagnetic signals and energy can be transported to a distance also using guided electromagnetic waves. Such waves are channeled through a guiding system composed of conductors and dielectrics. Guiding systems normally have a uniform cross section, and are classified into transmission lines, having two or more separate conductors, and waveguides, consisting of a single conductor or only dielectrics. In this chapter, we present a field analysis of two-conductor transmission lines, which is important for understanding physical processes that constitute the propagation and attenuation along a line of a given geometry and material composition. The principal result of the analysis is a set of parameters of a circuit model of an arbitrary line, in the form of a network of many cascaded equal small cells with lumped elements. This network is then solved, in the next chapter, using circuit-theory concepts and equations, as the starting point of the frequency-domain (complex-domain) and transient (time-domain) analysis of transmission lines as circuits with distributed parameters (circuit analysis of two-conductor transmission lines).

9.1 Field Analysis of Lossless Transmission Lines

Consider a transmission line consisting of two perfect conductors of arbitrary cross section, shown in Figure 9.1, in a homogeneous perfect dielectric of permittivity ε and permeability μ. We assume a time-harmonic variation of the electromagnetic field in the line, of frequency f (high frequency) and angular or radian frequency $\omega = 2\pi f$, and perform the analysis in the complex domain. The voltage $\underline{V}_{12}$ between the line conductors, that is, the difference between their potentials, $\underline{V}_1$ and $\underline{V}_2$, and the current $\underline{I}$ through the conductors are waves propagating along the z-axis in Figure 9.1,

$$\underline{V}_{12} = \underline{V}_{12}(z) = \underline{V}_{12}(0)\,\mathrm{e}^{-\mathrm{j}\beta z}\,, \quad \underline{I} = \underline{I}(z) = \underline{I}(0)\,\mathrm{e}^{-\mathrm{j}\beta z}\,, \quad \beta = \omega\sqrt{\varepsilon\mu}\,, \tag{9.1}$$

with the phase coefficient, β, being the same [see Eq. (7.9)] as for a uniform plane wave traveling in an unbounded medium with the same parameters as the dielectric of the transmission line. The phase velocity, v_p, and wavelength, $\lambda_z = 2\pi/\beta$, along the line (along the z-axis) are thus also the same as in Eqs. (7.3), (7.9) and (7.10).

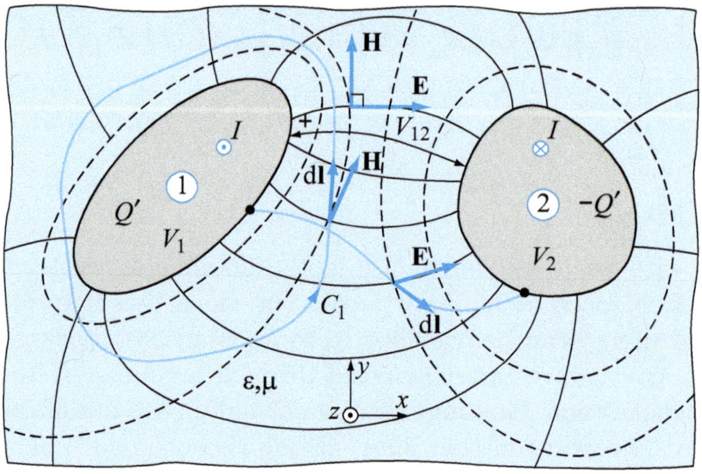

Figure 9.1 Cross section of a two-conductor transmission line with a homogeneous dielectric, carrying a TEM wave (in a high-frequency regime).

The electric and magnetic field lines in a cross section of the system (Figure 9.1) are as in electrostatics and magnetostatics, respectively, and both the electric and magnetic field vectors, $\underline{\mathbf{E}}$ and $\underline{\mathbf{H}}$, are transverse to the direction of wave propagation (i.e., to the line axis), constituting a TEM (transverse electromagnetic) wave. While $\underline{\mathbf{E}}$ and $\underline{\mathbf{H}}$ in any cross section of the transmission line undergo the respective electrostatic and magnetostatic field distributions (field dependences on transverse spatial coordinates, e.g., x and y in Figure 9.1), the field dependence, for both $\underline{\mathbf{E}}$ and $\underline{\mathbf{H}}$, on the z-coordinate is given by the propagation factor $\mathrm{e}^{-\mathrm{j}\beta z}$, as in Eq. (9.1) and as for a uniform plane wave in an unbounded dielectric medium. In addition, as for uniform plane waves, the field vectors are perpendicular to each other, and the ratio of their complex rms intensities equals a real constant, denoted by Z_{TEM} and called the wave impedance of TEM waves [it has the same value as the intrinsic impedance (η) in Eq. (7.6) of the medium of parameters ε and μ],

$$\underline{\mathbf{E}} \perp \underline{\mathbf{H}} \,, \quad \underline{\mathbf{E}}_z = 0 \,, \quad \underline{\mathbf{H}}_z = 0 \,, \quad Z_{\mathrm{TEM}} = \frac{\underline{E}}{\underline{H}} = \sqrt{\frac{\mu}{\varepsilon}} \quad \text{(TEM wave)} \,. \tag{9.2}$$

Note that TEM waves, with these same properties, can propagate on transmission lines consisting of an arbitrary number, M, of conductors, where $M \geq 2$.

With reference to Figure 9.1, the voltage, current, and charge per unit length ($\underline{Q}'$) of the transmission line can be computed as

$$\underline{V}_{12} = \underline{V}_1 - \underline{V}_2 = \int_1^2 \underline{\mathbf{E}} \cdot \mathrm{d}\mathbf{l} \,, \quad \underline{I} = \oint_{C_1} \underline{\mathbf{H}} \cdot \mathrm{d}\mathbf{l} \,, \quad \underline{V}_{12} = \frac{\underline{Q}'}{C'} \,, \quad \underline{I} = c\underline{Q}' = \frac{\underline{Q}'}{\sqrt{\varepsilon\mu}} \,, \tag{9.3}$$

where both the integration path between the conductors and the integration contour enclosing a conductor must lie entirely in a transverse plane (defined by a coordinate z), and C' is the capacitance per unit length of the line, Eq. (2.12).

Equations (9.1) tell us that the line voltage and current have the same exponential dependence on the z-coordinate, and hence their ratio is the same (a constant) for every cross section of the line. It is termed the characteristic impedance of the line and is designated by Z_0. Combining Eqs. (9.3) and (7.3), we obtain

$$Z_0 = \frac{\underline{V}_{12}}{\underline{I}} = \frac{\underline{V}_{12}}{c\underline{Q}'} = \frac{1}{cC'} = \frac{\sqrt{\varepsilon\mu}}{C'} \quad \text{(line characteristic impedance)} \ . \tag{9.4}$$

As a consequence of the proportionality between $\underline{E}$ and $\underline{H}$ in Eqs. (9.2), the following duality relationship between C' and the inductance per unit length of the line in Figure 9.1, L', holds true:

$$L'C' = \varepsilon\mu \quad \text{(duality of } L' \text{ and } C' \text{ of a transmission line)} \ , \tag{9.5}$$

with which the characteristic impedance in Eq. (9.4) can be written as $Z_0 = \sqrt{L'/C'}$.

CONCEPTUAL QUESTION 9.1 **Lossless transmission line with a homogeneous dielectric.** A time-harmonic TEM (transverse electromagnetic) wave of frequency f travels along a lossless transmission line consisting of two conductors of arbitrary cross section, in a homogeneous dielectric of permittivity ε and permeability μ, as shown in Figure 9.2. Consider the following basic propagation parameters of the line: the phase coefficient (β), the phase velocity (v_p), and the wavelength along the line (λ_z). Which of these parameters is the same as for a uniform plane wave of the same frequency, f, propagating in an unbounded medium having the same material parameters, ε and μ, as the dielectric of the transmission line?[1]

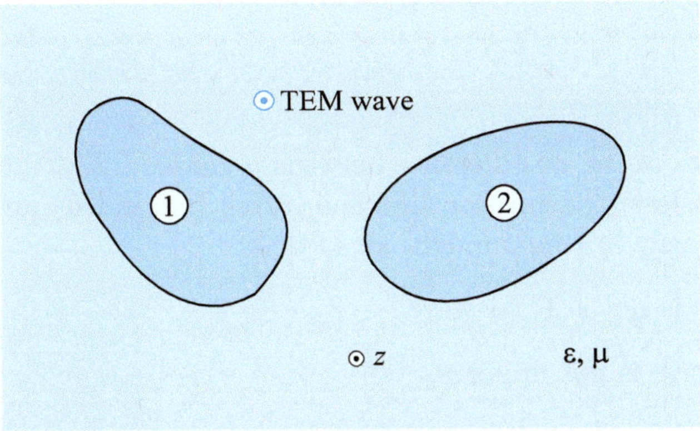

Figure 9.2 TEM wave propagating along a two-conductor transmission line with a homogeneous dielectric (cross section of the structure); for Conceptual Question 9.1.

[1] *For every conceptual question in this text, exactly one answer is correct.*

(A) β and v_p only.

(B) λ_z only.

(C) v_p and λ_z only.

(D) All three parameters.

(E) None of the parameters.

CONCEPTUAL QUESTION 9.2 Electric to magnetic field ratio in a transmission line. Consider the following two statements regarding the electromagnetic wave in a lossless two-conductor transmission line with a homogeneous dielectric of parameters ε and μ (Figure 9.2): (a) the electric and magnetic field vectors of the wave are perpendicular to each other at any instant of time and any point of the line dielectric; (b) the ratio of the electric and magnetic field intensities of the wave equals a constant, $\eta = \sqrt{\mu/\varepsilon}$, at any time and any location in the dielectric. Which of these statements is true?

(A) Statement (a) only.

(B) Statement (b) only.

(C) Both statements.

(D) Neither of the statements.

CONCEPTUAL QUESTION 9.3 Difference between guided TEM and free plane waves. A major difference between a guided TEM time-harmonic wave in an air-filled lossless transmission line and a free (unguided) uniform plane wave of the same frequency in (unbounded) air is that

(A) the free wave is not a TEM wave.

(B) the guided wave is not uniform.

(C) the velocity of the guided wave is frequency dependent.

(D) More than one of the above characterizations.

(E) None of the above characterizations.

CONCEPTUAL QUESTION 9.4 Energy transfer along a coaxial cable with a TEM wave. A TEM wave propagates along a coaxial cable, as shown in Figure 9.3. The cable conductors are perfectly conducting and the cable is filled with a lossless dielectric. The energy transfer along the cable is achieved completely through

(A) the inner conductor of the cable.

(B) the inner and outer conductors of the cable.

(C) the dielectric of the cable.

(D) the air surrounding the cable.

(E) the conductors, the dielectric, and the air.

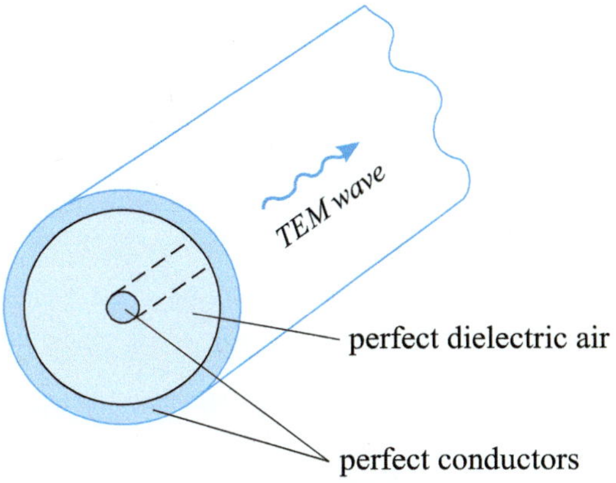

Figure 9.3 Coaxial cable carrying a TEM wave; for Conceptual Question 9.4.

CONCEPTUAL QUESTION 9.5 Electrostatic field distribution in a line cross section. A TEM wave propagates along a lossless transmission line with conductors of arbitrary cross section and a homogeneous dielectric. The distribution of the electric field vector, $\underline{\mathbf{E}}$, in a cross section of the line, namely, the dependence of $\underline{\mathbf{E}}$ on transverse coordinates,

(A) is as in electrostatics, and $\underline{\mathbf{E}}$ does not depend on the longitudinal coordinate.

(B) is as in electrostatics, but $\underline{\mathbf{E}}$ depends on the longitudinal coordinate.

(C) is not as in electrostatics, but $\underline{\mathbf{E}}$ is independent of the longitudinal coordinate.

(D) is not as in electrostatics, and $\underline{\mathbf{E}}$ depends on the longitudinal coordinate.

CONCEPTUAL QUESTION 9.6 **Quasistatic magnetic field distribution of a line.** The distribution of the magnetic field vector, $\underline{\mathbf{H}}$, in a cross section of a transmission line with a homogeneous dielectric and no losses carrying a TEM wave is the same as the quasistatic magnetic field distribution of the line.

(A) True.

(B) False.

CONCEPTUAL QUESTION 9.7 **Complex Poynting vector in a transmission line.** The complex Poynting vector of a TEM wave propagating along a lossless transmission line with a homogeneous dielectric is

(A) purely real

(B) purely imaginary

(C) zero

(D) with nonzero real and imaginary parts

at every point in the line dielectric.

CONCEPTUAL QUESTION 9.8 **Current distribution of a coaxial cable with a TEM wave.** For a lossless coaxial cable (filled with a homogeneous dielectric) with a TEM wave (in a high-frequency regime), the current of the cable conductors is distributed only over

(A) the surface of the inner conductor.

(B) the inner surface of the outer conductor.

(C) the outer surface of the outer conductor.

(D) the surface of the inner conductor and the inner surface of the outer conductor.

(E) the surface of the inner conductor and both surfaces of the outer conductor.

(F) the volume of both the inner and the outer conductors.

CONCEPTUAL QUESTION 9.9 **Current distribution of a coaxial cable at low frequencies.** Assuming that a coaxial cable with a homogeneous dielectric and no losses operates in a low-frequency regime, the current of the cable conductors is distributed only over

(A) the surface of the inner conductor.

(B) the inner surface of the outer conductor.

(C) the outer surface of the outer conductor.

(D) the surface of the inner conductor and the inner surface of the outer conductor.

(E) the surface of the inner conductor and both surfaces of the outer conductor.

(F) the volume of both the inner and the outer conductors.

CONCEPTUAL QUESTION 9.10 **Proportionality between surface current and charge densities.** In a coaxial cable with a traveling time-harmonic TEM wave in Figure 9.4, the complex rms surface current and charge densities, $\underline{J}_{\mathrm{s}}$ and $\underline{\rho}_{\mathrm{s}}$, are proportional to each other at

(A) all points

(B) some (but not all) points

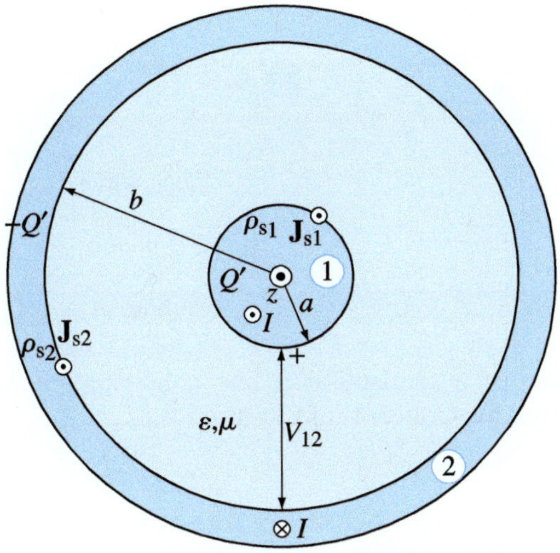

Figure 9.4 Cross section of a lossless coaxial cable with a homogeneous dielectric and TEM wave propagating in the positive z direction; for Conceptual Question 9.10.

(C) no points

of surfaces of conductors with a (nonzero) surface current.

CONCEPTUAL QUESTION 9.11 **Currents of a wire-plane transmission line.** A transmission line consists of a wire conductor and a grounded conducting plane, as shown in Figure 9.5. The medium above the plane is a homogeneous nonmagnetic perfect dielectric of permittivity ε, the wire is parallel to the plane, and both the wire and the plane are made of a perfect electric conductor. A time-harmonic TEM wave is established on the line. Denoting by $\underline{i}_{\text{wire}}$ the current intensity of the wire, the total intensity of the current flowing over the ground plane, computed with respect to the same reference direction as $\underline{i}_{\text{wire}}$ (positive z direction), amounts to

(A) $\underline{i}_{\text{ground}} = \underline{i}_{\text{wire}}$.

(B) $\underline{i}_{\text{ground}} = -\underline{i}_{\text{wire}}$.

(C) $\underline{i}_{\text{ground}} = 0$.

(D) $\underline{i}_{\text{ground}} \to \infty$.

(E) None of the above.

(F) Need more information.

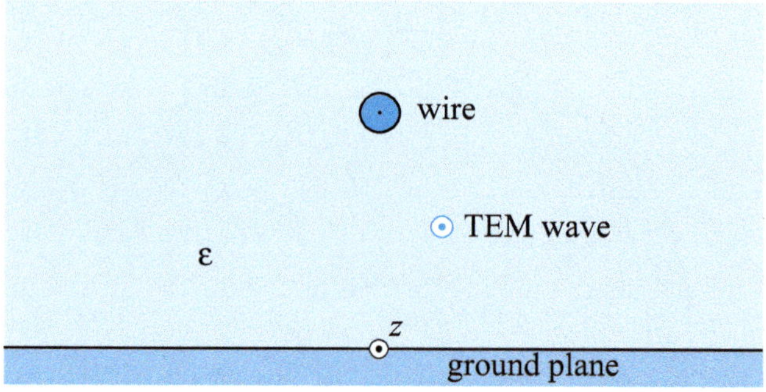

Figure 9.5 Cross section of a transmission line consisting of a wire conductor and a grounded conducting plane; for Conceptual Question 9.11.

CONCEPTUAL QUESTION 9.12 **Charges of a wire-plane transmission line.** If the complex rms charge per unit length of the wire conductor of the transmission line with a TEM wave in Figure 9.5 is doubled, i.e., it becomes $2\underline{Q}'$, with

$\underline{Q}'$ standing for the previous value of this charge, the total complex rms charge per unit length of the ground plane

(A) doubles as well.

(B) is halved.

(C) remains the same, and is nonzero and finite.

(D) remains the same, zero.

(E) remains the same, infinite.

(F) None of the above.

CONCEPTUAL QUESTION 9.13 **Changes to a coaxial cable that prevent TEM waves.** Consider a lossless coaxial cable with a homogeneous linear dielectric, of relative permittivity ε_r ($\varepsilon_r > 1$), and the following four changes to it: (a) removing the inner conductor of the cable, (b) removing the outer conductor of the cable, (c) inserting a thin cylindrical middle conductor coaxially with other conductors, and (d) removing the dielectric of the cable. How many of these changes, occurring separately, would prevent any TEM wave from propagating along the structure?

(A) None.

(B) One.

(C) Two.

(D) Three.

(E) All.

CONCEPTUAL QUESTION 9.14 **Possibility of TEM waves in optical fibers.** Can TEM waves propagate through an optical fiber [cylindrical dielectric structure, with a dielectric core and an outer dielectric layer (cladding)]?

(A) Yes.

(B) No.

CONCEPTUAL QUESTION 9.15 **Line integral of the electric field vector of a TEM wave.** A lossless two-conductor transmission line with a homogeneous linear dielectric carries a time-harmonic TEM wave. Consider two points, points 1 and 2, that belong to the same cross section of the line and to the surfaces of the first and second line conductor, respectively. The complex rms voltage between the conductors for this cross section, $\underline{V}_{12}$, equals the line integral of the complex rms electric field intensity vector, $\underline{\mathbf{E}}$, along an arbitrary path between points 1 and 2.

(A) True.

(B) False.

CONCEPTUAL QUESTION 9.16 **Circulation of the electric field vector in the line dielectric.** The line integral of $\underline{\mathbf{E}}$ along a closed path in the (homogeneous, linear, and perfect) dielectric of a transmission line with two perfect conductors of arbitrary cross section and a time-harmonic TEM wave is

(A) zero.

(B) nonzero.

(C) Need more information.

CONCEPTUAL QUESTION 9.17 **Circulation of the magnetic field vector of a TEM wave.** With $\underline{I}$ denoting the complex rms current intensity of a lossless coaxial cable with a homogeneous dielectric and TEM wave, the line integral of the complex rms magnetic field intensity vector, $\underline{\mathbf{H}}$, along a contour enclosing the inner conductor of the cable equals

(A) $\underline{I}$ or $-\underline{I}$, depending on the orientation of the contour.

(B) $\underline{I}/2$ or $-\underline{I}/2$, depending on the orientation of the contour.

(C) $2\underline{I}$.

(D) zero.

(E) None of the above.

(F) Need more information.

CONCEPTUAL QUESTION 9.18 **Voltage to current ratio for a traveling TEM wave.** A time-harmonic TEM wave travels along a lossless two-conductor transmission line with a homogeneous dielectric of permittivity ε and permeability μ. The ratio between the complex rms voltage, $\underline{V}_{12}$, and current intensity, $\underline{I}$, of the line

(A) equals $\sqrt{\mu/\varepsilon}$.

(B) equals $\varepsilon\mu$.

(C) depends on the geometry of the cross section of the line.

(D) depends on the longitudinal coordinate, z.

(E) depends on both the geometry of the line cross section and the coordinate z.

CONCEPTUAL QUESTION 9.19 **Characteristic impedance of a transmission line.** To find the characteristic impedance, Z_0, of an air-filled transmission line with perfect conductors, we need to know, in addition to possibly c_0 (speed of light in free space),

(A) both the capacitance C' and the inductance L' per unit length of the line.

(B) either C' or L'.

(C) both C' and L', and some other parameters of the line.

(D) either C' or L', and some other parameters of the line.

9.2 Transmission Lines with Small Losses

All real transmission lines have some losses, which, in general, consist of losses in conductors and losses in the dielectric of the line. However, for lines used in engineering practice, these losses, evaluated per unit length of the line, are small. Simply, the conductors and dielectrics in practical transmission lines, if not perfect, are good – by design, such that, having in mind Eqs. (7.16) and (7.15), and denoting the conductivity of the line conductors by σ_c, and that of the line dielectric by σ_d, the following conditions are met:

$$\sigma_c \gg \omega\varepsilon_0 \quad \text{and} \quad \sigma_d \ll \omega\varepsilon \quad \text{(transmission line with small losses)} . \tag{9.6}$$

The losses in a transmission line result in the attenuation of TEM waves along the line, as in Eqs. (7.13), and hence the complex current intensity along the line in Eqs. (9.1), for instance, is now given by

$$\underline{\gamma} = \alpha + \mathrm{j}\beta \quad (\alpha = \alpha_\mathrm{c} + \alpha_\mathrm{d}) \quad \longrightarrow \quad \underline{I}(z) = \underline{I}(0)\,\mathrm{e}^{-\underline{\gamma} z} = \underline{I}(0)\,\mathrm{e}^{-\alpha z}\,\mathrm{e}^{-\mathrm{j}\beta z} \,, \quad (9.7)$$

and similarly for the voltage, field intensity vectors, and other z-dependent quantities in the analysis.

The attenuation coefficient (namely, the portion of α) corresponding to the losses in the conductors in the structure, α_c, is computed as

$$\alpha_\mathrm{c} = \frac{R'}{2Z_0} \,, \quad R' = \frac{1}{|\underline{I}|^2} \oint_{C_\mathrm{c}} R_\mathrm{s} |\underline{\mathbf{H}}_\mathrm{tang}|^2 \,\mathrm{d}l \,, \quad R_\mathrm{s} = \sqrt{\frac{\pi \mu_\mathrm{c} f}{\sigma_\mathrm{c}}} \quad \text{(for line conductors)} \quad (9.8)$$

(practically always, $\mu_\mathrm{c} = \mu_0$), where R' is the high-frequency resistance per unit length of the transmission line (in Ω/m) and C_c denotes the contour of both conductors in the line cross section (Figure 9.1), with $\mathrm{d}l$ being an elemental segment along C_c. R_s is the surface resistance of the conductors (with the skin effect pronounced – see Section 7.7), measured in Ω/square, which equals the real part of the complex intrinsic impedance of the conductors, $R_\mathrm{s} = \mathrm{Re}\{\underline{\eta}\}$, in Eqs. (7.16), and $\underline{\mathbf{H}}_\mathrm{tang}$ is the tangential component of the complex rms magnetic field intensity vector on the conductor surfaces, computed as if the conductors were perfect (perturbation method).

On the other side, the attenuation coefficient α_d in Eqs. (9.7), for the losses in the dielectric of the transmission line, amounts to

$$\alpha_\mathrm{d} = \frac{G'}{2Y_0} \quad \left(Y_0 = \frac{1}{Z_0}\right), \quad G' = \frac{\sigma_\mathrm{d}}{\varepsilon}\,C' \quad \text{(for line dielectric)} \,, \quad (9.9)$$

where Y_0 stands for the characteristic admittance of the line, while G' is the leakage conductance per unit length of the line (unit: S/m), obtained from the duality relationship in Eq. (3.16).

CONCEPTUAL QUESTION 9.20 **Low-loss vs. lossless transmission lines.** Transmission lines with small losses can generally be treated as transmission lines with no losses.

(A) True.

(B) False.

CONCEPTUAL QUESTION 9.21 Large wave attenuation on a line with small losses. The attenuation of a TEM wave on a transmission line with small losses can be prohibitively large for practical use of the line.

(A) True.

(B) False.

CONCEPTUAL QUESTION 9.22 Rate of attenuation of the time-average power flow. If α ($\alpha \neq 0$) is the attenuation coefficient of a transmission line with a TEM wave, the rate of attenuation of the time-average power flow (P) along the line is

(A) determined by this same α.

(B) determined by twice α.

(C) determined by α squared.

(D) determined by half of α.

(E) zero ($P = \text{const}$).

(F) Need more information.

CONCEPTUAL QUESTION 9.23 Wave attenuation vs. conductivity of line conductors. The larger the conductivity of the conductors of a transmission line, σ_c, the larger the magnitude of the current density of the conductors, losses in the line, and TEM-wave attenuation along the line.

(A) True.

(B) False.

CONCEPTUAL QUESTION 9.24 Surface resistance of a good conductor. If the frequency is doubled, the surface resistance, R_s, of a good conductor (with the skin effect pronounced)

(A) increases.

(B) decreases.

(C) remains the same.

(D) Need more information.

CONCEPTUAL QUESTION 9.25 **Surface resistance of a PEC.** The surface resistance (R_s) of a perfect electric conductor is

(A) zero.

(B) infinite.

(C) 377 Ω.

(D) a function of frequency.

(E) None of the above.

CONCEPTUAL QUESTION 9.26 **Surface density of ohmic power for aluminum conductors.** A time-harmonic TEM wave propagates along a transmission line with aluminum conductors. The skin effect is pronounced. Under these circumstances, the surface density of the time-average power of Joule's losses in any of the line conductors (power per unit area of the conductor surface) can be expressed in terms of the surface resistance of the conductor (R_s) and the magnitude of the complex rms surface current density vector ($\underline{\mathbf{J}}_s$) that exists on the surface of the conductor as follows:

(A) $\mathrm{d}(P_J)_{\text{ave}}/\mathrm{d}S = R_s|\underline{\mathbf{J}}_s|$,

(B) $\mathrm{d}(P_J)_{\text{ave}}/\mathrm{d}S = R_s|\underline{\mathbf{J}}_s|^2$,

(C) $\mathrm{d}(P_J)_{\text{ave}}/\mathrm{d}S = |\underline{\mathbf{J}}_s|/R_s$,

(D) $\mathrm{d}(P_J)_{\text{ave}}/\mathrm{d}S = |\underline{\mathbf{J}}_s|^2/R_s$,

(E) None of the above.

CONCEPTUAL QUESTION 9.27 **Electric field distribution in low-loss and lossless lines.** A transmission line with small losses has approximately the same distribution (dependence on respective spatial coordinates) of the electric field vector, $\underline{\mathbf{E}}$,

(A) in a cross section of the line and along the line

(B) in a cross section of the line, but not along the line,

(C) along the line, but not in a cross section of the line,

(D) in no cross sections and along no directions

as the same line with losses neglected.

CONCEPTUAL QUESTION 9.28 **Magnetic field distribution.** Distributions of the magnetic field vector, $\underline{\mathbf{H}}$, of the low-loss and lossless transmission lines are approximately the same

(A) in a cross section of the line and along the line.

(B) in a cross section of the line, but not along the line.

(C) along the line, but not in a cross section of the line.

(D) in no cross sections and along no directions.

CONCEPTUAL QUESTION 9.29 **High-frequency resistance p.u.l. of a coaxial cable.** Conductors of a coaxial cable are made from copper, and its dielectric is polyethylene. The radius of the inner conductor is a, whereas the inner and outer radii of the outer conductor are b and c, respectively ($a < b < c$), as shown in Figure 9.6. In computation of the high-frequency resistance per unit length of the cable, R', the contour of integration (of $R_{\text{s}}|\underline{\mathbf{H}}_{\text{tang}}|^2$), C_{c}, consists of

(A) circles of radii a, b, and c.

(B) circles of radii a and b.

(C) circles of radii a and c.

(D) circles of radii b and c.

(E) a circle of radius c.

(F) None of the above.

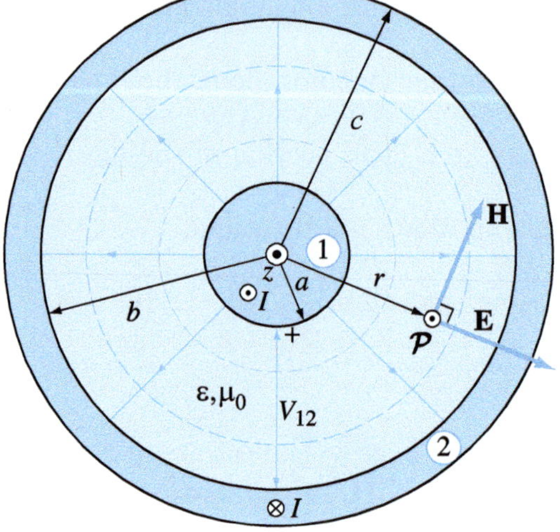

Figure 9.6 Computation of the high-frequency resistance per unit length of a coaxial cable with a TEM wave; for Conceptual Question 9.29.

CONCEPTUAL QUESTION 9.30 Magnetic field for computing R' of a coaxial cable. In computation of the resistance R' of a coaxial cable, with copper conductors of radii a (inner conductor) and b and $c > b$ (outer conductor) and polyethylene dielectric, carrying a TEM wave (Figure 9.6), the relevant expressions for magnetic field intensities to be squared and integrated (integration of $R_s|\underline{\mathbf{H}}_{\text{tang}}|^2$) are some of the following:

(A) $\underline{H}_1 = \underline{I}\,e^{-\alpha z}/(\pi a^2)$ and $\underline{H}_2 = \underline{I}\,e^{-\alpha z}/[\pi(c^2 - b^2)]$.

(B) $\underline{H}_1 = \underline{I}/(\pi a^2)$ and $\underline{H}_2 = \underline{I}/[\pi(c^2 - b^2)]$.

(C) $\underline{H}_1 = \underline{I}\,e^{-\alpha z}/(2\pi a)$, $\underline{H}_2 = \underline{I}\,e^{-\alpha z}/(2\pi b)$, and $\underline{H}_3 = \underline{I}\,e^{-\alpha z}/(2\pi c)$.

(D) $\underline{H}_1 = \underline{I}/(2\pi a)$, $\underline{H}_2 = \underline{I}/(2\pi b)$, and $\underline{H}_3 = \underline{I}/(2\pi c)$.

(E) None of the above.

CONCEPTUAL QUESTION 9.31 Frequency dependence of the attenuation coefficient for conductors. Consider the attenuation coefficient representing losses in conductors of a two-conductor transmission line, α_c, and its dependence on the frequency, f, of a propagating TEM wave. This dependence can be described as

(A) $\alpha_c \propto \sqrt{f}$.

(B) $\alpha_c \propto f$.

(C) $\alpha_c \propto f^2$.

(D) $\alpha_c \propto f^{-1/2}$.

(E) $\alpha_c \propto f^{-1}$.

(F) α_c does not depend on f.

CONCEPTUAL QUESTION 9.32 Practical importance of conductor vs. dielectric losses. Consider the attenuation coefficients for conductor and dielectric losses, α_c and α_d, respectively, of low-loss transmission lines. In typical applications and designs of such lines in engineering practice,

(A) α_c is of more concern than α_d.

(B) α_d is of more concern than α_c.

(C) the two coefficients are of equal concern.

(D) neither of the coefficients is of much concern.

9.3 Evaluation of Primary and Secondary Circuit Parameters of Transmission Lines

As we shall see in the next chapter, an arbitrary two-conductor transmission line with TEM waves can be analyzed as an electric circuit with distributed parameters, based on a representation of the line by a network of cascaded equal small cells, of length Δz, with lumped elements. These elements are characterized by per-unit-length parameters C', L', R', and G' of the line (studied in this and previous chapters), multiplied by Δz. As C', L', R', and G' are a basis for the circuit analysis of transmission lines (to be presented in the next chapter), they are referred to as primary circuit parameters of a line. The other parameters that will be used in the circuit analysis are the characteristic impedance, Z_0, phase coefficient, β, phase velocity, v_p, wavelength, λ_z, and attenuation coefficient, α, of the line. As these parameters can be derived from the primary parameters, they are called secondary circuit parameters of transmission lines. Moreover, once the secondary parameters are known for a given line, they suffice for the analysis (i.e., primary parameters are not needed).

In summary, the capacitance C' in Eqs. (9.3) is determined from a 2-D electrostatic analysis in the cross section of the line, in Figure 9.1. The inductance L' and conductance G' are then obtained from C' using the duality relationships in Eqs. (9.5) and (9.9), respectively, while the resistance R' is evaluated by means of Eqs. (9.8), based on a 2-D magnetostatic analysis in the line cross section (Figure 9.1). The impedance Z_0 is found from Eqs. (9.4), the coefficient β employing the expression in Eqs. (9.1) or as $\beta = \omega\sqrt{L'C'}$ [see Eq. (9.5)], the velocity v_p is given by Eq. (7.3) or $v_\text{p} = 1/\sqrt{L'C'}$, the wavelength along the line equals $\lambda_z = 2\pi/\beta$ [Eq. (7.9)], and the coefficient α is computed from R', G', and Z_0 using Eqs. (9.7)–(9.9).

CONCEPTUAL QUESTION 9.33 **Two transmission lines with equal primary circuit parameters.** Two transmission lines with small losses have the same capacitance, inductance, resistance, and conductance per unit of their length, at the same frequency. However, at this same frequency, some of the following parameters are different for the two lines: the characteristic impedance, phase coefficient, phase velocity, wavelength, and attenuation coefficient. Is this possible?

(A) Yes.

(B) No.

CONCEPTUAL QUESTION 9.34 **Two lines with equal secondary circuit parameters.** At a given frequency, two low-loss transmission lines have the same characteristic impedance, phase coefficient, phase velocity, wavelength, and attenuation coefficient (secondary circuit parameters). However, one or more primary circuit parameters, the per-unit-length capacitance, inductance, resistance, and conductance, are not equal for the two lines. Is this possible?

(A) Yes.

(B) No.

CONCEPTUAL QUESTION 9.35 **Obtaining primary circuit parameters from one another.** Consider a transmission line with small losses, and its primary circuit parameters at high frequencies (for TEM waves), namely, the capacitance (C'), inductance (L'), resistance (R'), and conductance (G') per unit length of the line. If the dielectric of the line is homogeneous (and linear), with permittivity ε and permeability μ, some of the parameters can be found from others. In particular,

(A) C' and L' can be obtained from R' and G'.

(B) L' can be obtained from C', and R' from G'.

(C) L' can be obtained from R', and G' from C'.

(D) L' and G' can be obtained from C'.

(E) L', R', and G' can be obtained from C'.

(F) None of the above.

CONCEPTUAL QUESTION 9.36 Relationship between p.u.l. inductance and capacitance. At an angular (radian) frequency ω of a TEM wave, the inductance and capacitance per unit length of a lossless transmission line with a homogeneous dielectric, whose intrinsic impedance is η and intrinsic phase velocity (velocity of waves in an unbounded medium with the same electromagnetic parameters) is c, are related as follows:

(A) $L'C' = 1/\omega^2$.

(B) $L'/C' = \eta^2$.

(C) $L'C' = c^2$.

(D) $L'C' = 1/c^2$.

(E) more than one of the above relationships hold true.

(F) L' and C' are not related to each other.

CONCEPTUAL QUESTION 9.37 2-D electrostatic analysis of a transmission line. Which of the primary circuit parameters of a transmission line with small losses and a homogeneous dielectric in a high-frequency regime, C', L', R', and G', can be determined based on a 2-D electrostatic analysis in the cross section of the line, for the given geometry and material properties of the line, and the frequency of a TEM wave?

(A) C' only.

(B) C' and L' only.

(C) C' and G' only.

(D) C', L', and G' only.

(E) All parameters, C', L', R', and G'.

(F) None of the parameters.

CONCEPTUAL QUESTION 9.38 **Line resistance and conductance at high frequencies.** Consider the per-unit-length resistance (R') and conductance (G') of a low-loss transmission line in a high-frequency regime. If the total length of the line is l, which one of the following relationships is always satisfied?

(A) $R' = 1/G'$.

(B) $R' = l^2/G'$.

(C) $R'l^2 = 1/G'$.

(D) None.

CONCEPTUAL QUESTION 9.39 **Low- and high-frequency resistance and conductance.** Which of the two parameters, R' and G', of a transmission line is approximately the same whether the line is considered in a low-frequency or high-frequency regime?

(A) R' only.

(B) G' only.

(C) Both R' and G'.

(D) Neither R' nor G'.

CONCEPTUAL QUESTION 9.40 **Transmission line with perfect conductors and dielectric.** In a transmission line with perfect conductors and a perfect dielectric, we have the following for its p.u.l. resistance and conductance, R' and G', respectively, at high frequencies:

(A) $R' = 0$ and $G' = 0$.

(B) $R' = 0$ and $G' \to \infty$.

(C) $R' \to \infty$ and $G' = 0$.

(D) $R' \to \infty$ and $G' \to \infty$.

9.4 Transmission Lines with Inhomogeneous Dielectrics

Consider a two-conductor transmission line with an inhomogeneous, lossless ($\sigma_\mathrm{d} = 0$), and nonmagnetic ($\mu = \mu_0$) dielectric [examples are the coaxial cable with a four-piece dielectric in Figure 2.17 and a microstrip transmission line, Figure 2.9(e), but with fringing effects included, i.e., with field lines also existing in air above the dielectric substrate, as illustrated in Figure 9.7]. We define the effective relative permittivity of the line as

$$\varepsilon_\mathrm{reff} = \frac{C'}{C'_0} \quad \text{(effective relative permittivity of a transmission line; dimensionless)}, \tag{9.10}$$

where C'_0 stands for the p.u.l. capacitance of the same line if air-filled [for the first example mentioned, the coaxial cable in Figure 2.9(b) with $\varepsilon = \varepsilon_0$]. Note that for lines with homogeneous dielectrics ($\varepsilon_\mathrm{r} = \mathrm{const}$), $\varepsilon_\mathrm{reff} = \varepsilon_\mathrm{r}$. The phase coefficient [Eqs. (9.1)] of the actual line can now be computed as $\beta = \omega\sqrt{\varepsilon_\mathrm{reff}}/c_0$ [c_0 is the wave velocity in free space, Eq. (7.4)]. The inductance L' of the line is the same

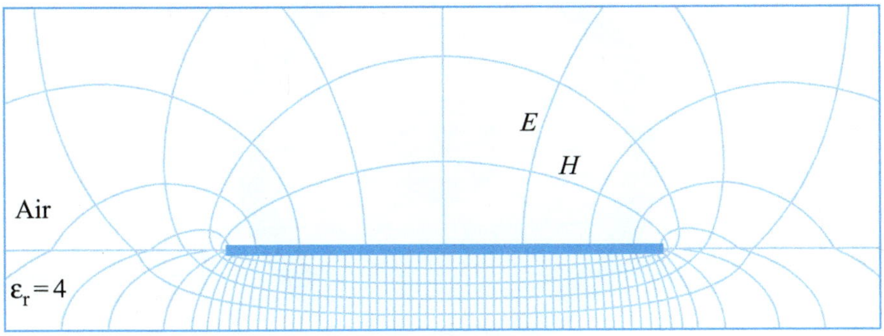

Figure 9.7 Electric (E) and magnetic (H) field lines in a cross section of a microstrip line with the strip width to substrate height ratio $w/h = 5.4$ and substrate relative permittivity $\varepsilon_\mathrm{r} = 4$; field pattern plots are obtained by numerical analysis.

as if the dielectric were air, so that Eq. (9.5) gives $L' = L'_0 = \varepsilon_0\mu_0/C'_0$, and the line characteristic impedance is obtained from $Z_0 = \sqrt{L'/C'}$ [Eqs. (9.4) and (9.5)]. Finally, the resistance R' of the line is also the same as for the air-filled line, $R' = R'_0$.

CONCEPTUAL QUESTION 9.41 **Meaning of the effective relative permittivity.** The effective relative permittivity of a transmission line with an inhomogeneous dielectric, $\varepsilon_{\rm reff}$, can be interpreted as the relative permittivity of an equivalent homogeneous dielectric material which, if occupying the space between the conductors of the line, would give the same capacitance per unit length, C', as the inhomogeneous dielectric of the actual line.

(A) True.

(B) False.

CONCEPTUAL QUESTION 9.42 **Coaxial cable half filled with a liquid dielectric.** Consider a horizontally laid coaxial cable half filled with a liquid dielectric of relative permittivity $\varepsilon_{\rm r}$ ($\varepsilon_{\rm r} \neq 1$), so that the upper half of the space between the cable conductors is air-filled, as shown in Figure 9.8. We can write the following for the effective relative permittivity of the cable, $\varepsilon_{\rm reff}$:

(A) $0 < \varepsilon_{\rm reff} < 1$.

(B) $1 < \varepsilon_{\rm reff} < \varepsilon_{\rm r}$.

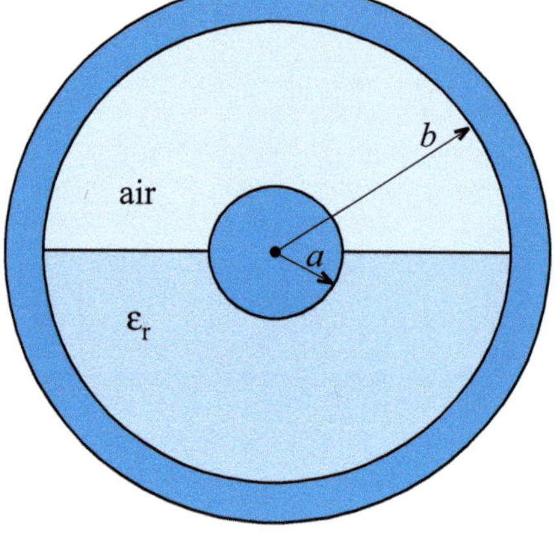

Figure 9.8 Cross section of a coaxial cable half filled with a liquid dielectric of relative permittivity $\varepsilon_{\rm r}$; for Conceptual Question 9.42.

(C) $\varepsilon_r < \varepsilon_{reff} < 2\varepsilon_r$.

(D) $\varepsilon_{reff} = 2\varepsilon_r$.

(E) $\varepsilon_{reff} > 2\varepsilon_r$.

CONCEPTUAL QUESTION 9.43 **Microstrip line with and without fringing effects.** Effective relative permittivity (ε_{reff}) of a microstrip line, shown in Figure 9.9, analyzed with fringing effects taken into account (as in Figure 9.7) is

(A) larger than

(B) the same as

(C) smaller than

that of the same line with fringing effects neglected [as in Figure 2.9(e)].

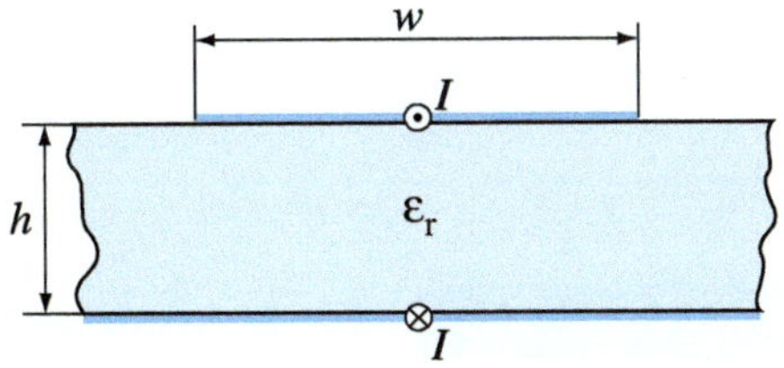

Figure 9.9 Cross section of a microstrip line; for Conceptual Question 9.43.

CONCEPTUAL QUESTION 9.44 **Strip line with and without fringing effects.** Considering the effective relative permittivity of a strip line, shown in Figure 9.10, analyzed including and neglecting the fringing effects, respectively, ε_{reff} in the former case (with fringing) comes out to be

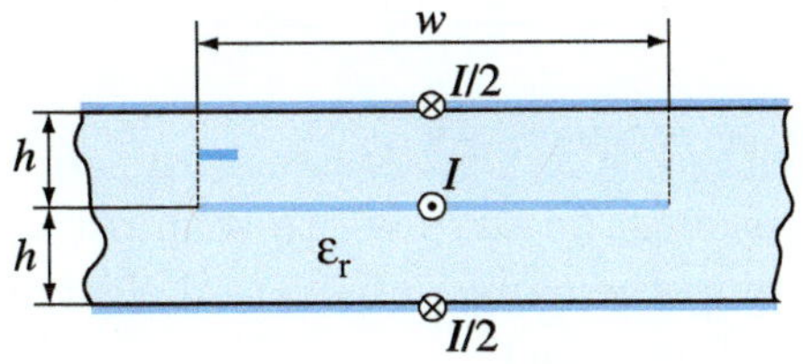

Figure 9.10 Cross section of a strip line; for Conceptual Question 9.44.

(A) larger than

(B) the same as

(C) smaller than

$\varepsilon_{\text{reff}}$ in the latter case [analysis as in Figure 2.9(f)].

CONCEPTUAL QUESTION 9.45 **Effective relative permittivity of microstrip and strip lines.** Consider a microstrip line and a strip line with the same ratio of the strip width (w) and substrate thickness (h), equal to $w/h = 1$ (fringing effects cannot be neglected), and the same dielectric, of relative permittivity $\varepsilon_{\text{r}} = 4$. The effective relative permittivity, $\varepsilon_{\text{reff}}$, of the microstrip line is

(A) larger than

(B) the same as

(C) smaller than

that of the strip line.

CONCEPTUAL QUESTION 9.46 **Is a microstrip line "faster" or "slower" than a strip line?** For a microstrip line and a strip line both with $w/h = 1$ (fringing effects pronounced) and $\varepsilon_{\text{r}} = 4$, signals along the microstrip line travel

(A) faster than

(B) at the same speed as

(C) slower than

along the strip line.

CONCEPTUAL QUESTION 9.47 **Increasing the w to h ratio of a microstrip line.** Upon increasing the strip width to substrate height ratio, w/h, of a microstrip line, while keeping the same substrate dielectric (same ε_{r}), the effective relative permittivity of the line ($\varepsilon_{\text{reff}}$)

(A) increases.

(B) decreases.

(C) remains the same.

CONCEPTUAL QUESTION 9.48 Increasing the w to h ratio of a strip line. Consider the dependence of $\varepsilon_{\rm reff}$ on the w/h ratio of a strip line. If w/h is increased, for the same $\varepsilon_{\rm r}$ of the substrate, $\varepsilon_{\rm reff}$ of the line

(A) increases.

(B) decreases.

(C) remains the same.

CONCEPTUAL QUESTION 9.49 Inductance p.u.l. of a line with an inhomogeneous dielectric. The inductance per unit length of a transmission line with an inhomogeneous nonmagnetic dielectric can be obtained from the capacitance per unit length of

(A) this transmission line.

(B) the same line if air-filled.

(C) the same line if filled with a homogeneous nonmagnetic dielectric (not air).

(D) the same line if filled with a homogeneous dielectric of parameters ε and μ.

(E) more than one of the structures above.

(F) none of the structures above.

CONCEPTUAL QUESTION 9.50 L'–C' relationship, line with an inhomogeneous dielectric. Consider a transmission line with an inhomogeneous nonmagnetic dielectric at an angular frequency ω. The effective relative permittivity and phase velocity of the line are $\varepsilon_{\rm reff}$ and $v_{\rm p}$, respectively. The inductance and capacitance per unit length of the line are related as

(A) $L'C' = 1/\omega^2$.

(B) $L'/C' = \eta_0^2/\varepsilon_{\rm reff}$ $(\eta_0 = \sqrt{\mu_0/\varepsilon_0})$.

(C) $L'C' = \varepsilon_{\rm reff}\varepsilon_0\mu_0$.

(D) $L'C' = 1/v_{\rm p}^2$.

(E) more than one of the above relationships hold true.

(F) L' and C' are not related to each other.

CONCEPTUAL QUESTION 9.51 Characteristic impedance, line with an inhomogeneous dielectric. Denoting by Z_0 and $Z_0^{(\rm air)}$, respectively, the characteristic impedance of a transmission line with an inhomogeneous nonmagnetic dielectric and effective relative permittivity $\varepsilon_{\rm reff}$ and that of the same line if air-filled, we have that

(A) $Z_0 = Z_0^{(\rm air)}$.

(B) $Z_0 = \varepsilon_{\rm reff} Z_0^{(\rm air)}$.

(C) $Z_0 = \sqrt{\varepsilon_{\rm reff}}\, Z_0^{(\rm air)}$.

(D) $Z_0 = Z_0^{(\rm air)}/\varepsilon_{\rm reff}$.

(E) $Z_0 = Z_0^{(\rm air)}/\sqrt{\varepsilon_{\rm reff}}$.

(F) None of the above.

CONCEPTUAL QUESTION 9.52 Resistance p.u.l. of a line with an inhomogeneous dielectric. The high-frequency resistance per unit length of a transmission line with an inhomogeneous nonmagnetic dielectric equals the high-frequency resistance p.u.l. of the same line if filled with a homogeneous nonmagnetic dielectric, at the same frequency.

(A) True.

(B) False.

CONCEPTUAL QUESTION 9.53 Attenuation coefficient, line with an inhomogeneous dielectric. A transmission line of effective relative permittivity

$\varepsilon_{\rm reff}$ has copper conductors, with small losses, while the losses in its inhomogeneous nonmagnetic dielectric can be neglected. The attenuation coefficient of this line, α, can be obtained using the attenuation coefficient of the same line if air-filled, $\alpha^{(\rm air)}$, as follows:

(A) $\alpha = \alpha^{(\rm air)}$.

(B) $\alpha = \varepsilon_{\rm reff}\, \alpha^{(\rm air)}$.

(C) $\alpha = \sqrt{\varepsilon_{\rm reff}}\, \alpha^{(\rm air)}$.

(D) $\alpha = \alpha^{(\rm air)}/\varepsilon_{\rm reff}$.

(E) $\alpha = \alpha^{(\rm air)}/\sqrt{\varepsilon_{\rm reff}}$.

(F) None of the above.

10 CIRCUIT ANALYSIS OF TRANSMISSION LINES

Introduction

This chapter takes the primary and secondary circuit parameters of transmission lines computed in the field analysis of lines in the previous chapter, and uses them to solve for the voltage and current along lossless and lossy lines, with various excitations and load terminations. Most importantly, this is a circuit analysis of transmission lines, using only pure circuit-theory concepts to develop the complete frequency-domain and transient analysis of lines as circuits with distributed parameters whose per-unit-length characteristics are already known. The analysis is based on a circuit model of an arbitrary two-conductor transmission line in the form of a ladder network of elementary circuit cells with lumped elements, and on circuit differential equations for this network, termed telegrapher's equations, which can be easily solved for voltages and currents on the network. We also introduce and implement a graphical technique for the circuit analysis and design of transmission lines in the frequency domain based on the so-called Smith chart. Transient analysis of transmission lines will cover step and pulse excitations of lines and a variety of line terminations, including reactive loads, and both matched and unmatched conditions at either end of the line.

10.1 Telegrapher's Equations and Their Solution

Figure 10.1 (upper part) shows a circuit-theory representation of an arbitrary two-conductor lossy transmission line, where a pair of parallel horizontal thick lines in the schematic diagram, although resembling a two-wire transmission line, symbolizes a structure with conductors of completely arbitrary cross sections (Figure 9.1) and a generally inhomogeneous dielectric. We subdivide the line into short sections, of length Δz, so that, using primary circuit parameters C', L', R', and G' of the line, studied in the previous chapter, each such section can be represented by a circuit cell shown in Figure 10.1. From Kirchhoff's laws for the cells and current–voltage characteristics (element laws) for their elements, we obtain, in the limit of $\Delta z \to 0$, transmission-line equations or telegrapher's equations for the complex rms voltage and current on the line:

$$\frac{\mathrm{d}\underline{V}}{\mathrm{d}z} = -\underbrace{(R' + \mathrm{j}\omega L')}_{\underline{Z}'}\,\underline{I}\,, \quad \frac{\mathrm{d}\underline{I}}{\mathrm{d}z} = -\underbrace{(G' + \mathrm{j}\omega C')}_{\underline{Y}'}\,\underline{V} \quad \text{(telegrapher's equations)}\,, \tag{10.1}$$

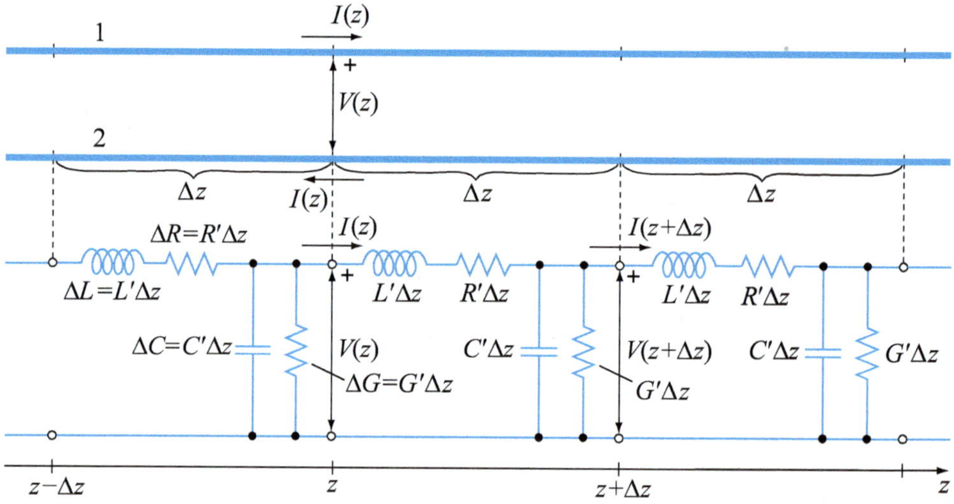

Figure 10.1 Circuit model of a two-conductor lossy transmission line in an ac regime.

whose general solutions are complex exponential functions in z given by

$$\underline{V}(z) = \underbrace{\underline{V}_{i0}\,e^{-\underline{\gamma}z}}_{\text{incident wave}} + \underbrace{\underline{V}_{r0}\,e^{\underline{\gamma}z}}_{\text{reflected wave}}, \quad \underline{I}(z) = \underbrace{\frac{\underline{V}_{i0}}{\underline{Z}_0}\,e^{-\underline{\gamma}z}}_{\text{incident wave}} + \underbrace{\left(-\frac{\underline{V}_{r0}}{\underline{Z}_0}\,e^{\underline{\gamma}z}\right)}_{\text{reflected wave}},$$

$$\underline{\gamma} = \sqrt{\underline{Z}'\underline{Y}'} = \alpha + j\beta, \quad \underline{Z}_0 = \sqrt{\frac{\underline{Z}'}{\underline{Y}'}} = |\underline{Z}_0|\,e^{j\phi} \quad \text{(solutions for } \underline{V} \text{ and } \underline{I} \text{ on a line)}. \tag{10.2}$$

As expected, the total voltage and current waves along the line are, in general, sums of two oppositely directed traveling waves, an incident (forward) wave, propagating in the positive z direction, and a reflected (backward) wave, progressing in the negative z direction [analogously to Eqs. (8.6) and (8.7)].

For transmission lines with small losses [see Eqs. (9.6)], we have $R' \ll \omega L'$ and $G' \ll \omega C'$, and hence the characteristic impedance, $\underline{Z}_0$, and the attenuation and phase coefficients, α and β, of the line can be computed (approximately) using Eqs. (9.4), (9.5), (9.8), (9.9), and (9.1), namely, as

$$\underline{Z}_0 = Z_0 = \sqrt{\frac{L'}{C'}}, \quad \alpha = \frac{R'}{2Z_0} + \frac{G'}{2Y_0}, \quad \beta = \omega\sqrt{L'C'} \quad \text{(line with small losses)}. \tag{10.3}$$

CONCEPTUAL QUESTION 10.1 Series resistor in a circuit model of a transmission line. Figure 10.2 shows a cell in a circuit model of an arbitrary two-conductor lossy transmission line in an ac regime, representing a short section, with length Δz, of the line. What losses are modeled by the series resistor, of resistance $\Delta R = R'\Delta z$?[1]

(A) Losses in conductors of the line.

(B) Losses in the dielectric of the line.

(C) Losses at lower frequencies.

(D) Losses at higher frequencies.

(E) All possible losses in the line at any frequency.

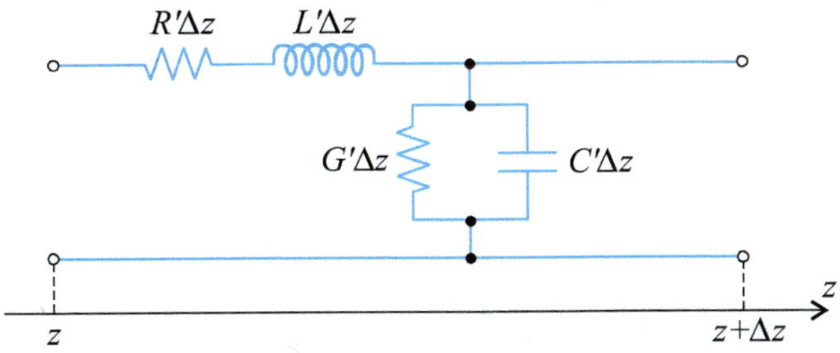

Figure 10.2 Elementary cell in a circuit model of a transmission line; for Conceptual Question 10.1.

CONCEPTUAL QUESTION 10.2 Shunt resistor in a circuit model of a line. What losses are represented by the shunt (parallel) resistor, of conductance $\Delta G = G'\Delta z$, in the circuit model in Figure 10.2?

(A) Losses in conductors of the line.

(B) Losses in the dielectric of the line.

(C) Losses at lower frequencies.

(D) Losses at higher frequencies.

(E) All possible losses in the line at any frequency.

[1] *For every conceptual question in this text, exactly one answer is correct.*

CONCEPTUAL QUESTION 10.3 **Capacitor and inductor in a dc circuit model of a line.** In a dc regime, with time-constant voltages and currents on a transmission line, we have the following for the capacitor, of capacitance $\Delta C = C'\Delta z$, and inductor, of inductance $\Delta L = L'\Delta z$, in the circuit model of an arbitrary lossy transmission line in an ac regime (in Figure 10.2):

(A) The capacitor is a short circuit and the inductor is an open circuit.

(B) The capacitor is an open circuit and the inductor is a short circuit.

(C) Both the capacitor and the inductor are short circuits.

(D) The capacitor is a voltage generator and the inductor is a current generator.

(E) The capacitor and the inductor switch places in the model.

(F) None of the above.

CONCEPTUAL QUESTION 10.4 **Complex voltage and current on a lossless transmission line.** A transmission line is connected at one end to an ideal time-harmonic voltage source, while its other end is open-circuited. Both the conductors and the dielectric of the line can be treated as perfect (lossless). Considering the complex rms voltage between the conductors, $\underline{V}$, and the complex rms current through them, $\underline{I}$, the following of the two quantities is (are) constant along the line:

(A) $\underline{V}$ only.

(B) $\underline{I}$ only.

(C) both.

(D) neither.

CONCEPTUAL QUESTION 10.5 **Voltage drops in a circuit model of a transmission line.** The voltage drop $\Delta\underline{V} = \underline{V}(z) - \underline{V}(z + \Delta z)$ across a circuit cell in Figure 10.3, in a circuit model of an arbitrary transmission line, amounts to

(A) $R'\Delta z\underline{I}(z)$,

(B) $\mathrm{j}\omega L'\Delta z\underline{I}(z)$,

(C) $R'\Delta z\underline{I}(z) + \mathrm{j}\omega L'\Delta z\underline{I}(z)$,

(D) $R'\Delta z\underline{I}(z) + \underline{I}(z)/(G'\Delta z)$,

(E) $\mathrm{j}\omega L'\Delta z\underline{I}(z) + \underline{I}(z)/(\mathrm{j}\omega C'\Delta z)$,

(F) $R'\Delta z\underline{I}(z) + \mathrm{j}\omega L'\Delta z\underline{I}(z) + \underline{I}(z)/(G'\Delta z) + \underline{I}(z)/(\mathrm{j}\omega C'\Delta z)$,

where ω ($\omega = 2\pi f$) is the angular (radian) frequency of the voltage $\underline{V}$ and current $\underline{I}$ on the line.

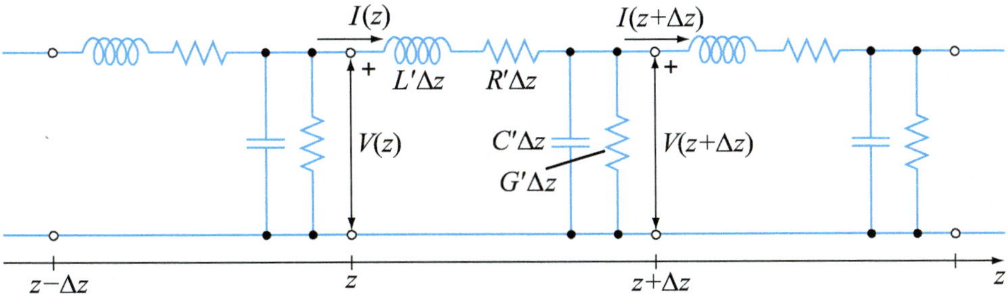

Figure 10.3 Finding the voltage and current drops across an elementary cell in a circuit model of an arbitrary two-conductor lossy transmission line in an ac regime; for Conceptual Question 10.5.

CONCEPTUAL QUESTION 10.6 **Current drops in a circuit model of a transmission line.** The current drop $\Delta\underline{I} = \underline{I}(z) - \underline{I}(z + \Delta z)$ across a circuit cell in Figure 10.3 equals

(A) $V(z)/(R'\Delta z) + G'\Delta z\underline{V}(z + \Delta z)$.

(B) $V(z)/(\mathrm{j}\omega L'\Delta z) + \mathrm{j}\omega C'\Delta z\underline{V}(z + \Delta z)$.

(C) $V(z)/(R'\Delta z) + V(z)/(\mathrm{j}\omega L'\Delta z) + G'\Delta z\underline{V}(z + \Delta z) + \mathrm{j}\omega C'\Delta z\underline{V}(z + \Delta z)$.

(D) $G'\Delta z\underline{V}(z + \Delta z)$.

(E) $\mathrm{j}\omega C'\Delta z\underline{V}(z + \Delta z)$.

(F) $G'\Delta z\underline{V}(z + \Delta z) + \mathrm{j}\omega C'\Delta z\underline{V}(z + \Delta z)$.

CONCEPTUAL QUESTION 10.7 **p.u.l. complex impedance and admittance of a line.** Consider the per-unit-length complex impedance and admittance of a transmission line, given by $\underline{Z}' = R' + \mathrm{j}\omega L'$ and $\underline{Y}' = G' + \mathrm{j}\omega C'$, where R', L', G', and C' are primary circuit parameters (Figure 10.3) and ω is the operating angular frequency of the line. Moreover, consider the following four relationships between $\underline{Z}'$ and $\underline{Y}'$: (a) $\underline{Z}' = 1/\underline{Y}'$, (b) $\underline{Z}' = \underline{\gamma}^2/\underline{Y}'$, (c) $R' + \mathrm{j}\omega L' = 1/G' + \mathrm{j}/(\omega C')$, and (d) $\underline{Z}' = \underline{Z}_0^2\underline{Y}'$, with $\underline{\gamma}$ and $\underline{Z}_0$ standing for the complex propagation coefficient and characteristic impedance, respectively, of the line. How many of these relationships hold true?

(A) None.

(B) One.

(C) Two.

(D) Three.

(E) All.

CONCEPTUAL QUESTION 10.8 Voltage to current ratio for incident, reflected, and total waves. On a transmission line, let $\underline{V}_\mathrm{i}$, $\underline{V}_\mathrm{r}$, and $\underline{V}$ denote the complex rms voltages of the incident, reflected, and resultant (total) waves, respectively, all relative to the same reference direction, as shown in Figure 10.4. Similarly, let $\underline{I}_\mathrm{i}$, $\underline{I}_\mathrm{r}$, and $\underline{I}$ be the complex rms current intensities of the three waves, again all given with respect to the same reference direction (Figure 10.4). In the line cross section defined by a coordinate z,

(A) $\underline{V}_\mathrm{i}(z)/\underline{I}_\mathrm{i}(z) = \underline{V}_\mathrm{r}(z)/\underline{I}_\mathrm{r}(z) = \underline{V}(z)/\underline{I}(z)$.

(B) $\underline{V}_\mathrm{i}(z)/\underline{I}_\mathrm{i}(z) = -\underline{V}_\mathrm{r}(z)/\underline{I}_\mathrm{r}(z) = \underline{V}(z)/\underline{I}(z)$.

(C) $\underline{V}_\mathrm{i}(z)/\underline{I}_\mathrm{i}(z) = \underline{V}_\mathrm{r}(z)/\underline{I}_\mathrm{r}(z) = -\underline{V}(z)/\underline{I}(z)$.

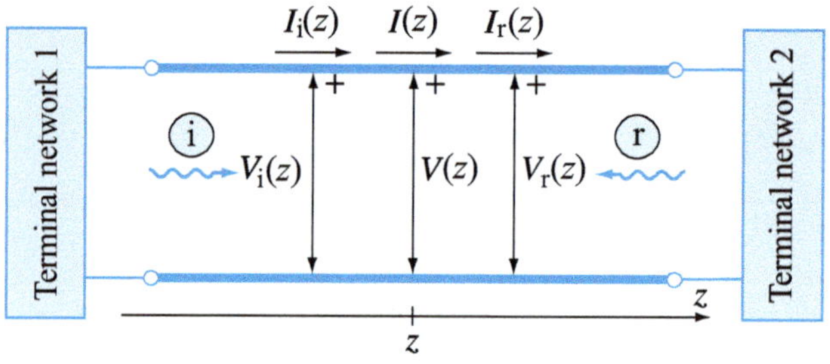

Figure 10.4 Incident, reflected, and total voltages and currents on a transmission line; for Conceptual Question 10.8.

(D) $\underline{V}_\mathrm{i}(z)/\underline{I}_\mathrm{i}(z) = \underline{V}(z)/\underline{I}(z) \neq \underline{V}_\mathrm{r}(z)/\underline{I}_\mathrm{r}(z)$.

(E) $\underline{V}_\mathrm{i}(z)/\underline{I}_\mathrm{i}(z) = \underline{V}_\mathrm{r}(z)/\underline{I}_\mathrm{r}(z) \neq \underline{V}(z)/\underline{I}(z)$.

(F) $\underline{V}_\mathrm{i}(z)/\underline{I}_\mathrm{i}(z) = -\underline{V}_\mathrm{r}(z)/\underline{I}_\mathrm{r}(z) \neq \underline{V}(z)/\underline{I}(z)$.

CONCEPTUAL QUESTION 10.9 **Low-loss and lossless transmission lines.** Which of the following secondary circuit parameters, the attenuation coefficient (α), the phase coefficient (β), the magnitude of the complex characteristic impedance ($|\underline{Z}_0|$), and the phase angle of the characteristic impedance (ϕ), for a transmission line with small losses are practically the same as those for the same line with losses neglected, at the same frequency?

(A) β only.

(B) β, $|\underline{Z}_0|$, and ϕ only.

(C) $|\underline{Z}_0|$ and ϕ only.

(D) α, β, and $|\underline{Z}_0|$ only.

(E) All parameters.

(F) None of the parameters.

10.2 Reflection Coefficient for Transmission Lines

Let the terminal network at the beginning of a transmission line be a voltage generator of complex rms electromotive force (open-circuit voltage) $\underline{\mathcal{E}}$ and complex internal (series) impedance $\underline{Z}_\mathrm{g}$, as shown in Figure 10.5. In general, such a generator represents the Thévenin equivalent generator (circuit), with respect to the line input terminals, of an arbitrary input network. In addition, let the other end of the line be terminated in a load of complex impedance $\underline{Z}_\mathrm{L}$, which, in general, is an equivalent (input) impedance of an arbitrary passive (with no generators) output network. Finally, we adopt the origin of the z-axis to be at the output terminals of the line (i.e., at the load), so that, denoting the length of the line by l, the location of the line input terminals (generator) is defined by $z = -l$ (Figure 10.5).

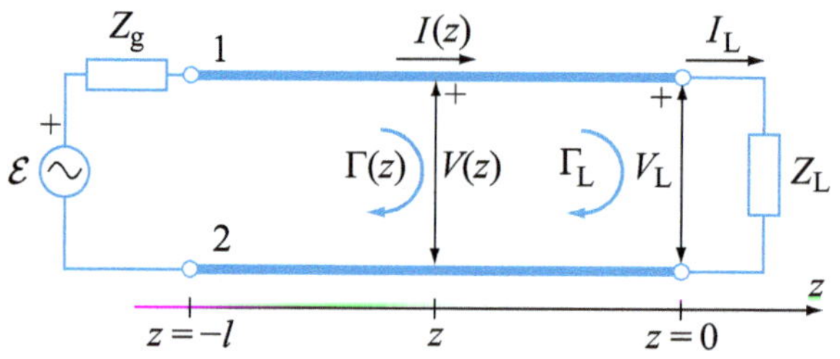

Figure 10.5 Transmission line of Figure 10.1 with a voltage generator (at $z = -l$) and complex impedance load (at $z = 0$) as terminal networks.

Boundary conditions at the load terminals ($z = 0$) in Figure 10.5 for the total voltage and current of the line, given by Eqs. (10.2), result, analogously to Eq. (8.10), in the following solution for the ratio of $\underline{V}_{r0}$ and $\underline{V}_{i0}$:

$$\underline{\Gamma}_L = \frac{\underline{V}_{r0}}{\underline{V}_{i0}} = \frac{\underline{Z}_L - \underline{Z}_0}{\underline{Z}_L + \underline{Z}_0} \quad \text{(load voltage reflection coefficient)}\,, \tag{10.4}$$

which we term the load voltage reflection coefficient of the line. Note that, from Eqs. (10.2), the load reflection coefficient of the line for currents,

$$\underline{\Gamma}_{\text{for currents}} = \frac{\underline{I}_r(0)}{\underline{I}_i(0)} = -\frac{\underline{V}_{r0}}{\underline{V}_{i0}} = -\underline{\Gamma}_L\,, \tag{10.5}$$

comes out to be just opposite to the voltage coefficient. Note also that the complex $\underline{\Gamma}_L$ can be written in the exponential form:

$$\underline{\Gamma}_L = |\underline{\Gamma}_L|\,e^{j\psi_L} \quad (0 \le |\underline{\Gamma}_L| \le 1\,;\ -180^\circ < \psi_L \le 180^\circ)\,, \tag{10.6}$$

where ψ_L denotes its phase angle. Finally, having in mind Eqs. (10.2), we generalize the concept of the line voltage reflection coefficient at the load, Eq. (10.4), to that at an arbitrary position (defined by the coordinate z) along the line, Figure 10.5,

$$\underline{\Gamma}(z) = \frac{\underline{V}_r(z)}{\underline{V}_i(z)} = \frac{\underline{V}_{r0}\,e^{\underline{\gamma} z}}{\underline{V}_{i0}\,e^{-\underline{\gamma} z}} = |\underline{\Gamma}_L|\,e^{2\alpha z}\,e^{j(2\beta z + \psi_L)} \quad \text{(generalized reflection coefficient)}\,. \tag{10.7}$$

With the use of the coefficient $\underline{\Gamma}(z)$, the total voltage and current along the line, Eqs. (10.2), can now be written as

$$\underline{V}(z) = \underline{V}_i(z)\,[1 + \underline{\Gamma}(z)]\,, \quad \underline{I}(z) = \underline{I}_i(z)\,[1 - \underline{\Gamma}(z)]\,. \tag{10.8}$$

From the analysis of the expression for $\underline{V}(z)$ given by Eqs. (10.8) and (10.7), the voltage maxima on a lossless transmission line are, in analogy to the electric field maxima in Figure 8.9(b),

$$|\underline{V}|_{\max} = V_{i0}\,(1 + |\underline{\Gamma}_L|) \quad \text{at} \quad z_{\max} = -m\frac{\lambda_z}{2} - \frac{\psi_L}{2\beta}\,, \quad \begin{cases} m = 0, 1, 2, \ldots \text{ if } \psi_L \ge 0 \\ m = 1, 2, \ldots \text{ if } \psi_L < 0 \end{cases}$$
$$\text{(voltage maxima on a transmission line)}\,, \tag{10.9}$$

where $m \geq 1$ for $\psi_L < 0$ to ensure that $z_{max} \leq 0$, and $\lambda_z = 2\pi/\beta$ is the wavelength along the line. The minima are [also see Figure 8.9(b)]

$$|\underline{V}|_{min} = |\underline{V}_{i0}|\,(1 - |\underline{\Gamma}_L|) \quad \text{at} \quad z_{min} = -(2m+1)\frac{\lambda_z}{4} - \frac{\psi_L}{2\beta}\,, \quad m = 0, 1, 2, \ldots,$$
$$\text{and } z_{min} = 0 \text{ if } \psi_L = 180^\circ \quad \text{(voltage minima)}\,, \quad (10.10)$$

i.e., they are shifted by $\lambda_z/4$ with respect to the adjacent maxima. From the expression for $\underline{I}(z)$ in Eqs. (10.8), the current maxima occur at the locations of the voltage minima, and vice versa, and we have that $|\underline{I}|_{max} = |\underline{V}|_{max}/Z_0$ and $|\underline{I}|_{min} = |\underline{V}|_{min}/Z_0$ [see the magnetic field standing wave pattern in Figure 8.9(b)]. The standing wave ratio (SWR) of a transmission line, for a given load impedance, defined as the ratio of voltage or current maxima to minima on the line, turns out to be

$$s = \frac{|\underline{V}|_{max}}{|\underline{V}|_{min}} = \frac{|\underline{I}|_{max}}{|\underline{I}|_{min}} = \frac{1 + |\underline{\Gamma}_L|}{1 - |\underline{\Gamma}_L|} \quad \text{(standing wave ratio)}\,. \quad (10.11)$$

CONCEPTUAL QUESTION 10.10 **Load voltage reflection coefficient, purely resistive load.** A lossless transmission line is terminated in a purely resistive load. The load voltage reflection coefficient of the line, $\underline{\Gamma}_L$, for this case is

(A) purely real and positive.

(B) purely real and negative.

(C) zero.

(D) None of the above.

(E) Need more information.

CONCEPTUAL QUESTION 10.11 **Load voltage transmission coefficient.** The load voltage transmission coefficient of a transmission line for a given load, defined as $\underline{\tau}_L = \underline{V}_L/\underline{V}_{i0}$, where $\underline{V}_L$ is the load voltage (complex rms) and $\underline{V}_{i0}$ is the incident complex rms voltage at the load, is related to the load voltage reflection coefficient of the line, $\underline{\Gamma}_L$, as follows:

(A) $\underline{\tau}_L = 1 + \underline{\Gamma}_L$.

(B) $\underline{\tau}_L = 1 - \underline{\Gamma}_L$.

(C) $\underline{\tau}_L = 2 - \underline{\Gamma}_L$.

(D) $\underline{\tau}_\mathrm{L} = -\underline{\Gamma}_\mathrm{L}$.

(E) $\underline{\tau}_\mathrm{L} = -1 + \underline{\Gamma}_\mathrm{L}$.

CONCEPTUAL QUESTION 10.12 **Load reflection coefficient for currents.** When compared to the load voltage reflection coefficient of a transmission line, the load reflection coefficient of the line for currents (defined as the ratio of the reflected and incident complex rms current intensities)

(A) is the same.

(B) has the same magnitude and by 180° larger phase angle.

(C) has a different magnitude and the same phase angle.

(D) has a different magnitude and by 180° larger phase angle.

(E) None of the above.

(F) Need more information.

CONCEPTUAL QUESTION 10.13 **Generalized voltage reflection coefficient along a lossless line.** For the generalized voltage reflection coefficient, $\underline{\Gamma}(z)$, at an arbitrary position defined by the coordinate z along a lossless transmission line, we have that

(A) its magnitude is an exponential function of z and its phase is zero.

(B) both its magnitude and its phase are exponential functions of z.

(C) its magnitude is a constant and its phase is an exponential function of z.

(D) its magnitude is a constant and its phase is a linear function of z.

(E) its magnitude is a constant and its phase is zero.

CONCEPTUAL QUESTION 10.14 **Limits for the SWR of a transmission line.** The range of possible values of the standing wave ratio (SWR) of a transmission line is as follows:

(A) $0 \le s \le 1$.

(B) $-1 \le s \le 1$.

(C) $0 \le s < \infty$.

(D) $1 \le s < \infty$.

(E) $-\infty < s < \infty$.

CONCEPTUAL QUESTION 10.15 **SWR for a 0-dB reflection coefficient.** The load voltage reflection coefficient – in decibels – of a lossless transmission line is found to be 0 dB. The corresponding standing wave ratio of the line is

(A) $s = 1$.

(B) $s = 0$.

(C) $s = 2$.

(D) $s \to \infty$.

(E) not defined.

(F) Need more information.

CONCEPTUAL QUESTION 10.16 **Measuring frequency by a slotted line.** Measurements with a lossless air-filled slotted coaxial line connected to an unknown load impedance show that the distance between two adjacent voltage maxima is $\Delta l = 15$ cm. From this, the unknown frequency of a generator feeding the line is found to be

(A) $f = 1$ GHz.

(B) $f = 2$ GHz.

(C) $f = 500$ MHz.

(D) $f = 4$ GHz.

(E) None of the above frequencies.

(F) Need more information.

CONCEPTUAL QUESTION 10.17 **Slotted-line measurement of the load reflection coefficient.** Using a lossless slotted line terminated in an unknown load, the measured standing wave ratio of the line turns out to be $s = 2$. The magnitude of the load reflection coefficient is

(A) $|\underline{\Gamma}_{\rm L}| = 1/2$.

(B) $|\underline{\Gamma}_{\rm L}| = 2$.

(C) $|\underline{\Gamma}_{\rm L}| = 1/3$.

(D) $|\underline{\Gamma}_{\rm L}| = 2/3$.

(E) None of the above values.

(F) Need more information.

CONCEPTUAL QUESTION 10.18 **Voltage maximum at load terminals.** A time-harmonic wave propagates along a lossless transmission line with a characteristic impedance of 50 Ω and partially reflects from a load whose impedance is unknown. It was found that the resultant wave on the line exhibits standing wave properties and that a voltage maximum is located at the load terminals. From this, it can be concluded that the impedance of the load is

(A) purely real and greater than 50 Ω.

(B) purely real, nonzero, and less than 50 Ω.

(C) equal to 50 Ω.

(D) purely imaginary.

(E) zero.

(F) infinite.

CONCEPTUAL QUESTION 10.19 **Voltage minimum at load terminals.** Assuming that a voltage minimum, which is not zero, of the standing wave is identified at the load terminals of a lossless transmission line with $Z_0 = 50\ \Omega$, we can conclude that the impedance of the load is

(A) purely real and greater than 50 Ω.

(B) purely real, nonzero, and less than 50 Ω.

(C) equal to 50 Ω.

(D) purely imaginary.

(E) zero.

(F) infinite.

CONCEPTUAL QUESTION 10.20 Voltage and current standing wave patterns, purely resistive load. A time-harmonic wave of rms voltage V_{i0} propagates with phase coefficient β along a lossless transmission line of characteristic impedance Z_0. The line is terminated in a load with complex impedance $\underline{Z}_L = 2Z_0 + j0$. Under these circumstances, the voltage and current standing wave patterns are as sketched in

(A) Figure 10.6(a).

(B) Figure 10.6(b).

(C) Figure 10.6(c).

(D) Figure 10.6(d).

(E) Figure 10.6(e).

CONCEPTUAL QUESTION 10.21 Standing wave patterns, another purely resistive load. Assuming that a lossless transmission line is terminated in a load of impedance $\underline{Z}_L = Z_0/4 + j0$, Z_0 being the characteristic impedance of the line, which one of the five offered sketches in Figure 10.6 represents the voltage and current standing wave patterns on this line?

(A) The sketch in Figure 10.6(a).

(B) The sketch in Figure 10.6(b).

(C) The sketch in Figure 10.6(c).

(D) The sketch in Figure 10.6(d).

(E) The sketch in Figure 10.6(e).

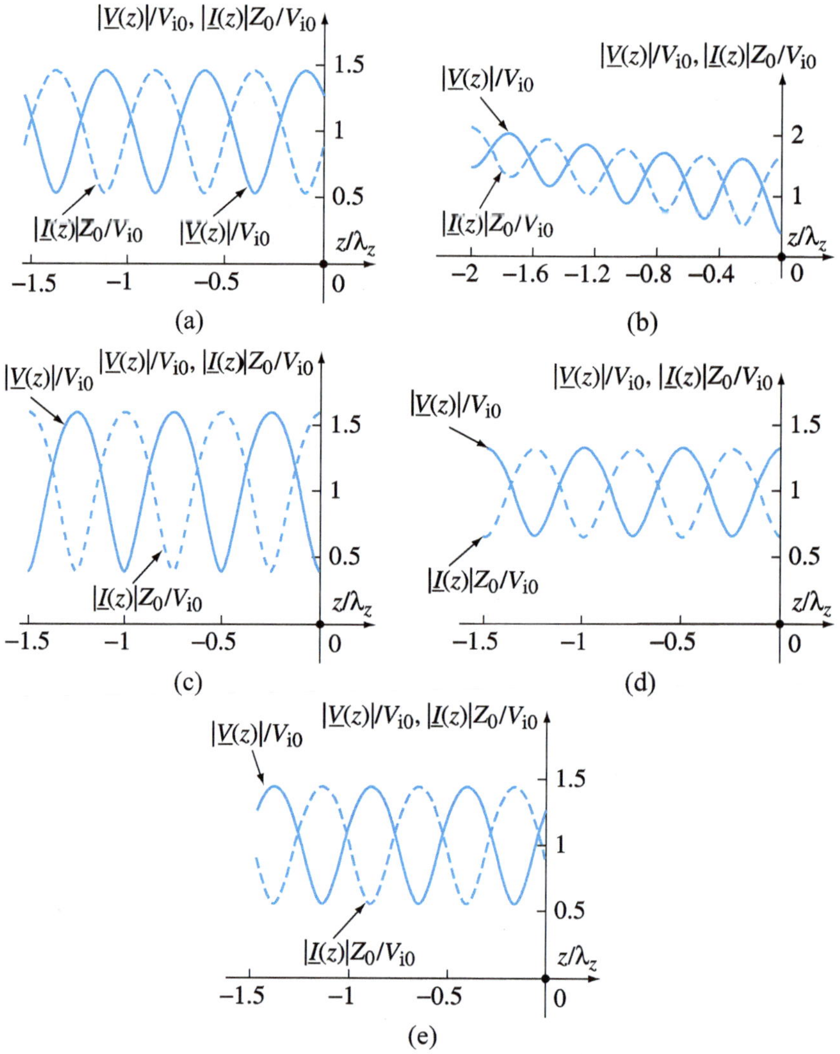

Figure 10.6 Five offered sketches of normalized voltage and current standing wave patterns [$|\underline{V}(z)|/V_{i0}$ and $|\underline{I}(z)|Z_0/V_{i0}$, as functions of z/λ_z ($\lambda_z = 2\pi/\beta$)] for transmission lines with certain impedance loads; for Conceptual Question 10.20.

CONCEPTUAL QUESTION 10.22 **Standing wave patterns, inductive complex load.** If the load impedance for a lossless transmission line of characteristic impedance Z_0 is given by $\underline{Z}_L = Z_0(1 + j)$, the voltage and current standing wave patterns on the line are as shown in

(A) Figure 10.6(a).

(B) Figure 10.6(b).

(C) Figure 10.6(c).

(D) Figure 10.6(d).

(E) Figure 10.6(e).

CONCEPTUAL QUESTION 10.23 Standing wave patterns, capacitive complex load. Which one of the sketches in Figure 10.6 is for the impedance $\underline{Z}_L = Z_0(1 - j)$ as load terminating a lossless transmission line with characteristic impedance Z_0?

(A) The one in Figure 10.6(a).

(B) The one in Figure 10.6(b).

(C) The one in Figure 10.6(c).

(D) The one in Figure 10.6(d).

(E) The one in Figure 10.6(e).

CONCEPTUAL QUESTION 10.24 Reflected complex power on a transmission line. For the notation and reference directions of voltages and currents given in Figure 10.7, the reflected complex power [flowing in the backward (negative z) direction in Figure 10.7], that is, the complex power carried by the reflected wave, is computed as

(A) $\underline{S}_r = \underline{V}_r(z)\underline{I}_r(z)$.

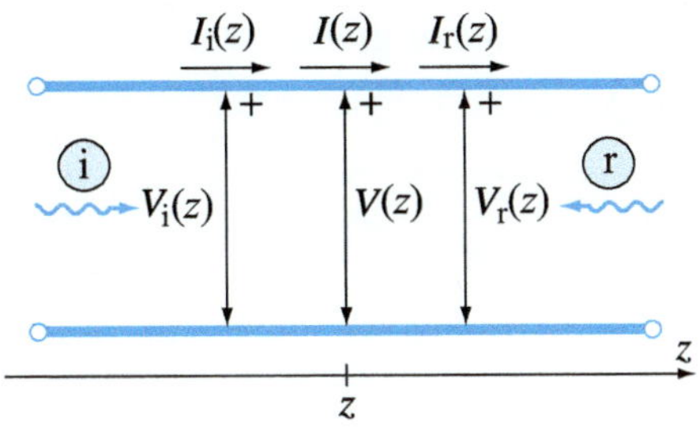

Figure 10.7 Incident and reflected waves on a transmission line; for Conceptual Question 10.24.

(B) $\underline{S}_r = -\underline{V}_r(z)\underline{I}_r(z)$.

(C) $\underline{S}_r = \underline{V}_r(z)[\underline{I}_r(z)]^*$.

(D) $\underline{S}_r = \underline{V}_r(z)[-\underline{I}_r(z)]^*$.

(E) None of the above.

CONCEPTUAL QUESTION 10.25 **Time-average powers carried by incident and reflected waves.** A time-harmonic wave of rms voltage V_{i0} propagates with phase coefficient β along a lossless transmission line of characteristic admittance Y_0. The line is terminated in a complex load whose voltage reflection coefficient has magnitude $|\underline{\Gamma}_L|$ $(0 < |\underline{\Gamma}_L| < 1)$ and phase angle ψ_L ($\psi_L \neq 0$ and $\psi_L \neq 180°$). The time-average powers carried by the incident and reflected waves in the positive and negative z directions (Figure 10.7), respectively, are given by

(A) $(P_i)_{ave} = Y_0 V_{i0}^2 \cos^2 \beta z$ and $(P_r)_{ave} = Y_0 |\underline{\Gamma}_L|^2 V_{i0}^2 \sin^2 \beta z$.

(B) $(P_i)_{ave} = Y_0 V_{i0}^2 \cos \psi_L$ and $(P_r)_{ave} = Y_0 |\underline{\Gamma}_L|^2 V_{i0}^2 \sin \psi_L$.

(C) $(P_i)_{ave} = Y_0 V_{i0}^2$ and $(P_r)_{ave} = Y_0 |\underline{\Gamma}_L|^2 V_{i0}^2 \cos \psi_L$.

(D) $(P_i)_{ave} = Y_0 V_{i0}^2$ and $(P_r)_{ave} = Y_0 |\underline{\Gamma}_L|^2 V_{i0}^2$.

(E) $(P_i)_{ave} = Y_0 V_{i0}^2$ and $(P_r)_{ave} = |\underline{\Gamma}_L| V_{i0}^2$.

(F) $(P_i)_{ave} = 0$ and $(P_r)_{ave} = 0$.

CONCEPTUAL QUESTION 10.26 **Time-average power dissipated in a complex load.** For a lossless transmission line of characteristic admittance Y_0 carrying a time-harmonic incident wave of rms voltage V_{i0} and its load described by the voltage reflection coefficient of magnitude $|\underline{\Gamma}_L|$ and phase angle ψ_L, the time-average power dissipated in the load is given by

(A) $P_L = Y_0 V_{i0}^2 (1 - |\underline{\Gamma}_L|^2) \cos \psi_L$.

(B) $P_L = Y_0 V_{i0}^2 (1 - |\underline{\Gamma}_L|^2)$.

(C) $P_L = Y_0 V_{i0}^2 |\underline{\Gamma}_L|^2$.

(D) $P_L = Y_0 V_{i0}^2$.

(E) $P_L = 0$.

CONCEPTUAL QUESTION 10.27 **Power flow on a low-loss line with a matched load.** A time-harmonic wave travels with attenuation and phase coefficients α and β, respectively, along a transmission line with small losses. The rms voltage of the wave at the load terminals is V_{i0}. The characteristic impedance of the line can be taken to be purely real, equal to Z_0. The line is terminated in a matched load, so that there is no reflected wave on the line. In a cross section of the line that is at a distance l ($l > 0$) from the load terminals, the time-average power carried by the wave equals

(A) $P = V_{i0}^2/Z_0$.

(B) $P = (V_{i0}^2/Z_0)\ \mathrm{e}^{2\alpha l}$.

(C) $P = (V_{i0}^2/Z_0)\ \mathrm{e}^{-2\alpha l}$.

(D) $P = (V_{i0}^2/Z_0)\ \mathrm{e}^{4\alpha l}$.

(E) $P = (V_{i0}^2/Z_0)\ \mathrm{e}^{-4\alpha l}$.

(F) $P = (V_{i0}^2/Z_0)\ \mathrm{e}^{-2\alpha l}\cos\beta l$.

10.3 Transmission-Line Impedance

As can be seen in Eqs. (10.2), the proportionality between the voltage and current of a transmission line (Figure 10.5) via its characteristic impedance, $\underline{Z}_0$, takes place only for a single traveling (forward or backward) wave, and not for a general solution for $\underline{V}(z)$ and $\underline{I}(z)$, on the line. Combining Eqs. (10.8), (10.2), and (10.7), the total voltage to current ratio expressed in terms of $\underline{Z}_0$ and either the generalized voltage reflection coefficient, $\underline{\Gamma}(z)$, or the load voltage reflection coefficient, $\underline{\Gamma}_\mathrm{L}$, and the complex propagation coefficient, $\underline{\gamma}$, of the line amounts to

$$\underline{Z}(z) = \frac{\underline{V}(z)}{\underline{I}(z)} = \underline{Z}_0\,\frac{1+\underline{\Gamma}(z)}{1-\underline{\Gamma}(z)} = \underline{Z}_0\,\frac{1+\underline{\Gamma}_\mathrm{L}\,\mathrm{e}^{2\underline{\gamma}z}}{1-\underline{\Gamma}_\mathrm{L}\,\mathrm{e}^{2\underline{\gamma}z}} \quad (-l \le z \le 0)\,, \quad \underline{Z}_\mathrm{in} = \underline{Z}(-l)$$

(transmission-line impedance, equivalent input impedance) , (10.12)

with $\underline{\Gamma}_\mathrm{L}$ given by Eq. (10.4). This ratio, denoted simply by $\underline{Z}(z)$, represents the so-called transmission-line impedance, seen at a line cross section defined by the coordinate z looking toward the load. In other words, $\underline{Z}(z)$ equals the complex impedance of an equivalent load that can be used to completely replace the portion of the line beyond this cross section, including the (original) load of impedance $\underline{Z}_\mathrm{L}$, with respect to the rest of the line (and the generator). For $z = -l$, $\underline{Z}$ at the generator terminals represents the equivalent input impedance, relative to the generator, of the entire line in Figure 10.5, and we mark it as $\underline{Z}_{\text{in}}$.

CONCEPTUAL QUESTION 10.28 **Voltage at generator terminals of a transmission line.** Figure 10.8 shows a lossless transmission line of characteristic impedance $Z_0 = 50\ \Omega$ fed by a time-harmonic voltage generator of frequency $f = 750$ MHz, rms emf $\mathcal{E} = 5$ V, and internal resistance $R_\text{g} = 50\ \Omega$. At the other end, the line is terminated in a purely resistive load of resistance $R_\text{L} = 100\ \Omega$. Under these circumstances, the rms voltage at the beginning (at generator terminals) of the line, V_g, in the steady state

(A) equals 5 V.

(B) equals 3.33 V.

(C) equals 2.5 V.

(D) equals 2 V.

(E) depends on the length of the line.

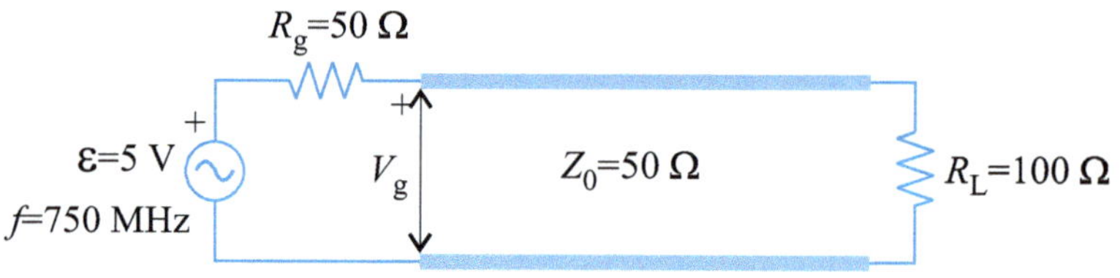

Figure 10.8 Lossless transmission line fed by a time-harmonic voltage generator with a purely resistive internal impedance and terminated in a purely resistive load; for Conceptual Question 10.28.

CONCEPTUAL QUESTION 10.29 **Input impedance of a line with a purely resistive load.** Which of the following changes would not affect the input impedance of a lossless transmission line terminated in a purely resistive load?

(A) The length of the line doubled.

(B) The operating frequency of the line doubled.

(C) Both the length and the frequency doubled.

(D) The length doubled and the frequency halved.

(E) More than one of the above changes, (A)–(D).

(F) None of the above changes.

CONCEPTUAL QUESTION 10.30 **Input impedance of a line with a complex load.** For a complex load consisting of a resistor and an inductor connected in series – as the termination of a lossless transmission line – the following change would not affect the input impedance of the line:

(A) The length of the line doubled.

(B) The operating frequency of the line doubled.

(C) Both the length and the frequency doubled.

(D) The length doubled and the frequency halved.

(E) More than one of the above changes, (A)–(D).

(F) None of the above changes.

CONCEPTUAL QUESTION 10.31 **Current of a voltage generator driving a lossy line.** A transmission line of length l and characteristic impedance Z_0 is driven by a time-harmonic voltage generator. The emf of the generator has rms value $\mathcal{E}$ and zero initial phase; its internal impedance is purely real and equal to R_g. At the other end, the line is open-circuited. The line conductors have small losses described by the high-frequency resistance per unit length of the line R', whereas the line dielectric is lossless. The input impedance of the line is $\underline{Z}_\mathrm{in}$ and the complex propagation coefficient along the line is $\underline{\gamma}$. Under these circumstances, the complex current intensity of the generator is given by

(A) $\underline{I}_\mathrm{g} = \mathcal{E}\,\mathrm{e}^{-\underline{\gamma}l}/(\underline{Z}_\mathrm{in} + R'l + R_\mathrm{g})$.

(B) $\underline{I}_\mathrm{g} = \mathcal{E}\,\mathrm{e}^{-\underline{\gamma}l}/(\underline{Z}_\mathrm{in} + R_\mathrm{g})$.

(C) $\underline{I}_\mathrm{g} = \mathcal{E}/(Z_0 + R_\mathrm{g})$.

(D) $\underline{I}_\mathrm{g} = \mathcal{E}/(\underline{Z}_\mathrm{in} + R_\mathrm{g})$.

(E) $\underline{I}_\mathrm{g} = \mathcal{E}\,\mathrm{e}^{-\underline{\gamma}l}/[(\underline{Z}_\mathrm{in} + R_\mathrm{g})(1 + \mathrm{e}^{-2\underline{\gamma}l})]$.

(F) $\underline{I}_\mathrm{g} = 0$.

CONCEPTUAL QUESTION 10.32 **Total ohmic power in line conductors.** Consider the open-circuited transmission line with length l, characteristic impedance Z_0, high-frequency p.u.l. resistance R', $G' = 0$, input impedance $\underline{Z}_{\text{in}}$, and propagation coefficient $\underline{\gamma}$, and its excitation by a time-harmonic voltage generator of rms emf $\mathcal{E}$, zero initial phase, and internal resistance R_g. Assume that the current $\underline{I}_\text{g}$ of the voltage generator is also known. What is the total time-average power of Joule's losses in the line conductors?

(A) $P_{\text{line losses}} = R'l|\underline{I}_\text{g}|^2$.

(B) $P_{\text{line losses}} = R'l|\underline{I}_\text{g}|^2\,\text{e}^{2\text{Re}\{\underline{\gamma}\}l}$.

(C) $P_{\text{line losses}} = \text{Re}\{\underline{Z}_{\text{in}}\}|\underline{I}_\text{g}|^2$.

(D) $P_{\text{line losses}} = \text{Re}\{\mathcal{E}\underline{I}_\text{g}^*\}$.

(E) $P_{\text{line losses}} = \text{Re}\{\mathcal{E}\underline{I}_\text{g}^*\,\text{e}^{-2\underline{\gamma}l}\}$.

(F) $P_{\text{line losses}} = \mathcal{E}^2\,\text{e}^{2\text{Re}\{\underline{\gamma}\}l}/(R'l + R_\text{g})$.

10.4 Short-Circuited, Open-Circuited, and Matched Transmission Lines

This section discusses three important special cases of load terminations of a transmission line, in Figure 10.5: a short-circuited, open-circuited, and matched line. If the line is terminated in a short circuit (sc), i.e., if its output terminals (at $z = 0$) are galvanically connected together, as shown in Figure 10.9(a), the load voltage ($\underline{V}_\text{L}$) is zero, and so is the load impedance ($\underline{Z}_\text{L}$). Therefore, the load voltage reflection coefficient ($\underline{\Gamma}_\text{L}$), Eq. (10.4), equals $(0 - Z_0)/(0 + Z_0) = -1$, and Eqs. (10.2), (10.4), (10.12), and (6.20) give the following for the total complex rms voltage ($\underline{V}_\text{sc}$) and current ($\underline{I}_\text{sc}$) on the line and the input impedance ($\underline{Z}_\text{sc}$) of the line (assuming that it is lossless, $\gamma = \text{j}\beta$):

$$\underline{Z}_\text{L} = 0 \quad \longrightarrow \quad \underline{\Gamma}_\text{L} = -1\,, \quad \underline{V}_\text{sc}(z) = -2\text{j}\underline{V}_{\text{i0}}\sin\beta z\,, \quad \underline{I}_\text{sc}(z) = 2\frac{\underline{V}_{\text{i0}}}{Z_0}\cos\beta z\,,$$

$$\underline{Z}_\text{sc} = \underline{Z}_{\text{in}}\big|_{\underline{Z}_\text{L}=0} = \text{j}Z_0\tan\beta l \quad \text{(short-circuited transmission line)}\,. \qquad (10.13)$$

On the other side, if the line is terminated in an open circuit (oc), that is, if the output terminals ($z = 0$) in Figure 10.5 are left open, as in Figure 10.9(b), then the load current, $\underline{I}_\text{L}$, is forced to be zero, and hence an infinite load impedance. This, in turn, gives [in Eq. (10.4)] a unity load voltage reflection coefficient of the line [it

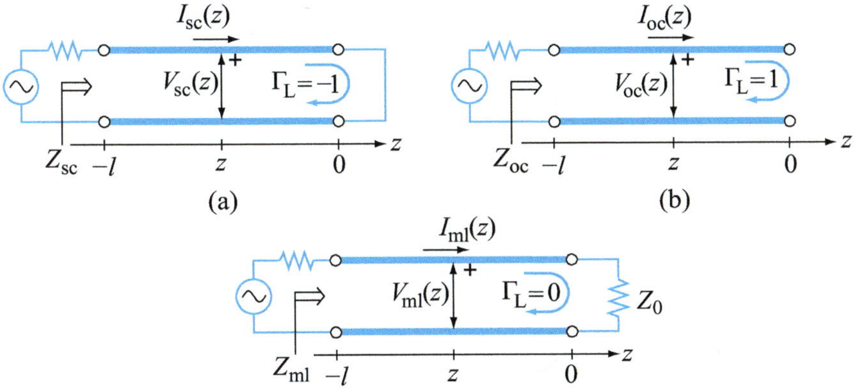

Figure 10.9 Three important special cases of the impedance load termination of a transmission line: (a) short circuit, (b) open circuit, and (c) impedance-matched load.

equals $(\underline{Z}_L - 0)/(\underline{Z}_L - 0) = 1$, since Z_0 can be treated as a zero value in comparison with the infinitely large $|\underline{Z}_L|$]. Using the same equations as for the shorted line, we now obtain

$$|\underline{Z}_L| \to \infty \quad \longrightarrow \quad \Gamma_L = 1\,, \quad \underline{V}_{oc}(z) = 2\underline{V}_{i0}\cos\beta z\,, \quad \underline{I}_{oc}(z) = -2j\frac{\underline{V}_{i0}}{Z_0}\sin\beta z\,,$$

$$\underline{Z}_{oc} = \underline{Z}_{in}\big|_{|\underline{Z}_L|\to\infty} = -jZ_0\cot\beta l \quad \text{(open-circuited transmission line)}\,. \qquad (10.14)$$

As the last special case, if the line is terminated in a load whose impedance is equal to the line characteristic impedance (Z_0), Figure 10.9(c), the load voltage reflection coefficient is zero, we say that the load is matched (by its impedance) to the line – a matched load (ml) – and also refer to the line itself as a matched line. Since there is no reflected wave on the line, we have

$$\underline{Z}_L = Z_0 \quad \longrightarrow \quad \Gamma_L = 0\,, \quad \underline{V}_{ml}(z) = \underline{V}_{i0}\,e^{-j\beta z}\,, \quad \underline{I}_{ml}(z) = \frac{\underline{V}_{i0}}{Z_0}\,e^{-j\beta z}\,,$$

$$\underline{Z}_{ml} = \underline{Z}_{in}\big|_{\underline{Z}_L = Z_0} = Z_0 \quad \text{(matched transmission line)}\,. \qquad (10.15)$$

CONCEPTUAL QUESTION 10.33 **Voltage and current standing wave patterns, short-circuited line.** The voltage and current standing wave patterns of a short-circuited lossless transmission line are as sketched in

(A) Figure 10.10(a).

(B) Figure 10.10(b).

(C) Figure 10.10(c).

(D) none of the above figures.

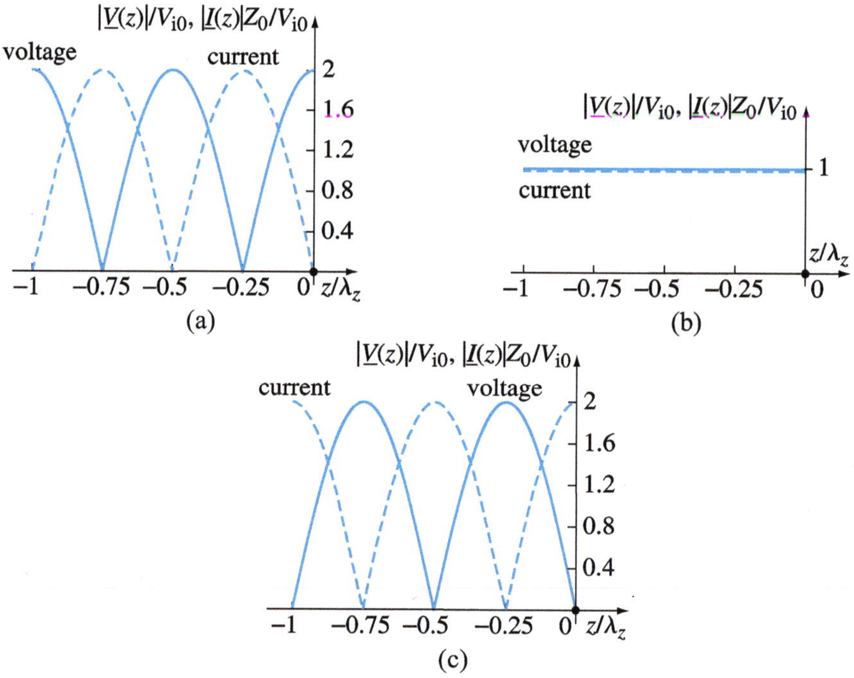

Figure 10.10 Three offered sketches of normalized voltage and current standing wave patterns, $|\underline{V}|/V_{i0}$ and $|\underline{I}|Z_0/V_{i0}$ against z/λ_z, for a lossless transmission line and three different load terminations; for Conceptual Question 10.33.

CONCEPTUAL QUESTION 10.34 **Standing wave patterns, open-circuited line.** Assuming that a lossless transmission line is open-circuited, which one of the three offered sketches in Figure 10.10 may represent the voltage and current standing wave patterns of the line?

(A) The one in Figure 10.10(a).

(B) The one in Figure 10.10(b).

(C) The one in Figure 10.10(c).

(D) None.

CONCEPTUAL QUESTION 10.35 **Standing wave patterns, matched line.** The voltage and current standing wave patterns of a matched lossless transmission line are as in

(A) Figure 10.10(a).

(B) Figure 10.10(b).

(C) Figure 10.10(c).

(D) none of the above figures.

CONCEPTUAL QUESTION 10.36 **Current pattern from a voltage pattern, relative to line loads.** The current at an arbitrary given location on an open-circuited lossless transmission line equals $1/Z_0$ times the voltage at the same location of that same line

(A) if matched (with an impedance-matched load).

(B) with the same load (open circuit).

(C) if short-circuited.

(D) None of the above.

CONCEPTUAL QUESTION 10.37 **Another voltage–current relationship.** The voltage at an arbitrary given location on a matched lossless transmission line (line with a matched load) equals Z_0 times the current at the same location of that same line

(A) with the same (matched) load.

(B) if short-circuited.

(C) if open-circuited.

(D) None of the above.

CONCEPTUAL QUESTION 10.38 **Shifting standing wave patterns along a transmission line.** The voltage and current standing wave patterns of an open-circuited lossless transmission line can be obtained by shifting along the line (along the z-axis) the corresponding patterns of the line when short-circuited by

(A) $\lambda_z/4$ (λ_z is the wavelength along the line).

(B) $\lambda_z/2$.

(C) π.

(D) 2π.

(E) 0.

(F) The two patterns cannot be obtained one from the other.

CONCEPTUAL QUESTION 10.39 **Phase difference between voltage and current, shorted line.** At every location along a short-circuited lossless transmission line, the voltage and current are

(A) in phase.

(B) in counter-phase (180° out of phase with respect to each other).

(C) in time-phase quadrature (±90° out of phase with respect to each other).

(D) out of phase by a constant angle different from 0, 180°, and ±90°.

(E) out of phase by an angle that depends on the distance from the load.

CONCEPTUAL QUESTION 10.40 **Phase difference between voltage and current, open line.** Assuming that a lossless transmission line is open-circuited, the voltage and current at every location along the line are

(A) in phase.

(B) in counter-phase (180° out of phase with respect to each other).

(C) in time-phase quadrature (±90° out of phase with respect to each other).

(D) out of phase by a constant angle different from 0, 180°, and ±90°.

(E) out of phase by an angle that depends on the distance from the load.

CONCEPTUAL QUESTION 10.41 **Phase difference between voltage and current, matched line.** For a lossless transmission line with an impedance-matched load, the voltage and current along the line are

(A) in phase.

(B) in counter-phase (180° out of phase with respect to each other).

(C) in time-phase quadrature (±90° out of phase with respect to each other).

(D) out of phase by a constant angle different from 0, 180°, and ±90°.

(E) out of phase by an angle that depends on the distance from the load.

CONCEPTUAL QUESTION 10.42 **Voltage zero at a quarter-wavelength from a load.** Consider a time-harmonic incident wave on a lossless transmission line with a characteristic impedance of 50 Ω that partially reflects from a load with an unknown impedance. It was found that a voltage zero of the resultant wave on the line, which exhibits standing wave properties, appears in a cross section of the line that is at a distance equal to $\lambda_z/4$ from the load terminals, with λ_z being the wavelength along the line. From this, we can conclude that the impedance of the load is

(A) purely real and greater than 50 Ω.

(B) purely real, nonzero, and less than 50 Ω.

(C) equal to 50 Ω.

(D) purely imaginary.

(E) zero.

(F) infinite.

CONCEPTUAL QUESTION 10.43 **Voltage zero at a half-wavelength from a load.** If the distance of a voltage zero of the standing wave from the load terminals amounts to $\lambda_z/2$ on a lossless transmission line with $Z_0 = 50\ \Omega$, the load impedance is

(A) purely real and greater than 50 Ω.

(B) purely real, nonzero, and less than 50 Ω.

(C) equal to 50 Ω.

(D) purely imaginary.

(E) zero.

(F) infinite.

CONCEPTUAL QUESTION 10.44 **What load makes a transmission line appear as infinitely long?** If terminated in the following load, a lossless transmission line appears as if it were infinitely long, i.e., the load impedance may be considered as an equivalent of an infinite extension of the line beyond the load terminals (to the right, for $0 \leq z < \infty$, in Figure 10.5):

(A) short circuit.

(B) open circuit.

(C) matched load.

(D) None of the above loads.

CONCEPTUAL QUESTION 10.45 **Voltage at generator terminals for a matched load.** In the steady state, the rms voltage V_g at generator terminals of the lossless transmission line shown in Figure 10.11

(A) equals 5 V.

(B) equals 3.33 V.

(C) equals 2.5 V.

(D) equals 2 V.

(E) depends on the length of the line.

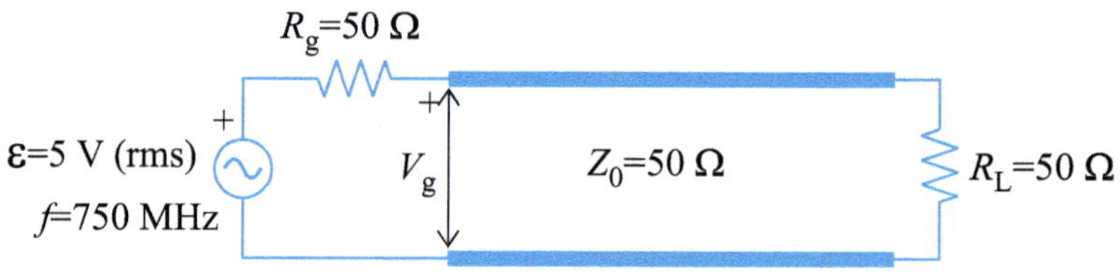

Figure 10.11 Lossless transmission line fed by a time-harmonic real (nonideal) voltage generator (emf and resistive internal impedance in series) and terminated in a matched load; for Conceptual Question 10.45.

CONCEPTUAL QUESTION 10.46 Open-circuited section of a coaxial cable. The input impedance of an open-circuited section of a lossless coaxial cable that is $l = 2\lambda_z/3$ long (λ_z is the wavelength along the cable) comes out to be

(A) nonzero, finite, and purely real.

(B) nonzero, finite, and purely imaginary.

(C) zero.

(D) infinite.

(E) None of the above.

(F) Need more information.

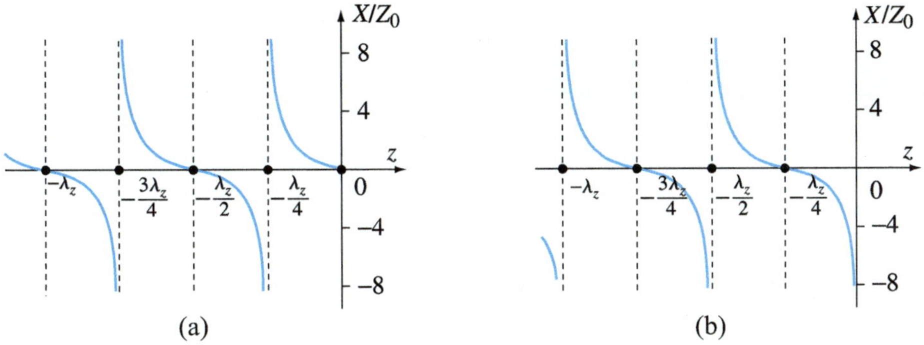

Figure 10.12 Two offered line impedance (reactance) plots for a lossless transmission line and two different load terminations; for Conceptual Question 10.47.

CONCEPTUAL QUESTION 10.47 Impedance plots for short- and open-circuited lines. Which one of the following graphs represents a plot of the transmission-line impedance (reactance) for a short-circuited lossless transmission line, and which one is for an open-circuited line?

(A) Figure 10.12(a) – short-circuited line, Figure 10.12(b) – open-circuited line.

(B) Figure 10.12(a) – open-circuited line, Figure 10.12(b) – short-circuited line.

(C) Figure 10.12(a) – short-circuited line, Figure 10.12(b) – neither of the two cases.

(D) Figure 10.12(a) – open-circuited line, Figure 10.12(b) – neither of the two cases.

(E) Figure 10.12(a) – neither of the two cases, Figure 10.12(b) – open-circuited line.

(F) Figure 10.12(a) – neither of the cases, Figure 10.12(b) – neither of the cases.

CONCEPTUAL QUESTION 10.48 **Input impedance of an open-circuited line.** The input impedance of an open-circuited lossless transmission line is

(A) purely resistive.

(B) purely inductive.

(C) purely capacitive.

(D) inductive complex (has also a resistive part).

(E) capacitive complex.

(F) Need more information.

CONCEPTUAL QUESTION 10.49 **Varying the length of a short-circuited line.** By varying the length of a short-circuited lossless transmission line, for a given operating frequency of the structure, it is possible to realize any input

(A) resistance.

(B) reactance.

(C) complex impedance.

(D) None of the above.

CONCEPTUAL QUESTION 10.50 **Varying the operating frequency of a short-circuited line.** Assume that the operating frequency of a short-circuited lossless transmission line is varied, for a fixed line length. Such a variation may result in any input

(A) resistance.

(B) reactance.

(C) complex impedance.

(D) None of the above.

CONCEPTUAL QUESTION 10.51 **Input impedance of a quarter-wave short-circuited line.** A quarter-wave (quarter-wavelength long, $l = \lambda_z/4$) short-circuited lossless transmission line appears at its input terminals as

(A) a short circuit.

(B) an open circuit.

(C) a resistor of resistance equal to the characteristic impedance of the line.

(D) None of the above.

CONCEPTUAL QUESTION 10.52 **Quarter-wave section of a coaxial cable.** Assume that the length of an open-circuited section of a lossless coaxial cable, with the wavelength (along the line) λ_z, is $l = \lambda_z/4$. The input impedance of this cable section is

(A) nonzero, finite, and purely real.

(B) nonzero, finite, and purely imaginary.

(C) zero.

(D) infinite.

(E) None of the above.

(F) Need more information.

CONCEPTUAL QUESTION 10.53 **Half-wave cable section.** The input impedance of a half-wave section ($l = \lambda_z/2$) of an open-circuited lossless coaxial cable is

(A) nonzero, finite, and purely real.

(B) nonzero, finite, and purely imaginary.

(C) zero.

(D) infinite.

(E) None of the above.

(F) Need more information.

CONCEPTUAL QUESTION 10.54 **Impedance inverting property of a quarter-wave line.** A quarter-wave ($\lambda_z/4$ long) section of a lossless transmission line transforms

(A) an open to a short (but not vice versa).

(B) a short to an open (but not vice versa).

(C) both an open to a short and a short to an open.

(D) None of the above transformations takes place.

CONCEPTUAL QUESTION 10.55 **Transmission line as a lumped capacitor.** Consider a lossless coaxial cable of characteristic impedance $Z_0 = 50\ \Omega$, for which the phase velocity of propagating waves equals $v_p = 2 \times 10^8$ m/s. Is it possible to find a length of an open-circuited section of this cable such that it is equivalent to a lumped capacitor of capacitance $C_{eq} = 10$ pF at a frequency of $f = 500$ MHz?

(A) Yes.

(B) No.

CONCEPTUAL QUESTION 10.56 **Admittance-matching by a shunt short-circuited stub, part 1.** To admittance-match a load of complex admittance $\underline{Y}_L$ to a lossless transmission line of characteristic admittance Y_0 ($Y_0 = 1/Z_0$) using the circuit shown in Figure 10.13, a shunt (connected in parallel) short-circuited transmission line segment (called a stub) should be connected at a location on a transmission line where

(A) the real part of the complex line impedance ($\underline{Z}_{line} = 1/\underline{Y}_{line}$) equals $1/Y_0$.

(B) the imaginary part of the complex line admittance ($\underline{Y}_{line}$) equals Y_0.

(C) the real part of the complex line admittance is zero.

(D) the real part of the complex line admittance equals Y_0.

(E) Any location on the line would enable the matching.

(F) No location on the line would enable the matching.

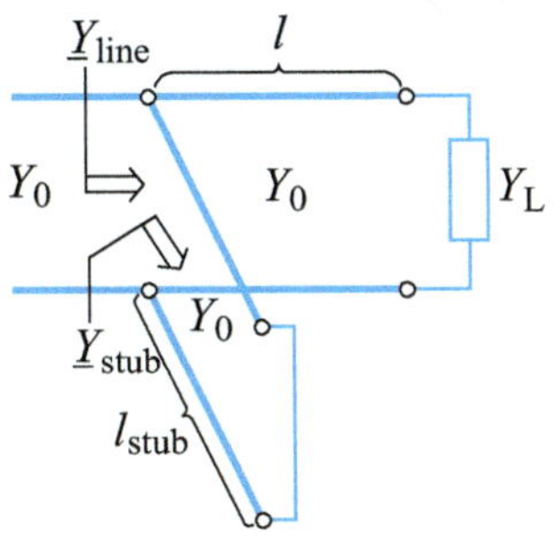

Figure 10.13 Admittance-matching using a shunt short-circuited stub (matching is performed by ensuring that the combined input admittance of the line of length l and the stub of length l_{stub} equals Y_0); for Conceptual Question 10.56.

CONCEPTUAL QUESTION 10.57 Admittance-matching by a shunt stub, part 2. In the admittance-matching circuit in Figure 10.13, the electrical length of the stub should be adjusted so that the stub input admittance

(A) is equal to Y_0.

(B) is equal to $\mathrm{j}Y_0$.

(C) annuls the line admittance ($\underline{Y}_{\text{line}}$).

(D) compensates (cancels) the real part of the line admittance.

(E) compensates the imaginary part of the line admittance.

(F) The proposed circuit configuration cannot enable the matching.

CONCEPTUAL QUESTION 10.58 Admittance-matching by a shunt open-circuited stub. In a circuit for admittance-matching of a complex-admittance load using a shunt stub, the stub is open-circuited, as shown in Figure 10.14. To enable matching, the location of this stub on the transmission line should be chosen such that

(A) the real part of the complex line impedance ($\underline{Z}_{\text{line}} = 1/\underline{Y}_{\text{line}}$) equals $1/Y_0$.

(B) the imaginary part of the complex line admittance ($\underline{Y}_{\text{line}}$) equals Y_0.

(C) the real part of the complex line admittance is zero.

(D) the real part of the complex line admittance equals Y_0.

(E) Any location on the line would enable the matching.

(F) No location on the line would enable the matching.

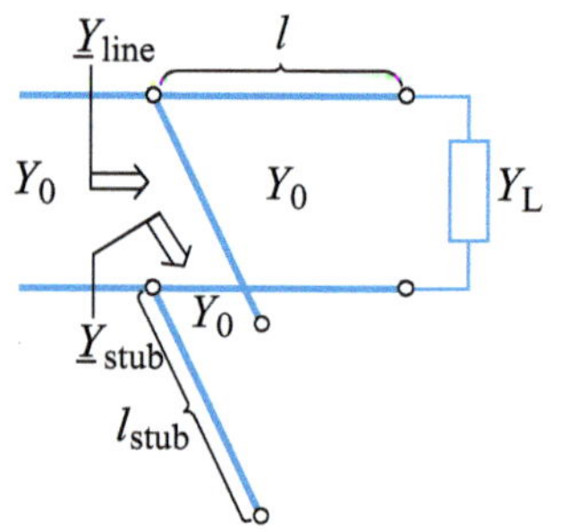

Figure 10.14 Admittance-matching transmission-line circuit with an open-circuited stub (all line sections are lossless); for Conceptual Question 10.58.

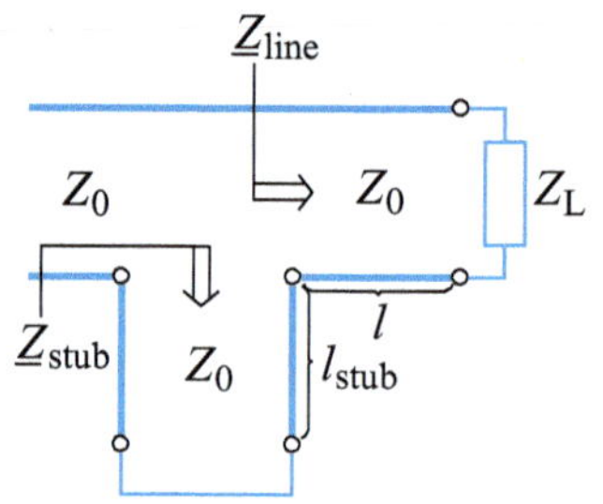

Figure 10.15 Impedance-matching using a series short-circuited stub (the combined input impedance of the line of length l and the stub of length l_{stub} should equal Z_0); for Conceptual Question 10.59.

CONCEPTUAL QUESTION 10.59 Impedance-matching by a series short-circuited stub, part 1. To impedance-match a load of complex impedance $\underline{Z}_L$ to a lossless transmission line of characteristic impedance Z_0 using the circuit shown in Figure 10.15, a series short-circuited stub should be connected at a location on a transmission line where

(A) the real part of the complex line impedance ($\underline{Z}_{\text{line}}$) equals Z_0.

(B) the imaginary part of the complex line impedance equals Z_0.

(C) the real part of the complex line impedance is zero.

(D) the real part of the complex line admittance ($\underline{Y}_{\text{line}} = 1/\underline{Z}_{\text{line}}$) equals $1/Z_0$.

(E) Any location on the line would enable the matching.

(F) No location on the line would enable the matching.

CONCEPTUAL QUESTION 10.60 **Impedance-matching by a series stub, part 2.** In the impedance-matching circuit in Figure 10.15, the electrical length of the stub should be adjusted so that the stub input impedance

(A) is equal to Z_0.

(B) is equal to $\mathrm{j}Z_0$.

(C) annuls the line impedance ($\underline{Z}_{\text{line}}$).

(D) compensates (cancels) the real part of the line impedance.

(E) compensates the imaginary part of the line impedance.

(F) The proposed circuit configuration cannot enable the matching.

CONCEPTUAL QUESTION 10.61 **Transmission-line resonator.** Considering a short-circuited lossless transmission line, there are some locations on the line where we can also short-circuit (galvanically connect together) the line conductors, and nothing will change in the entire structure, namely, the newly added short circuit will not affect the voltage and current on the line. We can then remove the part of the transmission line on the other side of the new short circuit, to obtain a self-contained structure (shorted at both ends), with a standing wave trapped between the two short circuits – this structure represents a transmission-line resonator. With λ_z denoting the wavelength along the line, the possible distances from the load end at which the new short circuit is added, namely, the possible lengths of the resonator, are

(A) $l = m\lambda_z/2$ $(m = 1, 2, \ldots)$.

(B) $l = m\lambda_z/4$ $(m = 1, 2, \ldots)$.

(C) $l = m\lambda_z/8$ $(m = 1, 2, \ldots)$.

(D) $l = (2m + 1)\lambda_z/4$ $(m = 0, 1, 2, \ldots)$.

(E) any l, i.e., any location on the line.

(F) There is no such l, i.e., no such location on the line.

10.5 The Smith Chart

As the last topic in the circuit theory of two-conductor transmission lines in a time-harmonic regime, in this section we present an alternative, graphical, technique for the circuit analysis and design of transmission lines in the frequency domain. The technique is based on the so-called Smith chart. This chart is, essentially, a polar plot of the generalized voltage reflection coefficient, $\underline{\Gamma}(z)$, given by Eq. (10.7), along a transmission line (in Figure 10.5), lying in the complex plane of $\underline{\Gamma} = \Gamma_r + j\Gamma_i$, i.e., the Γ_r–Γ_i plane, as depicted in Figure 10.16(a). Confining our analysis to the lossless case, the magnitude of $\underline{\Gamma}$ is constant along the line (for $-l \leq z \leq 0$), which corresponds to a circle of radius $|\underline{\Gamma}|$ in Figure 10.16(a), referred to as the s circle or SWR circle [see Eq. (10.11)]. Since $0 \leq |\underline{\Gamma}| \leq 1$, Eq. (10.6), the Smith chart is bounded by the circle defined by $|\underline{\Gamma}| = 1$, called the unit circle. Impedances are displayed on the chart using their normalized (to Z_0) values, $\underline{z}_n$. By means of Eq. (10.12),

$$\underline{z}_n = \frac{\underline{Z}}{Z_0} = \frac{R}{Z_0} + j\frac{X}{Z_0} = r + jx = \frac{1+\underline{\Gamma}}{1-\underline{\Gamma}} \quad \text{(normalized line impedance)}, \quad (10.16)$$

where r and x are the normalized line resistance and reactance, respectively. From this relationship between $\underline{\Gamma}$ and $\underline{z}_n$, r = const is an equation of a circle, r circle, in the Γ_r–Γ_i plane, while x = const gives an x arc, as shown in Figure 10.16(b).

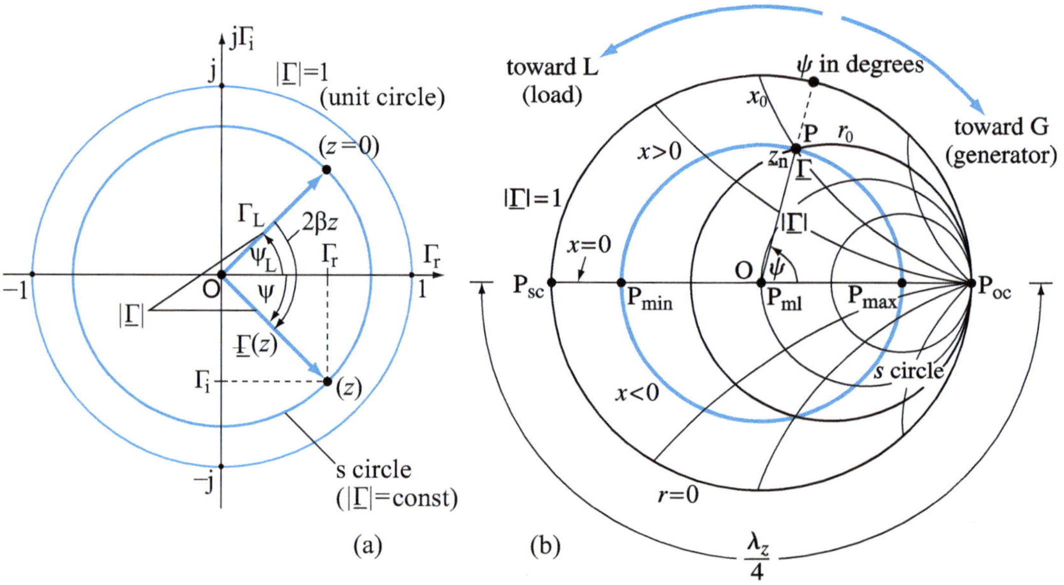

Figure 10.16 (a) Graphical representation in the complex plane of the generalized voltage reflection coefficient $\underline{\Gamma}(z)$, Eq. (10.7), along a lossless transmission line (Figure 10.5), as a basis for construction of the Smith chart. (b) Highlighting several key features of the Smith chart and its use.

Superposing the r circles and x arcs for a large number of respective values of r and x we obtain the Smith chart – provided in Figure 10.17. On the chart, a normalized line impedance given by $\underline{z}_{\rm n} = r_0 + {\rm j}x_0$ corresponds to the point of intersection of the $r = r_0$ circle and $x = x_0$ arc. Overall, the Smith chart simultaneously displays values of $\underline{z}_{\rm n}$ and $\underline{\Gamma}$, according to the relationship in Eq. (10.16), in a convenient format – for graphical calculations on transmission lines and/or visualization of analysis data.

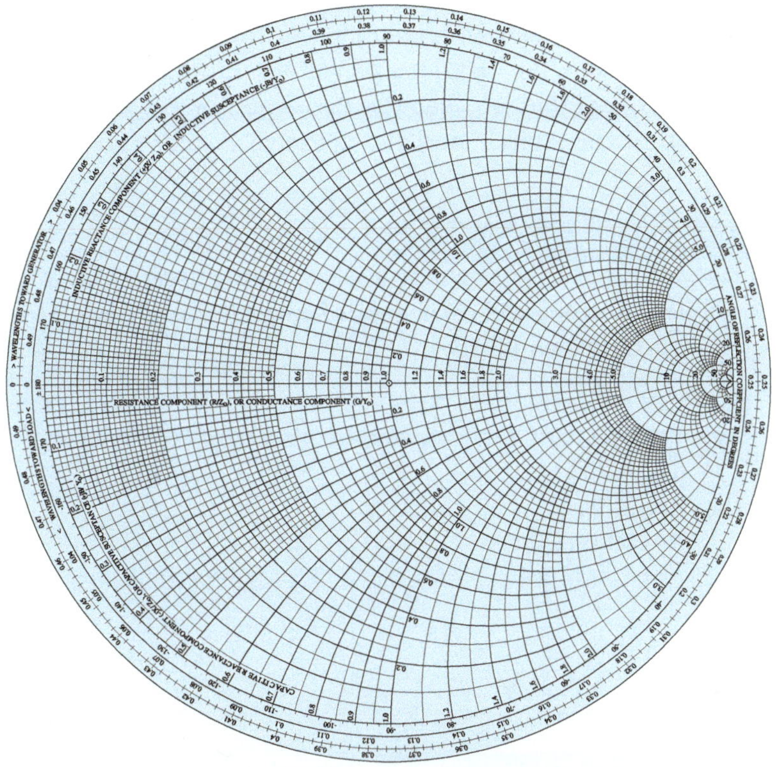

Figure 10.17 The Smith chart.

Figure 10.16(b) highlights several key features of the chart in Figure 10.17. For instance, the three important special cases of load terminations of a transmission line in Figure 10.9, a short circuit, open circuit, and matched load, are marked in Figure 10.16(b): points $\rm P_{sc}$, $\rm P_{oc}$, and $\rm P_{ml}$, respectively. The figure also emphasizes that the upper (lower) half of the chart corresponds to inductive (capacitive) line impedances. From Eq. (10.7), the change in the coordinate z due to a movement along the line (with no losses), in Figure 10.5, results in the following change of the phase angle ψ of the reflection coefficient $\underline{\Gamma}(z)$: $\Delta\psi = 2\beta\Delta z$ ($\beta = 2\pi/\lambda_z$). Hence, a clockwise rotation on the chart in Figure 10.16(b) ($\Delta\psi < 0$) corresponds to a movement toward the generator (G) along the line in Figure 10.5 ($\Delta z < 0$), while moving toward the load (L) in Figure 10.5 gives the counterclockwise direction of rotation on the chart. In movements in either direction, one complete rotation around the chart corresponds to a half-wave shift along the line,

$$360^\circ \text{ around the Smith chart} \quad \longleftrightarrow \quad \frac{\lambda_z}{2} \text{ along the transmission line}\,. \qquad (10.17)$$

With the use of Eq. (10.8), we realize that voltage maxima ($|\underline{V}| = |\underline{V}|_{\text{max}}$) on the line in Figure 10.5 occur at locations where the generalized voltage reflection coefficient is purely real and positive, and this corresponds to the point P_{max} on the Smith chart [Figure 10.16(b)] where the s circle intersects the positive Γ_r-axis (i.e., the line $\overline{OP_{\text{oc}}}$). On the other side, $\underline{\Gamma}$ at locations of voltage minima ($|\underline{V}| = |\underline{V}|_{\text{min}}$) on the line is purely real and negative, so that the corresponding point P_{min} on the chart is at the intersection of the s circle and the negative Γ_r-axis ($\overline{OP_{\text{sc}}}$).

CONCEPTUAL QUESTION 10.62 **Transforming a purely reactive load impedance.** By varying the length of a lossless transmission line of characteristic impedance $Z_0 = 50\ \Omega$ terminated in a load of impedance $\underline{Z}_L = \text{j}50\ \Omega$, we cannot obtain the input impedance equal to

(A) $50\ \Omega$.

(B) that of an open circuit.

(C) that of a short circuit.

(D) $\text{j}50\ \Omega$.

(E) $-\text{j}50\ \Omega$.

CONCEPTUAL QUESTION 10.63 **Transforming a purely resistive load impedance.** By varying the length of a lossless transmission line of characteristic impedance $Z_0 = 50\ \Omega$, it is possible to transform a load impedance of $\underline{Z}_L = (100 + \text{j}0)\ \Omega$ to an input impedance equal to

(A) $(100 + \text{j}100)\ \Omega$.

(B) $\text{j}100\ \Omega$.

(C) $15\ \Omega$.

(D) $25\ \Omega$.

(E) $150\ \Omega$.

(F) More than one of the above impedances.

CONCEPTUAL QUESTION 10.64 **Load resistance greater than the line characteristic impedance.** A lossless transmission line of characteristic impedance $Z_0 = 100\ \Omega$ and length equal to one eighth of the wavelength along the line is terminated in a load of impedance $\underline{Z}_L = (200 + \mathrm{j}0)\ \Omega$. The input impedance of the line is

(A) purely resistive, greater than 100 Ω.

(B) purely resistive, less than 100 Ω.

(C) purely resistive, equal to 100 Ω.

(D) purely reactive.

(E) inductive complex (has also a resistive part).

(F) capacitive complex.

CONCEPTUAL QUESTION 10.65 **Load resistance less than the line characteristic impedance.** Assuming that the load impedance terminating a lossless transmission line with $Z_0 = 100\ \Omega$ and $l = \lambda_z/8$ amounts to $\underline{Z}_L = (50+\mathrm{j}0)\ \Omega$, the input impedance of the line comes out to be

(A) purely resistive, greater than 100 Ω.

(B) purely resistive, less than 100 Ω.

(C) purely resistive, equal to 100 Ω.

(D) purely reactive.

(E) inductive complex (has also a resistive part).

(F) capacitive complex.

CONCEPTUAL QUESTION 10.66 **Load resistance equal to the line characteristic impedance.** The input impedance of a lossless transmission line with $Z_0 = 100\ \Omega$, $l = \lambda_z/8$, and $\underline{Z}_L = (100 + \mathrm{j}0)\ \Omega$ (load) is

(A) purely resistive, greater than 100 Ω.

(B) purely resistive, less than 100 Ω.

(C) purely resistive, equal to 100 Ω.

(D) purely reactive.

(E) inductive complex (has also a resistive part).

(F) capacitive complex.

CONCEPTUAL QUESTION 10.67 **Finding load impedance using the Smith chart.** Assume that a lossless transmission line of characteristic impedance $Z_0 = 100\ \Omega$ and length equal to 1/8 of the wavelength along the line is terminated in an unknown load. If it is determined that the input impedance of the line amounts to $\underline{Z}_{\text{in}} = (200 + \text{j}0)\ \Omega$, what can be said about the load impedance? The impedance $\underline{Z}_{\text{L}}$ is

(A) purely resistive, greater than 100 Ω.

(B) purely resistive, less than 100 Ω.

(C) purely resistive, equal to 100 Ω.

(D) purely reactive.

(E) inductive complex (has also a resistive part).

(F) capacitive complex.

CONCEPTUAL QUESTION 10.68 **Change of the line length.** If the lossless transmission line with $Z_0 = 100\ \Omega$ is five eighths of the wavelength long and its input impedance is $\underline{Z}_{\text{in}} = (200 + \text{j}0)\ \Omega$, the load impedance ($\underline{Z}_{\text{L}}$) for the line turns out to be

(A) purely resistive, greater than 100 Ω.

(B) purely resistive, less than 100 Ω.

(C) purely resistive, equal to 100 Ω.

(D) purely reactive.

(E) inductive complex (has also a resistive part).

(F) capacitive complex.

CONCEPTUAL QUESTION 10.69 **Slotted line measurement and the Smith chart.** Based on measurements on a lossless air-filled slotted coaxial line connected to an unknown load impedance showing that the distance between two adjacent voltage maxima is $\Delta l = 15$ cm, we can say – with the help of the Smith chart – that the frequency of the generator is

(A) $f = 1$ GHz.

(B) $f = 2$ GHz.

(C) $f = 500$ MHz.

(D) $f = 4$ GHz.

(E) None of the above frequencies.

(F) Need more information.

CONCEPTUAL QUESTION 10.70 **Voltage maximum in the Smith chart.** For a lossless transmission line with a characteristic impedance of 50 Ω carrying a time-harmonic wave that partially reflects from a load with an unknown impedance, the resultant wave on the line exhibits standing wave properties and a voltage maximum is identified at the load terminals. From this, it can be concluded, using the Smith chart, that the impedance of the load ($\underline{Z}_L$) is

(A) purely real and greater than 50 Ω.

(B) purely real, nonzero, and less than 50 Ω.

(C) equal to 50 Ω.

(D) purely imaginary.

(E) zero.

(F) infinite.

CONCEPTUAL QUESTION 10.71 **Voltage minimum in the Smith chart.** If a voltage minimum, which is not zero, of the resultant standing wave occurs at the load terminals of a lossless transmission line with $Z_0 = 50$ Ω, we can conclude, with the help of the Smith chart, that the impedance $\underline{Z}_L$ is

(A) purely real and greater than 50 Ω.

(B) purely real, nonzero, and less than 50 Ω.

(C) equal to 50 Ω.

(D) purely imaginary.

(E) zero.

(F) infinite.

CONCEPTUAL QUESTION 10.72 **Transforming an open circuit in the Smith chart.** Consider an open-circuited section of a lossless coaxial cable that is $l = 2\lambda_z/3$ in length, λ_z standing for the wavelength along the cable. With the use of the Smith chart, the input impedance of this cable section can be characterized as being

(A) nonzero, finite, and purely real.

(B) nonzero, finite, and purely imaginary.

(C) zero.

(D) infinite.

(E) None of the above.

(F) Need more information.

CONCEPTUAL QUESTION 10.73 **Quarter-wave shift along a line, and the Smith chart.** Viewing the Smith chart, the input impedance of a $\lambda_z/4$ long open-circuited section of a lossless coaxial cable can be determined to be

(A) nonzero, finite, and purely real.

(B) nonzero, finite, and purely imaginary.

(C) zero.

(D) infinite.

(E) None of the above.

(F) Need more information.

CONCEPTUAL QUESTION 10.74 Half-wave movement and the Smith chart. Using the Smith chart, the input impedance of a half-wave ($\lambda_z/2$ long) open-circuited lossless coaxial-cable section comes out to be

(A) nonzero, finite, and purely real.

(B) nonzero, finite, and purely imaginary.

(C) zero.

(D) infinite.

(E) None of the above.

(F) Need more information.

CONCEPTUAL QUESTION 10.75 Impedance inverting in the Smith chart. With the help of the Smith chart, we realize that a quarter-wave ($l = \lambda_z/4$) section of a lossless transmission line transforms

(A) an open to a short (but not vice versa).

(B) a short to an open (but not vice versa).

(C) both an open to a short and a short to an open.

(D) None of the above transformations takes place.

CONCEPTUAL QUESTION 10.76 Transmission lines with equal input impedances. Two lossless 50-Ω (characteristic impedance) transmission lines of different lengths operate at the same frequency, and the wavelength along the lines is the same, $\lambda_z = 1$ m. The lines are terminated in equal complex loads, of impedance $\underline{Z}_L = (100 - j50)$ Ω, and the input impedances of the lines are also the same. Which of the following combinations of values for l_1 and l_2 are possible lengths of the two lines?

(A) $l_1 = 1.2$ m and $l_2 = 0.2$ m.

(B) $l_1 = 1.2$ m and $l_2 = 0.7$ m.

(C) $l_1 = 0.95$ m and $l_2 = 0.7$ m.

(D) $l_1 = 0.5$ m and $l_2 = 0.25$ m.

(E) More than one of the above combinations.

(F) None of the above combinations.

CONCEPTUAL QUESTION 10.77 **Numbers of voltage maxima and minima on a transmission line.** A lossless transmission line of length $l = 2\lambda_z$ (λ_z is the wavelength along the line) is terminated in a load with normalized impedance $\underline{z}_{\rm L} = \underline{Z}_{\rm L}/Z_0 = 1 + {\rm j}1$ (Z_0 is the characteristic impedance of the line). What are the total numbers of voltage maxima and minima, respectively, on the line?

(A) 2 voltage maxima and 2 voltage minima.

(B) 2 voltage maxima and 3 voltage minima.

(C) 3 voltage maxima and 2 voltage minima.

(D) 4 voltage maxima and 4 voltage minima.

(E) 4 voltage maxima and 5 voltage minima.

(F) 5 voltage maxima and 4 voltage minima.

CONCEPTUAL QUESTION 10.78 **Counting voltage maxima and minima for a different line length.** Assuming that the length of a lossless transmission line with the normalized load impedance of $\underline{z}_{\rm L} = \underline{Z}_{\rm L}/Z_0 = 1 + {\rm j}1$ is $l = 2.25\lambda_z$, there are a total of

(A) 2 voltage maxima and 2 voltage minima

(B) 2 voltage maxima and 3 voltage minima

(C) 3 voltage maxima and 2 voltage minima

(D) 4 voltage maxima and 4 voltage minima

(E) 4 voltage maxima and 5 voltage minima

(F) 5 voltage maxima and 4 voltage minima

on the line.

CONCEPTUAL QUESTION 10.79 Normalized input impedance of a transmission line. What is the normalized input impedance ($\underline{z}_{\text{in}} = \underline{Z}_{\text{in}}/Z_0$) of a lossless transmission line of length $l = 2\lambda_z$ terminated in a load with $\underline{z}_{\text{L}} = \underline{Z}_{\text{L}}/Z_0 = 1 + \text{j}1$?

(A) $\underline{z}_{\text{in}} = 1 + \text{j}1$.

(B) $\underline{z}_{\text{in}} = 1 - \text{j}1$.

(C) $\underline{z}_{\text{in}} = 1 - \text{j}2$.

(D) $\underline{z}_{\text{in}} = 2 + \text{j}2$.

(E) $\underline{z}_{\text{in}} = 1$.

(F) $\underline{z}_{\text{in}} = \text{j}1$.

10.6 Transient Analysis of Transmission Lines with Step Excitations

Temporary variations of voltages and currents on transmission lines in establishing the steady state of time-harmonic or any other forms of signals in the structure are called transients. A notable example is transients produced by step-like (on or off) abrupt changes of the input voltage or current at the beginning of a transmission line, which corresponds to establishing a time-constant (dc) voltage and current along the line. To start our transient analysis of transmission lines, let us assume that the voltage generator at the beginning of the line in Figure 10.5 is given by its time-varying emf $e(t)$ and internal resistance R_{g}, and that the load is purely resistive, of resistance R_{L}. In addition, for the convenience of the study in this and the sections to follow, we adopt the origin of the z-axis ($z = 0$) to be at the generator terminals, as shown in Figure 10.18(a).

Let the emf in Figure 10.18(a) be a step time function, defined as $e(t) = 0$ for $t < 0$ and $e(t) = \mathcal{E} = \text{const}$ for $t > 0$. Prior to the time $t = 0$, the voltage and current are identically zero at every position along the line. At $t = 0$, the generator launches an incident voltage v_{i1}, and the accompanying current i_{i1}, to propagate toward the load, as indicated in Figure 10.18(b). The incident wavefront reaches the load at instant $t = T$, where T designates the full one-way transit or delay time along the line,

$$T = \frac{l}{v_{\text{p}}} \quad \text{(one-way delay time of a transmission line)} \qquad (10.18)$$

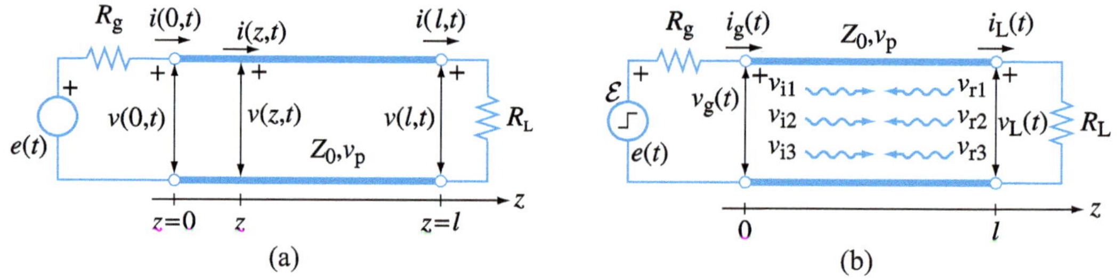

Figure 10.18 (a) Transmission line fed by a time-varying (nonsinusoidal) voltage generator with a purely resistive internal impedance and terminated in a purely resistive load. (b) Determining the step response of a lossless transmission line with a purely resistive load by the multiple reflection tracking method: components of the incident and reflected voltage waves on the line.

(v_p is the phase velocity along the line). Prior to the return of any backward propagating wave reflected from the load, the only wave on the line is the incident one. The dynamic (for transients) line impedance at the generator terminals, that is, the input dynamic impedance of the line, amounts to

$$(Z_{in})_{dynamic} = \frac{v_{i1}}{i_{i1}} = Z_0 \quad \text{(input dynamic impedance of a transmission line)} . \tag{10.19}$$

In other words, the generator sees, looking into the line, a purely resistive impedance equal to the characteristic impedance of the line, Z_0, and from the voltage divider formed by Z_0 and R_g,

$$v_{i1} = \frac{Z_0}{Z_0 + R_g}\mathcal{E} \quad \text{(initial incident voltage on a line)} . \tag{10.20}$$

In the general case for Figure 10.18(b), the load is not matched to the line ($R_L \neq Z_0$), so that a reflected wave, of voltage $v_{r1} = \Gamma_L v_{i1}$, is generated at the load. The load voltage reflection coefficient (Γ_L) is given by Eq. (10.4) with $\underline{Z}_L$ replaced by R_L. As the reflected wave travels away from the load, in Figure 10.18(b), its wavefront arrives at the generator at $t = 2T$. The travel of these forward and backward voltage waves on the transmission line and how they are superposed to each other can be viewed in Figure 10.19. If the generator is not matched to the line, that is, if $R_g \neq Z_0$, reflection (namely, re-reflection) of the reflected voltage v_{r1} from the generator occurs. The expression for the voltage reflection coefficient at the generator (Γ_g) is the same as for the load coefficient but with R_g in place of the load resistance, so we have

$$\Gamma_g = \frac{R_g - Z_0}{R_g + Z_0}, \quad \Gamma_L = \frac{R_L - Z_0}{R_L + Z_0} \quad \text{(generator and load reflection coefficients)} . \tag{10.21}$$

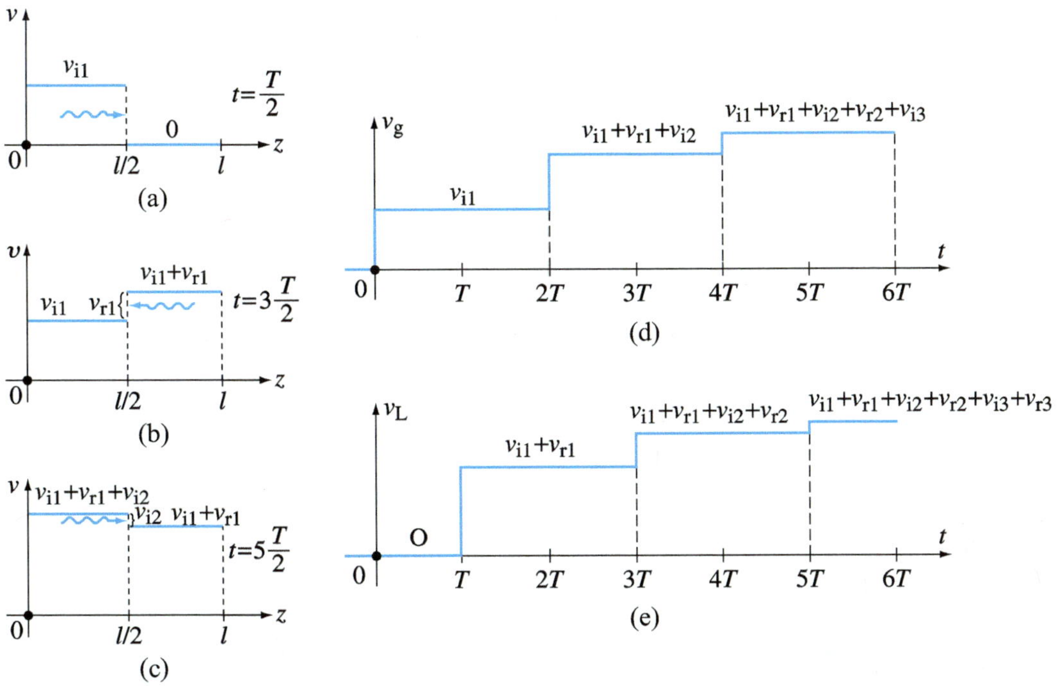

Figure 10.19 Total voltage distribution, $v(z,t)$, $0 \leq z \leq l$, on the transmission line in Figure 10.18(b) for R_L, $R_g > Z_0$ at times (a) $t = T/2$, (b) $t = 3T/2$, and (c) $t = 5T/2$, and voltage waveforms at (d) the generator ($z = 0$) and (e) the load ($z = l$) for $0 \leq t \leq 6T$.

The voltage of this new incident (forward) wave in Figure 10.18(b) is $v_{i2} = \Gamma_g v_{r1}$. Next, v_{i2} bounces off the load, with the reflection coefficient Γ_L, at $t = 3T$, so that the new reflected voltage [Figure 10.18(b)] is $v_{r2} = \Gamma_L v_{i2}$, and so on ($v_{i3} = \Gamma_g v_{r2}$, $v_{r3} = \Gamma_L v_{i3}$, ...). Figure 10.19 shows snapshots (scans) of the total voltage along the line, $v(z,t)$, at three characteristic instants of time and voltage waveforms at both ends of the line, $v_g(t) = v(0,t)$ and $v_L(t) = v(l,t)$, within an interval of time.

CONCEPTUAL QUESTION 10.80 **One-way wave travel time along two coaxial cables.** A pulse signal is launched to propagate along two coaxial cables of the same dimensions but with dielectrics of different permittivities, as shown in Figure 10.20. Both dielectrics are nonmagnetic and lossless. The wave travel time from the generator to the load along the cable with the relative permittivity $\varepsilon_r = 2$ is

(A) four times

(B) a quarter of

(C) two times

(D) a half of

(E) equal to

the travel time along the cable with $\varepsilon_r = 8$.

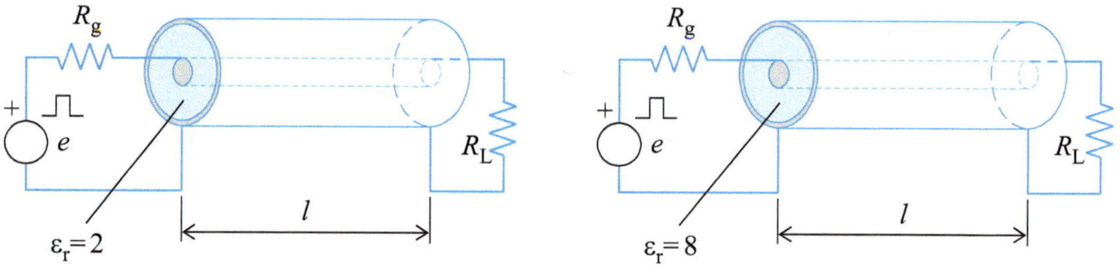

Figure 10.20 Pulse travel along two coaxial cables with different dielectrics; for Conceptual Question 10.80.

CONCEPTUAL QUESTION 10.81 Initial transient voltage at generator terminals of a line. Assume that a lossless transmission line of characteristic impedance $Z_0 = 50\ \Omega$ is driven by a voltage generator of step emf $\mathcal{E} = 5$ V applied at $t = 0$, as shown in Figure 10.21. At $t = 0^+$, the transient voltage at the beginning (at generator terminals) of the line, $v_g(t)$,

(A) equals 5 V.

(B) equals 3.33 V.

(C) equals 2.5 V.

(D) equals 2 V.

(E) depends on the length of the line.

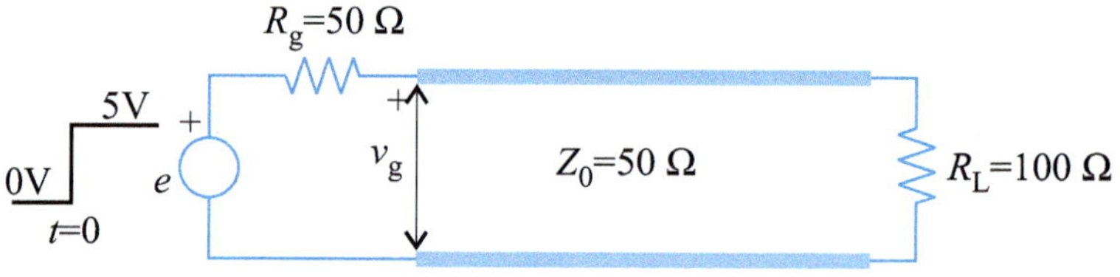

Figure 10.21 Circuit from Figure 10.8 but with a step excitation (step emf applied at $t = 0$); for Conceptual Question 10.81.

CONCEPTUAL QUESTION 10.82 **Step response for a matched load.** Assuming that the load terminating a lossless transmission line with a step excitation (as in Figure 10.21) is matched to the line, the transitional regime on the line lasts only one one-way wave travel time along the line, T, after which a steady state is established on the entire line,

(A) only if the generator is matched to the line as well.

(B) only if the generator is not matched to the line.

(C) regardless of whether the generator is matched to the line or not.

CONCEPTUAL QUESTION 10.83 **Step response for a matched generator and unmatched load.** In the transmission-line circuit in Figure 10.21, a steady state is established at the generator terminals, for the voltage $v_{\mathrm{g}}(t)$, at instant

(A) $t = 0$,

(B) $t = T$,

(C) $t = 2T$,

(D) $t = 3T$,

(E) $t = 4T$,

(F) $t \to \infty$,

where T is the one-way delay period of the line.

CONCEPTUAL QUESTION 10.84 **Step response for unmatched generator and load.** For the lossless transmission line with a step excitation and a transit period T shown in Figure 10.22, the voltage $v_{\mathrm{g}}(t)$ reaches a steady state at instant

(A) $t = 0$.

(B) $t = T$.

(C) $t = 2T$.

(D) $t = 3T$.

(E) $t = 4T$.

(F) $t \to \infty$.

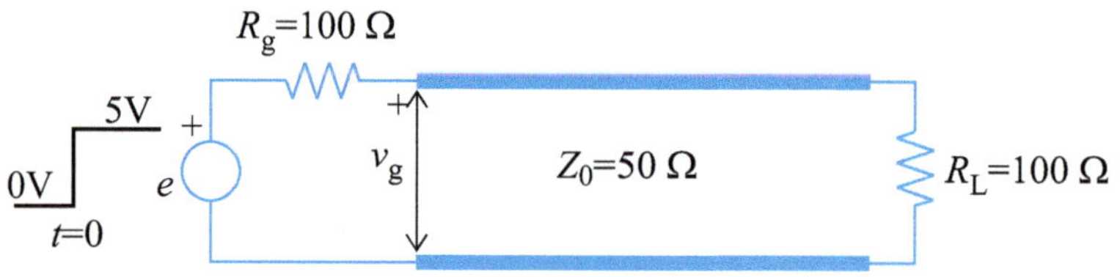

Figure 10.22 Another transmission-line circuit with transient voltages and currents; for Conceptual Question 10.84.

CONCEPTUAL QUESTION 10.85 **Is a steady state for step response analysis actually a dc regime?** A steady state for a structure excited by a voltage step generator (as the circuit in Figure 10.22) is actually a dc regime, in which a lossless transmission line, if present in the structure, can be considered simply as a pair of ideal short-circuiting conductors.

(A) True.

(B) False.

CONCEPTUAL QUESTION 10.86 **Steady state in a model with distributed capacitors and inductors.** Consider a lossless transmission line with a step excitation and unmatched generator and load (e.g., the line in Figure 10.22), and the distributed shunt capacitors and series inductors in a circuit model of the line in the form of a ladder network of elementary circuit cells with lumped elements (see Figure 10.23). In a steady state on the (lossless) line, we have the following for voltages of the capacitors and for the currents of the inductors:

(A) all voltages are zero; all currents are zero.

(B) all voltages are the same, nonzero; all currents are the same, nonzero.

(C) all voltages are the same, nonzero; all currents are zero.

(D) all voltages are zero; all currents are the same, nonzero.

(E) all voltages are the same, nonzero; currents are not the same.

(F) voltages are not the same; currents are not the same.

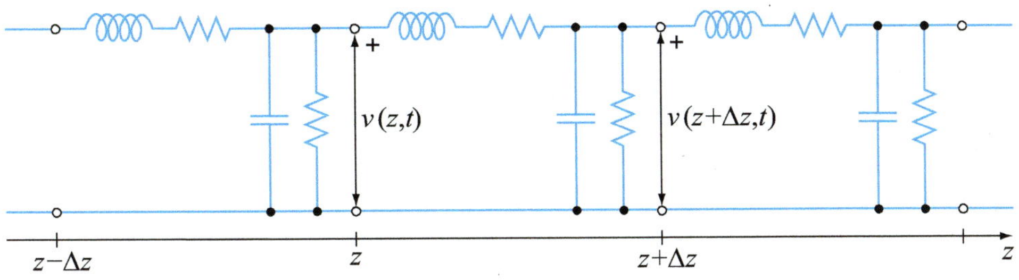

Figure 10.23 Circuit model of an arbitrary (lossy) transmission line with transient voltages and currents; for Conceptual Question 10.86.

CONCEPTUAL QUESTION 10.87 Steady-state voltage on a transmission line. For the transmission-line circuit shown in Figure 10.24 ($\mathcal{E} = 10$ V, $R_g = R_L = 100\ \Omega$, and $Z_0 = 400\ \Omega$), the voltage across the generator terminals, v_g, in the steady state

(A) equals 10 V.

(B) equals 2 V.

(C) equals 8 V.

(D) equals 5 V.

(E) is zero.

(F) depends on the length of the line.

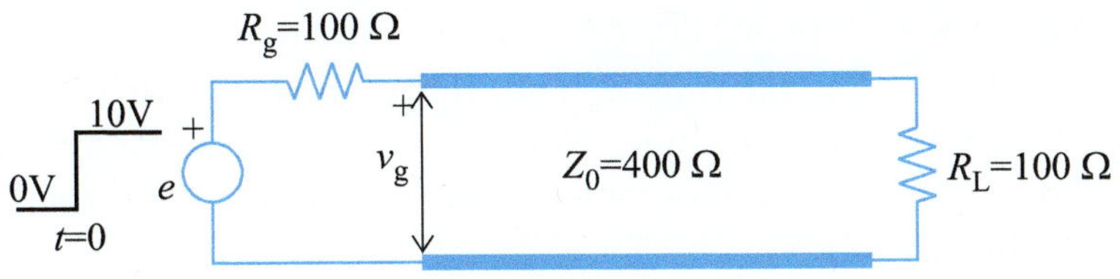

Figure 10.24 Evaluation of the steady-state voltage on a lossless transmission line with an unmatched generator and an unmatched load, excited by a step emf; for Conceptual Question 10.87.

CONCEPTUAL QUESTION 10.88 **Time-domain reflectometry.** A time-domain reflectometer (TDR) consisting of a step voltage generator and an oscilloscope is used to determine the resistance of an unknown load resistor connected to a lossless transmission line. The characteristic impedance of the line is $Z_0 = 50\ \Omega$, and so is the internal resistance of the generator. If the line is fed by this generator, with emf $\mathcal{E} = 4$ V applied at $t = 0$, the oscilloscope displays voltage $v_g(t)$ in Figure 10.25 at the generator end of the line. What is the load resistance?

(A) $R_L = 0$.

(B) $R_L = 15\ \Omega$.

(C) $R_L = 30\ \Omega$.

(D) $R_L = 50\ \Omega$.

(E) $R_L = 150\ \Omega$.

(F) $R_L \to \infty$.

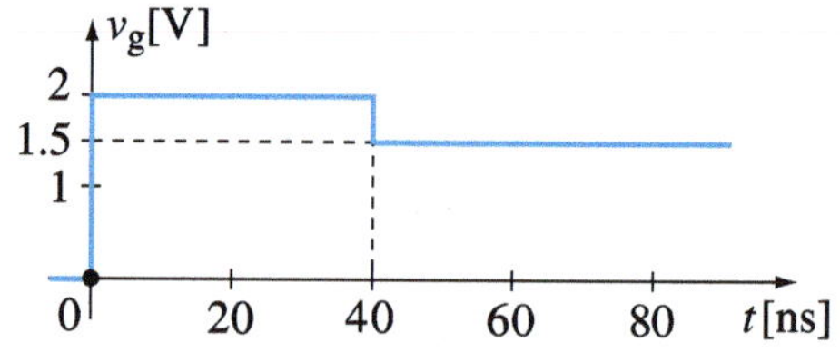

Figure 10.25 Measuring unknown load termination of a transmission line by time-domain reflectometry: oscilloscope display of the voltage step response of the line at its generator end ($R_g = Z_0$), from which the load resistance, R_L, is determined; for Conceptual Question 10.88.

CONCEPTUAL QUESTION 10.89 **Determining the line length from a TDR display.** Assume that the time-domain reflectometer is used to determine the unknown length of a transmission line, which is filled with polyethylene ($\varepsilon_r = 2.25$ and $\mu_r = 1$). From the TDR oscilloscope display in Figure 10.25, the length of the line turns out to be

(A) $l = 3$ m.

(B) $l = 4$ m.

(C) $l = 6$ m.

(D) $l = 8$ m.

(E) $l = 12$ m.

(F) $l = 16$ m.

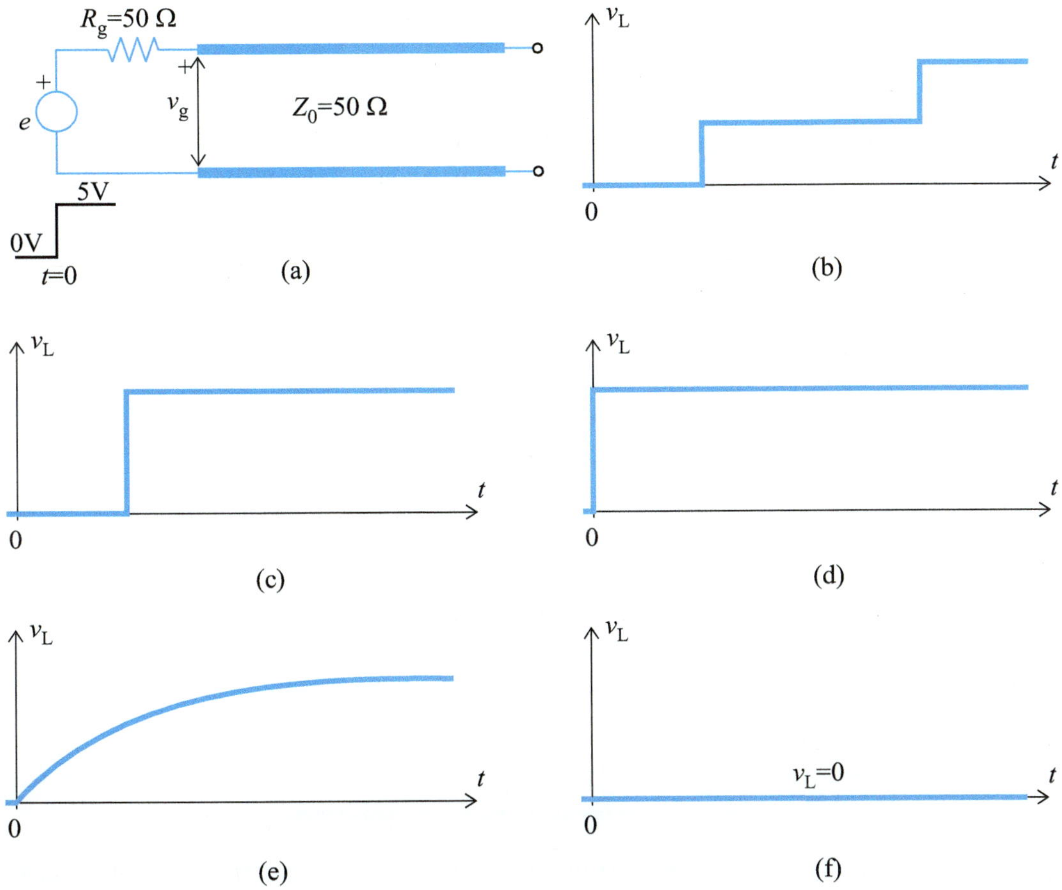

Figure 10.26 (a) Open-circuited transmission line with a step excitation (step emf $\mathcal{E} = 5$ V applied at $t = 0$). (b)–(f) Step response of the line: five offered sketches for the voltage waveform at the load; for Conceptual Question 10.90.

CONCEPTUAL QUESTION 10.90 **Step response of an open-circuited line.** Consider a lossless transmission line of characteristic impedance $Z_0 = 50\ \Omega$ driven by a voltage generator of step emf $\mathcal{E} = 5$ V and internal resistance $R_g = 50\ \Omega$ and open-circuited at its load end ($R_L \to \infty$), as shown in Figure 10.26(a). The voltage waveform at the load, $v_L(t)$, is as sketched in

(A) Figure 10.26(b).

(B) Figure 10.26(c).

(C) Figure 10.26(d).

(D) Figure 10.26(e).

(E) Figure 10.26(f).

CONCEPTUAL QUESTION 10.91 **Step response of a short-circuited line.** Assuming that the line in Figure 10.26(a) is short-circuited at the load end ($R_L = 0$), as shown in Figure 10.27, which one of the five offered sketches in Figure 10.26 may represent the voltage waveform at the load, $v_L(t)$?

(A) The one in Figure 10.26(b).

(B) The one in Figure 10.26(c).

(C) The one in Figure 10.26(d).

(D) The one in Figure 10.26(e).

(E) The one in Figure 10.26(f).

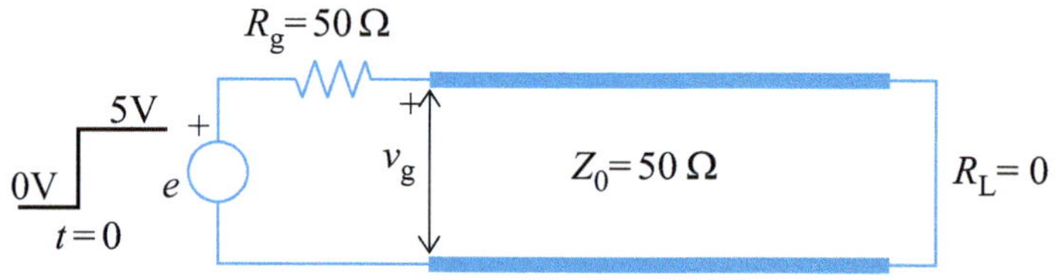

Figure 10.27 Short-circuited lossless transmission line with a step excitation; for Conceptual Question 10.91.

CONCEPTUAL QUESTION 10.92 **Reflection coefficient for an ideal voltage generator.** A lossless transmission line of characteristic impedance $Z_0 = 100\ \Omega$ is driven by an ideal voltage generator with a step emf applied at $t = 0$. The other end of the line is terminated in a purely resistive load of resistance $R_L = 200\ \Omega$. Under these circumstances, the transient partial line voltages traveling toward the generator

(A) do not reflect from it (reflection coefficient is $\Gamma_g = 0$).

(B) reflect from it, with reflection coefficient $\Gamma_g = 0.5$.

(C) reflect from it, with reflection coefficient $\Gamma_g = -0.5$.

(D) reflect from it, with reflection coefficient $\Gamma_g = 1$.

(E) reflect from it, with reflection coefficient $\Gamma_g = -1$.

(F) Need more information.

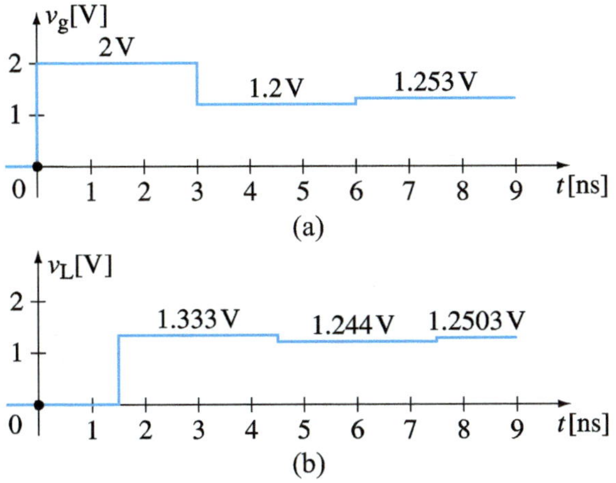

Figure 10.28 Voltage waveforms at the generator (a) and load (b) for a lossless coaxial cable with a step excitation; for Conceptual Question 10.93.

CONCEPTUAL QUESTION 10.93 **Voltage waveforms at both ends of a coaxial cable.** A lossless coaxial cable with characteristic impedance $Z_0 = 50\ \Omega$ and one-way time delay $T = 1.5$ ns is fed by a voltage generator of step emf $\mathcal{E} = 5$ V applied at $t = 0$ and internal resistance R_g. The other end of the cable is terminated in a purely resistive load of resistance R_L. Figure 10.28 shows voltage waveforms at both ends of the cable within a time interval $0 \leq t \leq 9$ ns. Based on these diagrams, we can conclude that

(A) $R_L > Z_0$ and $R_g > Z_0$.

(B) $R_L < Z_0$ and $R_g < Z_0$.

(C) $R_L > Z_0$ and $R_g < Z_0$.

(D) $R_L < Z_0$ and $R_g > Z_0$.

(E) $R_L > Z_0$ and $R_g = Z_0$.

(F) $R_L < Z_0$ and $R_g = Z_0$.

CONCEPTUAL QUESTION 10.94 **Another combination of load and generator resistances.** A lossless transmission line of characteristic impedance $Z_0 = 100\ \Omega$ and one-way transit time $T = 2$ ns, terminated in a purely resistive load, is driven by a voltage generator of step emf $\mathcal{E} = 4$ V applied at $t = 0$. The voltage waveforms at both the generator and load within a time interval $0 \leq t \leq 12$ ns are shown in Figure 10.29. What can be concluded about the load resistance, R_L, and the internal resistance of the generator, R_g, relative to Z_0?

(A) $R_L > Z_0$ and $R_g > Z_0$.

(B) $R_L < Z_0$ and $R_g < Z_0$.

(C) $R_L > Z_0$ and $R_g < Z_0$.

(D) $R_L < Z_0$ and $R_g > Z_0$.

(E) $R_L > Z_0$ and $R_g = Z_0$.

(F) $R_L < Z_0$ and $R_g = Z_0$.

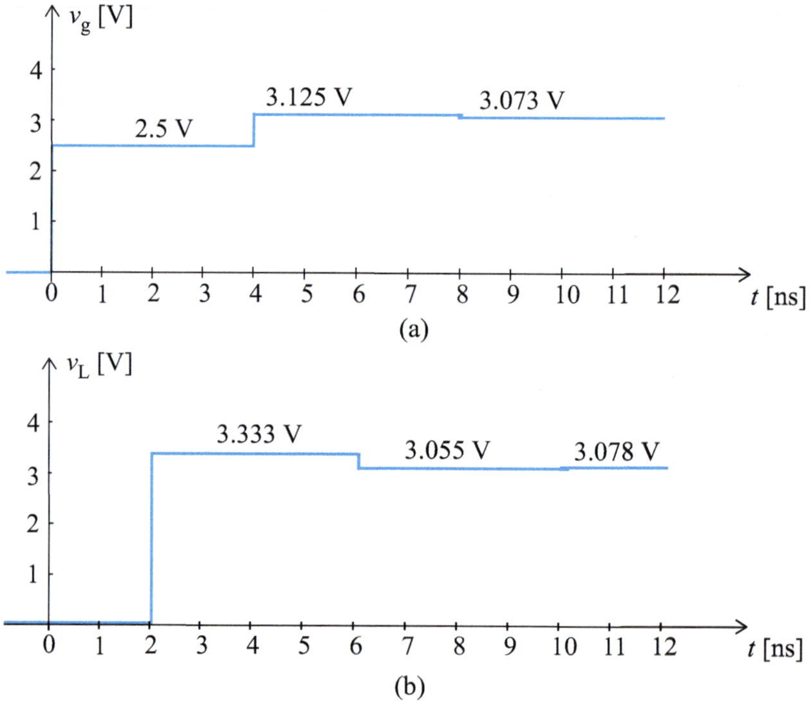

Figure 10.29 Step transient analysis of a lossless transmission line: input (a) and output (b) voltage waveforms; for Conceptual Question 10.94.

CONCEPTUAL QUESTION 10.95 **One more combination of load and generator resistances.** If the voltage waveforms for a lossless transmission line with $Z_0 = 100\ \Omega$, $T = 2$ ns, voltage generator of step emf $\mathcal{E} = 4$ V applied at $t = 0$ and internal resistance R_g, and a purely resistive load of resistance R_L are given in Figure 10.30, we can conclude that

(A) $R_L > Z_0$ and $R_g > Z_0$.

(B) $R_L < Z_0$ and $R_g < Z_0$.

(C) $R_L > Z_0$ and $R_g < Z_0$.

(D) $R_L < Z_0$ and $R_g > Z_0$.

(E) $R_L > Z_0$ and $R_g = Z_0$.

(F) $R_L < Z_0$ and $R_g = Z_0$.

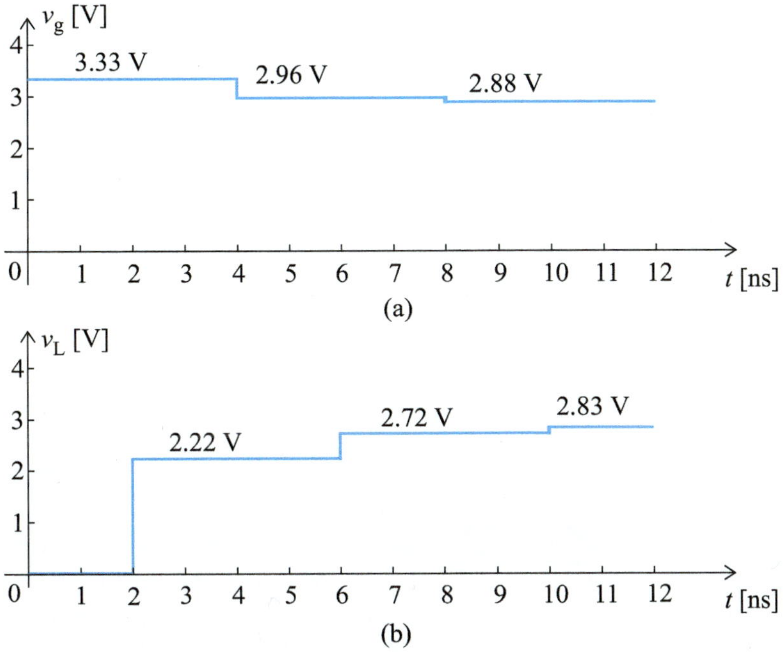

Figure 10.30 Voltage waveforms for a transmission line at (a) the generator and (b) the load; for Conceptual Question 10.95.

CONCEPTUAL QUESTION 10.96 **Voltage snapshots along a line at two time instants.** Figure 10.31 shows snapshots (scans) of the total voltage along a lossless transmission line with $Z_0 = 100\ \Omega$, $T = 2$ ns, and a step emf

$\mathcal{E} = 4$ V excitation applied at $t = 0$, $v(z,t)$, $0 \leq z \leq l$, at two instants of time. In particular, Figure 10.31(a) and Figure 10.31(b) show $v(z,t)$ at times

(A) $t = 1$ ns and $t = 3$ ns,

(B) $t = 3$ ns and $t = 1$ ns,

(C) $t = 3$ ns and $t = 5$ ns,

(D) $t = 5$ ns and $t = 3$ ns,

(E) $t = 5$ ns and $t = 7$ ns,

(F) $t = 7$ ns and $t = 5$ ns,

respectively.

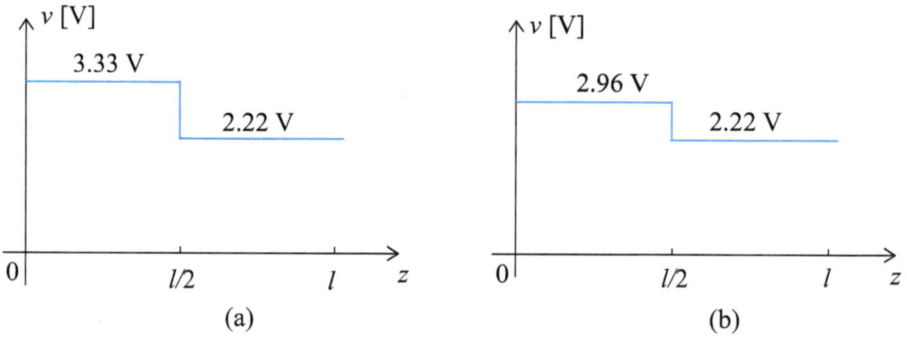

Figure 10.31 Voltage snapshots, $v(z,t)$, along a transmission line with a step excitation (input/output voltage waveforms of the line are those in Figure 10.30) at two instants of time; for Conceptual Question 10.96.

CONCEPTUAL QUESTION 10.97 Determining the generator resistance from voltage waveforms. Figure 10.32 shows the input and output waveforms, $v_{\rm g}(t)$ and $v_{\rm L}(t)$, for a lossless transmission line of characteristic impedance Z_0 fed by a voltage generator of step emf $\mathcal{E}$ applied at $t = 0$. The internal resistance of the generator is

(A) $R_{\rm g} = Z_0$.

(B) $R_{\rm g} = 2Z_0$.

(C) $R_{\rm g} = Z_0/2$.

(D) $R_{\rm g} = 0$.

(E) $R_g \to \infty$.

(F) Need more information.

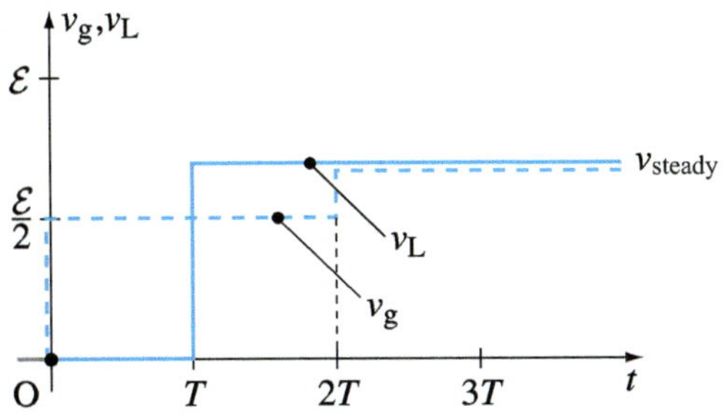

Figure 10.32 Input and output voltage waveforms, $v_g(t)$ and $v_L(t)$, for a lossless transmission line with a step excitation; for Conceptual Question 10.97.

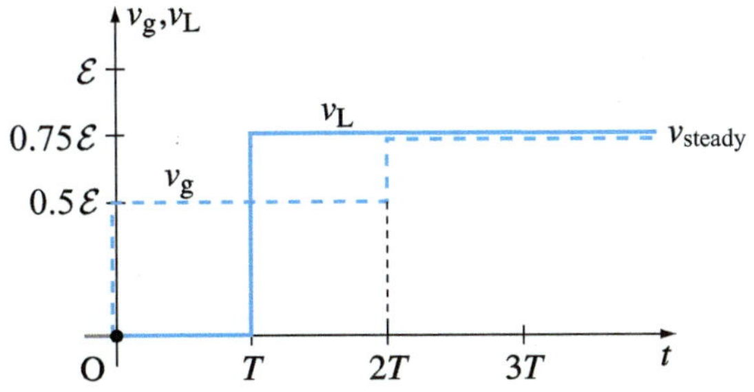

Figure 10.33 Step response of a transmission line: voltage waveforms at the generator and the load; for Conceptual Question 10.98.

CONCEPTUAL QUESTION 10.98 Determining the load reflection coefficient from voltage waveforms. If $v_{steady} = 0.75\,\mathcal{E}$ for a lossless transmission line excited by a step emf $\mathcal{E}$ applied at $t = 0$, as shown in Figure 10.33, the load voltage reflection coefficient of the line amounts to

(A) $\Gamma_L = -1$.

(B) $\Gamma_L = -0.5$.

(C) $\Gamma_L = 0$.

(D) $\Gamma_L = 0.25$.

(E) $\Gamma_L = 0.5$.

(F) $\Gamma_L = 1$.

CONCEPTUAL QUESTION 10.99 **Another determination of the load reflection coefficient.** Assuming that $v_{\text{steady}} = \mathcal{E}$ for a lossless transmission line with a step emf $\mathcal{E}$ as excitation, as in Figure 10.34, the load voltage reflection coefficient of the line is

(A) $\Gamma_{\text{L}} = -1$.

(B) $\Gamma_{\text{L}} = -0.5$.

(C) $\Gamma_{\text{L}} = 0$.

(D) $\Gamma_{\text{L}} = 0.25$.

(E) $\Gamma_{\text{L}} = 0.5$.

(F) $\Gamma_{\text{L}} = 1$.

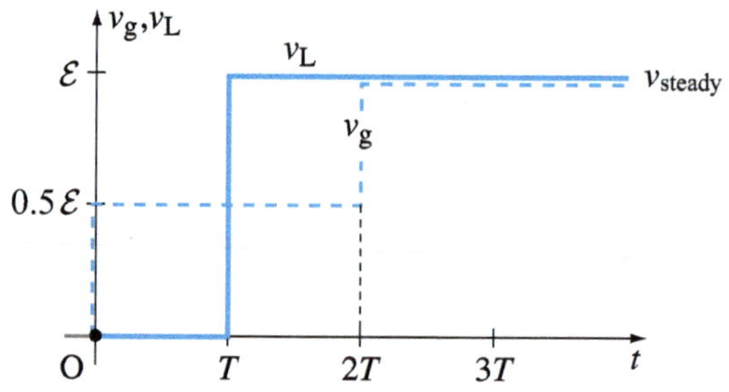

Figure 10.34 Step response of a transmission line with another load; for Conceptual Question 10.99.

10.7 Analysis of Transmission Lines with Pulse Excitations

Let us now assume that the emf in Figure 10.18(a) is a rectangular pulse function of time, of magnitude $\mathcal{E}$ and duration t_0, triggered at $t = 0$, as shown in Figure 10.35(a). It is obvious from Figure 10.35(b) that $e(t)$ can be viewed as a superposition of two step functions, in Figure 10.18(b), with opposite polarities and a time shift t_0 between them, so that the response of the transmission line in Figure 10.18(a) to $e(t)$ can be computed combining the individual responses to the two step inputs if applied alone. Namely, marking by $v_{\text{L1}}(t)$ the line output response (load voltage) to the input $e_1(t)$ [step emf applied at $t = 0$, in Figure 10.35(b)] alone, the resultant output response to the combined excitation is obtained as

$$e(t) = e_1(t) - e_1(t - t_0) \quad \longrightarrow \quad v_{\text{L}}(t) = v_{\text{L1}}(t) - v_{\text{L1}}(t - t_0) \quad \text{(pulse response)}, \tag{10.22}$$

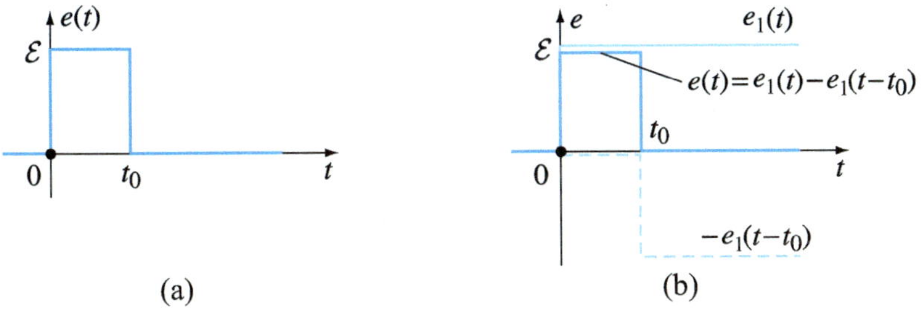

Figure 10.35 Pulse excitation of a lossless transmission line [Figure 10.18(a)]: (a) rectangular pulse emf function in time and (b) its representation using two step functions, in Figure 10.18(b).

i.e., the load voltage is the same superposition of $v_{L1}(t)$ and its flipped-over (multiplied by -1) and delayed (by t_0) version – as for the excitation. Analogous transformations can also be applied for other signals on the line.

CONCEPTUAL QUESTION 10.100 Pulse response of a line with unmatched load and generator. A lossless transmission line of length $l = 45$ cm, for which the one-way delay period of the line comes out to be $T = 3$ ns, and characteristic impedance $Z_0 = 50\ \Omega$ is driven by an ideal voltage generator of rectangular pulse emf with magnitude $\mathcal{E} = 10$ V and width $t_0 = 2$ ns, applied at $t = 0$. At its other end, the line is terminated in a purely resistive load of resistance $R_L = 200\ \Omega$. The step and pulse responses of the line at the load terminals, $v_{L1}(t)$ and $v_L(t)$, are shown, for $0 \le t \le 20$ ns, in Figures 10.36(a) and (b), respectively, where

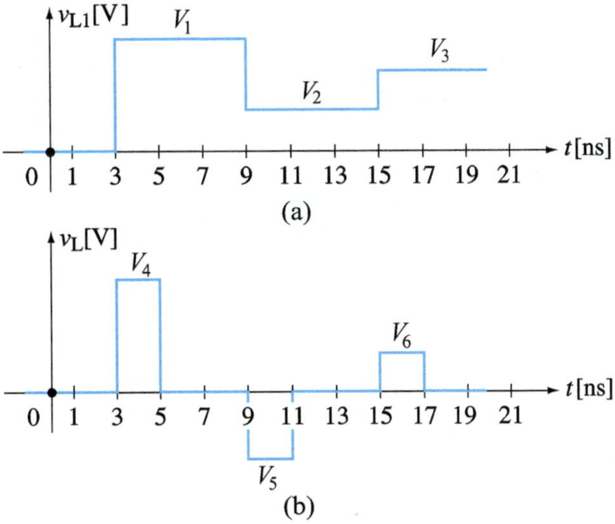

Figure 10.36 Output step (a) and pulse (b) response of a lossless transmission line, in an interval of time; for Conceptual Question 10.100.

(A) $V_4 = V_1$, $V_5 = -V_2$, and $V_6 = V_3$.

(B) $V_4 = V_1$, $V_5 = V_1 - V_2$, and $V_6 = V_2 - V_3$.

(C) $V_4 = V_1$, $V_5 = V_1 - V_2$, and $V_6 = V_1 - V_3$.

(D) $V_4 = V_1 - V_2$, $V_5 = -V_2$, and $V_6 = V_3 - V_2$.

(E) $V_4 = V_1$, $V_5 = V_2 - V_1$, and $V_6 = V_3 - V_2$.

(F) None of the above.

CONCEPTUAL QUESTION 10.101 **Pulse response of a line with a matched generator.** If the generator is matched ($R_g = Z_0$) to a lossless transmission line (with $l = 45$ cm, $T = 3$ ns, $Z_0 = 50\ \Omega$, and $R_L = 200\ \Omega$), the output response of the line to the pulse excitation (emf $\mathcal{E} = 10$ V, pulse width $t_0 = 2$ ns, applied at $t = 0$), $v_L(t)$, consists of

(A) an infinite series of pulses with alternating polarity.

(B) an infinite series of pulses with the same polarity.

(C) only one pulse.

(D) two pulses with different polarities.

(E) two pulses with the same polarity.

(F) None of the above.

CONCEPTUAL QUESTION 10.102 **Pulse response of a line with a matched load.** Assuming that the load is matched to the transmission line (with $T = 3$ ns), $R_L = Z_0$, while the generator is an ideal one, $R_g = 0$, the pulse (with $t_0 = 2$ ns) response of the line at the load, $v_L(t)$, consists of

(A) an infinite series of pulses with alternating polarity.

(B) an infinite series of pulses with the same polarity.

(C) only one pulse.

(D) two pulses with different polarities.

(E) two pulses with the same polarity.

(F) None of the above.

CONCEPTUAL QUESTION 10.103 **Overlapping pulses at the load terminals.** Figure 10.37 shows the output voltage response of a transmission line to a rectangular pulse excitation. With T denoting the one-way delay period of the line and t_0 the duration of the pulse emf, we have that

(A) $t_0 < T$.

(B) $t_0 = T$.

(C) $T < t_0 < 2T$.

(D) $t_0 = 2T$.

(E) $t_0 > 2T$.

(F) Need more information.

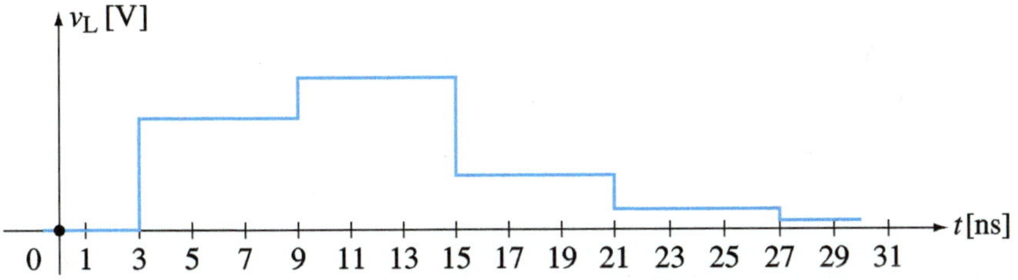

Figure 10.37 Pulse response of a transmission line; for Conceptual Question 10.103.

10.8 Transient Response for Reactive Terminations

Often, transmission-line terminations involve reactive lumped elements, inductors and capacitors. As an example of the transient analysis, consider a lossless transmission line terminated in an ideal inductor, of inductance L, and excited by a matched step generator, as in Figure 10.38(a). The incident voltage is $v_\text{i} = v_\text{i1} = \mathcal{E}/2$ [Eq. (10.20)]. At $t = T$, with T being the one-way time delay of the line, Eq. (10.18), the wavefront of the signal v_{i1} arrives at the load terminals, and the voltage of the inductor is abruptly changed. In general, for rapid variations of an applied voltage, an inductor behaves as an open circuit (as it does for time-harmonic signals of very high frequencies), and its current is zero. So, at this time the incident voltage is reflected from the load in Figure 10.38(a) as in the case of an open-circuited transmission line, and the total voltage of the load jumps to

$$v_{\rm L}(T) = v_{i1} + v_{r1} = 2v_{i1} = \mathcal{E} \quad (\text{inductor} \to \text{open circuit}, \Gamma_{\rm L} = 1) . \tag{10.23}$$

In the steady state, as in a dc regime (zero frequency), the inductor can be considered as a short circuit, and hence

$$v_{\rm L}(t \to \infty) = v_{\rm steady} = 0 \quad (\text{inductor} \to \text{short circuit. } \Gamma_{\rm L} = -1) . \tag{10.24}$$

Between the time $t = T$ and $t \to \infty$, the change (decrease) of the inductor voltage is an exponential one, as sketched in Figure 10.38(b). Since the input dynamic impedance that the load sees looking into the transmission line equals the line characteristic impedance, Z_0, the time constant of this exponential change is

$$\tau = \frac{L}{Z_0} \quad (\text{time constant of a line with a purely inductive load}) . \tag{10.25}$$

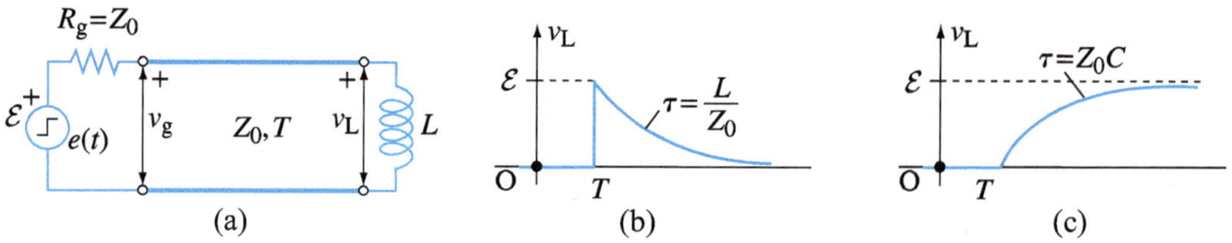

Figure 10.38 Step transient analysis of a lossless transmission line with a purely inductive load and matched generator: (a) circuit schematic diagram and (b) voltage waveform at the load. (c) The same as in (b) but for a purely capacitive load.

Transient analysis of a transmission line with an ideal capacitor, of capacitance C, as load is performed in a similar fashion, having in mind that, just opposite to an inductor, a capacitor acts as a short circuit for rapid variations of an applied signal, whereas as an open circuit in the steady (dc) state, and the time constant of the exponential variation between $t = T$ and $t \to \infty$ is $\tau = Z_0 C$, Figure 10.38(c). In addition, similar evaluations of transients can be performed on transmission lines with various combined resistive and reactive terminations, such as series or parallel combinations of a resistor and an inductor (capacitor).

CONCEPTUAL QUESTION 10.104 Resistor and inductor in series at the load end of a line. A lossless transmission line is fed by a voltage generator of step emf, applied at $t = 0$. The characteristic impedance of the line, Z_0, and internal resistance of the generator, R_g, are the same (the generator is matched to the line). At its other end, the line is terminated in a load in the form of a series connection of a resistor of resistance R and an inductor of inductance L. The voltage waveform across the load, $v_L(t)$, is as sketched in

(A) Figure 10.39(a).

(B) Figure 10.39(b).

(C) Figure 10.39(c).

(D) Figure 10.39(d).

(E) Figure 10.39(e).

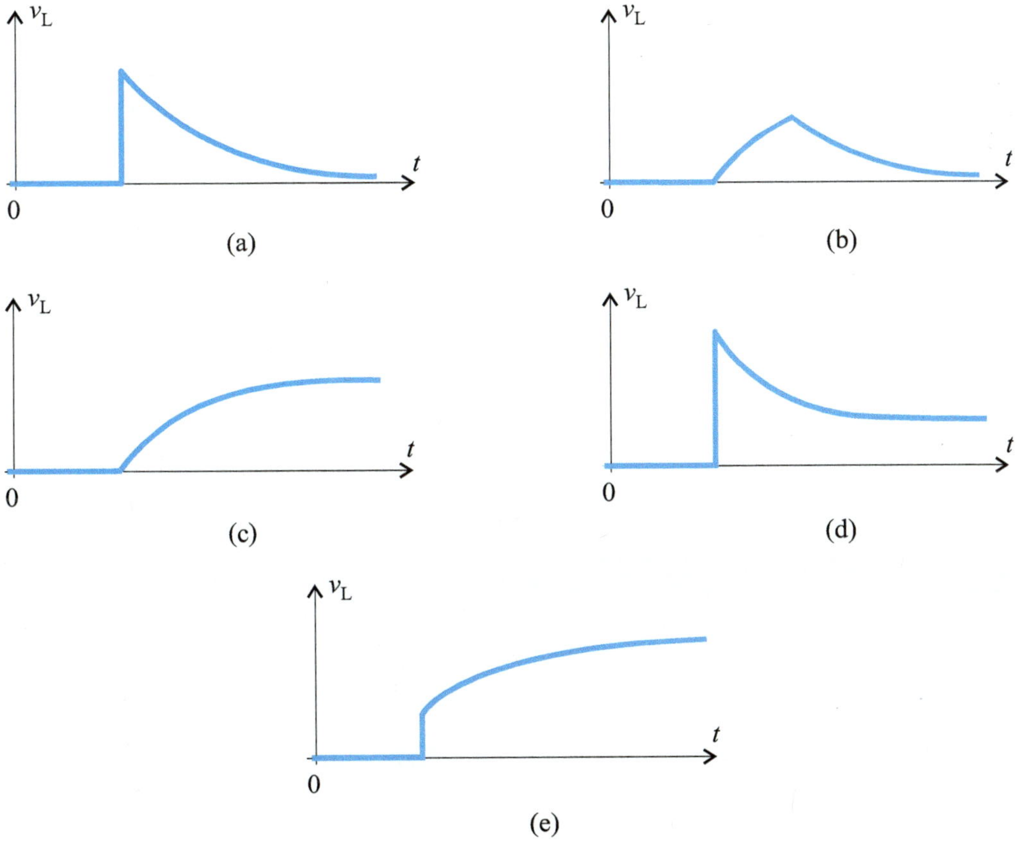

Figure 10.39 Step response of a lossless transmission line with $R_g = Z_0$ (matched generator) terminated in a combined resistive and reactive load: five offered sketches for the voltage waveform across the load; for Conceptual Question 10.104.

CONCEPTUAL QUESTION 10.105 **Resistor and capacitor in series at the load end.** Assuming that the transmission line with a step excitation and matched generator is terminated in a series connection of a resistor of resistance R and a capacitor of capacitance C, the voltage waveform at the load is as in

(A) Figure 10.39(a).

(B) Figure 10.39(b).

(C) Figure 10.39(c).

(D) Figure 10.39(d).

(E) Figure 10.39(e).

CONCEPTUAL QUESTION 10.106 **Resistor and inductor in parallel as load.** For a parallel connection of a resistor of resistance R and an inductor of inductance L at the load end of the line, which one of the five offered sketches in Figure 10.39 may represent the step response of the line?

(A) The one in Figure 10.39(a).

(B) The one in Figure 10.39(b).

(C) The one in Figure 10.39(c).

(D) The one in Figure 10.39(d).

(E) The one in Figure 10.39(e).

CONCEPTUAL QUESTION 10.107 **Resistor and capacitor in parallel as load.** For a parallel connection of a resistor of resistance R and a capacitor of capacitance C as load, the step response of the line is as in

(A) Figure 10.39(a).

(B) Figure 10.39(b).

(C) Figure 10.39(c).

(D) Figure 10.39(d).

(E) Figure 10.39(e).

CONCEPTUAL QUESTION 10.108 **Time constant for resistor and inductor in series as the line load.** Consider the transmission-line circuit consisting of a lossless transmission line excited by a matched voltage generator of step emf, applied at $t = 0$, and terminated at the other end in a load in the form of a series connection of a resistor of resistance R and an inductor of inductance L. What is the time constant of the exponential change of the load voltage in the transitional period, from the time $t = T$ to $t \to \infty$, for this circuit?

(A) $\tau = L/R$.

(B) $\tau = RZ_0L/(R + Z_0)$.

(C) $\tau = (R + Z_0)L/(RZ_0)$.

(D) $\tau = L/(R + Z_0)$.

(E) $\tau = L/(R + Z_0 + R_g)$.

(F) $\tau = (R + Z_0 + R_g)L/[R(Z_0 + R_g)]$.

CONCEPTUAL QUESTION 10.109 **Time constant for resistor and inductor in parallel as load.** For the transmission-line circuit with a parallel connection of a resistor of resistance R and an inductor of inductance L at the load end of the line, the time constant of the exponential change of the load voltage in the transitional period, from $t = T$ to $t \to \infty$, is given by

(A) $\tau = L/R$.

(B) $\tau = RZ_0L/(R + Z_0)$.

(C) $\tau = (R + Z_0)L/(RZ_0)$.

(D) $\tau = L/(R + Z_0)$.

(E) $\tau = L/(R + Z_0 + R_g)$.

(F) $\tau = (R + Z_0 + R_g)L/[R(Z_0 + R_g)]$.

CONCEPTUAL QUESTION 10.110 **Time constant for resistor and capacitor in series as load.** What is the time constant for the transitional process on the transmission line with a step excitation and matched generator terminated in a series connection of a resistor of resistance R and a capacitor of capacitance C?

(A) $\tau = RC$.

(B) $\tau = RZ_0C/(R + Z_0)$

(C) $\tau = (R + Z_0)C/(RZ_0)$.

(D) $\tau = (R + Z_0)C$.

(E) $\tau = (R + Z_0 + R_g)C$.

(F) $\tau = R(Z_0 + R_g)C/(R + Z_0 + R_g)$.

CONCEPTUAL QUESTION 10.111 **Time constant for resistor and capacitor in parallel as load.** For the line with a parallel connection of a resistor of resistance R and a capacitor of capacitance C as load, the time constant of the exponential change of the load voltage comes out to be

(A) $\tau = RC$.

(B) $\tau = RZ_0C/(R + Z_0)$

(C) $\tau = (R + Z_0)C/(RZ_0)$.

(D) $\tau = (R + Z_0)C$.

(E) $\tau = (R + Z_0 + R_g)C$.

(F) $\tau = R(Z_0 + R_g)C/(R + Z_0 + R_g)$.

CONCEPTUAL QUESTION 10.112 **Rate of the transitional process versus L and C values.** In a transmission line whose load contains an inductor of inductance L or a capacitor of capacitance C, the higher L and C

(A) the faster

(B) the slower

(C) no difference

the transitional process on the line.

CONCEPTUAL QUESTION 10.113 **Meaning of the time constant for a circuit with a reactive element.** In the transitional process depicted in Figure 10.40, the soonest time by which the load voltage acquires 99% of its steady-state value is the closest to

(A) 0.05τ

(B) 0.5τ

(C) 5τ

(D) 50τ

(E) 500τ

after the start of the process, where τ is the time constant of the exponential variation of the voltage.

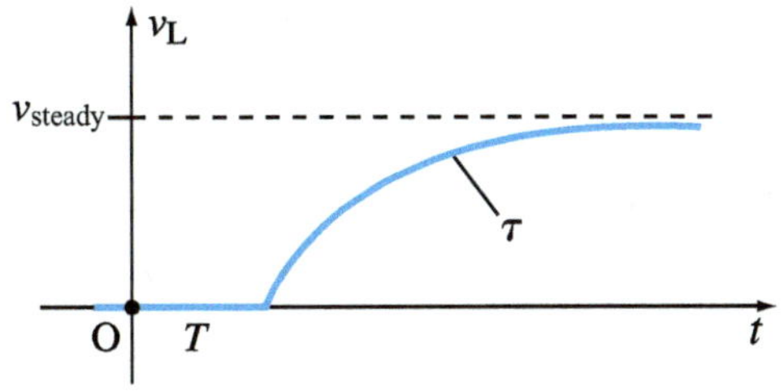

Figure 10.40 Step response of a lossless transmission line with a matched generator and a combined resistive and reactive load; for Conceptual Question 10.113.

CONCEPTUAL QUESTION 10.114 **Pulse response of a transmission line with a series RL load.** A lossless transmission line, for which the one-way wave travel time is T, is fed by a voltage generator of rectangular pulse emf with magnitude $\mathcal{E}$ and width t_0, applied at $t = 0$. The characteristic impedance of the line and internal resistance of the generator are the same, $Z_0 = R_\text{g}$. The other end of the line is terminated in a load consisting of a resistor of resistance R and inductor of inductance L connected in series. The waveform of the voltage across the load, $v_\text{L}(t)$, is as sketched in

(A) Figure 10.41(a).

(B) Figure 10.41(b).

(C) Figure 10.41(c).

(D) Figure 10.41(d).

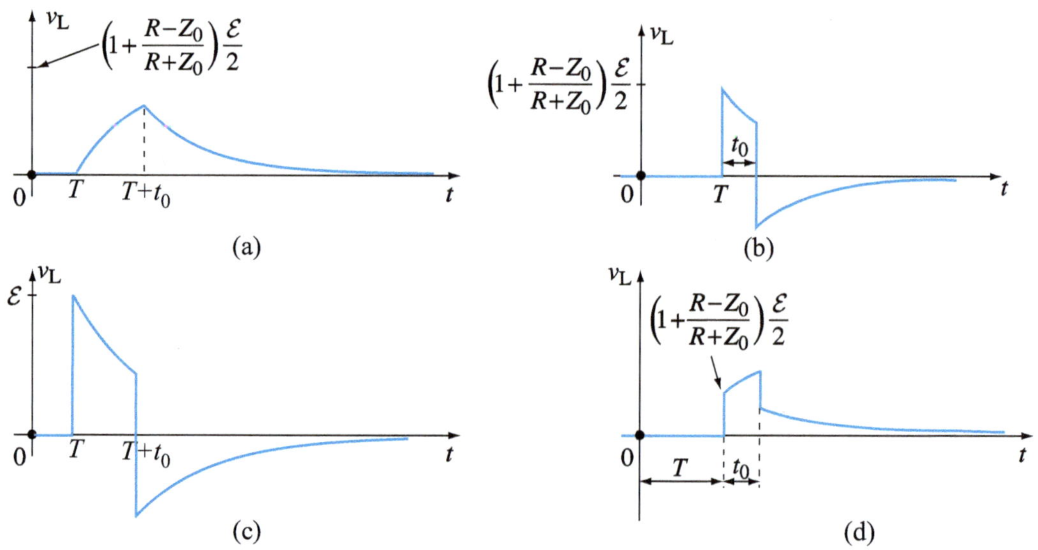

Figure 10.41 Pulse (emf $\mathcal{E}$, width t_0, applied at $t = 0$) response of a lossless transmission line, with one-way wave travel time T and $Z_0 = R_g$ (matched generator), terminated in a combined resistive and reactive load: four offered sketches for the voltage waveform across the load; for Conceptual Question 10.114.

CONCEPTUAL QUESTION 10.115 Pulse response of a line with a series RC load. Assuming that the line load is a series connection of a resistor of resistance R and a capacitor of capacitance C, the load voltage waveform for the pulse excitation of the line is as sketched in

(A) Figure 10.41(a).

(B) Figure 10.41(b).

(C) Figure 10.41(c).

(D) Figure 10.41(d).

CONCEPTUAL QUESTION 10.116 Pulse response of a line with a parallel RL load. For a transmission line terminated in a parallel connection of a resistor of resistance R and an inductor of inductance L, which one of the four offered sketches in Figure 10.41 may represent the pulse response of the line?

(A) The one in Figure 10.41(a).

(B) The one in Figure 10.41(b).

(C) The one in Figure 10.41(c).

(D) The one in Figure 10.41(d).

CONCEPTUAL QUESTION 10.117 **Pulse response of a line with a parallel RC load.** The pulse response of the line with a parallel connection of a resistor of resistance R and a capacitor of capacitance C at the load end is as in

(A) Figure 10.41(a).

(B) Figure 10.41(b).

(C) Figure 10.41(c).

(D) Figure 10.41(d).

CONCEPTUAL QUESTION 10.118 **Series-parallel combination of two resistors and an inductor as load.** A lossless transmission line for which the one-way wave travel time is $T = 1$ ns and whose characteristic impedance is $Z_0 = 100\ \Omega$ is driven by a matched voltage generator of step emf $\mathcal{E} = 2$ V applied at $t = 0$. Consider the series-parallel combination of two resistors of resistance $R = 100\ \Omega$ and an inductor of inductance $L = 50$ nH shown in Figure 10.42(a) as a load termination for this line. The voltage waveform at the load end of the line is as sketched in

(A) Figure 10.42(b).

(B) Figure 10.42(c)

(C) Figure 10.42(d).

(D) Figure 10.42(e).

CONCEPTUAL QUESTION 10.119 **Another combination of two resistors and an inductor as load.** Assuming that the load of the line with

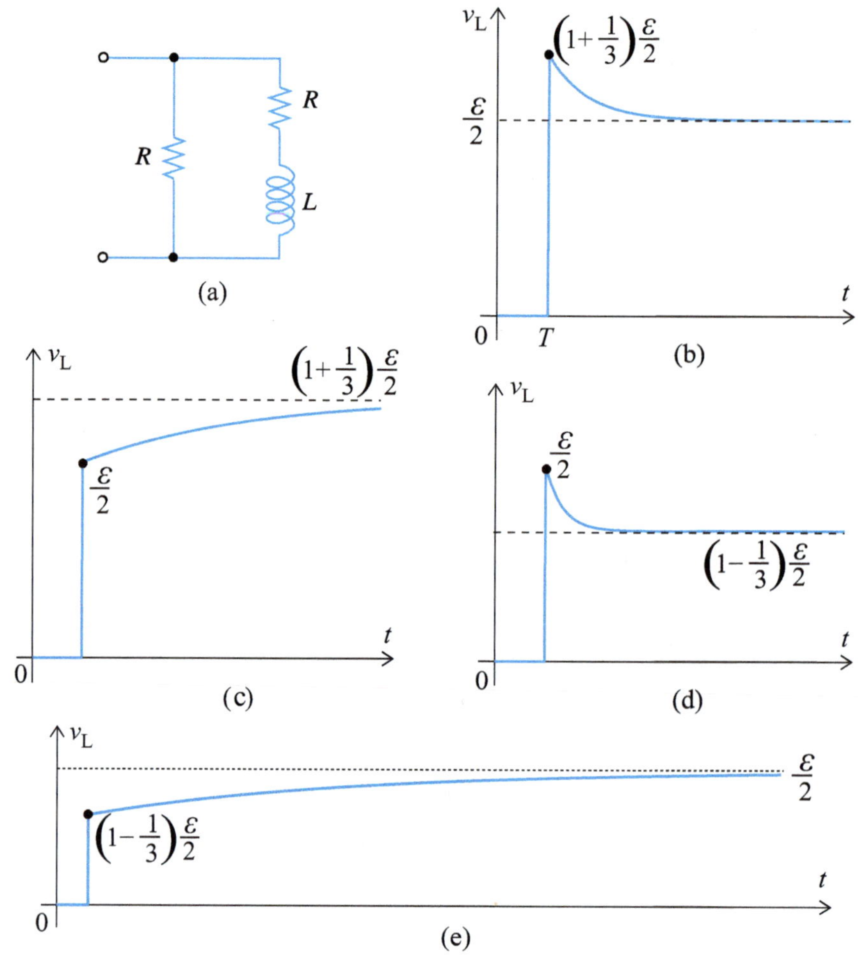

Figure 10.42 (a) Series-parallel combination of two resistors and an inductor as a load termination of a transmission line driven by a matched step voltage generator. (b)–(e) Four offered sketches for the voltage waveform across the load; for Conceptual Question 10.118.

$Z_0 = 100\ \Omega$, $R_g = Z_0$, and a step excitation is the series-parallel combination of two resistors of resistance $R = 100\ \Omega$ and an inductor of inductance $L = 50$ nH shown in Figure 10.43, the load voltage waveform is as sketched in

(A) Figure 10.42(b).

(B) Figure 10.42(c)

(C) Figure 10.42(d).

(D) Figure 10.42(e).

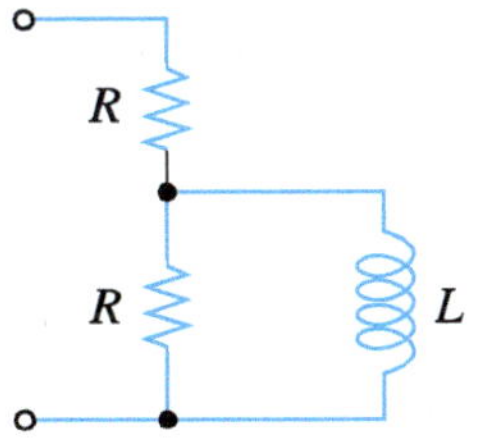

Figure 10.43 Another series-parallel combination of two resistors and an inductor as a load of a transmission line with a step excitation; for Conceptual Question 10.119.

CONCEPTUAL QUESTION 10.120 **Series-parallel combination of two resistors and a capacitor as load.** For the series-parallel combination of two resistors of resistance $R = 100\ \Omega$ and a capacitor of capacitance $C = 50$ pF shown in Figure 10.44 as a load termination for the line ($Z_0 = 100\ \Omega$, $R_g = Z_0$), the step response of the line is as in

(A) Figure 10.42(b).

(B) Figure 10.42(c)

(C) Figure 10.42(d).

(D) Figure 10.42(e).

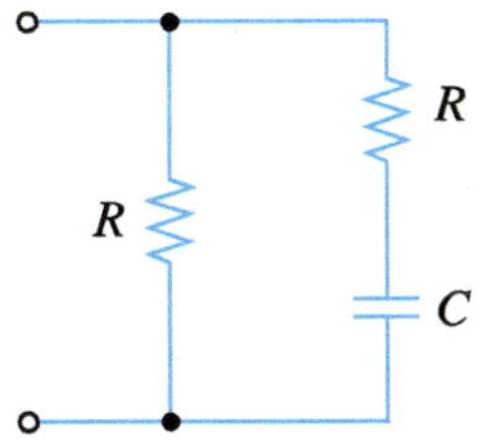

Figure 10.44 Series-parallel combination of two resistors and a capacitor as a load of a transmission line fed by a matched step voltage generator; for Conceptual Question 10.120.

CONCEPTUAL QUESTION 10.121 **Another combination of two resistors and a capacitor as load.** The step response of a transmission line ($Z_0 = 100\ \Omega$, $R_g = Z_0$) terminated in the series-parallel combination of two resistors of resistance $R = 100\ \Omega$ and a capacitor of capacitance $C = 50$ pF shown in Figure 10.45 is as in

(A) Figure 10.42(b).

(B) Figure 10.42(c)

(C) Figure 10.42(d).

(D) Figure 10.42(e).

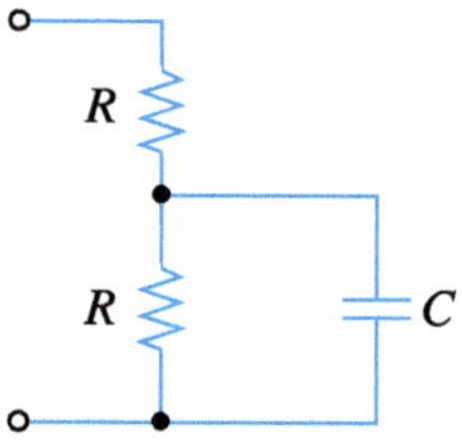

Figure 10.45 Another series-parallel combination of two resistors and a capacitor as the line load; for Conceptual Question 10.121.

11 WAVEGUIDES AND CAVITY RESONATORS

Introduction

At frequencies in the microwave region, waveguides in the form of metallic tubes are used for energy and information transfer in electromagnetic devices and systems. Essentially, electromagnetic waves travel along such tubes by means of multiple reflections from the metallic walls, through the dielectric filling the tube (most frequently, air), so the waves are guided by the tube conductor. Waveguides carry non-TEM waves, which include transverse electric (TE) and transverse magnetic (TM) waves. TE waves have a zero electric and nonzero magnetic field component along the waveguide axis, whereas the situation for TM waves is just opposite (magnetic field vector is in a transverse plane, perpendicular to the axis). In addition to metallic waveguides for energy/information transmission, waveguide sections closed at both ends, thus forming rectangular metallic cavities, represent microwave resonators – also with widespread applications. Although arbitrary cross sections of metallic tubes and cavities are theoretically possible, our focus here will be on rectangular metallic waveguides and cavity resonators, which are involved most frequently in practical microwave devices and systems.

11.1 Rectangular Waveguide Analysis Based on Multiple Reflections of Plane Waves

Consider an infinitely long uniform rectangular metallic waveguide with cross-sectional interior dimensions a and b, filled by a homogeneous dielectric of permittivity ε and permeability μ, as shown in Figure 11.1. We assume that the waveguide is lossless, i.e., that its walls are made of a perfect electric conductor (PEC), and that

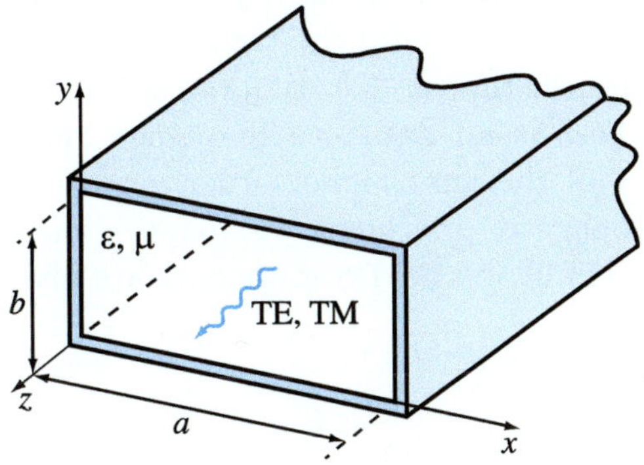

Figure 11.1 Rectangular waveguide with a TE or TM wave.

the dielectric is also perfect. We would like to find a solution for a time-harmonic electromagnetic wave, of frequency f (and angular frequency $\omega = 2\pi f$), that propagates inside the waveguide, along the z-axis. One such solution is a normally (or TE – transverse electric) polarized uniform plane wave obliquely incident on a PEC boundary in Figure 8.15(a). The wave propagates in the positive z direction by bouncing back and forth, at an incident angle $\theta_{\rm i}$, between the walls at $x = 0$ and at $x = a$ in Figure 11.1 [note that the coordinate axes in Figure 11.1 are set up differently from Figure 8.15(a)], where the boundary condition stipulating that the tangential component of the total electric field vector be zero at the second plane gives

$$\sin\beta_{\rm n}a = 0\,, \quad \beta_{\rm n} = \frac{\omega\cos\theta_{\rm i}}{c} \quad \longrightarrow \quad \theta_{\rm i} = \arccos\left(m\,\frac{c}{2af}\right) \quad (m = 1, 2, \ldots) \tag{11.1}$$

($\arccos \equiv \cos^{-1}$), with $\beta_{\rm n}$ standing for the phase coefficient of the wave travel in the direction normal to the PEC boundary in Figure 8.15(a) and c for the intrinsic phase velocity of the waveguide dielectric, that is, the velocity of electromagnetic waves in the medium of parameters ε and μ, Eq. (7.3). Using Eqs. (11.1), the longitudinal phase coefficient (in the z direction), β_z, of the waveguide in Figure 11.1, which is the principal phase coefficient, β, for the structure, comes out to be

$$\beta = \beta_z = \frac{\omega\sin\theta_{\rm i}}{c} = \frac{\omega}{c}\sqrt{1-\cos^2\theta_{\rm i}} = \frac{\omega}{c}\sqrt{1-\frac{f_{\rm c}^2}{f^2}}\,, \quad f_{\rm c} = (f_{\rm c})_{m0} = m\,\frac{c}{2a} \quad (n = 0)$$
$$(\beta \text{ – waveguide phase coefficient})\,, \tag{11.2}$$

where the frequency $f_{\rm c}$ is called the cutoff frequency of the waveguide – for a particular, TE_{m0}, mode. Namely, each integer value of m determines a possible field solution in the waveguide, and these distinct waves that can exist in a waveguide are referred to as modes. As we shall see in the next section, waveguide modes with both m and n being arbitrary nonnegative integers are also possible (if properly excited) in a waveguide (in Figure 11.1). The frequency $f_{\rm c}$ has the same role as the plasma frequency ($f_{\rm p}$), in Eq. (7.18). Analogously to a plasma medium, the waveguide in Figure 11.1 behaves like a high-pass filter, letting only waves whose frequency is higher than the cutoff frequency, $f > f_{\rm c}$, propagate through it.

As it is customary to always denote the transverse dimensions of a rectangular waveguide (Figure 11.1) such that $a \geq b$, the lowest waveguide mode is TE_{10}, for $m = 1$ and $n = 0$. There is an exclusive frequency range, that between $(f_{\rm c})_{10}$ and the cutoff frequency of the next higher order mode, in which only one mode, the TE_{10} mode, can propagate, and hence its name – the dominant mode. From Figure 8.15(a) and Eqs. (11.2), its field components are given by

$$\mathrm{TE}_{10}\colon\ \underline{E}_y = -\mathrm{j}\omega\mu\,\frac{a}{\pi}\,\underline{H}_0\sin\left(\frac{\pi}{a}\,x\right)\mathrm{e}^{-\mathrm{j}\beta z}\,, \quad \underline{H}_x = \mathrm{j}\beta\,\frac{a}{\pi}\,\underline{H}_0\sin\left(\frac{\pi}{a}\,x\right)\mathrm{e}^{-\mathrm{j}\beta z}\,,$$
$$\underline{H}_z = \underline{H}_0\cos\left(\frac{\pi}{a}\,x\right)\mathrm{e}^{-\mathrm{j}\beta z}\,, \quad \underline{H}_0 = -\frac{2\pi\underline{E}_{\rm i0}}{\omega\mu a}\,, \quad (f_{\rm c})_{10} = \frac{c}{2a} \quad \text{(dominant mode)} \tag{11.3}$$

($\underline{E}_x$, $\underline{E}_z$, $\underline{H}_y = 0$). Of course, this is a transverse electric (TE) wave, since $\underline{E}_z = 0$ and $\underline{H}_z \neq 0$.

CONCEPTUAL QUESTION 11.1 **Ray paths of several TE modes in a rectangular waveguide.** Consider an air-filled rectangular metallic waveguide with the larger transverse dimension amounting to $a = 6$ cm. At an operating frequency of the guide of $f = 10$ GHz, Figure 11.2 shows the ray paths, traced by a uniform plane wave bouncing back and forth at angles θ_i between the waveguide walls, corresponding to the first four TE_{m0} modes ($m = 1, 2, 3, 4$) in the structure. In particular, the paths in Figures 11.2(a), 11.2(b), 11.2(c), and 11.2(d) are of[1]

(A) TE_{40}, TE_{30}, TE_{20}, and TE_{10}

(B) TE_{30}, TE_{20}, TE_{10}, and TE_{40}

(C) TE_{10}, TE_{20}, TE_{30}, and TE_{40}

(D) TE_{20}, TE_{30}, TE_{40}, and TE_{10}

modes, respectively.

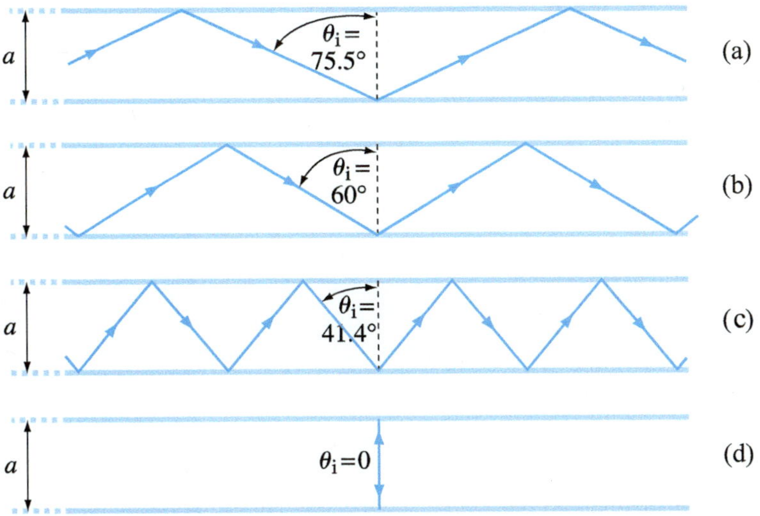

Figure 11.2 Ray paths of the first four TE_{m0} modes ($m = 1, 2, 3, 4$) in an air-filled rectangular metallic waveguide [the order (a)–(d) does not necessarily coincide with $m = 1, 2, 3, 4$]; for Conceptual Question 11.1.

[1] *For every conceptual question in this text, exactly one answer is correct.*

CONCEPTUAL QUESTION 11.2 **Number of propagating modes from the ray path diagrams.** Based on Figure 11.2, how many different TE_{m0} modes, out of all theoretically possible such modes ($m = 1, 2, \ldots$), can propagate along the air-filled metallic waveguide with $a = 6$ cm at the specified frequency ($f = 10$ GHz)?

(A) Zero.

(B) One.

(C) Two.

(D) Three.

(E) Four.

(F) Infinite number of modes.

CONCEPTUAL QUESTION 11.3 **Waveguide phase coefficient below the cutoff frequency.** Figure 11.3 shows a lossless rectangular metallic waveguide of transverse dimensions a and b, filled with a homogeneous dielectric of permittivity ε and permeability μ. The cutoff frequency of this waveguide for a given wave mode is f_c. At a frequency f such that $f < f_c$, the associated phase coefficient of the waveguide, β, is

(A) nonzero, finite, and purely real.

(B) nonzero, finite, and purely imaginary.

(C) zero.

(D) infinite.

(E) None of the above.

(F) Need more information.

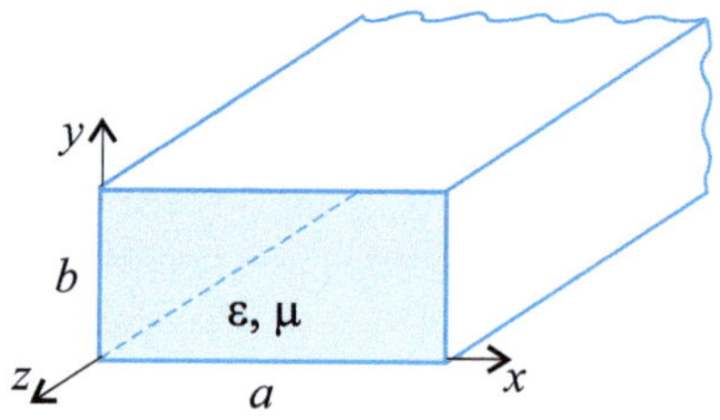

Figure 11.3 Rectangular metallic waveguide with a homogeneous dielectric; for Conceptual Question 11.3.

CONCEPTUAL QUESTION 11.4 **Waveguide phase coefficient above the cutoff.** At a frequency f satisfying the condition $f > f_c$, with f_c standing for the cutoff frequency of a lossless rectangular metallic waveguide (Figure 11.3) for a given wave mode, the phase coefficient β of the waveguide for this mode is

(A) nonzero, finite, and purely real.

(B) nonzero, finite, and purely imaginary.

(C) zero.

(D) infinite.

(E) None of the above.

(F) Need more information.

CONCEPTUAL QUESTION 11.5 **Phase coefficient at the cutoff.** At the frequency $f = f_c$, the waveguide phase coefficient (β) for the wave mode with this cutoff frequency (f_c) comes out to be

(A) nonzero, finite, and purely real.

(B) nonzero, finite, and purely imaginary.

(C) zero.

(D) infinite.

(E) None of the above.

(F) Need more information.

CONCEPTUAL QUESTION 11.6 **Propagating wave modes in a waveguide.** At a given operating frequency (f) of a lossless rectangular metallic waveguide with a homogeneous dielectric, all wave modes whose cutoff frequency (f_c) satisfies the condition

(A) $f_c < f$

(B) $f_c \leq f$

(C) $f_c > f$

(D) $f_c \geq f$

(E) $f_c = f$

(F) $f_c \neq f$

can propagate along the structure.

CONCEPTUAL QUESTION 11.7 **TE, TM, and TEM wave types on a waveguide.** Consider the possibilities that waves of different types, namely, of transverse electric (TE), transverse magnetic (TM), and transverse electromagnetic (TEM) types, propagate along a rectangular metallic waveguide. Denoting by f the operating frequency of a time-harmonic electromagnetic wave on the waveguide, we have that

(A) TE waves of all frequencies f

(B) TM waves of all frequencies f

(C) TEM waves of all frequencies f

(D) TE and TM waves of some frequencies f

(E) TE, TM, and TEM waves of some frequencies f

can travel along the structure.

CONCEPTUAL QUESTION 11.8 **Cutoff frequency of TEM waves on a coaxial cable.** The cutoff frequency of TEM waves on a lossless coaxial cable, with a homogeneous dielectric of permittivity ε and permeability μ_0, and conductor radii a and b ($a < b$), is

(A) $f_c = 1/(2\sqrt{\varepsilon\mu_0}\, a)$.

(B) $f_c = 1/(2\sqrt{\varepsilon\mu_0}\, b)$.

(C) $f_c = 1/[\sqrt{\varepsilon\mu_0}\,(a + b)]$.

(D) $f_c = 1/[\sqrt{\varepsilon_0\mu_0}\,(a + b)]$.

(E) $f_c = 0$.

(F) $f_c \to \infty$.

CONCEPTUAL QUESTION 11.9 **Field components of the dominant waveguide mode.** For the waveguide shown in Figure 11.4, assuming that $a > b$, the following is a complete list of nonzero electric and magnetic field components of the dominant mode:

(A) E_x, E_z, and H_y.

(B) E_x and H_y.

(C) E_y, H_x, and H_z.

(D) E_y, H_y, and H_z.

(E) E_x, E_z, H_y, and H_z.

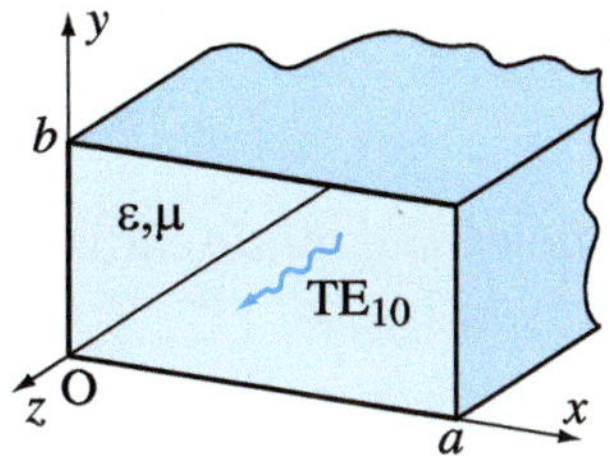

Figure 11.4 Metallic waveguide with a homogeneous dielectric carrying the dominant wave mode; for Conceptual Question 11.9.

CONCEPTUAL QUESTION 11.10 **Standing-wave patterns of the dominant mode.** With reference to the notation in Figure 11.4, the electric and magnetic fields of the dominant mode traveling through an air-filled rectangular waveguide with $a = 1.5b$ exhibit standing-wave patterns in

(A) the x direction only.

(B) the y direction only.

(C) the x and y directions only.

(D) the x and z directions only.

(E) all three directions.

(F) none of the directions.

11.2 Arbitrary TE and TM Modes in a Rectangular Waveguide

From Maxwell's equations for a perfect dielectric of parameters ε and μ (waveguide dielectric), Eqs. (6.13), and boundary conditions at the waveguide (PEC) walls, Eqs. (6.15), the field components of a TE_{mn} mode in a rectangular waveguide, Figure 11.1, are found to be

$$\underline{E}_x = \frac{\mathrm{j}\omega\mu}{k^2}\frac{n\pi}{b}\underline{H}_0\cos\left(\frac{m\pi}{a}x\right)\sin\left(\frac{n\pi}{b}y\right)\mathrm{e}^{-\mathrm{j}\beta z}, \quad \underline{E}_y = -\frac{\mathrm{j}\omega\mu}{k^2}\frac{m\pi}{a}\underline{H}_0\sin\left(\frac{m\pi}{a}x\right)$$

$$\times\cos\left(\frac{n\pi}{b}y\right)\mathrm{e}^{-\mathrm{j}\beta z}, \quad \underline{E}_z = 0, \quad \underline{H}_x = \frac{\mathrm{j}\beta}{k^2}\frac{m\pi}{a}\underline{H}_0\sin\left(\frac{m\pi}{a}x\right)\cos\left(\frac{n\pi}{b}y\right)\mathrm{e}^{-\mathrm{j}\beta z},$$

$$\underline{H}_y = \frac{\mathrm{j}\beta}{k^2}\frac{n\pi}{b}\underline{H}_0\cos\left(\frac{m\pi}{a}x\right)\sin\left(\frac{n\pi}{b}y\right)\mathrm{e}^{-\mathrm{j}\beta z}, \quad \underline{H}_z = \underline{H}_0\cos\left(\frac{m\pi}{a}x\right)\cos\left(\frac{n\pi}{b}y\right)$$

$$\times\,\mathrm{e}^{-\mathrm{j}\beta z}, \quad k^2 = \left(\frac{m\pi}{a}\right)^2 + \left(\frac{n\pi}{b}\right)^2 \quad (\mathrm{TE}_{mn}\text{ mode; } m, n = 0, 1, 2, \ldots), \tag{11.4}$$

with the restriction that only one of the mode indices can be zero. The electromagnetic field of a TM_{mn} mode is given by

$$\underline{E}_x = -\frac{\mathrm{j}\beta}{k^2}\frac{m\pi}{a}\underline{E}_0\cos\left(\frac{m\pi}{a}x\right)\sin\left(\frac{n\pi}{b}y\right)\mathrm{e}^{-\mathrm{j}\beta z}, \quad \underline{E}_y = -\frac{\mathrm{j}\beta}{k^2}\frac{n\pi}{b}\underline{E}_0\sin\left(\frac{m\pi}{a}x\right)$$

$$\times\cos\left(\frac{n\pi}{b}y\right)\mathrm{e}^{-\mathrm{j}\beta z}, \quad \underline{E}_z = \underline{E}_0\sin\left(\frac{m\pi}{a}x\right)\sin\left(\frac{n\pi}{b}y\right)\mathrm{e}^{-\mathrm{j}\beta z},$$

$$\underline{H}_x = \frac{\mathrm{j}\omega\varepsilon}{k^2}\frac{n\pi}{b}\underline{E}_0\sin\left(\frac{m\pi}{a}x\right)\cos\left(\frac{n\pi}{b}y\right)\mathrm{e}^{-\mathrm{j}\beta z}, \quad \underline{H}_y = -\frac{\mathrm{j}\omega\varepsilon}{k^2}\frac{m\pi}{a}\underline{E}_0\cos\left(\frac{m\pi}{a}x\right)$$

$$\times\sin\left(\frac{n\pi}{b}y\right)\mathrm{e}^{-\mathrm{j}\beta z}, \quad \underline{H}_z = 0 \quad (\mathrm{TM}_{mn}\text{ mode; } m, n = 1, 2, \ldots), \tag{11.5}$$

where the parameter k (or k^2) is the same as in Eqs. (11.4). Note that this is a TM (transverse magnetic) wave because $\underline{H}_z = 0$ and $\underline{E}_z \neq 0$. Note also that the lowest TM mode is TM_{11}. The expression for the waveguide phase coefficient, β, is that in Eqs. (11.2) in both TE and TM cases, with the cutoff frequency of the TE_{mn} or TM_{mn} mode computed as

$$f_\mathrm{c} = (f_\mathrm{c})_{mn} = \frac{c}{2}\sqrt{\left(\frac{m}{a}\right)^2 + \left(\frac{n}{b}\right)^2} \quad (\text{cutoff frequency, } \mathrm{TE}_{mn}\text{ or } \mathrm{TM}_{mn}\text{ mode}). \tag{11.6}$$

CONCEPTUAL QUESTION 11.11 Field configuration of an arbitrary TE mode. Consider the field configuration of an arbitrary TE_{mn} wave mode propagating along a rectangular metallic waveguide. With reference to the notation given in Figure 11.5, the integers m and n (if nonzero) equal the number of

(A) wavelengths along the x- and y-axes that fit into a and b, respectively.

(B) half-wavelengths along the x- and y-axes that fit into a and b, respectively.

(C) wavelengths along the x-axis for electric and magnetic fields, respectively.

(D) half-wavelengths along the x-axis for electric and magnetic fields, respectively.

(E) wavelengths along the x-axis for the electric and y-axis for the magnetic field.

(F) None of the above.

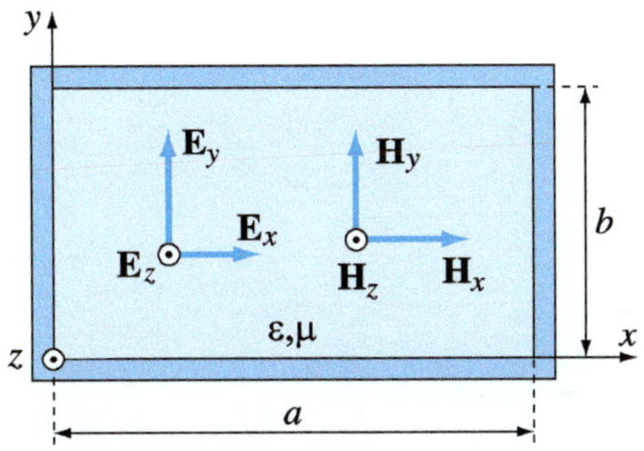

Figure 11.5 Rectangular metallic waveguide of transverse dimensions a and b ($a \geq b$). Shown are possible electric and magnetic field components; for Conceptual Question 11.11.

CONCEPTUAL QUESTION 11.12 Field components of the lowest TM waveguide mode. A complete list of nonzero electric and magnetic field components of the TM_{11} wave mode propagating along a rectangular metallic waveguide of transverse dimensions a and b, where $a > b$, in Figure 11.5, is as follows:

(A) E_y, H_x, and H_z.

(B) E_x, E_z, and H_y.

(C) E_x, E_y, H_x, and H_y.

(D) E_x, E_y, H_x, H_y, and H_z.

(E) E_x, E_y, E_z, H_x, and H_y.

(F) E_x, E_y, E_z, H_x, H_y, and H_z.

CONCEPTUAL QUESTION 11.13 **Field components of an arbitrary TM wave mode.** For an arbitrary TM_{mn} propagating wave mode in a rectangular metallic waveguide (Figure 11.5), the following is a complete list of nonzero electric and magnetic field components:

(A) E_y, H_x, and H_z.

(B) E_x, E_z, and H_y.

(C) E_x, E_y, H_x, and H_y.

(D) E_x, E_y, H_x, H_y, and H_z.

(E) E_x, E_y, E_z, H_x, and H_y.

(F) E_x, E_y, E_z, H_x, H_y, and H_z.

CONCEPTUAL QUESTION 11.14 **Field pattern of the dominant waveguide mode.** Field configuration of the dominant mode, TE_{10}, in a cross section of a rectangular waveguide, Figure 11.5, is as shown in

(A) Figure 11.6(a).

(B) Figure 11.6(b).

(C) Figure 11.6(c).

(D) Figure 11.6(d).

(E) Figure 11.6(e).

(F) Figure 11.6(f).

CONCEPTUAL QUESTION 11.15 **Field pattern of the TE_{01} wave mode.** Which one of the six offered field patterns in Figure 11.6 may represent the electric and magnetic field configuration, in a transverse plane, of the TE_{01} waveguide mode?

(A) The one in Figure 11.6(a).

(B) The one in Figure 11.6(b).

(C) The one in Figure 11.6(c).

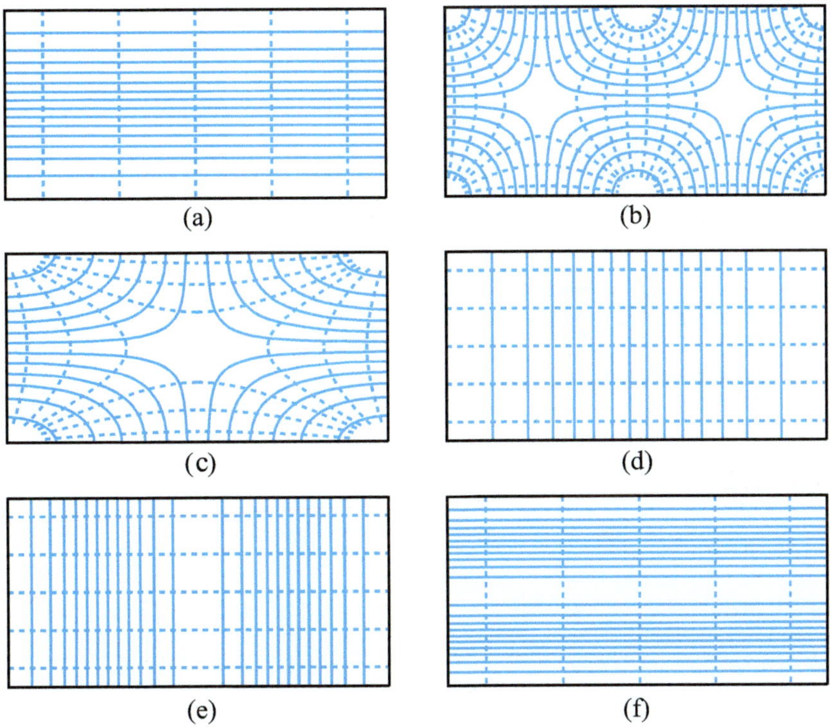

Figure 11.6 Field lines representing selected TE_{mn} modal field distributions in a cross section of a rectangular waveguide (Figure 11.5) with $a = 2b$: electric field – solid line, magnetic field – dashed line (recall that, in general, field lines are lines to which a field vector is tangential at all points, as well as that the magnitude of a field vector at a point is proportional to the density of field lines at that point); for Conceptual Question 11.14.

(D) The one in Figure 11.6(d).

(E) The one in Figure 11.6(e).

(F) The one in Figure 11.6(f).

CONCEPTUAL QUESTION 11.16 **Field pattern of the TE_{20} wave mode.** Which pattern in Figure 11.6 corresponds to the transverse E and H field distribution of the TE_{20} waveguide mode?

(A) The one in Figure 11.6(a).

(B) The one in Figure 11.6(b).

(C) The one in Figure 11.6(c).

(D) The one in Figure 11.6(d).

(E) The one in Figure 11.6(e).

(F) The one in Figure 11.6(f).

CONCEPTUAL QUESTION 11.17 **Field pattern of the TE_{11} wave mode.** The transverse electromagnetic field pattern of the TE_{11} mode in a rectangular waveguide is as shown in

(A) Figure 11.6(a).

(B) Figure 11.6(b).

(C) Figure 11.6(c).

(D) Figure 11.6(d).

(E) Figure 11.6(e).

(F) Figure 11.6(f).

CONCEPTUAL QUESTION 11.18 **Field pattern of the TE_{21} wave mode.** For the TE_{21} mode, the field configuration in a transverse plane is as in

(A) Figure 11.6(a).

(B) Figure 11.6(b).

(C) Figure 11.6(c).

(D) Figure 11.6(d).

(E) Figure 11.6(e).

(F) Figure 11.6(f).

CONCEPTUAL QUESTION 11.19 **Field pattern of the TE_{02} wave mode.** The transverse field lines of the TE_{02} waveguide mode are as in

(A) Figure 11.6(a).

(B) Figure 11.6(b).

(C) Figure 11.6(c).

(D) Figure 11.6(d).

(E) Figure 11.6(e).

(F) Figure 11.6(f).

CONCEPTUAL QUESTION 11.20 **Phase coefficients for TE and TM waves.** The phase coefficient for a TE_{mn} wave and that for a TM_{mn} wave at the same propagating frequency in a lossless air-filled rectangular waveguide are the same.

(A) True.

(B) False.

CONCEPTUAL QUESTION 11.21 **Dominant frequency range of a rectangular waveguide.** The dominant frequency range of a rectangular metallic waveguide of transverse dimensions a and b (Figure 11.5) is defined as the largest possible range of frequencies in which (at each of the frequencies) only the dominant mode (TE_{10}) can propagate along the structure. For a waveguide with $a = 2b$, which is referred to as a standard waveguide, the dominant range is given by

(A) $c/(2a) \leq f < \infty$,

(B) $c/(2b) \leq f < c/b$,

(C) $c/(2b) < f \leq c/b$,

(D) $c/(2a) \leq f < c/a$,

(E) $c/(2a) < f \leq c/a$,

(F) $c/(2a) < f \leq c\sqrt{1/a^2 + 1/b^2}/2$,

where c is the intrinsic phase velocity of the waveguide dielectric.

CONCEPTUAL QUESTION 11.22 **Waveguide aspect ratio that maximizes the dominant range.** Consider a rectangular metallic waveguide, in Figure 11.5, with $a \geq b$ and assume that the dimension a is given (fixed). Out of all possible values for the side ratio a/b, called the waveguide aspect ratio, the dominant frequency range of the waveguide (the largest possible frequency range in which only the dominant mode can propagate) is maximal for

(A) $a/b = 1$.

(B) $1 \leq a/b < 1.5$.

(C) $1 \leq a/b < 2$.

(D) $a/b \geq 2$.

(E) any a/b.

(F) None of the above.

CONCEPTUAL QUESTION 11.23 **Maximum relative size of the dominant range.** Denoting the lower and upper limits of the dominant frequency range of a rectangular metallic waveguide (Figure 11.5) by f_1 and f_2, respectively, the maximum possible value of the relative size of the dominant range defined as f_2/f_1 amounts to

(A) $(f_2/f_1)_{\rm max} = 1.5$.

(B) $(f_2/f_1)_{\rm max} = 2$.

(C) $(f_2/f_1)_{\rm max} = 4$.

(D) $(f_2/f_1)_{\rm max} \to \infty$.

(E) None of the above.

(F) Need more information.

CONCEPTUAL QUESTION 11.24 **Dominant frequency range of a square waveguide.** What is the relative size f_2/f_1 of the dominant frequency range (bounded by frequencies f_1 and f_2) of a square waveguide ($a = b$)?

(A) $f_2/f_1 = 1$.

(B) $f_2/f_1 = 1.5$.

(C) $f_2/f_1 = 2$.

(D) $f_2/f_1 \to \infty$.

(E) None of the above.

(F) Need more information.

CONCEPTUAL QUESTION 11.25 Relative size of the dominant range of a standard waveguide. For a standard waveguide ($a = 2b$), the relative size of the dominant frequency range comes out to be

(A) $f_2/f_1 = 1$.

(B) $f_2/f_1 = 1.5$.

(C) $f_2/f_1 = 2$.

(D) $f_2/f_1 \to \infty$.

(E) None of the above.

(F) Need more information.

CONCEPTUAL QUESTION 11.26 List of possible propagating modes in a standard waveguide. For an air-filled standard ($a = 2b$) rectangular metallic waveguide (see Figure 11.5), $a = 5$ cm and $b = 2.5$ cm, so that the cutoff frequency of the dominant mode amounts to $(f_c)_{10} = 3$ GHz. The following is a complete list of modes that can propagate along this waveguide at an operating frequency of $f = 4$ GHz:

(A) TE_{10}.

(B) TE_{10} and TM_{10}.

(C) TE_{10}, TE_{01}, and TE_{20}.

(D) TE_{10}, TE_{01}, TM_{10}, and TM_{01}.

(E) TE_{10}, TE_{01}, TE_{20}, TE_{11}, and TM_{11}.

(F) None.

CONCEPTUAL QUESTION 11.27 **List of propagating modes at a higher frequency.** At an operating frequency of $f = 6.2$ GHz, a complete list of propagating modes for a metallic waveguide with $a = 2b = 5$ cm, $\varepsilon = \varepsilon_0$, and $\mu = \mu_0$ [$(f_c)_{10} = 3$ GHz] consists of the following waveguide modes:

(A) TE_{10}.

(B) TE_{10} and TM_{10}.

(C) TE_{10}, TE_{01}, and TE_{20}.

(D) TE_{10}, TE_{01}, TM_{10}, and TM_{01}.

(E) TE_{10}, TE_{01}, TE_{20}, TE_{11}, and TM_{11}.

(F) None.

CONCEPTUAL QUESTION 11.28 **Dielectric-filled waveguide.** Assuming that the waveguide with $a = 2b = 5$ cm is filled with a lossless dielectric of relative permittivity $\varepsilon_r = 2.25$ and relative permeability $\mu_r = 1$ (polyethylene), a complete list of propagating modes at an operating frequency of $f = 4$ GHz is as follows:

(A) TE_{10}.

(B) TE_{10} and TM_{10}.

(C) TE_{10}, TE_{01}, and TE_{20}.

(D) TE_{10}, TE_{01}, TM_{10}, and TM_{01}.

(E) TE_{10}, TE_{01}, TE_{20}, TE_{11}, and TM_{11}.

(F) None.

CONCEPTUAL QUESTION 11.29 **Four times larger waveguide.** Consider an air-filled standard waveguide with transverse dimensions $a = 20$ cm and $b = 10$ cm, at an operating frequency of $f = 1$ GHz. All wave modes that can propagate along this waveguide are

(A) TE_{10}.

(B) TE_{10} and TM_{10}.

(C) TE_{10}, TE_{01}, and TE_{20}.

(D) TE_{10}, TE_{01}, TM_{10}, and TM_{01}.

(E) TE_{10}, TE_{01}, TE_{20}, TE_{11}, and TM_{11}.

(F) None.

CONCEPTUAL QUESTION 11.30 **FM and AM radio waves in a railway tunnel.** A railway tunnel can be approximated by a rectangular waveguide with transverse dimensions $a = 7$ m and $b = 4$ m and nonpenetrable walls. How many different wave modes can propagate inside the tunnel at an FM radio frequency of 100 MHz and at an AM radio frequency of 1 MHz, respectively?

(A) One mode at each of the two radio frequencies.

(B) More than one mode at each of the two frequencies.

(C) No modes at either of the frequencies.

(D) One mode at the FM frequency and no modes at the AM frequency.

(E) More than one mode at the FM frequency and one mode at the AM frequency.

(F) More than one mode at the FM frequency and no modes at the AM frequency.

11.3 Wave Impedances of TE and TM Waves

From Eqs. (11.4) and (11.2), the ratio of the electric and magnetic transverse complex field intensities, $\underline{E}_\text{t}$ and $\underline{H}_\text{t}$, for an arbitrary TE wave in a rectangular metallic waveguide comes out to be independent of the coordinates in Figure 11.1, and purely real, so a real constant, equal to

$$Z_\text{TE} = \left(\frac{\underline{E}_\text{t}}{\underline{H}_\text{t}}\right)_\text{TE} = \frac{\underline{E}_x}{\underline{H}_y} = -\frac{\underline{E}_y}{\underline{H}_x} = \frac{\omega\mu}{\beta} = \frac{\eta}{\sqrt{1 - f_\text{c}^2/f^2}} \quad \text{(TE wave impedance)}\,, \tag{11.7}$$

since $\mu c = \eta$, where $c = 1/\sqrt{\varepsilon\mu}$ and $\eta = \sqrt{\mu/\varepsilon}$ are the intrinsic phase velocity and impedance, respectively, of the waveguide dielectric. The ratio $\underline{E}_\text{t}/\underline{H}_\text{t}$, in turn, defines the wave impedance of a TE wave, namely, a TE_{mn} mode, analogously to the TEM case (for a transmission line), in Eqs. (9.2). Similarly, Eqs. (11.5) and

(11.2) give the following expression for the wave impedance of an arbitrary TM wave (TM_{mn} mode) in Figure 11.1:

$$Z_{\mathrm{TM}} = \left(\frac{\underline{E}_{\mathrm{t}}}{\underline{H}_{\mathrm{t}}}\right)_{\mathrm{TM}} = \frac{\beta}{\omega\varepsilon} = \eta\sqrt{1 - \frac{f_{\mathrm{c}}^2}{f^2}} \quad \text{(TM wave impedance)} . \tag{11.8}$$

CONCEPTUAL QUESTION 11.31 TE wave impedance, dependence on coordinates. There is a single traveling TE_{mn} wave on a lossless rectangular waveguide, as shown in Figure 11.7. Consider the wave impedance of this wave, Z_{TE}, defined as the ratio of transverse electric and magnetic complex field intensities, $\underline{E}_{\mathrm{t}}$ and $\underline{H}_{\mathrm{t}}$, in the structure, and whether it is a function of spatial coordinates, x, y, and/or z. The impedance Z_{TE}

(A) depends on x and y, but not on z.

(B) depends on x, but not on y and z.

(C) depends on z, but not on x and y.

(D) depends on x, y, and z.

(E) does not depend on any of the coordinates.

(F) Need more information.

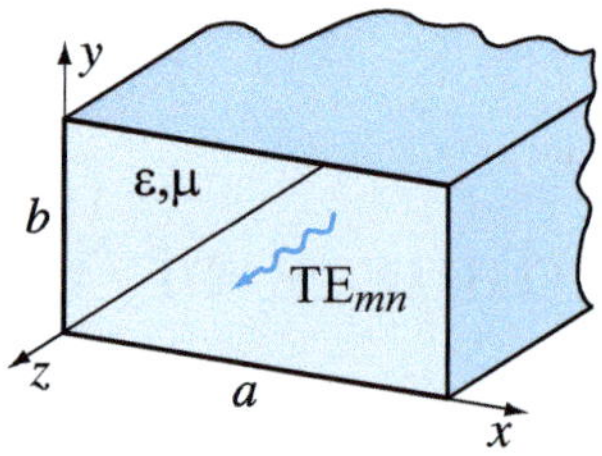

Figure 11.7 Lossless waveguide with a single traveling TE_{mn} wave; for Conceptual Question 11.31.

CONCEPTUAL QUESTION 11.32 Dependence on frequency and/or mode indices. For the wave impedance $Z_{\mathrm{TE}} = \underline{E}_{\mathrm{t}}/\underline{H}_{\mathrm{t}}$ of a TE_{mn} wave propagating along a lossless rectangular waveguide, consider whether it is a function of the operating frequency of the wave, f, and of mode indices, m and n, respectively. This impedance

(A) depends on f, but not on (m, n).

(B) depends on (m, n), but not on f.

(C) depends on both f and (m, n).

(D) does not depend on any of the above parameters.

(E) Need more information.

CONCEPTUAL QUESTION 11.33 **TE and TM wave impedances.** In a lossless air-filled rectangular waveguide, the wave impedance of a TE_{mn} wave and that of a TM_{mn} wave at the same operating frequency, in the propagating frequency region for that wave mode,

(A) have very different magnitudes.

(B) are close together in magnitude.

(C) are exactly the same.

(D) None of the above.

(E) Need more information.

CONCEPTUAL QUESTION 11.34 **TE and TM waves similar to TEM waves.** Both TE_{mn} and TM_{mn} waves in an air-filled rectangular waveguide have characteristics of TEM waves in free space

(A) at propagating frequencies close to the cutoff.

(B) at frequencies far above the cutoff.

(C) at intermediate frequencies between those in (A) and (B).

(D) at none of the propagating frequencies.

CONCEPTUAL QUESTION 11.35 **TE and TM wave impedances below the cutoff.** At frequencies below the cutoff, both TE and TM wave impedances of a rectangular waveguide become

(A) zero.

(B) infinite.

(C) nonzero, finite, and purely real (resistive).

(D) nonzero, finite, and purely imaginary (reactive).

(E) None of the above.

(F) Need more information.

CONCEPTUAL QUESTION 11.36 **Comparing TE, TM, and TEM wave impedances.** In the entire propagating frequency region (for $f > f_c$), we have the following for TE and TM wave impedances of a rectangular waveguide, as compared to the wave impedance of a TEM wave in a transmission line with the same dielectric, Z_{TEM}:

(A) $Z_{TE} > Z_{TEM}$ and $Z_{TM} > Z_{TEM}$.

(B) $Z_{TE} > Z_{TEM}$ and $Z_{TM} < Z_{TEM}$.

(C) $Z_{TE} < Z_{TEM}$ and $Z_{TM} > Z_{TEM}$.

(D) $Z_{TE} \gg Z_{TEM}$ and $Z_{TM} \ll Z_{TEM}$.

(E) $Z_{TE} \ll Z_{TEM}$ and $Z_{TM} \gg Z_{TEM}$.

(F) None of the above.

CONCEPTUAL QUESTION 11.37 **Relationship between TE, TM, and TEM wave impedances.** The relationship $Z_{TE}Z_{TM} = Z_{TEM}^2$ between the TE and TM wave impedances of a rectangular waveguide and the TEM wave impedance of a transmission line (with the same dielectric) holds true

(A) at all propagating frequencies ($f > f_c$) of the waveguide and the given mode.

(B) only at propagating frequencies close to the cutoff.

(C) only at frequencies far above the cutoff.

(D) only at intermediate frequencies between those in (B) and (C).

(E) at none of the propagating frequencies.

11.4 Waveguides with Small Losses

To take into account conductor and dielectric losses in a rectangular waveguide (Figure 11.1), with a TE or TM wave, we assume that these losses are small, i.e., that the conditions in Eqs. (9.6) are satisfied. Accordingly, we assume that the field distributions in every cross section of the waveguide are practically the same as if there were no losses, while the difference is in the axial (z) direction, in which the fields attenuate as $\mathrm{e}^{-\alpha z}$, α being the attenuation coefficient of the structure. In place of Eqs. (9.8), the attenuation coefficient for the waveguide conductor is given by

$$\alpha_{\mathrm{c}} = \frac{P'_{\mathrm{c}}}{2P} , \quad P'_{\mathrm{c}} = \oint_{C_{\mathrm{c}}} R_{\mathrm{s}} |\underline{\mathbf{H}}_{\mathrm{tang}}|^2 \, \mathrm{d}l \quad \text{(losses in waveguide walls)} , \tag{11.9}$$

where P'_{c} is the time-average power of Joule's losses in the conductor (i.e., in four waveguide walls) per unit length of the structure and P is the time-average power transmitted along the guide, equal to the real part of the complex power carried by the wave along the z-axis, i.e., of the flux of the complex Poynting vector of the wave through a cross section of the guide dielectric, S_{d}. On the other side, the attenuation coefficient for the waveguide dielectric is, as a generalization of Eqs. (9.9), obtained as

$$\alpha_{\mathrm{d}} = \frac{P'_{\mathrm{d}}}{2P} , \quad P'_{\mathrm{d}} = \int_{S_{\mathrm{d}}} \sigma_{\mathrm{d}} |\underline{\mathbf{E}}|^2 \, \mathrm{d}S \quad \text{(losses in waveguide dielectric)} . \tag{11.10}$$

Most importantly, both α_{c} and α_{d} are computed using the no-loss field distributions (perturbation method).

CONCEPTUAL QUESTION 11.38 Field components not contributing to the power flow. For an arbitrary TE or TM wave mode propagating along a lossless rectangular waveguide (Figure 11.8), the following electric and magnetic field components either are zero or, if nonzero, do not contribute to the power transfer along the waveguide:

(A) E_x, E_y, H_x, and H_y.

(B) E_x, E_z, H_y, and H_z.

(C) E_y, H_y, and H_z.

(D) E_x and H_z.

(E) E_z and H_z.

(F) None of the components.

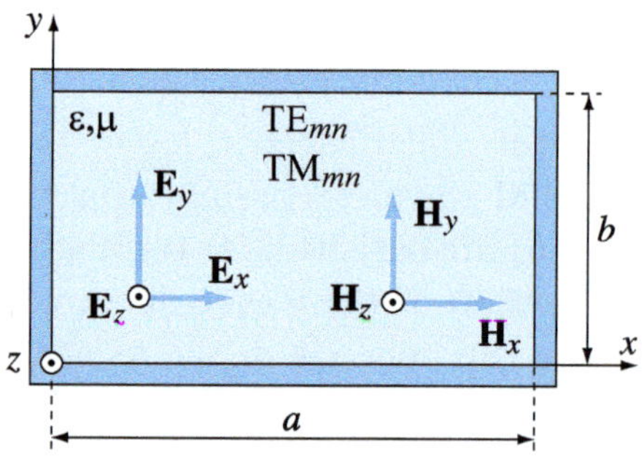

Figure 11.8 For the discussion of electric and magnetic field components of an arbitrary TE or TM wave mode propagating along a rectangular waveguide; for Conceptual Question 11.38.

CONCEPTUAL QUESTION 11.39 Flux of the Poynting vector into waveguide walls. In a lossless rectangular metallic waveguide with the dominant (TE_{10}) mode only, given the notation in Figure 11.8 – the flux of the complex Poynting vector, $\underline{\mathcal{P}}$, into the following waveguide walls is zero:

(A) walls defined by $x = 0$ and $x = a$ only.

(B) walls defined by $x = 0$ and $y = 0$ only.

(C) walls defined by $y = 0$ and $y = b$ only.

(D) all four walls.

(E) none of the walls.

CONCEPTUAL QUESTION 11.40 Transmitted power of the dominant mode. Consider the time-average (real) power flow, P, associated with a traveling TE_{10} wave through a rectangular waveguide, Figure 11.8, and the following two statements: (a) P can be written as a constant times $\int_{x=0}^{a} |\underline{E}_y|^2 \, dx$; (b) P can be written as a constant times $\int_{x=0}^{a} |\underline{H}_x|^2 \, dx$. Which of the statements is true?

(A) Statement (a) only.

(B) Statement (b) only.

(C) Both statements.

(D) Neither of the statements.

CONCEPTUAL QUESTION 11.41 **Low-loss vs. lossless waveguides.** Waveguides with small losses can generally be treated as those with no losses.

(A) True.

(B) False.

CONCEPTUAL QUESTION 11.42 **Large wave attenuation on a waveguide with small losses.** The attenuation of a TE or TM wave on a waveguide with small losses can be prohibitively large for practical use of the structure.

(A) True.

(B) False.

CONCEPTUAL QUESTION 11.43 **Electric-field distribution in low-loss and lossless waveguides.** A waveguide with the dominant (TE_{10}) wave mode and small losses has approximately the same distribution (dependence on respective spatial coordinates) of the electric field vector, $\underline{\mathbf{E}}$,

(A) in a cross section of the guide and along the guide

(B) in a cross section of the guide, but not along the guide,

(C) along the guide, but not in a cross section of the guide,

(D) in no cross sections and along no directions

as the same waveguide with losses neglected.

CONCEPTUAL QUESTION 11.44 **Magnetic field of low- and no-loss waveguides.** Distributions of the magnetic field vector, $\underline{\mathbf{H}}$, of the low-loss and lossless waveguides carrying the dominant wave mode are approximately the same

(A) in a cross section of the guide and along the guide.

(B) in a cross section of the guide, but not along the guide.

(C) along the guide, but not in a cross section of the guide.

(D) in no cross sections and along no directions.

CONCEPTUAL QUESTION 11.45 Contour of integration for evaluating conductor losses. In evaluating the per-unit-length time-average power of Joule's losses in the waveguide conductor (four waveguide walls), $P'_{\rm c}$, needed to obtain the attenuation coefficient $\alpha_{\rm c}$ for an arbitrary TM wave mode, the contour of integration of the associated surface power density $R_{\rm s}|\underline{\mathbf{H}}_{\rm tang}|^2$, $C_{\rm c}$, represents

(A) the interior contour of the conductor.

(B) the exterior contour of the conductor.

(C) both the interior and the exterior contours of the conductor.

(D) None of the above.

CONCEPTUAL QUESTION 11.46 Magnetic field components in the conductor loss power integral. A TE_{10} wave (and no other wave modes) propagates through an air-filled rectangular waveguide with copper walls, as shown in Figure 11.9. In the integral along the contour $C_{\rm c}$ needed for obtaining the p.u.l. ohmic power and the attenuation coefficient of the waveguide, the following nonzero magnetic field components contribute, through the surface power density $R_{\rm s}|\underline{\mathbf{H}}_{\rm tang}|^2$, to the result for $P'_{\rm c}$:

(A) H_x only.

(B) H_z only.

(C) both H_x and H_z.

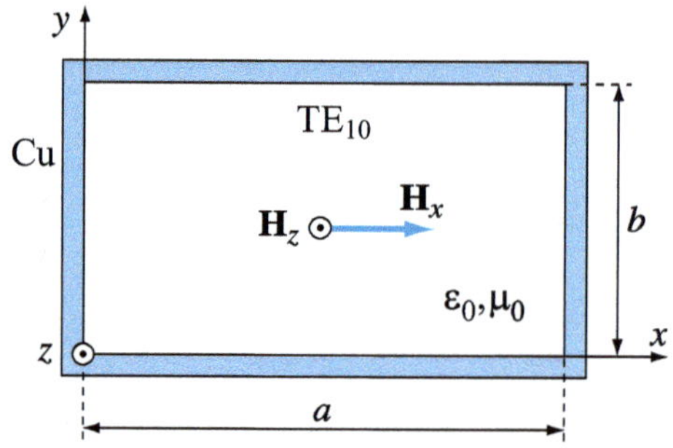

Figure 11.9 Air-filled rectangular waveguide with copper (Cu) walls carrying a TE_{10} wave; for Conceptual Question 11.46.

CONCEPTUAL QUESTION 11.47 Frequency dependence of the guide attenuation coefficient. An air-filled rectangular aluminum waveguide carries a TE_{10} wave, in the dominant frequency range (only the dominant mode can propagate). The attenuation coefficient of the waveguide, $\alpha = \alpha_c$,

(A) is proportional to the square root of frequency.

(B) depends on frequency but in a more complex way than $\sqrt{f}$.

(C) does not depend on frequency.

(D) Need more information.

CONCEPTUAL QUESTION 11.48 Electric field components in the dielectric loss power integral. Consider an arbitrary TM_{mn} wave in a rectangular waveguide filled with a low-loss dielectric, as shown in Figure 11.10. For the mode, all three Cartesian components of the electric field vector are nonzero. In the integral over the surface of the waveguide dielectric in a cross section of the structure, S_d, needed to determine the time-average power of Joule's losses in the dielectric per unit length of the structure, P'_d, and the associated attenuation coefficient, α_d, for the waveguide, the following electric field components contribute, through the volume power density $\sigma_d|\mathbf{E}|^2$, to the result for P'_d:

(A) E_x and E_y only.

(B) E_z only.

(C) all components, E_x, E_y, and E_z.

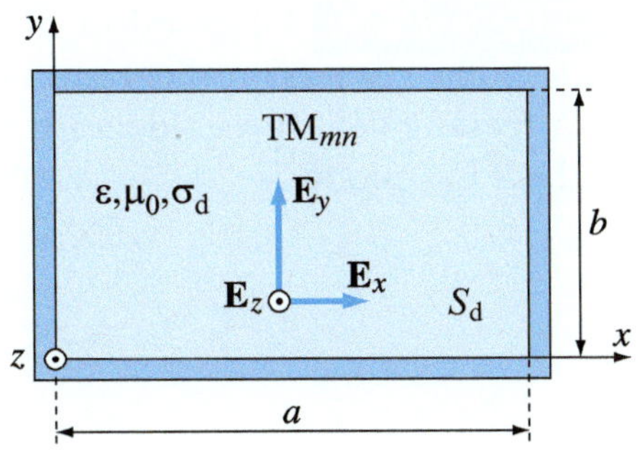

Figure 11.10 Rectangular waveguide filled with a low-loss dielectric; for Conceptual Question 11.48.

11.5 Waveguide Dispersion and Wave Velocities

Since the phase coefficient β in Eq. (11.2), of the rectangular waveguide in Figure 11.1, is a nonlinear function of the angular frequency, ω, of a propagating TE or TM wave, the phase velocity of the wave, $v_{\rm p}$, is frequency dependent, and the waveguide represents a dispersive propagation medium. Combining Eqs. (7.19), (7.20), and (11.2), $v_{\rm p}$ and the group velocity (or energy velocity) along the waveguide, $v_{\rm g}$, are given by the following expressions (c is the intrinsic phase velocity of the waveguide dielectric):

$$v_{\rm p} = \frac{\omega}{\beta} = \frac{c}{\sqrt{1 - f_{\rm c}^2/f^2}} , \quad v_{\rm g} = \frac{1}{{\rm d}\beta/{\rm d}\omega} = c\sqrt{1 - \frac{f_{\rm c}^2}{f^2}} \quad \left(c = \frac{1}{\sqrt{\varepsilon\mu}}\right)$$

$$\text{(waveguide phase and group velocities)} . \quad (11.11)$$

CONCEPTUAL QUESTION 11.49 **Waveguide as a dispersive or nondispersive propagation medium.** A rectangular waveguide has walls made of a perfect electric conductor, which is a nonpenetrable medium, and is filled with a perfect dielectric, which is a nondispersive propagation medium. Above the cutoff frequency for the dominant wave mode, the waveguide as a whole represents

(A) a dispersive propagation medium

(B) a nondispersive propagation medium

for this mode.

CONCEPTUAL QUESTION 11.50 **Dispersion diagrams for two modes in an arbitrary waveguide.** The dispersion diagram, i.e., the β–ω relationship, for the TE_{20} wave mode in a rectangular waveguide with an arbitrary aspect ratio a/b is the same as that for the TE_{01} mode.

(A) True.

(B) False.

CONCEPTUAL QUESTION 11.51 **Dispersion diagrams for two modes in a standard waveguide.** For a standard waveguide, with the aspect ratio $a/b = 2$, β–ω diagrams for TE_{20} and TE_{01} modes are the same.

(A) True.

(B) False.

CONCEPTUAL QUESTION 11.52 **Phase and group velocities of the dominant mode.** With v_p and v_g being the phase and group velocities, respectively, of the dominant (TE_{10}) mode propagating along an air-filled rectangular waveguide with PEC walls and $c_0 = 3 \times 10^8$ m/s standing for the speed of light in free space, which of the following inequalities holds true?

(A) $v_\mathrm{p} > c_0$.

(B) $v_\mathrm{g} > c_0$.

(C) Both inequalities.

(D) Neither of the inequalities.

CONCEPTUAL QUESTION 11.53 **Comparison of values of phase and group velocities.** In a lossless air-filled rectangular waveguide, the phase velocity of a TE_{mn} wave, v_p, and the group velocity of that wave, v_g, at a given operating frequency, f, in the propagating frequency region for the (m, n) mode,

(A) are very different (in magnitude).

(B) are close together (in magnitude).

(C) are exactly the same.

(D) None of the above.

(E) Need more information.

CONCEPTUAL QUESTION 11.54 **Travel of signals with different carrier frequencies along a waveguide.** Two signals whose frequency spectra are confined to narrow bands around carrier frequencies f_1 and f_2 ($f_2 > f_1$), respectively, are launched at the same instant of time at one end of an air-filled rectangular waveguide with length l, to propagate along it. Both frequencies belong to the dominant frequency range of the waveguide (defined as the largest possible range of frequencies in which only the TE_{10} mode can propagate along the waveguide). At the frequency f_1, the phase and group velocities of the dominant mode are v_{p1} and v_{g1}, respectively, while these velocities equal v_{p2} and v_{g2} at the frequency f_2. The time lag between the two signals as they are received on the other end of the waveguide is

(A) nonzero and amounts to $\Delta t = |l/v_{p1} - l/v_{p2}|$.

(B) nonzero and equal to $\Delta t = |l/v_{g1} - l/v_{g2}|$.

(C) nonzero and given by $\Delta t = |l/v_{p1} - l/v_{p2}| = |l/v_{g1} - l/v_{g2}|$.

(D) zero.

11.6 Waveguide Couplers

In order to generate a particular TE_{mn} or TM_{mn} mode in a rectangular waveguide (Figure 11.1), we need an electromagnetic coupling mechanism that feeds external energy into the guide, and excites that particular modal field. This field then travels along the structure carrying the input signal away from the feed. Conversely, the same mechanism can be used, in the reversed process, to extract the energy (signal) carried by the wave (in the same mode), and deliver it to an external device or system. Such signal transmitters or receivers based on electromagnetic coupling to waveguide fields are generally referred to as waveguide couplers. Most frequently, couplers convert input power from a coaxial cable, attached, externally, to a guide wall, into waveguide modes, and vice versa (coax-to-waveguide couplers). An extension of the inner conductor of the cable, called the probe, is inserted into the guide dielectric (usually air), with the outer conductor being connected to the wall, as in Figure 11.11, showing a coupler with an electric probe and one with a magnetic probe. The electric probe, suitable for coupling to the electric field in the structure, is in the form of a short wire segment (straight extension of the cable conductor). The magnetic probe consists of a small wire loop (the conductor is folded and its tip connected back to the wall), and is better suited for magnetic field coupling. These probes are actually a short monopole wire antenna and a small loop antenna, respectively. In particular, electric probes aimed to launch or receive a TE or TM wave in the waveguide should be placed at the locations of the maxima of the guide electric field intensity, **E**, and directed in parallel to the electric field lines. On the other side, magnetic probes should be positioned at the magnetic field maxima in the waveguide, and oriented such that **H** is perpendicular to the loop plane.

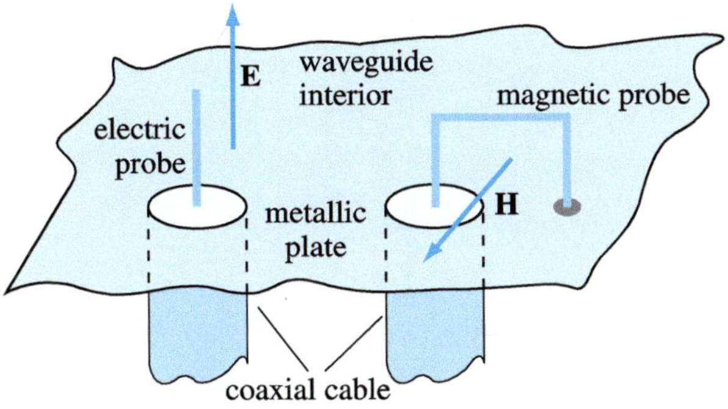

Figure 11.11 Coax-to-waveguide couplers in the form of an electric probe (short monopole antenna) and magnetic probe (small loop antenna) used to excite or receive wave modes in a rectangular metallic waveguide.

CONCEPTUAL QUESTION 11.55 **Evaluation of the emf induced in magnetic probes in a waveguide.** A TE_{10} wave of angular frequency ω propagates along a lossless air-filled rectangular metallic waveguide, and it is to be received by a small wire loop (magnetic probe), of the loop surface area S, attached to one of the guide walls. Consider loops attached to the conductor surface at (a) $x = 0$ (left wall) and (b) $y = 0$ (bottom wall), respectively, as shown in Figure 11.12. If $\underline{H}_x$ and $\underline{H}_z$ designate the x- and z-components, respectively, of the complex rms magnetic field intensity vector of the wave at their maxima in a cross section of the waveguide, the rms electromotive force (emf) induced in loop (a), $\mathcal{E}_{\rm ind}^{(\rm a)}$, and that in loop (b), $\mathcal{E}_{\rm ind}^{(\rm b)}$, are given by

(A) $\mathcal{E}_{\rm ind}^{(\rm a)} = \omega\mu_0|\underline{H}_x|S$ and $\mathcal{E}_{\rm ind}^{(\rm b)} = \omega\mu_0|\underline{H}_z|S$.

(B) $\mathcal{E}_{\rm ind}^{(\rm a)} = \omega\mu_0|\underline{H}_z|S$ and $\mathcal{E}_{\rm ind}^{(\rm b)} = \omega\mu_0|\underline{H}_x|S$.

(C) $\mathcal{E}_{\rm ind}^{(\rm a)} = \omega\mu_0|\underline{H}_x|S$ and $\mathcal{E}_{\rm ind}^{(\rm b)} = 0$.

(D) $\mathcal{E}_{\rm ind}^{(\rm a)} = 0$ and $\mathcal{E}_{\rm ind}^{(\rm b)} = \omega\mu_0|\underline{H}_z|S$.

(E) $\mathcal{E}_{\rm ind}^{(\rm a)} = |\omega\mu_0\underline{H}_xS + {\rm j}\omega\mu_0\underline{H}_zS|$ and $\mathcal{E}_{\rm ind}^{(\rm b)} = |\omega\mu_0\underline{H}_xS - {\rm j}\omega\mu_0\underline{H}_zS|$.

(F) $\mathcal{E}_{\rm ind}^{(\rm a)} = \mathcal{E}_{\rm ind}^{(\rm b)} = 0$.

CONCEPTUAL QUESTION 11.56 **Magnetic-probe coupling above the dominant range.** Consider an air-filled waveguide with transverse dimensions $a = 16.51$ cm and $b = 8.255$ cm, at a frequency of $f = 2$ GHz, and assume that

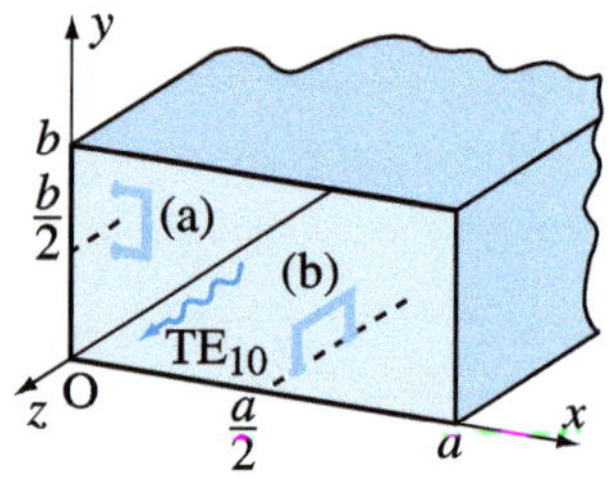

Figure 11.12 Magnetic probes (small loop antennas) for receiving a dominant (TE_{10}) wave in a rectangular metallic waveguide; for Conceptual Question 11.55.

all possible propagating modes, which turn out to be TE_{10}, TE_{20}, and TE_{01}, are established in the structure. With reference to the coordinate system in Figure 11.12, let a small wire loop be attached to the left wall of the waveguide (wall at $x = 0$) such that it lies in the plane $y = b/2$. Under these circumstances, the loop couples to the magnetic field of

(A) only one

(B) only two

(C) all three

(D) none

of the established modes.

CONCEPTUAL QUESTION 11.57 Unidirectional electric-probe coupler for the dominant mode. Figure 11.13 depicts a coupling configuration designed to transmit a TE_{10} wave mode in only one direction along the waveguide. To prevent the propagation in the other direction, the guide is closed (short-circuited), by inserting a metallic plate at one end. The distance of the electric probe (short monopole antenna) from the plate, d, is chosen such that the backward propagating wave launched by the probe, after its reflection from the plate (PEC surface) and its round trip (from the probe to the plate and back), adds constructively with the forward propagating wave (Figure 11.13). Hence, the two waves are in phase as they propagate to the right, away from the probe. With $\lambda_z = 2\pi/\beta$ being the wavelength along the structure (measured along the z-axis in Figure 11.13), a choice for this distance resulting in the constructive addition of the waves is

(A) $d = \lambda_z/8$.

(B) $d = \lambda_z/4$.

(C) $d = \lambda_z/2$.

(D) $d = \lambda_z$.

(E) $d = 2\lambda_z$.

(F) More than one of the above values.

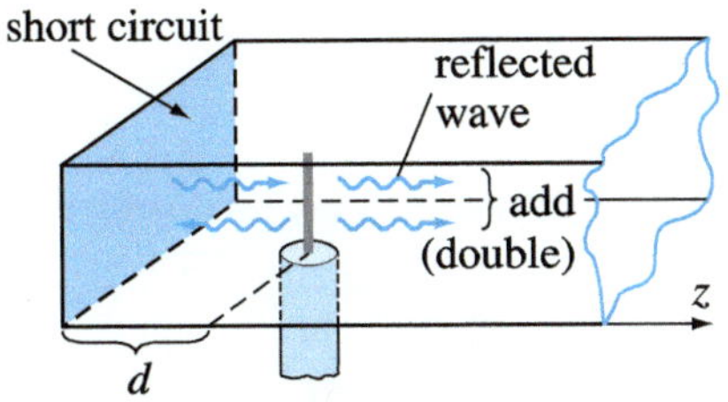

Figure 11.13 Electric-probe coupler transmitting the TE_{10} mode in only one direction along the waveguide; for Conceptual Question 11.57.

CONCEPTUAL QUESTION 11.58 Unidirectional coupler in the receiving mode of operation. Considering the waveguide coupling configuration in Figure 11.13 (with a proper choice of the distance d), the same constructive addition of the direct (incident) and reflected (from the plate) waves occurs in the receiving mode of operation of the waveguide, when the electric probe in Figure 11.13 is used to receive the signal (and deliver it to the coaxial cable) from a TE_{10} wave propagating in the negative z direction (toward the probe).

(A) True.

(B) False.

CONCEPTUAL QUESTION 11.59 Electric-probe waveguide coupler for the TE_{20} mode. Which one of the four offered coupling configurations with electric probes in Figure 11.14 can be used for excitation/reception of the TE_{20} waveguide mode?

(A) Coupler in Figure 11.14(a).

(B) Coupler in Figure 11.14(b).

(C) Coupler in Figure 11.14(c).

(D) Coupler in Figure 11.14(d).

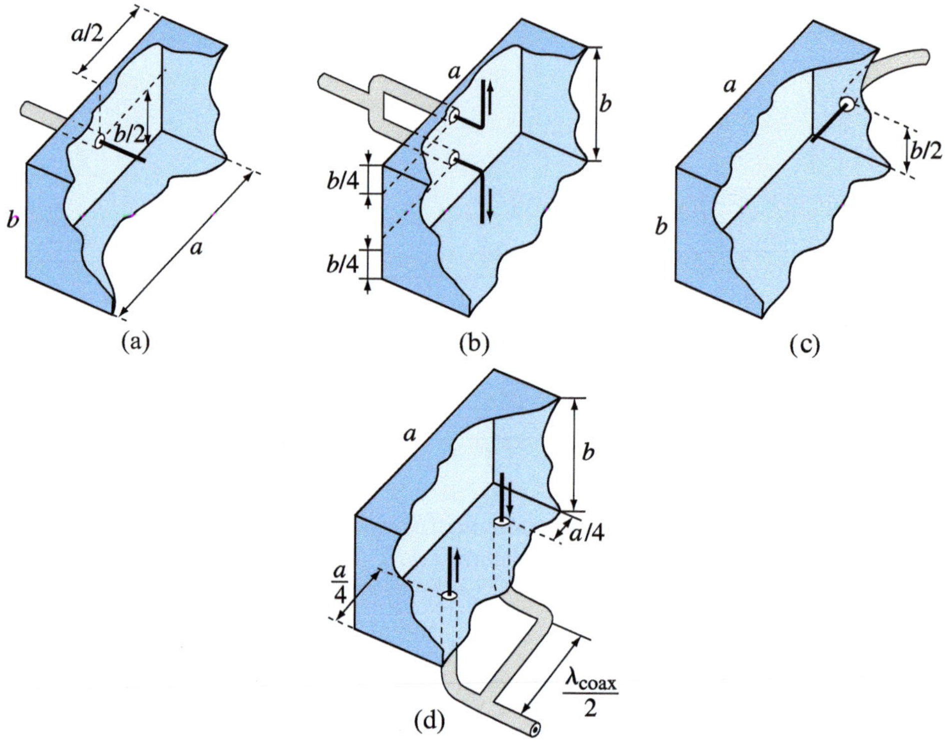

Figure 11.14 Coupling configurations using electric probes for four different higher-order modes (modes higher than the dominant) in a rectangular waveguide (each configuration is designed for a particular mode); for Conceptual Question 11.59.

CONCEPTUAL QUESTION 11.60 **Electric-probe waveguide coupler for the TE_{01} mode.** Which coupling configuration out of the four offered in Figure 11.14 can be used for excitation/reception of the TE_{01} waveguide mode?

(A) Coupler in Figure 11.14(a).

(B) Coupler in Figure 11.14(b).

(C) Coupler in Figure 11.14(c).

(D) Coupler in Figure 11.14(d).

CONCEPTUAL QUESTION 11.61 **Electric-probe waveguide coupler for the TE_{11} mode.** Which one in Figure 11.14 is a coupling configuration for the TE_{11} waveguide mode?

(A) The one in Figure 11.14(a).

(B) The one in Figure 11.14(b).

(C) The one in Figure 11.14(c).

(D) The one in Figure 11.14(d).

CONCEPTUAL QUESTION 11.62 **Electric-probe waveguide coupler for the TM_{11} mode.** A coupling configuration for the TM_{11} waveguide mode is as shown in

(A) Figure 11.14(a).

(B) Figure 11.14(b).

(C) Figure 11.14(c).

(D) Figure 11.14(d).

11.7 Rectangular Cavity Resonators

Next, we study electromagnetic resonators made from rectangular metallic waveguides with TE or TM waves. A section of a waveguide closed at both ends with new transverse conducting walls, thus forming a rectangular metallic box (cavity), as shown in Figure 11.15, represents a three-dimensional resonant wave structure (at certain resonant frequencies), called a rectangular cavity resonator. In the cavity, a standing TE or TM wave is formed by the incident and reflected waves bouncing back and forth between the two transverse walls, and the resultant field pattern along the z-axis looks the same as in Figure 8.1(b). The dimension d of the cavity must be such that the tangential component of the electric field vector of the resultant wave is zero at the PEC boundary $z = -d$ in Figure 11.15, which gives $\sin\beta d = 0$, as in Figure 8.1(b) and Eqs. (8.4), that is, $\beta d = p\pi$ ($p = 1, 2, \ldots$). Combining this zero-field condition with the expressions for the waveguide phase coefficient, β, in Eq. (11.2), and the cutoff frequency of an arbitrary TE_{mn} or TM_{mn} mode, $f_c = (f_c)_{mn}$, in Eq. (11.6), we obtain the following expression for the resonant frequency (f_{res}) of a mode (m, n, p), i.e., TE_{mnp} or TM_{mnp}, in a cavity of dimensions a, b, and d and dielectric parameters ε and μ (Figure 11.15):

$$f = f_{\mathrm{res}} = (f_{\mathrm{res}})_{mnp} = \frac{c}{2}\sqrt{\left(\frac{m}{a}\right)^2 + \left(\frac{n}{b}\right)^2 + \left(\frac{p}{d}\right)^2} \quad (m, n, p = 0, 1, 2, \ldots)$$

$$(\text{cavity resonance, } \mathrm{TE}_{mnp} \text{ or } \mathrm{TM}_{mnp} \text{ mode}) . \quad (11.12)$$

If the cavity dimensions are not all the same and coordinate axes (x, y, and z) in Figure 11.15 are chosen such that $a > b$ and $d > b$, out of all solutions the TE_{101} mode ($m = p = 1$, $n = 0$) has the lowest frequency, given by

$$(f_{\mathrm{res}})_{101} = \frac{c}{2}\sqrt{\frac{1}{a^2} + \frac{1}{d^2}} \quad (\mathrm{TE}_{101},\ \text{dominant cavity mode})\ , \tag{11.13}$$

and is hence termed the dominant cavity mode.

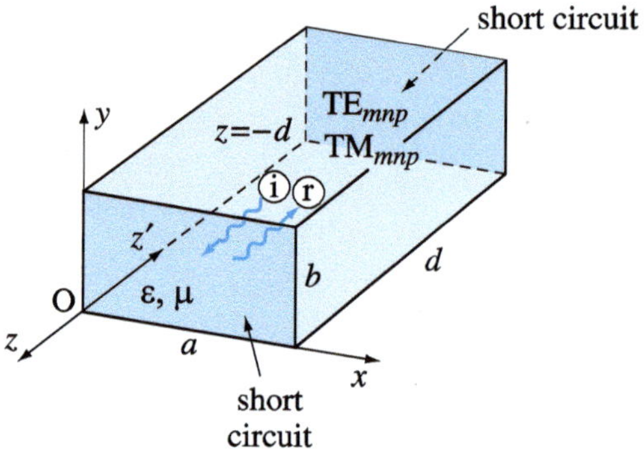

Figure 11.15 Rectangular cavity resonator, obtained by short-circuiting the rectangular metallic waveguide in Figure 11.1 in two transverse planes.

Considering a TE_{10} wave (dominant waveguide mode) described by Eqs. (11.3) as the incident wave that reflects at PEC boundaries $z = 0$ and $z = -d$ in Figure 11.15, and computing the resultant (incident plus reflected) wave, as in Eqs. (8.4), we get the expressions for the electric and magnetic fields of a TE_{101} wave (dominant cavity mode) in the cavity:

$$\underline{E}_y = -2\omega\mu\,\frac{a}{\pi}\,\underline{H}_0 \sin\left(\frac{\pi}{a}\,x\right)\sin\left(\frac{\pi}{d}\,z\right)\,, \quad \underline{H}_x = 2\mathrm{j}\,\frac{a}{d}\,\underline{H}_0 \sin\left(\frac{\pi}{a}\,x\right)\cos\left(\frac{\pi}{d}\,z\right)\,,$$

$$\underline{H}_z = -2\mathrm{j}\underline{H}_0 \cos\left(\frac{\pi}{a}\,x\right)\sin\left(\frac{\pi}{d}\,z\right) \quad (\text{fields of the } \mathrm{TE}_{101} \text{ cavity mode}) \tag{11.14}$$

($\underline{E}_x\,,\ \underline{E}_z\,,\ \underline{H}_y = 0$), where $\omega = 2\pi f$, with $f = (f_{\mathrm{res}})_{101}$ being the resonant frequency in Eq. (11.13). Field expressions of arbitrary TE_{mnp} and TM_{mnp} resonance modes in the cavity are obtained in a similar fashion.

CONCEPTUAL QUESTION 11.63 **Operating frequencies for individual resonant-cavity modes.** Consider a lossless rectangular metallic cavity of dimensions a, b, and d, filled with a homogeneous dielectric of permittivity ε and permeability μ. For fixed a, b, d, ε, and μ, each cavity mode, namely, TE_{mnp} or TM_{mnp} mode for given (m, n, p), can exist

(A) at a single frequency only.

(B) at a discrete set of frequencies.

(C) in a continuous range of frequencies.

(D) in multiple (separated) frequency ranges.

CONCEPTUAL QUESTION 11.64 **Operating frequencies for corresponding waveguide modes.** Considering the corresponding (infinitely long) waveguide of transverse dimensions a and b: for fixed a, b, ε, and μ, what is the type of frequencies (frequency ranges) at (in) which each waveguide mode, namely, TE_{mn} or TM_{mn} mode, for given (m, n), can exist?

(A) A single frequency only.

(B) A discrete set of frequencies.

(C) A continuous range of frequencies.

(D) Multiple (separated) frequency ranges.

CONCEPTUAL QUESTION 11.65 **Cavity resonant frequency versus waveguide cutoff frequency.** For an air-filled rectangular metallic cavity of dimensions a, b, and d, the resonant frequency of the TE_{101} cavity mode is

(A) higher than

(B) equal to

(C) lower than

the cutoff frequency of the corresponding TE_{10} waveguide mode, for an air-filled waveguide with transverse dimensions a and b.

CONCEPTUAL QUESTION 11.66 **Standing-wave patterns of the dominant cavity mode.** Assume that the TE_{101} wave mode (dominant cavity mode) is established in an air-filled cavity resonator of dimensions a, b, and d. Considering the three mutually orthogonal directions parallel to the respective cavity edges, the electric and magnetic field components of this mode exhibit a standing-wave variation in

(A) one of the directions only.

(B) two of the directions only.

(C) all three directions.

(D) none of the directions.

CONCEPTUAL QUESTION 11.67 **Field expressions for an arbitrary TE mode in a resonant cavity.** In the expressions for the electric and magnetic fields of an arbitrary TE_{mnp} wave mode in a lossless rectangular cavity resonator, in Figure 11.16, the field dependence, for all nonzero electric and magnetic field components, on the coordinate z is given by

(A) $\sin(p\pi z/d)$

(B) $\cos(p\pi z/d)$

(C) $\sin[p\pi z/(2d)]$

(D) $\sin(p\pi z/d)$ or $\cos(p\pi z/d)$

(E) $\mathrm{e}^{-\mathrm{j}p\pi z/d}$

(F) const (no dependence on z)

$(p = 1, 2, \ldots)$.

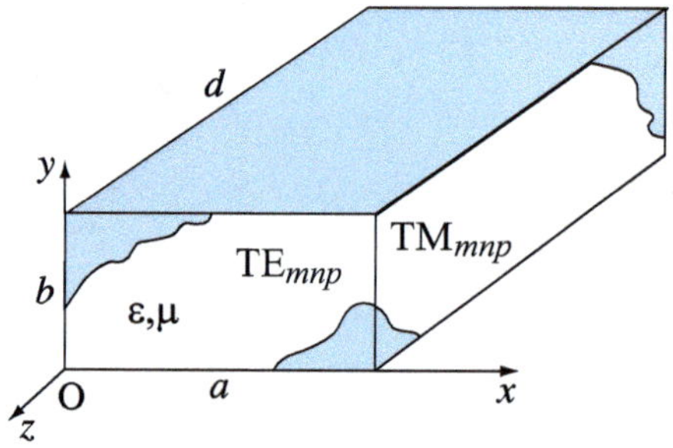

Figure 11.16 Rectangular cavity resonator with a TE_{mnp} or TM_{mnp} wave mode; for Conceptual Question 11.67.

CONCEPTUAL QUESTION 11.68 **Field of an arbitrary TM cavity mode.** For an arbitrary TM_{mnp} resonance mode in a lossless rectangular cavity (Figure 11.16), we have the following dependence on the coordinate z of all nonzero electric and magnetic field components:

(A) $\sin(p\pi z/d)$

(B) $\cos(p\pi z/d)$

(C) $\sin[p\pi z/(2d)]$

(D) $\sin(p\pi z/d)$ or $\cos(p\pi z/d)$

(E) $e^{-jp\pi z/d}$

(F) const (no dependence on z)

($p = 1, 2, \ldots$).

CONCEPTUAL QUESTION 11.69 **Boundary conditions for an arbitrary TE resonance mode.** The following is a full list of electric and magnetic field components of an arbitrary TE_{mnp} wave mode in a waveguide cavity resonator with PEC walls in Figure 11.16 that must be zero in planes $z = 0^-$ and $z = -d^+$:

(A) E_z and H_z.

(B) E_x, E_y, H_x, and H_y.

(C) E_x and E_y.

(D) E_x, E_y, and H_z.

(E) E_x, E_y, E_z, and H_z.

(F) None of the components.

CONCEPTUAL QUESTION 11.70 **Boundary conditions for an arbitrary TM mode.** For an arbitrary TM_{mnp} resonance mode in the PEC cavity (Figure 11.16), a complete list of electric and magnetic field components that must be zero in planes $z = 0^-$ and $z = -d^+$ is as follows:

(A) E_z and H_z.

(B) E_x, E_y, H_x, and H_y.

(C) E_x and E_y.

(D) E_x, E_y, and H_z.

(E) E_x, E_y, E_z, and H_z.

(F) None of the components.

CONCEPTUAL QUESTION 11.71 **Wave impedance in a short-circuited waveguide.** A TE_{10} wave propagates, at a frequency f, in the positive z direction along a lossless rectangular metallic waveguide of transverse dimensions a and b and dielectric parameters ε and μ, and is incident on a short-circuiting PEC plate placed in the plane $z = 0$. The wave impedance of the resultant wave in the waveguide, $\underline{Z}_{\text{standing}} = -\underline{E}_y/\underline{H}_x$, is

(A) purely real

(B) purely imaginary

(C) with nonzero real and imaginary parts

at every point in the waveguide, in front of the PEC plate.

CONCEPTUAL QUESTION 11.72 **Poynting vector in a short-circuited waveguide.** Considering a lossless rectangular metallic waveguide carrying an incident TE_{10} wave that reflects from a PEC plate closing the waveguide at $z = 0$, the longitudinal component (z-component) of the complex Poynting vector of the resultant wave in the waveguide, $\underline{\mathcal{P}}_z$, is

(A) purely real

(B) purely imaginary

(C) with nonzero real and imaginary parts

at every point in the waveguide, in front of the PEC plate.

CONCEPTUAL QUESTION 11.73 **Poynting vector inside a cavity resonator.** Consider an air-filled rectangular PEC cavity resonator of dimensions a, b, and d, in Figure 11.16, and assume that a dominant (TE_{101}) standing wave is established in the cavity. The complex Poynting vector of the wave ($\underline{\mathcal{P}}$) comes out to be

(A) purely real at every point in the cavity.

(B) purely imaginary at every point in the cavity.

(C) purely real at some points and purely imaginary at other points in the cavity.

(D) None of the above.

CONCEPTUAL QUESTION 11.74 **Surface currents on cavity walls.** Consider the distribution of surface currents on interior surfaces of all six sides of an air-filled PEC cavity (Figure 11.16) with a TE_{101} wave. How many sides (cavity walls) are current-free ($\underline{\mathbf{J}}_s = 0$ at every point of the surface)?

(A) Zero.

(B) One.

(C) Two.

(D) Three.

(E) Four.

(F) Six.

CONCEPTUAL QUESTION 11.75 **Surface charges on cavity walls.** How many inner wall surfaces, out of the six walls, of an air-filled PEC cavity with a TE_{101} wave are charge-free, namely, with a zero surface charge density ($\underline{\rho}_s = 0$) at every point of the surface?

(A) Zero.

(B) One.

(C) Two.

(D) Three.

(E) Four.

(F) Six.

11.8 Quality Factor of Rectangular Cavities with Small Losses

Neglecting the losses in a cavity resonator, its electromagnetic energy, $W_{\rm em}(t)$, once established remains the same indefinitely (to $t \to \infty$). In a real (lossy) resonator, on the other hand, $W_{\rm em}(t)$ decreases exponentially with time, as $W_{\rm em}(t) = W_{\rm em}(0)\,{\rm e}^{-2t/\tau}$, where the time constant τ is proportional to the so-called quality factor, Q, of the structure, which is given by $Q = \pi\tau/T_{\rm res}$, with $T_{\rm res} = 1/f_{\rm res}$ being the time period at resonance of time-harmonic variation of the electromagnetic field in the cavity. The Q factor also defines the bandwidth of a resonator, so the higher the Q of a resonant structure the sharper the resonance and higher the frequency selectivity of the device. For cavity resonators with small losses, Q can be obtained as follows:

$$Q = \frac{Q_{\rm c}Q_{\rm d}}{Q_{\rm c}+Q_{\rm d}}, \quad Q_{\rm c} = \omega_{\rm res}\frac{W_{\rm em}}{P_{\rm c}}, \quad Q_{\rm d} = \omega_{\rm res}\frac{W_{\rm em}}{P_{\rm d}} \quad \text{(quality factor)} \quad (11.15)$$

($\omega_{\rm res} = 2\pi f_{\rm res}$), where the time-average power of Joule's losses in the cavity conductor (metallic walls), $P_{\rm c}$, and that for the imperfect dielectric in Figure 11.15, $P_{\rm d}$, are computed similarly to the evaluation of conductor and dielectric per-unit-length losses in a waveguide in Eqs. (11.9) and (11.10), respectively. Moreover, Q (under the low-loss assumption) is determined using the field distributions for the lossless case (perturbation method).

CONCEPTUAL QUESTION 11.76 **Phase shift between electric and magnetic fields in a cavity.** The instantaneous total electric and magnetic fields of the $\rm TE_{101}$ mode in a lossless rectangular cavity resonator are in time-phase quadrature (are shifted in phase by 90° with respect to each other)

(A) at every point of the cavity.

(B) at some (but not all) locations in the cavity.

(C) at no locations in the cavity.

CONCEPTUAL QUESTION 11.77 **Instantaneous electric and magnetic field vectors in a cavity.** The instantaneous electric and magnetic field intensity vectors of the dominant cavity mode ($\rm TE_{101}$) in a lossless waveguide resonator can be written – at an arbitrary point inside the cavity – as follows [$\mathbf{A}_1$ and $\mathbf{A}_2$ are purely real vectors (that depend on the coordinates of the point), and $\underline{H}_0 = |\underline{H}_0|\,{\rm e}^{{\rm j}\xi}$ in the expressions for complex electric and magnetic field vectors of the dominant mode]:

(A) $\mathbf{E}(t) = |\underline{H}_0|\mathbf{A}_1 \cos(\omega t + \xi)$ and $\mathbf{H}(t) = -|\underline{H}_0|\mathbf{A}_2 \cos(\omega t + \xi)$.

(B) $\mathbf{E}(t) = |\underline{H}_0|\mathbf{A}_1 \sin(\omega t + \xi)$ and $\mathbf{H}(t) = -|\underline{H}_0|\mathbf{A}_2 \sin(\omega t + \xi)$.

(C) $\mathbf{E}(t) = |\underline{H}_0|\mathbf{A}_1 \cos(\omega t + \xi)$ and $\mathbf{H}(t) = -|\underline{H}_0|\mathbf{A}_2 \sin(\omega t + \xi)$.

(D) $\mathbf{E}(t) = |\underline{H}_0|\mathbf{A}_1 \cos \omega t$ and $\mathbf{H}(t) = -|\underline{H}_0|\mathbf{A}_2 \cos \omega t$.

(E) $\mathbf{E}(t) = |\underline{H}_0|\mathbf{A}_1 \sin \omega t$ and $\mathbf{H}(t) = -|\underline{H}_0|\mathbf{A}_2 \sin \omega t$.

(F) None of the above (there are no such real vectors $\mathbf{A}_1$ and $\mathbf{A}_2$).

CONCEPTUAL QUESTION 11.78 **Polarization state of field vectors in a cavity.** Consider the polarization state of the electric field vector, **E**, and magnetic field vector, **H**, of the TE_{101} wave mode in a lossless rectangular cavity resonator. At an arbitrary point inside the cavity,

(A) both **E** and **H** are elliptically polarized.

(B) both **E** and **H** are circularly polarized.

(C) both **E** and **H** are linearly polarized.

(D) **E** is linearly polarized and **H** is elliptically polarized.

(E) **E** is linearly polarized and **H** is circularly polarized.

(F) None of the above combinations.

CONCEPTUAL QUESTION 11.79 **Maximum electric energy of a cavity resonator.** At instants when the electric energy stored in a lossless rectangular metallic cavity resonator with the dominant resonance wave mode (TE_{101}) is at its maximum value, the magnetic field intensity is

(A) at its maximum value at all points in the cavity.

(B) at $1/\sqrt{2}$ of its maximum value at all points in the cavity.

(C) at $1/2$ of its maximum value at all points in the cavity.

(D) at its maximum and minimum values periodically in equidistant planes.

(E) zero at all points in the cavity.

CONCEPTUAL QUESTION 11.80 **Energy of a cavity when it is all electric or all magnetic.** Consider a lossless waveguide cavity resonator with a TE_{101} wave and the following two statements: (a) There are instants of time at which the instantaneous electromagnetic energy of the cavity, $W_{\mathrm{em}}(t)$, is all electric. (b) There are times when the energy $W_{\mathrm{em}}(t)$ is all magnetic. Which of the statements is true?

(A) Statement (a) only.

(B) Statement (b) only.

(C) Both statements.

(D) Neither of the statements.

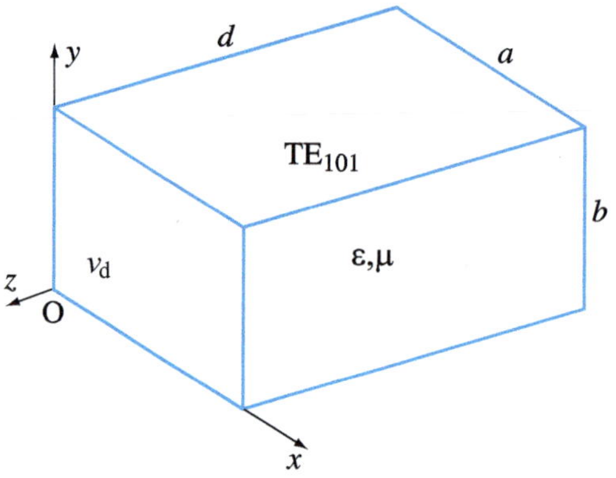

Figure 11.17 Rectangular metallic cavity resonator with the dominant resonance wave mode (TE_{101}); for Conceptual Question 11.81.

CONCEPTUAL QUESTION 11.81 **Volume integral to find the stored energy in a cavity.** A TE_{101} wave is established in a lossless rectangular metallic cavity of dimensions a, b, and d, filled with a homogeneous dielectric of permittivity ε and permeability μ, Figure 11.17. The complex rms electric field intensity of the wave is $\underline{E}_y$ (it is a function of coordinates x and z in Figure 11.17). The electromagnetic energy stored in the cavity can be obtained by the following integration throughout the volume of the cavity dielectric, v_{d} (Figure 11.17), that is, the entire cavity interior:

(A) $W_{\mathrm{em}} = (\varepsilon/4) \int_{v_{\mathrm{d}}} |\underline{E}_y|^2 \,\mathrm{d}v$.

(B) $W_{\mathrm{em}} = (\varepsilon/2) \int_{v_{\mathrm{d}}} |\underline{E}_y|^2 \,\mathrm{d}v$.

(C) $W_{\text{em}} = \varepsilon \int_{v_\text{d}} |\underline{E}_y|^2 \,\mathrm{d}v$.

(D) $W_{\text{em}} = 2\varepsilon \int_{v_\text{d}} |\underline{E}_y|^2 \,\mathrm{d}v$.

(E) $W_{\text{em}} = 4\varepsilon \int_{v_\text{d}} |\underline{E}_y|^2 \,\mathrm{d}v$.

(F) None of the above.

CONCEPTUAL QUESTION 11.82 **Quality factor versus damping rate of a resonator.** The larger the quality factor of a waveguide cavity resonator

(A) the faster

(B) the slower

(C) no difference

the damping (discharge) of the resonator, i.e., the decay of the energy stored in the resonator.

CONCEPTUAL QUESTION 11.83 **Quality factor of an ideal resonator.** The Q factor of an ideal resonator, with no losses, is

(A) $Q = 0$.

(B) $Q = 1$.

(C) $Q \to \infty$.

(D) not defined.

CONCEPTUAL QUESTION 11.84 **Quality factors for the conductor and dielectric of a resonator.** If the quality factor of a cavity resonator associated with conductor losses, Q_c, and the factor representing the losses in the dielectric, Q_d, are such that $0 < Q_\text{c} < Q_\text{d} < \infty$, the following holds true for the overall Q factor of the resonator:

(A) $Q = Q_\text{c}$.

(B) $Q = Q_\text{d}$.

(C) $Q \to \infty$.

(D) $0 < Q < Q_\mathrm{c}$.

(E) $Q_\mathrm{d} < Q < \infty$.

CONCEPTUAL QUESTION 11.85 **Cavity resonator with a perfect dielectric.** The quality factor for the conductor (metallic walls) of a waveguide cavity resonator is $Q_\mathrm{c} = 1000$, while that for the resonator dielectric is $Q_\mathrm{d} \to \infty$. The total quality factor of the resonator is then

(A) $Q = 1000$.

(B) $Q = 2000$.

(C) $Q = 500$.

(D) $Q \to \infty$.

(E) $Q = 0$.

CONCEPTUAL QUESTION 11.86 **Quality factor of a brass cavity, air- vs. teflon-filled.** Consider an air-filled cubical cavity with edge length $a = 30$ cm and brass ($\sigma_\mathrm{c} = 15$ MS/m and $\mu_\mathrm{c} = \mu_0$) walls. Let Q_c and Q_d denote the quality factors for the conductor and dielectric, respectively, of this resonator – for its dominant (TE_{101}) mode of operation. If the cavity is then filled with teflon ($\varepsilon_\mathrm{r} = 2.1$ and $\sigma_\mathrm{d} = 5 \times 10^{-6}$ S/m),

(A) Q_c changes and Q_d remains the same.

(B) Q_d changes and Q_c remains the same.

(C) both Q_c and Q_d change.

(D) both Q_c and Q_d remain the same.

CONCEPTUAL QUESTION 11.87 **Integral over inner cavity walls to find conductor losses.** A TE_{101} field oscillates in an air-filled rectangular metallic cavity (Figure 11.17) with copper walls. In the integral over the inner surface of the walls needed for obtaining the time-average ohmic power in the walls, P_c, and the quality factor Q_c of the cavity, the following nonzero magnetic field components contribute, through the surface power density $R_s|\underline{\mathbf{H}}_{\rm tang}|^2$, to the result for P_c:

(A) H_x only.

(B) H_z only.

(C) both H_x and H_z.

CONCEPTUAL QUESTION 11.88 **Volume integral to find the dielectric loss power in a cavity.** Consider a rectangular PEC cavity with a TE_{101} field oscillating at an angular (radian) frequency $\omega_{\rm res}$. The dielectric filling the cavity is a low-loss one, of permittivity ε and conductivity $\sigma_{\rm d}$. Comparing the volume integrals of $|\underline{E}_y|^2$ needed to obtain the time-average ohmic power in the dielectric, $P_{\rm d}$, and the electromagnetic energy stored in the cavity, $W_{\rm em}$, respectively, we conclude that the quality factor $Q_{\rm d}$ of the cavity amounts to

(A) $Q_{\rm d} = \omega_{\rm res}\,\varepsilon/(4\sigma_{\rm d})$.

(B) $Q_{\rm d} = \omega_{\rm res}\,\varepsilon/(2\sigma_{\rm d})$.

(C) $Q_{\rm d} = \omega_{\rm res}\,\varepsilon\,\sigma_{\rm d}/2$.

(D) $Q_{\rm d} = \omega_{\rm res}\,\varepsilon\,\sigma_{\rm d}$.

(E) $Q_{\rm d} = \omega_{\rm res}\,\varepsilon/\sigma_{\rm d}$.

(F) None of the above.

CONCEPTUAL QUESTION 11.89 **Quality factor versus bandwidth of a device.** Consider a device with a very small bandwidth (BW), where BW is defined as the range of frequencies, $\mathrm{BW} = f_2 - f_1$, across which the performance of the device with respect to some parameter conforms to a specified standard, i.e., across which the device operates properly. In principle, the narrow-band operation of the device can be made more broadband (BW can be increased) by

(A) making the device more lossy (e.g., adding resistors in a circuit).

(B) making the device less lossy.

(C) neither of the above.

CONCEPTUAL QUESTION 11.90 **Bandwidth of an ideal resonator.** The bandwidth of an ideal (lossless) resonator is

(A) $\mathrm{BW} = 0$.

(B) $\mathrm{BW} \to \infty$.

(C) not defined.

12 ANTENNAS AND WIRELESS COMMUNICATION SYSTEMS

Introduction

Although any conductor with a rapidly time-varying (e.g., high-frequency time-harmonic) current (Chapter 6) radiates electromagnetic energy into the surrounding space, some conductor configurations are specially designed to maximize electromagnetic radiation, in desired directions at given frequencies. Such systems of conductors, which sometimes also include dielectric parts, are called antennas. In other words, antennas are electromagnetic devices designed and built to provide a means of efficient transmitting or receiving of radio waves. More precisely, they provide transition from a guided electromagnetic wave, in a transmission line (Chapters 9 and 10) or waveguide (Chapter 11) feeding the antenna, to a radiated unbounded electromagnetic wave (in free space or other ambient medium) in the transmitting (radiating) mode of operation, and vice versa for an antenna operating in the receiving mode. In many discussions, we shall study not only antennas but wireless communication systems with antennas at the two ends. In this, we shall use concepts and equations describing the propagation of uniform plane electromagnetic waves, from Chapters 7 and 8.

12.1 Electromagnetic Field due to a Hertzian Dipole

Consider the simplest antenna, a so-called Hertzian dipole, which is an electrically short ($l \ll \lambda$) straight metallic wire segment with a rapidly time-varying current that does not change along the wire, as shown in Figure 12.1. Assuming a time-harmonic regime of the dipole, let it be fed at its center by a lumped generator, of frequency f, and let its complex rms current intensity be $\underline{I}$. As the current is nonzero at the wire ends, it must be terminated by charges [see Eq. (3.4)] $\underline{Q}$ and $-\underline{Q}$ that accumulate on a pair of small metallic spheres (Figure 12.1), or conductors of other shapes, attached to these ends. The complex magnetic vector potential at the point P in Figure 12.1, for which $r \gg l$ (short wire), has the form of a spherical electromagnetic wave emanating from the dipole center [see Figure 6.42(a)], and it equals

$$\underline{\mathbf{A}} = \frac{\mu \underline{I}\, l\, \mathrm{e}^{-\mathrm{j}\beta r}}{4\pi r}\,\hat{\mathbf{z}} \quad (\beta = \omega\sqrt{\varepsilon\mu}) \quad \text{(magnetic potential of a Hertzian dipole)}\,, \tag{12.1}$$

where β is the phase coefficient (wavenumber) for the ambient medium and given operating frequency, Eq. (7.9). Namely, $\mathbf{J}(t - R/c)\,\mathrm{d}v$ or $i(t - R/c)\,\mathrm{d}\mathbf{l}$ in the ex-

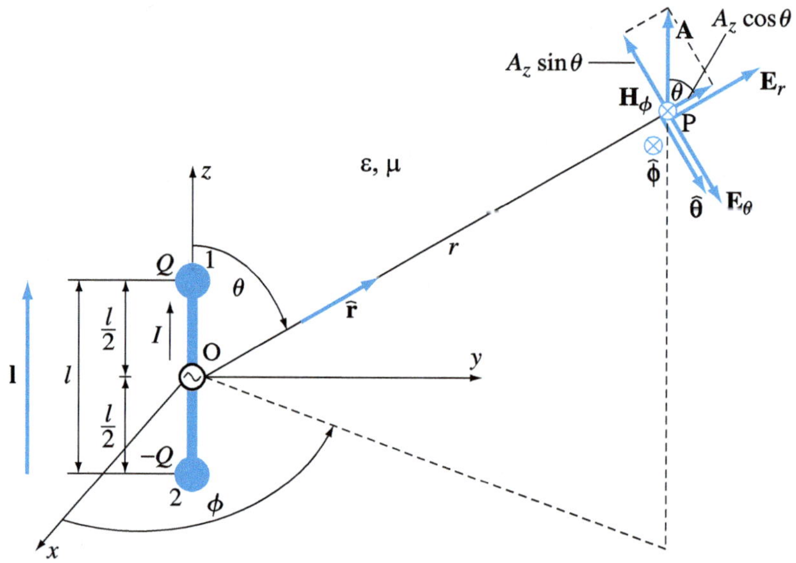

Figure 12.1 Hertzian dipole ($\underline{I} = \text{const}$, $l \ll \lambda$, and $l \ll r$).

pression for $\mathbf{A}(t)$ in Eqs. (6.22) is converted to the complex domain in the same way $E_x(t)$ in Eqs. (7.2) is transformed to $\underline{E}_x$ in Eqs. (7.11), so we have $\underline{I}\,\mathrm{d}\mathbf{l}\,\mathrm{e}^{-\mathrm{j}\beta R}/R$ in the integral, and then, since the antenna in Figure 12.1 can be treated as an infinitesimal dipole, $\mathrm{d}\mathbf{l} \to l\,\hat{\mathbf{z}}$, there is essentially no integration in Eqs. (6.22) and $R = r$.

Using Eqs. (4.17) and (5.5), we next obtain, from $\underline{\mathbf{A}}$, the complex magnetic field intensity vector of the antenna, $\underline{\mathbf{H}}$, from which, in turn, Maxwell's second equation in Eqs. (6.13) gives $\underline{\mathbf{E}} = -\mathrm{j}\,\nabla \times \underline{\mathbf{H}}/(\omega\varepsilon)$, and hence the following field expressions in the spherical coordinate system in Figure 12.1:

$$\underline{\mathbf{E}} = -\frac{\eta\beta^2 \underline{I}\,l\,\mathrm{e}^{-\mathrm{j}\beta r}}{4\pi}\left\{\left[\frac{1}{(\mathrm{j}\beta r)^2} + \frac{1}{(\mathrm{j}\beta r)^3}\right] 2\cos\theta\,\hat{\mathbf{r}} + \left[\frac{1}{\mathrm{j}\beta r} + \frac{1}{(\mathrm{j}\beta r)^2} + \frac{1}{(\mathrm{j}\beta r)^3}\right]\sin\theta\,\hat{\boldsymbol{\theta}}\right\},$$

$$\underline{\mathbf{H}} = -\frac{\beta^2 \underline{I}\,l\,\mathrm{e}^{-\mathrm{j}\beta r}\sin\theta}{4\pi}\left[\frac{1}{\mathrm{j}\beta r} + \frac{1}{(\mathrm{j}\beta r)^2}\right]\hat{\boldsymbol{\phi}} \quad \text{(field vectors of a dipole)}\,, \qquad (12.2)$$

where $\eta = \sqrt{\mu/\varepsilon}$ is the intrinsic impedance of the medium, Eq. (7.6). On the other side, the same result for the electric field can alternatively be obtained as $\underline{\mathbf{E}} = -\mathrm{j}\omega\underline{\mathbf{A}} - \nabla\underline{V}$ [see Eqs. (6.2) and (1.16)], with $\underline{V}$ being the complex electric scalar potential at the point P due to point charges $\underline{Q}$ and $-\underline{Q}$ of the Hertzian dipole in Figure 12.1, computed based on Eqs. (6.22) as well.

Finally, combining Eqs. (6.26) and (12.2), the complex Poynting vector at the point P in Figure 12.1 is

$$\underline{\boldsymbol{\mathcal{P}}} = \eta\left(\frac{\beta^2 I\,l}{4\pi}\right)^2\left\{\left[\frac{\mathrm{j}}{(\beta r)^3} + \frac{\mathrm{j}}{(\beta r)^5}\right]\sin 2\theta\,\hat{\boldsymbol{\theta}} + \left[\frac{1}{(\beta r)^2} - \frac{\mathrm{j}}{(\beta r)^5}\right]\sin^2\theta\,\hat{\mathbf{r}}\right\}, \qquad (12.3)$$

where $\underline{I}\underline{I}^* = |\underline{I}|^2 = I^2$, so that the time average of the instantaneous Poynting vector due to the antenna, Eq. (6.27), comes out to be

$$\mathcal{P}_{\text{ave}} = \text{Re}\{\underline{\mathcal{P}}\} = \frac{\eta\beta^2 I^2 l^2 \sin^2\theta}{16\pi^2 r^2}\,\hat{\mathbf{r}} \quad \text{(time-average Poynting vector)}\,. \qquad (12.4)$$

CONCEPTUAL QUESTION 12.1 **Current-charge relationship for a Hertzian dipole.** Figure 12.2 shows a Hertzian dipole, of length l and complex rms current intensity $\underline{I}$, radiating in free space at an angular (radian) frequency ω. The complex rms charge of the dipole, that is, the charge of the upper small metallic sphere in Figure 12.2, is given by[1]

(A) $\underline{Q} = \mathrm{j}\underline{I}/\omega$.

(B) $\underline{Q} = -\mathrm{j}\underline{I}/\omega$.

(C) $\underline{Q} = \mathrm{j}\omega\underline{I}$.

(D) $\underline{Q} = -\underline{I}/\omega$.

(E) $\underline{Q} = 0$.

(F) $\underline{I}$ and $\underline{Q}$ are not related to each other.

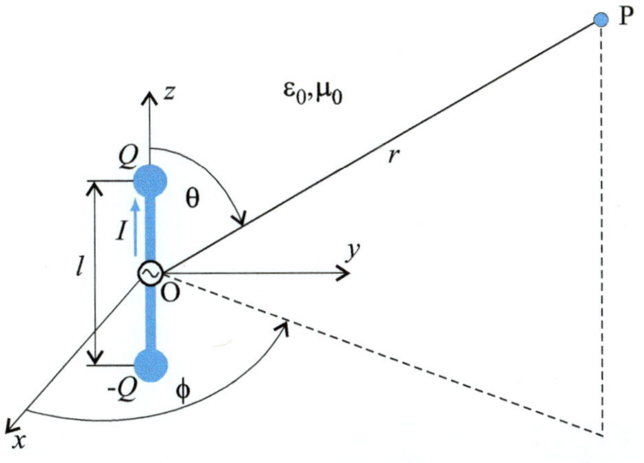

Figure 12.2 Hertzian dipole radiating in free space; for Conceptual Question 12.1.

[1] *For every conceptual question in this text, exactly one answer is correct.*

CONCEPTUAL QUESTION 12.2 **Dipole current from charge in time domain.** If the complex rms charge of the Hertzian dipole in Figure 12.2 is $\underline{Q} = Q\,e^{j0}$ (the operating angular frequency is ω), then its instantaneous current intensity amounts to

(A) $i(t) = -\omega Q\sqrt{2}\cos\omega t$.

(B) $i(t) = -Q\sqrt{2}\cos\omega t$.

(C) $i(t) = (Q\sqrt{2}/\omega)\sin\omega t$.

(D) $i(t) = -\omega Q\sqrt{2}\sin\omega t$.

(E) $i(t) = \omega Q$.

(F) Need more information.

CONCEPTUAL QUESTION 12.3 **Electrical length of a Hertzian dipole.** Which of the following combinations of the length, l, and frequency, f, of the dipole in Figure 12.2 implies that this antenna cannot be considered as a Hertzian dipole?

(A) $l = 1$ cm and $f = 300$ MHz.

(B) $l = 1$ cm and $f = 30$ GHz.

(C) $l = 1$ m and $f = 3$ MHz.

(D) $l = 1$ m and $f = 300$ kHz.

(E) $l = 10$ m and $f = 300$ kHz.

(F) More than one of the combinations above.

CONCEPTUAL QUESTION 12.4 **Distance of an observation point from the dipole center.** Which of the combinations of the length and frequency indicates that the antenna in Figure 12.2 can be treated as a Hertzian dipole if observed from a field point, P, at the distance $r = 1$ m from the dipole center?

(A) $l = 1$ cm and $f = 300$ MHz.

(B) $l = 1$ cm and $f = 30$ GHz.

(C) $l = 1$ m and $f = 3$ MHz.

(D) $l = 1$ m and $f = 300$ kHz.

(E) $l = 10$ m and $f = 300$ kHz.

(F) More than one of the combinations above.

CONCEPTUAL QUESTION 12.5 **Electric fields due to the current and charge of a Hertzian dipole.** Let $\underline{\mathbf{E}}_{\text{current}}$ denote the complex electric field intensity vector at an arbitrary observation (field) point (P) due to the current, $\underline{I}$, of a Hertzian dipole radiating in free space, and let $\underline{\mathbf{E}}_{\text{charge}}$ stand for the field vector due to the charge, $\underline{Q}$ (and $-\underline{Q}$), of the dipole, at the same point. In other words, $\underline{\mathbf{E}}_{\text{current}}$ is the field that would be obtained assuming that $\underline{Q} = 0$ for the dipole, and $\underline{\mathbf{E}}_{\text{charge}}$ is the field with the assumption that $\underline{I} = 0$. The total electric field, $\underline{\mathbf{E}}$, at this point can (exactly or approximately) be computed

(A) as $\underline{\mathbf{E}} = \underline{\mathbf{E}}_{\text{current}}$.

(B) as $\underline{\mathbf{E}} = \underline{\mathbf{E}}_{\text{charge}}$.

(C) as $\underline{\mathbf{E}} = \underline{\mathbf{E}}_{\text{current}} = \underline{\mathbf{E}}_{\text{charge}}$.

(D) as $\underline{\mathbf{E}} = \underline{\mathbf{E}}_{\text{current}} + \underline{\mathbf{E}}_{\text{charge}}$.

(E) using more than one of the expressions above.

CONCEPTUAL QUESTION 12.6 **Electric and magnetic potentials of a Hertzian dipole.** Let $\underline{V}$ and $\underline{\mathbf{A}}$ denote the complex electric scalar potential and magnetic vector potential, respectively, of a Hertzian dipole in free space. At an arbitrary observation point, we have that (exactly or approximately)

(A) $\underline{V} = 0$ (and $\underline{\mathbf{A}} \neq 0$).

(B) $\underline{\mathbf{A}} = 0$ (and $\underline{V} \neq 0$).

(C) $\underline{V} = 0$ and $\underline{\mathbf{A}} = 0$.

(D) None of the above.

CONCEPTUAL QUESTION 12.7 Components of field vectors radiated by a Hertzian dipole. Let $\underline{\mathbf{E}}$ and $\underline{\mathbf{H}}$ stand, respectively, for the complex electric and magnetic field vectors of a Hertzian dipole radiating in free space. Referring to the spherical coordinate system with the z-axis along the dipole axis and origin (O) at the dipole center, determine which of the components of these vectors are nonzero. At an arbitrary observation point (P),

(A) $\underline{\mathbf{E}}$ has only an r-component, and $\underline{\mathbf{H}}$ has only a ϕ-component,

(B) $\underline{\mathbf{E}}$ has only a θ-component, and $\underline{\mathbf{H}}$ has only a ϕ-component,

(C) $\underline{\mathbf{E}}$ has only a ϕ-component, and $\underline{\mathbf{H}}$ has only a θ-component,

(D) $\underline{\mathbf{E}}$ has only a θ- and a ϕ-component, and the same for $\underline{\mathbf{H}}$,

(E) $\underline{\mathbf{E}}$ has only an r- and a θ-component, and $\underline{\mathbf{H}}$ has only a ϕ-component,

(F) $\underline{\mathbf{E}}$ has all three components, and $\underline{\mathbf{H}}$ has only a θ- and a ϕ-component,

while the remaining field components are (exactly or approximately) zero.

CONCEPTUAL QUESTION 12.8 Complex Poynting vector due to a Hertzian dipole. The complex Poynting vector, $\underline{\boldsymbol{\mathcal{P}}}$, at an arbitrary field point due to a Hertzian dipole in free space is

(A) purely real.

(B) purely imaginary.

(C) with nonzero real and imaginary parts.

CONCEPTUAL QUESTION 12.9 Components of the time-average Poynting vector. Consider the time-average Poynting vector, $\boldsymbol{\mathcal{P}}_{\text{ave}}$, of a Hertzian dipole radiating in free space. At an arbitrary observation point, given in the spherical coordinate system with the z-axis along the dipole axis and origin at the dipole center, $\boldsymbol{\mathcal{P}}_{\text{ave}}$ has

(A) only an r-component.

(B) only a θ-component.

(C) only an r- and a θ-component.

(D) only an r- and a ϕ-component.

(E) all three components.

CONCEPTUAL QUESTION 12.10 **Azimuthal symmetry of a Hertzian dipole for some or all quantities?** In the analysis of radiation by a Hertzian dipole in free space, performed in a spherical coordinate system whose z-axis coincides with the dipole axis and origin with the dipole center, some quantities, but not all, do not depend on the azimuthal angle, ϕ.

(A) True.

(B) False.

CONCEPTUAL QUESTION 12.11 **Electrical distance of a field point from the dipole center.** Consider a Hertzian dipole transmitting in free space and the associated field, potential, and Poynting-vector dependences on the distance r of an observation point from the dipole center. Wherever we have r, it actually is βr in

(A) all dependences.

(B) some (but not all) dependences.

(C) no dependences.

CONCEPTUAL QUESTION 12.12 **Field picture in electrical units.** The relative spatial field distributions due to a Hertzian dipole radiating in free space do not change if we scale (increase or decrease) the dipole length, l, the distance of the field point from the dipole center, r, and the operating wavelength of the antenna, λ, by the same factor.

(A) True.

(B) False.

12.2 Far Field and Near Field

This section introduces an important special case of the electromagnetic field due to a Hertzian dipole (Figure 12.1): the far field, for observation locations that are electrically far away from the antenna. Specifically, in the far zone the distance r of the field point P in Figure 12.1 from the origin is much larger than the operating wavelength λ of the dipole (for the ambient medium), given in Eqs. (7.10) and (7.9), βr is much larger than unity, and we can write

$$r \gg \lambda \quad \longrightarrow \quad \beta r \gg 1 \quad \longrightarrow \quad \frac{1}{\beta r} \gg \frac{1}{(\beta r)^2} \gg \frac{1}{(\beta r)^3} \quad \text{(far zone)} . \tag{12.5}$$

In practice, a useful rule of thumb quantifying the far-field condition is $r > 10\lambda$. Therefore, the dominant terms in both field expressions in Eqs. (12.2) are those with the smallest inverse powers of r (or βr), that is, the $1/r$ terms. These expressions can thus be replaced by much simpler approximate ones as follows:

$$\underline{\mathbf{E}} \approx \frac{\mathrm{j}\eta\beta \underline{I}\, l\, \mathrm{e}^{-\mathrm{j}\beta r} \sin\theta}{4\pi r}\, \hat{\boldsymbol{\theta}}\, , \quad \underline{\mathbf{H}} \approx \frac{\mathrm{j}\beta \underline{I}\, l\, \mathrm{e}^{-\mathrm{j}\beta r} \sin\theta}{4\pi r}\, \hat{\boldsymbol{\phi}} \quad \text{(far fields, Hertzian dipole)} . \tag{12.6}$$

For electrically large antennas, the far-zone condition in Eqs. (12.5) must be combined with the condition that $r \gg D$, D being the maximum dimension of the antenna. In some cases, a more complex definition of the far-field region comparing the distance r and the value $2D^2/\lambda$ is used.

Just opposite to Eqs. (12.5) for the far field, r in the near zone in Figure 12.1 is, by definition, small relative to λ (in practice, $r < 0.1\lambda$), but still $r \gg l$, and βr satisfies the quasistatic condition discussed in Section 6.11, which leads to

$$r \ll \lambda \quad \longrightarrow \quad \beta r \ll 1 \quad \longrightarrow \quad \frac{1}{(\beta r)^3} \gg \frac{1}{(\beta r)^2} \gg \frac{1}{\beta r} \quad \text{(near zone)} , \tag{12.7}$$

so only the dominant terms with the largest inverse powers of r in the expressions for each of the field components in Eqs. (12.2) need to be retained. In addition, the retardation effect in this zone can be neglected, $\mathrm{e}^{-\mathrm{j}\beta r} \approx 1$.

CONCEPTUAL QUESTION 12.13 **Far electric fields due to the current and charge.** Consider the fields $\underline{\mathbf{E}}_{\text{current}}$ due to the current of a Hertzian dipole and $\underline{\mathbf{E}}_{\text{charge}}$ due to its charge at a far-field point, i.e., at an observation point in the far zone of the dipole, for which $r \gg \lambda_0$ (λ_0 being the operating free-space wavelength of the antenna). With such a notation, the total far electric field, $\underline{\mathbf{E}}$, at this point can (exactly or approximately) be obtained

(A) as $\underline{\mathbf{E}} = \underline{\mathbf{E}}_{\text{current}}$.

(B) as $\underline{\mathbf{E}} = \underline{\mathbf{E}}_{\text{charge}}$.

(C) as $\underline{\mathbf{E}} = \underline{\mathbf{E}}_{\text{current}} = \underline{\mathbf{E}}_{\text{charge}}$.

(D) as $\underline{\mathbf{E}} = \underline{\mathbf{E}}_{\text{current}} + \underline{\mathbf{E}}_{\text{charge}}$.

(E) using more than one of the expressions above.

CONCEPTUAL QUESTION 12.14 **Potentials and fields in the far zone.** Consider the potentials $\underline{V}$ and $\underline{\mathbf{A}}$ due to a Hertzian dipole radiating in free space. In the far zone (for $r \gg \lambda_0$), we have that (exactly or approximately)

(A) $\nabla\underline{V} = 0$ (and $\nabla \times \underline{\mathbf{A}} \neq 0$).

(B) $\nabla \times \underline{\mathbf{A}} = 0$ (and $\nabla\underline{V} \neq 0$).

(C) $\nabla\underline{V} = 0$ and $\nabla \times \underline{\mathbf{A}} = 0$.

(D) None of the above.

CONCEPTUAL QUESTION 12.15 **Field components in the far zone.** At a far-field point, we have the following for the vector components of fields $\underline{\mathbf{E}}$ and $\underline{\mathbf{H}}$ of a z-directed Hertzian dipole in the spherical coordinate system centered at the dipole center:

(A) $\underline{\mathbf{E}}$ has only an r-component, and $\underline{\mathbf{H}}$ has only a ϕ-component,

(B) $\underline{\mathbf{E}}$ has only a θ-component, and $\underline{\mathbf{H}}$ has only a ϕ-component,

(C) $\underline{\mathbf{E}}$ has only a ϕ-component, and $\underline{\mathbf{H}}$ has only a θ-component,

(D) $\underline{\mathbf{E}}$ has only a θ- and a ϕ-component, and the same for $\underline{\mathbf{H}}$,

(E) $\underline{\mathbf{E}}$ has only an r- and a θ-component, and $\underline{\mathbf{H}}$ has only a ϕ-component,

(F) $\underline{\mathbf{E}}$ has all three components, and $\underline{\mathbf{H}}$ has only a θ- and a ϕ-component,

with the other components being (exactly or approximately) zero.

CONCEPTUAL QUESTION 12.16 **Phase difference between E and H vectors in the far zone.** The instantaneous electric and magnetic field intensity vectors, $\mathbf{E}(t)$ and $\mathbf{H}(t)$, in the far zone of a radiating Hertzian dipole are

(A) in phase.

(B) in counter-phase (180° out of phase with respect to each other).

(C) in time-phase quadrature (±90° out of phase with respect to each other).

(D) out of phase by a constant angle different from 0, 180°, and ±90°.

(E) out of phase by an angle that depends on the distance from the dipole (r).

CONCEPTUAL QUESTION 12.17 **Complex electric to magnetic field ratio in the far zone.** The ratio of the electric and magnetic complex rms field intensities radiated by a Hertzian dipole in free space, $\underline{E}/\underline{H}$, at an arbitrary point in the far zone of the dipole is

(A) a purely real constant.

(B) a purely imaginary constant.

(C) a constant with nonzero real and imaginary parts.

(D) a purely real quantity that depends on spatial coordinates.

(E) a purely imaginary quantity that depends on spatial coordinates.

(F) a quantity with real and imaginary parts that depend on spatial coordinates.

CONCEPTUAL QUESTION 12.18 **Complex Poynting vector in the far zone.** At a far-field point, the complex Poynting vector, $\underline{\boldsymbol{\mathcal{P}}}$, of a Hertzian dipole comes out to be

(A) purely real.

(B) purely imaginary.

(C) with nonzero real and imaginary parts.

CONCEPTUAL QUESTION 12.19 **Time-average Poynting vector far and near the dipole.** Consider the magnitude of the time-average Poynting vector, $\mathcal{P}_{\text{ave}}$, due to a transmitting Hertzian dipole at a point in the far zone whose distance from the dipole center is r. The value $\mathcal{P}_{\text{ave}}$ computed using far-zone approximations (applicable if $r \gg \lambda$, λ being the operating wavelength of the dipole for the ambient medium) is

(A) larger than

(B) the same as

(C) smaller than

that obtained from the expression for $\mathcal{P}_{\text{ave}}$ holding true for any r.

CONCEPTUAL QUESTION 12.20 **Potentials and fields in the near zone.** The following is true (exactly or approximately) for the potentials $\underline{V}$ and $\underline{\mathbf{A}}$, and their gradient and curl, respectively, in the near zone ($r \ll \lambda_0$) of a Hertzian dipole:

(A) $\nabla \underline{V} = 0$ (and $\nabla \times \underline{\mathbf{A}} \neq 0$).

(B) $\nabla \times \underline{\mathbf{A}} = 0$ (and $\nabla \underline{V} \neq 0$).

(C) $\nabla \underline{V} = 0$ and $\nabla \times \underline{\mathbf{A}} = 0$.

(D) None of the above.

CONCEPTUAL QUESTION 12.21 **Field components in the near zone.** For the components of $\underline{\mathbf{E}}$ and $\underline{\mathbf{H}}$ due to a Hertzian dipole at a near-field point, we have that

(A) $\underline{\mathbf{E}}$ has only an r-component, and $\underline{\mathbf{H}}$ has only a ϕ-component.

(B) $\underline{\mathbf{E}}$ has only a θ-component, and $\underline{\mathbf{H}}$ has only a ϕ-component.

(C) $\underline{\mathbf{E}}$ has only a ϕ-component, and $\underline{\mathbf{H}}$ has only a θ-component.

(D) $\underline{\mathbf{E}}$ has only a θ- and a ϕ-component, and the same for $\underline{\mathbf{H}}$.

(E) $\underline{\mathbf{E}}$ has only an r- and a θ-component, and $\underline{\mathbf{H}}$ has only a ϕ-component.

(F) $\underline{\mathbf{E}}$ has all three components, and $\underline{\mathbf{H}}$ has only a θ- and a ϕ-component.

CONCEPTUAL QUESTION 12.22 **Phase shift between E and H vectors in the near zone.** Consider the phase difference between the instantaneous electric and magnetic field intensity vectors in the near zone of a Hertzian dipole. The vectors $\mathbf{E}(t)$ and $\mathbf{H}(t)$ are

(A) in phase.

(B) in counter-phase (180° out of phase with respect to each other).

(C) in time-phase quadrature (±90° out of phase with respect to each other).

(D) out of phase by a constant angle different from 0, 180°, and ±90°.

(E) out of phase by an angle that depends on the distance from the dipole (r).

CONCEPTUAL QUESTION 12.23 **Near electric field – equal to that of a quasistatic electric dipole?** The near electric field of a Hertzian dipole is equal to the electric field intensity vector of a quasistatic electric dipole, the same as an electrostatic dipole (a system consisting of two static point charges Q and $-Q$, with $\mathbf{l}$ being the position vector of Q with respect to $-Q$) except that the dipole charge is slowly oscillating in time (and not time-constant), whose complex moment is $\underline{\mathbf{p}} = \underline{Q}\mathbf{l}$.

(A) True.

(B) False.

CONCEPTUAL QUESTION 12.24 **Complex Poynting vector in the near zone.** The complex Poynting vector ($\underline{\mathcal{P}}$) in the near zone of a radiating Hertzian dipole is

(A) predominantly real.

(B) predominantly imaginary (reactive).

(C) neither of the above.

CONCEPTUAL QUESTION 12.25 **Orthogonality of electric and magnetic field vectors.** The field vectors $\underline{\mathbf{E}}$ and $\underline{\mathbf{H}}$ of a Hertzian dipole radiating in air are mutually orthogonal at

(A) every point of space (for every location of the field point).

(B) at far-field points only.

(C) at near-field points only.

(D) no points of space.

(E) Need more information.

CONCEPTUAL QUESTION 12.26 **Practical importance of far-field versus near-field calculations.** For most applications of antennas, evaluation of the radiated far field is

(A) more important than

(B) less important than

(C) equally important as

determination of the near field.

12.3 Steps in Far-Field Evaluation of an Arbitrary Antenna

Consider a straight wire antenna along the z-axis, shown in Figure 12.3. Under the far-field assumption, Eq. (12.5), we apply different approximations for the magnitude

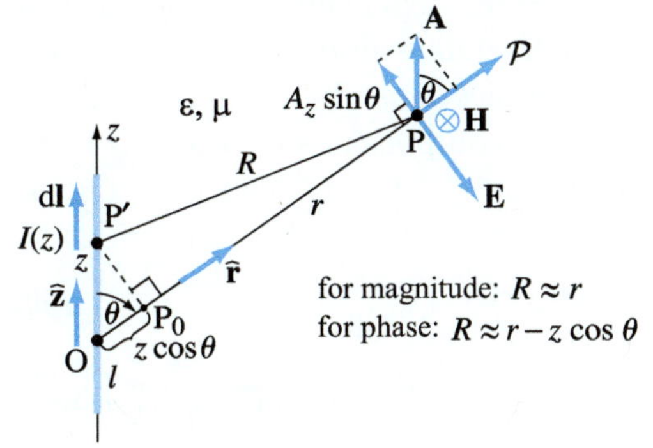

Figure 12.3 Straight wire antenna with an arbitrary current distribution: evaluation of the magnetic vector potential and electric and magnetic field vectors in the far zone.

and phase (as indicated in Figure 12.3) of the spherical-wave factor $\mathrm{e}^{-\mathrm{j}\beta R}/R$ in the integral for computing the magnetic vector potential, $\underline{\mathbf{A}}$, at an observation point P (field point) defined by (r, θ, ϕ), with R being the variable source-to-field distance for an arbitrary point P′ at the wire axis (source point). This integral (in terms of the coordinate z along the wire antenna) thus becomes

$$\underline{\mathbf{A}} = \frac{\mu}{4\pi} \int_l \frac{\underline{I}(z)\,\mathrm{d}l\,\mathrm{e}^{-\mathrm{j}\beta R}}{R} \approx \frac{\mu\,\mathrm{e}^{-\mathrm{j}\beta r}}{4\pi r}\,\hat{\mathbf{z}} \int_l \underline{I}(z)\,\mathrm{e}^{\mathrm{j}\beta z\cos\theta}\,\mathrm{d}z \quad \text{(radiation integral)}\,. \tag{12.8}$$

It is called the radiation integral, and its solution, for a given current distribution $\underline{I}(z)$, is the basis for analysis of the wire antenna in Figure 12.3. Similar integrals are in place for arbitrary (curvilinear) wire antennas, and for surface and volume antennas. Only for electrically short antennas, for which $l \ll \lambda$, $\beta z \approx 0$ in Eq. (12.8).

From Figure 12.1 and Eqs. (12.6) and (12.1), we can write for the far electric and magnetic field components of a z-directed Hertzian dipole: $\underline{E}_\theta = -\mathrm{j}\omega\underline{A}_\theta = \mathrm{j}\omega\underline{A}_z \sin\theta$ and $\underline{H}_\phi = \underline{E}_\theta/\eta$. Since the wire antenna in Figure 12.3 can be represented as a chain of many z-directed Hertzian dipoles, the same relationships hold true for that antenna as well. For an arbitrary transmitting antenna, the vectors $\underline{\mathbf{E}}$ and $\underline{\mathbf{H}}$ have both θ- and ϕ-components (but no radial component) in the far zone, so these relationships should be extended to

$$\underline{E}_\theta = -\mathrm{j}\omega\underline{A}_\theta\,, \quad \underline{E}_\phi = -\mathrm{j}\omega\underline{A}_\phi\,, \quad \underline{H}_\phi = \frac{\underline{E}_\theta}{\eta}\,, \quad \underline{H}_\theta = -\frac{\underline{E}_\phi}{\eta} \quad \text{(arbitrary antenna)}\,. \tag{12.9}$$

Once $\underline{\mathbf{A}}$ is obtained from Eq. (12.8), or a similar radiation integral, the remaining steps in the radiation analysis of the antenna are straightforward, and the same for any antenna type and geometry: $\underline{\mathbf{E}}$ and $\underline{\mathbf{H}}$ in the far zone are easily found from Eqs. (12.9).

CONCEPTUAL QUESTION 12.27 **Representation of a wire antenna by a chain of Hertzian dipoles.** An arbitrary transmitting wire metallic antenna is represented as a chain of Hertzian dipoles, as illustrated in Figure 12.4. In this representation, charges of dipoles must be such that charge $\underline{Q}$ at the end of one dipole in the chain completely compensates charge $-\underline{Q}$ of the next dipole, etc.; for example, $\underline{Q}_1$ compensates (annuls) $-\underline{Q}_2$, $\underline{Q}_2$ cancels $-\underline{Q}_3$, and so on.

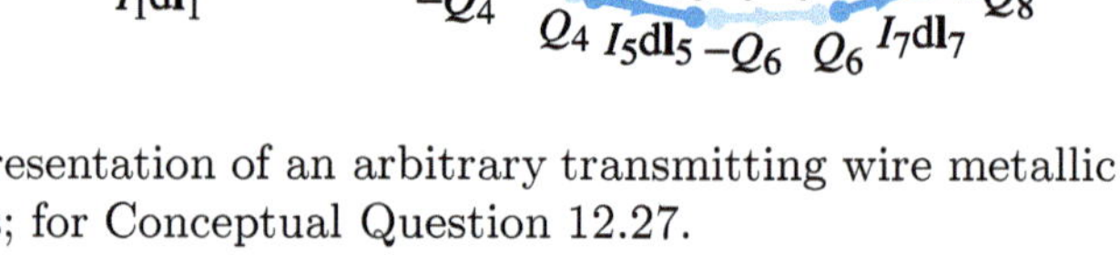

Figure 12.4 Representation of an arbitrary transmitting wire metallic antenna by a chain of Hertzian dipoles; for Conceptual Question 12.27.

(A) True.

(B) False.

CONCEPTUAL QUESTION 12.28 **Magnitude and phase approximations in radiation integrals.** In solving radiation integrals (in the far-field computation) for arbitrary transmitting antennas,

(A) the magnitude approximation is more stringent than the phase approximation.

(B) the phase approximation is more stringent than the magnitude approximation.

(C) the magnitude and phase approximations are equally stringent.

CONCEPTUAL QUESTION 12.29 **Approximations for electrically small antennas.** Which one of the following statements holds true when solving far-field radiation integrals for electrically small transmitting antennas?

(A) The magnitude approximation is more stringent than the phase approximation.

(B) The phase approximation is more stringent than the magnitude approximation.

(C) The magnitude and phase approximations are equally stringent.

CONCEPTUAL QUESTION 12.30 **Equivalent length of a short dipole with a triangular current.** Consider a nonloaded (with no metallic spheres at the wire ends) electrically short wire dipole antenna with length l and a triangular current distribution given by $\underline{I}(z) = \underline{I}_0(1 - 2|z|/l)$ $(-l/2 \leq z \leq l/2)$, as shown in Figure 12.5. There is an equivalent Hertzian dipole of length $l_{\text{equivalent}}$ and current $\underline{I}_0$, so with the same current (uniform along the wire) as the feed current of the original antenna, such that the magnetic potential of the two antennas (the original short antenna and the equivalent Hertzian dipole) at any point in the far zone is the same. This equivalent length of the short dipole with the triangular current, relative to the original length of the antenna, turns out to be

(A) $l_{\text{equivalent}} > l$.

(B) $l_{\text{equivalent}} = l$.

(C) $l_{\text{equivalent}} < l$.

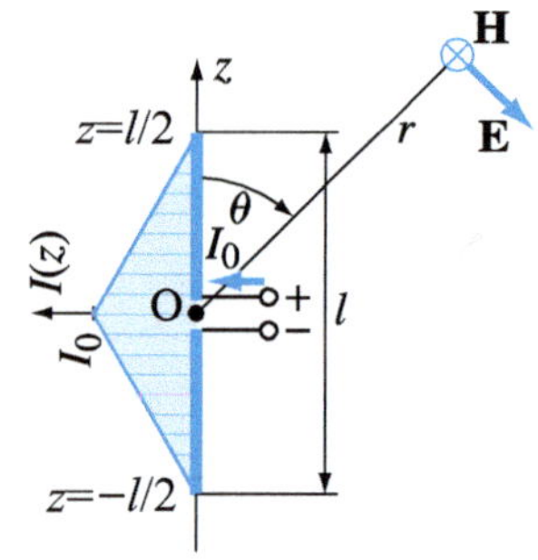

Figure 12.5 Radiation by a nonloaded electrically short symmetrical wire dipole antenna with a triangular current distribution; for Conceptual Question 12.30.

CONCEPTUAL QUESTION 12.31 **Far-zone Poynting vector from the magnetic potential.** If two transmitting antennas have the same magnetic vector potential, $\underline{\mathbf{A}}$, in the far zone, then they also have the same far-zone complex Poynting vector, $\underline{\mathcal{P}}$.

(A) True.

(B) False.

CONCEPTUAL QUESTION 12.32 **Antenna with a θ-directed magnetic potential.** At a far-field point, the magnetic vector potential of a transmitting antenna has only a θ-component in a spherical coordinate system attached to the antenna, so $\underline{\mathbf{A}} = \underline{A}\,\hat{\boldsymbol{\theta}}$. The magnetic field vector at that point is given by

(A) $\underline{\mathbf{H}} = (\mathrm{j}\omega\underline{A}/\eta)\,\hat{\boldsymbol{\theta}}$.

(B) $\underline{\mathbf{H}} = -(\mathrm{j}\omega\underline{A}/\eta)\,\hat{\boldsymbol{\theta}}$.

(C) $\underline{\mathbf{H}} = [(\mathrm{j}\omega\underline{A}\sin\theta)/\eta]\,\hat{\boldsymbol{\theta}}$.

(D) $\underline{\mathbf{H}} = (\mathrm{j}\omega\underline{A}/\eta)\,\hat{\boldsymbol{\phi}}$.

(E) $\underline{\mathbf{H}} = -(\mathrm{j}\omega\underline{A}/\eta)\,\hat{\boldsymbol{\phi}}$.

(F) $\underline{\mathbf{H}} = [(\mathrm{j}\omega\underline{A}\sin\theta)/\eta]\,\hat{\boldsymbol{\phi}}$.

CONCEPTUAL QUESTION 12.33 **Antenna with a ϕ-directed magnetic potential.** For an antenna having only a ϕ-component of the magnetic potential, $\underline{\mathbf{A}} = \underline{A}\,\hat{\boldsymbol{\phi}}$, at a far-field point, the associated magnetic field vector can be obtained as

(A) $\underline{\mathbf{H}} = (\mathrm{j}\omega \underline{A}/\eta)\,\hat{\boldsymbol{\theta}}$.

(B) $\underline{\mathbf{H}} = -(\mathrm{j}\omega \underline{A}/\eta)\,\hat{\boldsymbol{\theta}}$.

(C) $\underline{\mathbf{H}} = [(\mathrm{j}\omega \underline{A}\sin\theta)/\eta]\,\hat{\boldsymbol{\theta}}$.

(D) $\underline{\mathbf{H}} = (\mathrm{j}\omega \underline{A}/\eta)\,\hat{\boldsymbol{\phi}}$.

(E) $\underline{\mathbf{H}} = -(\mathrm{j}\omega \underline{A}/\eta)\,\hat{\boldsymbol{\phi}}$.

(F) $\underline{\mathbf{H}} = [(\mathrm{j}\omega \underline{A}\sin\theta)/\eta]\,\hat{\boldsymbol{\phi}}$.

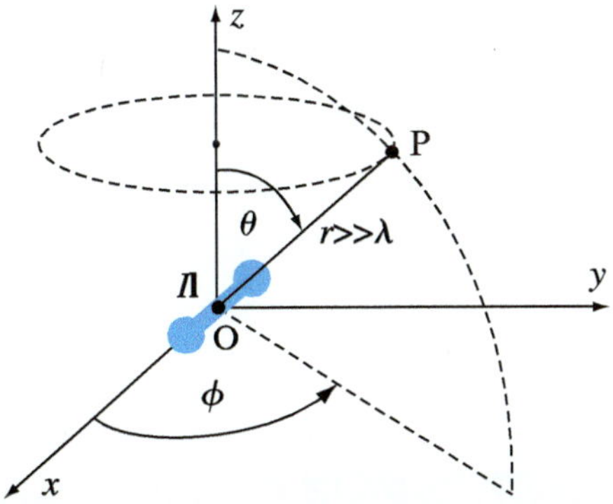

Figure 12.6 Evaluation of radiation in the spherical coordinate system of an x-directed Hertzian dipole; for Conceptual Question 12.34.

CONCEPTUAL QUESTION 12.34 **Hertzian dipole along the x-axis.** The far electric field intensity vector, $\underline{\mathbf{E}}$, of a Hertzian dipole placed at the coordinate origin along the x-axis of a Cartesian coordinate system, as shown in Figure 12.6, has (at the point P) the following components in the associated spherical coordinate system (in Figure 12.6):

(A) $\underline{E}_\theta$ only.

(B) $\underline{E}_\phi$ only.

(C) $\underline{E}_r$ and $\underline{E}_\theta$ only.

(D) $\underline{E}_r$ and $\underline{E}_\phi$ only.

(E) $\underline{E}_\theta$ and $\underline{E}_\phi$ only.

(F) $\underline{E}_r$, $\underline{E}_\theta$, and $\underline{E}_\phi$.

CONCEPTUAL QUESTION 12.35 **Vector relations between far E and H fields of an arbitrary antenna.** Using the intrinsic impedance of the medium, η, and the radial unit vector $\hat{\mathbf{r}}$ (in a spherical coordinate system attached to the antenna) defining the radiation direction, the following vector relations can be written between the far electric and magnetic field intensity vectors, $\underline{\mathbf{E}}$ and $\underline{\mathbf{H}}$, of an arbitrary transmitting antenna:

(A) $\underline{\mathbf{H}} = \hat{\mathbf{r}} \times \underline{\mathbf{E}}/\eta$ and $\underline{\mathbf{E}} = \eta\underline{\mathbf{H}} \times \hat{\mathbf{r}}$.

(B) $\underline{\mathbf{H}} = \hat{\mathbf{r}} \times \underline{\mathbf{E}}/\eta$ and $\underline{\mathbf{E}} = -\eta\underline{\mathbf{H}} \times \hat{\mathbf{r}}$.

(C) $\underline{\mathbf{H}} = -\hat{\mathbf{r}} \times \underline{\mathbf{E}}/\eta$ and $\underline{\mathbf{E}} = \eta\underline{\mathbf{H}} \times \hat{\mathbf{r}}$.

(D) $\underline{\mathbf{H}} = -\hat{\mathbf{r}} \times \underline{\mathbf{E}}/\eta$ and $\underline{\mathbf{E}} = -\eta\underline{\mathbf{H}} \times \hat{\mathbf{r}}$.

(E) More than one of the above pairs of relations.

(F) It is impossible to relate $\underline{\mathbf{E}}$ and $\underline{\mathbf{H}}$.

CONCEPTUAL QUESTION 12.36 **Complex Poynting vector from the far electric field of an antenna.** That the radiated complex Poynting vector at a point in the far zone of an antenna where the complex rms electric field is $\underline{\mathbf{E}}$ and the intrinsic impedance of the (homogeneous) medium is η equals $\underline{\boldsymbol{\mathcal{P}}} = |\underline{\mathbf{E}}|^2 \,\hat{\mathbf{r}}/\eta$ ($\hat{\mathbf{r}}$ is the unit vector in the direction of radiation) is true

(A) for arbitrary antennas.

(B) only for straight wire antennas.

(C) only for electrically small antennas.

(D) for none of the antennas.

CONCEPTUAL QUESTION 12.37 **Global spherical and local plane waves for an arbitrary antenna.** Consider the radiation of an arbitrary antenna in free space and the following two statements: (a) Globally, the radiated electric and magnetic fields constitute a spherical TEM wave, centered at the coordinate origin (at or near the antenna). (b) Locally, in the far zone, the fields can be considered to be uniform, and the wave can be treated as a uniform plane wave. Which of the statements is true?

(A) Statement (a) only.

(B) Statement (b) only.

(C) Both statements.

(D) Neither of the statements.

CONCEPTUAL QUESTION 12.38 **Field intensity dependence on the distance from an antenna.** At a distance of 6 km from a transmitting antenna operating at 100 MHz, the electric field intensity is 10 μV/m. The electric field intensity 12 km from the antenna is

(A) 1.25 μV/m.

(B) 1.67 μV/m.

(C) 2.5 μV/m.

(D) 5 μV/m.

(E) 40 μV/m.

12.4 Radiation and Ohmic Resistances of an Antenna, Antenna Input Impedance

To find the time-average power radiated by an antenna, P_{rad}, we compute the flux [see Eq. (6.25)] of the time-average Poynting vector, $\mathcal{P}_{\text{ave}}$, of the antenna, which – for a Hertzian dipole (Figure 12.1) – is given in Eq. (12.4), through a spherical surface S, of radius r, centered at the coordinate origin. This power is proportional to the magnitude of the feed current at the antenna input terminals, $\underline{I}_0$, squared. The constant of proportionality,

$$R_{\rm rad} = \frac{P_{\rm rad}}{I_0^2} = \frac{1}{I_0^2}\oint_S \boldsymbol{\mathcal{P}}_{\rm ave}\cdot {\rm d}\mathbf{S} \quad \text{(antenna radiation resistance)} \qquad (12.10)$$

($I_0 = |\underline{I}_0|$), is a resistance, called the radiation resistance of the antenna. For a Hertzian dipole ($\underline{I} = \underline{I}_0$) in free space (air), we obtain $R_{\rm rad} = 20(\beta l)^2\,\Omega$, where β is the free-space phase coefficient at a given operating frequency of the antenna.

In addition, all real antennas exhibit some ohmic (Joule's) losses in the lossy materials constituting the antenna body, because of the conduction current flow through the materials, and the time-average ohmic power of the antenna, $P_{\rm ohmic}$, can be written as $P_{\rm ohmic} = R_{\rm ohmic}I_0^2$, where $R_{\rm ohmic}$ is the high-frequency (with the skin effect pronounced) ohmic resistance of the antenna. On the other side, $P_{\rm ohmic}$ in the antenna metallic parts can be computed similarly to the integration in Eqs. (11.9). For a straight wire antenna extending along the z-axis, in Figure 12.3, the magnetic field on the surface of the wire, of radius a, equals $\underline{H} = \underline{I}(z)/2\pi a$ [from the generalized Ampère's law in integral form, Eq. (5.3)], so that $R_{\rm ohmic}$ can be found from the following integral:

$$R_{\rm ohmic} = \frac{P_{\rm ohmic}}{I_0^2} = \frac{1}{I_0^2}\int_l \frac{R_{\rm s}}{2\pi a}\,|\underline{I}(z)|^2\,{\rm d}z \quad \text{(antenna ohmic resistance)}\,, \qquad (12.11)$$

with $R_{\rm s}$ being the surface resistance of the antenna conductors, Eqs. (9.8), and $\underline{I}(z)$ the current intensity along the antenna.

The complex input impedance of an antenna is $\underline{Z}_{\rm A} = \underline{Z}_{\rm in} = R_{\rm A} + {\rm j}X_{\rm A} = R_{\rm rad} + R_{\rm ohmic} + {\rm j}X_{\rm A}$, where the antenna input reactance, $X_{\rm A}$, represents reactive power stored in the near field around the antenna, $X_{\rm A} = Q_{\rm near\ field}/I_0^2$.

CONCEPTUAL QUESTION 12.39 **Inverse square law for the real radiated Poynting vector.** That the magnitude of the radiated time-average Poynting vector is proportional to the inverse distance of an observation point from the coordinate origin squared, $\mathcal{P}_{\rm ave} \propto 1/r^2$, holds true

(A) for an arbitrary transmitting antenna.

(B) only for a Hertzian dipole.

(C) for none of the antennas.

CONCEPTUAL QUESTION 12.40 **Inverse square law for the far and near power flow densities.** If a transmitting antenna exhibits a dependence

$\mathcal{P}_{\rm ave} \propto 1/r^2$ in the far field, the same dependence on r must take place in the near field as well.

(A) True.

(B) False.

CONCEPTUAL QUESTION 12.41 **Time-average power radiated by an antenna.** The flux of the time-average Poynting vector, $\boldsymbol{\mathcal{P}}_{\rm ave}$, of an arbitrary transmitting antenna through a spherical surface of radius r, centered at the coordinate origin, depends on r.

(A) True.

(B) False.

CONCEPTUAL QUESTION 12.42 **Complex power flow through a spherical surface.** The flux of the complex Poynting vector, $\underline{\boldsymbol{\mathcal{P}}}$, radiated by an arbitrary antenna through a spherical surface centered at the coordinate origin depends the surface radius r.

(A) True.

(B) False.

CONCEPTUAL QUESTION 12.43 **Ohmic and radiation resistances of an arbitrary antenna.** The high-frequency ohmic resistance, $R_{\rm ohmic}$, and radiation resistance, $R_{\rm rad}$, of an arbitrary transmitting antenna represent Joule's (ohmic) losses in the antenna body and in the ambient medium, respectively.

(A) True.

(B) False.

CONCEPTUAL QUESTION 12.44 **Radiation resistance of a Hertzian dipole.** When compared to that of a half-wave (half-wavelength long) wire dipole antenna, the radiation resistance, $R_{\rm rad}$, of a Hertzian dipole is

(A) very small (much smaller).

(B) small.

(C) about the same.

(D) large.

(E) very large.

CONCEPTUAL QUESTION 12.45 Ohmic resistance of a Hertzian dipole. What is the high-frequency ohmic resistance of a Hertzian dipole of length l, if the rms intensity of its feed current is I_0, the skin-effect surface resistance of the wire is R_s, and wire radius is a?

(A) $R_{\rm ohmic} = R_s l/(\pi a I_0^2)$.

(B) $R_{\rm ohmic} = R_s l/(2\pi a I_0^2)$.

(C) $R_{\rm ohmic} = R_s l I_0^2/(2\pi a)$.

(D) $R_{\rm ohmic} = R_s l/(\pi a)$.

(E) $R_{\rm ohmic} = 2R_s l/(\pi a)$.

(F) $R_{\rm ohmic} = R_s l/(2\pi a)$.

CONCEPTUAL QUESTION 12.46 Real part of the antenna input impedance. The real part of the complex impedance that an arbitrary transmitting antenna presents to its input terminals equals

(A) the radiation resistance of the antenna.

(B) the high-frequency ohmic resistance of the antenna.

(C) the sum of the resistances in (A) and (B).

(D) none of the above.

CONCEPTUAL QUESTION 12.47 **Equivalent circuit for an arbitrary transmitting antenna.** An arbitrary transmitting antenna can be replaced, with respect to its input terminals (and to an arbitrary input network feeding into the antenna), by

(A) a real voltage generator (emf and impedance in series).

(B) an ideal voltage generator.

(C) a load with a complex impedance.

(D) a purely resistive load.

(E) a purely reactive load.

(F) an open circuit.

CONCEPTUAL QUESTION 12.48 **Impedance matching properties of a transmitting antenna.** An arbitrary transmitting antenna is fed by a lossless two-conductor transmission line of characteristic impedance Z_0. Radiation and ohmic resistances of the antenna are $R_{\rm rad}$ and $R_{\rm ohmic}$, respectively, and the antenna input reactance is $X_{\rm A}$, at a given operating frequency of the antenna. Out of these three antenna parameters, $R_{\rm rad}$, $R_{\rm ohmic}$, and $X_{\rm A}$, to maximize the transfer of power from the line to the antenna, it is necessary to consider and optimize, as much as possible, the following ones:

(A) $R_{\rm rad}$ only.

(B) $R_{\rm rad}$ and $R_{\rm ohmic}$ only.

(C) $R_{\rm rad}$ and $X_{\rm A}$ only.

(D) $X_{\rm A}$ only.

(E) all three parameters ($R_{\rm rad}$, $R_{\rm ohmic}$, and $X_{\rm A}$).

(F) none of the parameters.

CONCEPTUAL QUESTION 12.49 **Input power of a transmitting antenna.** A transmitting antenna whose radiation resistance is $R_{\rm rad}$, ohmic resistance is $R_{\rm ohmic}$, and input reactance is $X_{\rm A}$, is fed by a time-harmonic current of rms intensity I_0 and angular (radian) frequency ω. The time-average (real) input power, $P_{\rm in}$, delivered, at its input terminals, to the antenna amounts to

(A) $P_{\text{in}} = R_{\text{rad}} I_0^2$.

(B) $P_{\text{in}} = R_{\text{ohmic}} I_0^2$.

(C) $P_{\text{in}} = (R_{\text{rad}} + R_{\text{ohmic}}) I_0^2$.

(D) $P_{\text{in}} = (R_{\text{rad}} + R_{\text{ohmic}} + X_{\text{A}}) I_0^2$.

(E) $P_{\text{in}} = (R_{\text{rad}} + R_{\text{ohmic}} + \omega X_{\text{A}}) I_0^2$.

(F) Need more information.

CONCEPTUAL QUESTION 12.50 **Radiation efficiency of an antenna.** The radiation efficiency of an antenna is defined through the following ratio: $\eta_{\text{rad}} = P_{\text{rad}}/(P_{\text{rad}} + P_{\text{ohmic}})$, where P_{rad} and P_{ohmic} are the radiated and ohmic powers, respectively, of the antenna. The range of possible values of η_{rad} is

(A) $0 \le \eta_{\text{rad}} \le 1/2$.

(B) $0 \le \eta_{\text{rad}} \le 1$.

(C) $0 < \eta_{\text{rad}} \le 2$.

(D) $-1 < \eta_{\text{rad}} \le 1$.

(E) $0 < \eta_{\text{rad}} < \infty$.

CONCEPTUAL QUESTION 12.51 **Input reactances of a Hertzian dipole and a small loop antenna.** Considering the input reactance of a Hertzian dipole, $(X_{\text{A}})_{\text{dipole}}$, and that of an electrically small loop antenna, $(X_{\text{A}})_{\text{loop}}$, we have the following:

(A) $(X_{\text{A}})_{\text{dipole}} > 0$ (inductive) and $(X_{\text{A}})_{\text{loop}} > 0$ (inductive).

(B) $(X_{\text{A}})_{\text{dipole}} > 0$ (inductive) and $(X_{\text{A}})_{\text{loop}} < 0$ (capacitive).

(C) $(X_{\text{A}})_{\text{dipole}} < 0$ (capacitive) and $(X_{\text{A}})_{\text{loop}} > 0$ (inductive).

(D) $(X_{\text{A}})_{\text{dipole}} < 0$ (capacitive) and $(X_{\text{A}})_{\text{loop}} < 0$ (capacitive).

12.5 Antenna Radiation Patterns, Directivity, and Gain

The far electric field intensity vector, $\underline{\mathbf{E}}$, of an arbitrary antenna is proportional to the feed current at the antenna input terminals, $\underline{I}_0$. Having also in mind the spherical-wave dependence on r in Eq. (12.8), it is convenient to write $\underline{\mathbf{E}}$, given by Eqs. (12.9), in the following form:

$$\underline{\mathbf{E}}(r,\theta,\phi) = \frac{\mathrm{j}\eta}{2\pi}\,\underline{I}_0\,\frac{\mathrm{e}^{-\mathrm{j}\beta r}}{r}\,\underline{\mathbf{F}}(\theta,\phi) \quad (\underline{\mathbf{F}} - \text{characteristic radiation function})\,, \quad (12.12)$$

where $\underline{\mathbf{F}}(\theta,\phi)$ is termed the characteristic radiation function of the antenna (a dimensionless quantity); it represents the part of the field expression that is characteristic for individual antennas, i.e., that differs from antenna to antenna, while the remaining terms in the expression are the same for all antennas. Independent of r, and thus only a function of the direction of antenna radiation, defined by angles θ and ϕ, $\underline{\mathbf{F}}(\theta,\phi)$ determines the directional properties of the antenna. Its different aspects, presented graphically, give different radiation patterns of the antenna under consideration. Most frequently, we plot the normalized field pattern of the antenna, $f(\theta,\phi)$, defined as

$$f(\theta,\phi) = \frac{|\underline{\mathbf{F}}(\theta,\phi)|}{|\underline{\mathbf{F}}(\theta,\phi)|_{\max}} \qquad \{[f(\theta,\phi)]_{\max} = 1\} \qquad \text{(normalized field pattern)}\,, \quad (12.13)$$

and a typical pattern of a directional antenna – as a three-dimensional (3-D) polar plot – is shown in Figure 12.7. The pattern f can also be expressed in decibels, $f_{\mathrm{dB}}(\theta,\phi) = 20\log f(\theta,\phi)$, with the maximum pattern level now being 0 dB.

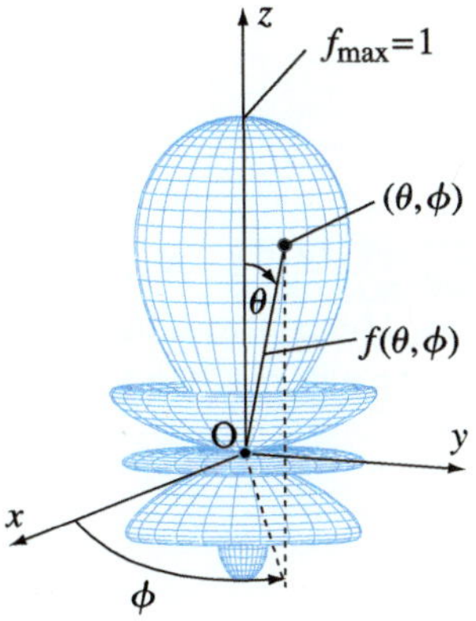

Figure 12.7 Typical normalized field pattern, Eq. (12.13), of a directional antenna.

To further describe and quantify directional properties of transmitting antennas, the directivity of an antenna in a given direction, $D(\theta,\phi)$, is defined via the far-zone time-average Poynting vector and the total time-average radiated power of the antenna [see Eqs. (12.4) and (12.10)] as

$$D(\theta,\phi) = \frac{4\pi r^2 \mathcal{P}_{\text{ave}}(r,\theta,\phi)}{P_{\text{rad}}} = \frac{\eta\,|\underline{\mathbf{F}}(\theta,\phi)|^2}{\pi R_{\text{rad}}} \quad \text{(antenna directivity)}\,, \qquad (12.14)$$

where R_{rad} is the radiation resistance of the antenna. Similarly, the gain of an antenna is computed as

$$G(\theta,\phi) = \frac{4\pi r^2 \mathcal{P}_{\text{ave}}(r,\theta,\phi)}{P_{\text{in}}} = \frac{\eta\,|\underline{\mathbf{F}}(\theta,\phi)|^2}{\pi(R_{\text{rad}}+R_{\text{ohmic}})} \quad \text{(antenna gain)}\,, \qquad (12.15)$$

with $P_{\text{in}} = P_{\text{rad}} + P_{\text{ohmic}}$ being the time-average total input power and R_{ohmic} the ohmic resistance of the antenna, Eq. (12.11). Frequently, D is used without specifying the direction of radiation, in which case the maximum directivity is implied, so $D \equiv D_{\text{max}} = [D(\theta,\phi)]_{\text{max}}$, and analogously $G \equiv G_{\text{max}}$.

CONCEPTUAL QUESTION 12.52 **Common dependence of the far field on some coordinates.** The dependence of the far electric and magnetic field vectors on the following coordinates (in a spherical coordinate system attached to the antenna) is the same for all transmitting antennas:

(A) r only.

(B) ϕ only.

(C) θ and ϕ only.

(D) none of the coordinates.

CONCEPTUAL QUESTION 12.53 **Antenna characteristic radiation function.** Two antennas having the same characteristic radiation function, $\underline{\mathbf{F}}$, at a point in the far zone have also the same electric field intensity vector, $\underline{\mathbf{E}}$, at that point.

(A) Always true.

(B) Sometimes true, sometimes false.

(C) Always false.

CONCEPTUAL QUESTION 12.54 **Antenna normalized field pattern.** Two antennas having the same normalized field pattern, f, at a far-zone point have also the same electric field vector at that point.

(A) Always true.

(B) Sometimes true, sometimes false.

(C) Always false.

CONCEPTUAL QUESTION 12.55 **Characteristic radiation function of a Hertzian dipole.** The characteristic radiation function of a z-directed Hertzian dipole, in Figure 12.8, comes out to be

(A) $\underline{\mathbf{F}}(\theta,\phi) = \beta l(-\cos\theta\cos\phi\,\hat{\boldsymbol{\theta}} + \sin\phi\,\hat{\boldsymbol{\phi}})/2$ ($\beta = 2\pi f\sqrt{\varepsilon_0\mu_0}$).

(B) $\underline{\mathbf{F}}(\theta) = \mathrm{j}\eta\beta l\,\mathrm{e}^{-\mathrm{j}\beta r}\sin\theta\,\hat{\boldsymbol{\theta}}/(4\pi r)$ ($\eta = \sqrt{\mu_0/\varepsilon_0}$).

(C) $\underline{\mathbf{F}}(\theta) = \mathrm{j}\eta\beta l\sin\theta\,\hat{\boldsymbol{\theta}}/(4\pi)$.

(D) $\underline{\mathbf{F}}(\theta) = \beta l\sin\theta\,\hat{\boldsymbol{\theta}}/2$.

(E) $\underline{\mathbf{F}}(\theta) = \sin\theta\,\hat{\boldsymbol{\theta}}$.

(F) $\underline{\mathbf{F}} = \beta l\,\hat{\boldsymbol{\theta}}/2$.

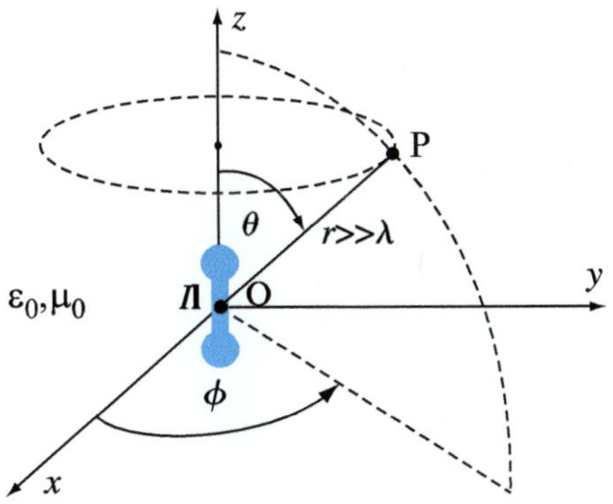

Figure 12.8 Hertzian dipole along the z-axis in the spherical coordinate system, radiating in free space; for Conceptual Question 12.55.

CONCEPTUAL QUESTION 12.56 **Normalized field pattern of a Hertzian dipole.** The normalized field pattern of a Hertzian dipole (Figure 12.8) is given by

(A) $f(\theta, \phi) = \sqrt{\cos^2\theta\cos^2\phi + \sin^2\phi}$ $(\beta = 2\pi f\sqrt{\varepsilon_0\mu_0})$.

(B) $f(\theta) = \eta\beta l \sin\theta/(4\pi r)$ $(\eta = \sqrt{\mu_0/\varepsilon_0})$.

(C) $f(\theta) = \sin\theta$.

(D) $f(\theta) = \beta l \sin\theta/2$.

(E) $f = \beta l/2$.

(F) $f = 1$.

CONCEPTUAL QUESTION 12.57 **3-D radiation pattern plot of a Hertzian dipole.** A 3-D polar plot of the normalized field pattern (f) of a Hertzian dipole, in Figure 12.8, is as shown in

(A) Figure 12.9(a).

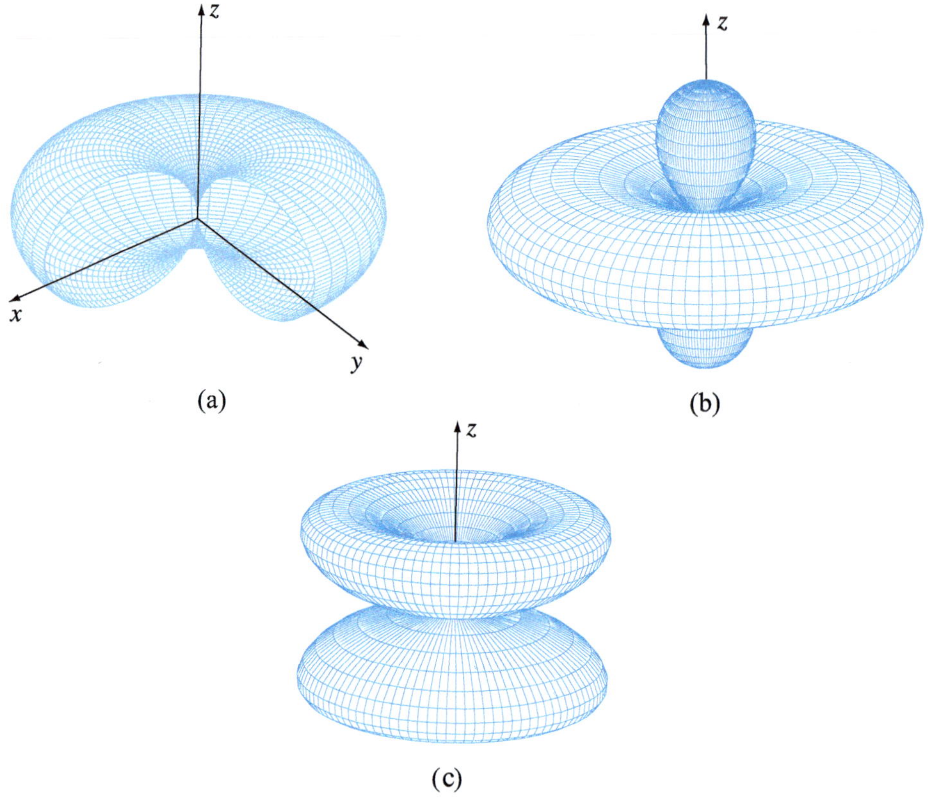

Figure 12.9 3-D polar plots of normalized field patterns of three different antennas; for Conceptual Question 12.57.

(B) Figure 12.9(b).

(C) Figure 12.9(c).

CONCEPTUAL QUESTION 12.58 **Radiation pattern of a Hertzian dipole in the equatorial plane.** The two-dimensional (2-D) cut of the normalized field pattern of a z-directed Hertzian dipole (Figure 12.8) in the equatorial plane, that is, the 2-D pattern for $\theta = 90°$ (xy-plane), is

(A) a circle.

(B) a curve composed of two circles.

(C) a point.

(D) none of the above.

CONCEPTUAL QUESTION 12.59 **Normalized power pattern of a transmitting antenna.** We sometimes plot the normalized power pattern of an antenna, $p(\theta, \phi)$, defined as the normalized magnitude of the associated far-zone Poynting vector for $r = r_0 =$ const, which comes out to be the square of the normalized field pattern of the antenna, $p(\theta, \phi) = f^2(\theta, \phi)$. Expressing the normalized power and field patterns in decibels, we have

(A) $p_{\mathrm{dB}}(\theta, \phi) = f_{\mathrm{dB}}^2(\theta, \phi)$.

(B) $p_{\mathrm{dB}}(\theta, \phi) = 2f_{\mathrm{dB}}(\theta, \phi)$.

(C) $p_{\mathrm{dB}}(\theta, \phi) = f_{\mathrm{dB}}(\theta, \phi)/2$.

(D) $p_{\mathrm{dB}}(\theta, \phi) = f_{\mathrm{dB}}(\theta, \phi) + 6$ dB.

(E) $p_{\mathrm{dB}}(\theta, \phi) = f_{\mathrm{dB}}(\theta, \phi) + 3$ dB.

(F) $p_{\mathrm{dB}}(\theta, \phi) = f_{\mathrm{dB}}(\theta, \phi)$.

CONCEPTUAL QUESTION 12.60 **Directivity of an isotropic radiator.** The directivity, D, of an isotropic radiator, with uniform radiation in all directions, is

(A) zero.

(B) unity.

(C) 1.5.

(D) 4π.

(E) infinite.

CONCEPTUAL QUESTION 12.61 **Directivity of a Hertzian dipole.** The maximum directivity of a z-directed Hertzian dipole radiating in free space (Figure 12.8), for $\theta_{\text{max}} = 90°$, amounts to (using the expressions for the characteristic radiation function, $\underline{\mathbf{F}}$, and radiation resistance, R_{rad}, of the dipole)

(A) $D_{\text{max}} = 0$.

(B) $D_{\text{max}} = 1$.

(C) $D_{\text{max}} = 1.5$.

(D) $D_{\text{max}} = 2$.

(E) $D_{\text{max}} = 3$.

(F) $D_{\text{max}} = 4$.

CONCEPTUAL QUESTION 12.62 **Antenna gain in decibels.** Antenna gain (G) in decibels (dB gain) is given by

(A) $G_{\text{dB}} = \log G$.

(B) $G_{\text{dB}} = 10 \log G$.

(C) $G_{\text{dB}} = 20 \log G$.

(D) $G_{\text{dB}} = \ln G$.

(E) $G_{\text{dB}} = 10 \ln G$.

(F) $G_{\text{dB}} = 20 \ln G$.

CONCEPTUAL QUESTION 12.63 **Gain vs. directivity of an antenna.** The maximum gain of an antenna, $G_{\rm max}$, is always

(A) smaller than or eventually equal to

(B) equal to

(C) larger than or eventually equal to

the maximum directivity of the same antenna, $D_{\rm max}$.

CONCEPTUAL QUESTION 12.64 **Antenna radiation pattern, directivity, and gain.** Antennas having the same 3-D radiation pattern (normalized field pattern) must also have the same

(A) maximum directivity, but may have different maximum gains.

(B) maximum gain, but may have different maximum directivities.

(C) maximum directivity and maximum gain, respectively.

(D) none of the above.

12.6 Wire Dipole Antennas of Arbitrary Lengths

We now consider a symmetrical (centrally fed) straight wire dipole antenna with an arbitrary length, l, and free ends (no capacitive loads at wire ends), as shown in Figure 12.10, and assume a sinusoidal (sine-wave) current distribution along the antenna,

$$\underline{I}(z) = \underline{I}_{\rm m} \sin\beta(h - |z|) \quad (-h \le z \le h) \quad \text{(sine-wave dipole current)}\,, \tag{12.16}$$

where β is the phase coefficient of the antenna radiation ($\beta = 2\pi f\sqrt{\varepsilon\mu}$) and $h = l/2$ is the length of each of the arms of the dipole. Note that the current at the antenna input terminals, $z = 0$, is $\underline{I}_0 = \underline{I}(0) = \underline{I}_{\rm m} \sin\beta h$. Using Eqs. (12.8), (12.16), (12.9), and (12.12), the characteristic radiation function of the dipole is found to be

$$\underline{\mathbf{F}}(\theta) = \frac{\cos(\beta h \cos\theta) - \cos\beta h}{\sin\beta h \sin\theta}\,\hat{\boldsymbol{\theta}} \quad \text{(radiation function, arbitrary dipole)}\,. \tag{12.17}$$

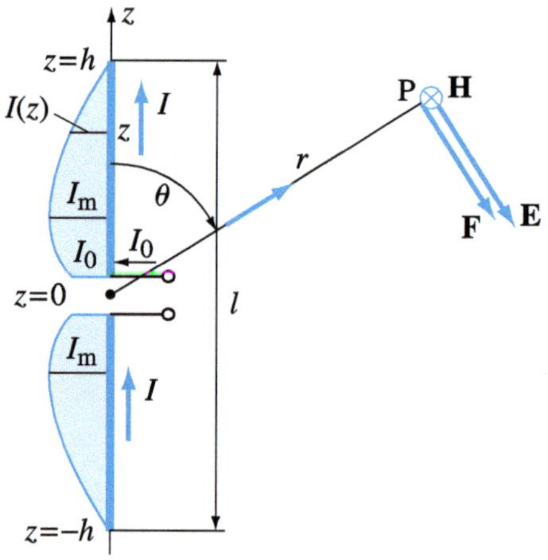

Figure 12.10 Symmetrical wire dipole antenna of arbitrary length with sinusoidal current approximation.

The most important dipole is by far that for $l = \lambda/2$ (in Figure 12.10). This simple wire antenna, known as a half-wave dipole, is, in fact, one of the most widely used of all antenna types. The dipole arm length being $h = \lambda/4$, we have $\beta h = 2\pi h/\lambda = \pi/2$, with which the characteristic radiation function in Eq. (12.17) becomes

$$\underline{\mathbf{F}}(\theta) = \frac{\cos\left(\frac{\pi}{2}\cos\theta\right)}{\sin\theta}\,\hat{\boldsymbol{\theta}} \quad \text{(radiation function, half-wave dipole)}\,. \tag{12.18}$$

For dipoles of various electrical lengths, l/λ, there is a clear correspondence between the current distributions along the wire, given by Eq. (12.16), and far-field distributions, i.e., radiation patterns, based on Eq. (12.17). Specifically, for dipoles with $l \leq \lambda$, because of the phase uniformity of the antenna current [the sine function in Eq. (12.16) does not change sign] along the whole wire, the radiation is strongest in a direction normal to the antenna ($\theta = 90°$). Namely, all parallel rays originating from current elements that constitute the current distribution of an antenna travel the same paths, and thus arrive in phase, at the far field at $\theta = 90°$ [$\cos\theta = 0$ in the phase correction term in the radiation integral, Eq. (12.8)]. However, when $l > \lambda$, multiple radiation lobes are formed in the pattern, due to cancelation effects of oppositely directed (counter-phase) currents on adjacent antenna sections, separated by the current zeros, caused by opposite signs of the sine function, in Eq. (12.16), on the two segments. On the other side, all patterns have nulls along the axis of the dipole ($\theta = 0$ or $\theta = 180°$), which comes from the factor $\sin\theta$ in obtaining the magnetic potential component (projection) $\underline{A}_\theta$ in Figure 12.3.

The high-frequency ohmic resistance ($R_{\rm ohmic}$), Eq. (12.11), of a half-wave and longer metallic (non-PEC), e.g., copper or aluminum, dipoles is typically very small relative to the radiation resistance ($R_{\rm rad}$), Eq. (12.10), so they are, as opposed to electrically short dipoles, very efficient radiators ($R_{\rm rad} \gg R_{\rm ohmic}$).

CONCEPTUAL QUESTION 12.65 **Constant and alternating phases of dipole currents.** Consider the complex current intensity, $\underline{I}(z)$, along a wire dipole antenna of length $l = \lambda/2$ and that along a dipole with $l = 3\lambda/2$, respectively, both radiating in free space, and whether $\underline{I}(z)$ has a constant phase or undergoes phase shifts of 180° along the wire, for $-l/2 \leq z \leq l/2$, for each of the antennas. We have

(A) constant phases for both $l = \lambda/2$ and $l = 3\lambda/2$.

(B) a constant phase for $l = \lambda/2$ and an alternating phase for $l = 3\lambda/2$.

(C) an alternating phase for $l = \lambda/2$ and a constant phase for $l = 3\lambda/2$.

(D) alternating phases for both $l = \lambda/2$ and $l = 3\lambda/2$.

CONCEPTUAL QUESTION 12.66 **Charge distribution of an arbitrary dipole antenna.** Consider a symmetrical wire dipole antenna with an arbitrary length, $l = 2h$, and sinusoidal current distribution, described as $\underline{I}(z) = \underline{I}_{\mathrm{m}} \sin\beta(h - |z|)$ $(-h \leq z \leq h)$. The operating angular frequency of the antenna is ω and the ambient medium is free space. The charge distribution (charge per unit length of the antenna) along the upper antenna arm, for $0 \leq z \leq h$, is given by

(A) $\underline{Q}'(z) = 0$.

(B) $\underline{Q}'(z) = -(\mathrm{j}\underline{I}_{\mathrm{m}}/\omega h) \sin\beta h$.

(C) $\underline{Q}'(z) = -(\mathrm{j}\underline{I}_{\mathrm{m}}/\omega) \sin\beta(h - z)$.

(D) $\underline{Q}'(z) = -(\mathrm{j}\beta\underline{I}_{\mathrm{m}}/\omega) \cos\beta(h - z)$.

(E) $\underline{Q}'(z) \to \infty$.

CONCEPTUAL QUESTION 12.67 **Short-dipole current from the expression for arbitrary dipole.** Consider a z-directed symmetrical wire dipole antenna with an arbitrary length, l, and sinusoidal current distribution, radiating in free space. For $l \ll \lambda$, the antenna current intensity, given by the sine function of the coordinate z, reduces to

(A) $\underline{I}(z) = \underline{I}_0$,

(B) $\underline{I}(z) = \underline{I}_0/2$,

(C) $\underline{I}(z) = \underline{I}_0(1 - |z|/h)$,

(D) $\underline{I}(z) = 0$,

(E) $\underline{I}(z) \to \infty$,

for $-h \leq z \leq h$, where $h = l/2$.

CONCEPTUAL QUESTION 12.68 **Maximum magnitude of the radiation function.** The magnitude of the characteristic radiation function, $\underline{\mathbf{F}}$, of a wire dipole antenna in the direction of maximum radiation is unity ($F_{\text{max}} = 1$) for

(A) a half-wave dipole.

(B) a dipole of any length.

(C) none of the dipoles.

CONCEPTUAL QUESTION 12.69 **Half-wave dipole antenna at different frequencies.** Radiation patterns in E-planes (planes containing the vector $\mathbf{E}$) of a half-wave dipole antenna at 1 GHz are

(A) more directive than

(B) less directive than

(C) the same as

those of a half-wave dipole antenna at 2 GHz.

CONCEPTUAL QUESTION 12.70 **Radiation pattern in the equatorial plane of a wire dipole.** For a wire dipole antenna of an arbitrary length, extending along the z-axis and being fed at the coordinate origin, the $\theta = 90°$ 2-D pattern cut comes out to be

(A) a circle.

(B) a curve composed of two circles.

(C) a point.

(D) none of the above.

CONCEPTUAL QUESTION 12.71 **Far-field vector components of an arbitrary dipole.** For the vector components of $\underline{\mathbf{E}}$ and $\underline{\mathbf{H}}$ at a far-field point due to a z-directed wire dipole antenna of an arbitrary length, we have the following:

(A) $\underline{\mathbf{E}}$ has only an r-component, and $\underline{\mathbf{H}}$ has only a ϕ-component

(B) $\underline{\mathbf{E}}$ has only a θ-component, and $\underline{\mathbf{H}}$ has only a ϕ-component

(C) $\underline{\mathbf{E}}$ has only a ϕ-component, and $\underline{\mathbf{H}}$ has only a θ-component

(D) $\underline{\mathbf{E}}$ has only a θ- and a ϕ-component, and the same for $\underline{\mathbf{H}}$,

(E) $\underline{\mathbf{E}}$ has only an r- and a θ-component, and $\underline{\mathbf{H}}$ has only a ϕ-component

(F) $\underline{\mathbf{E}}$ has all three components, and $\underline{\mathbf{H}}$ has only a θ- and a ϕ-component

(and the remaining field components are, exactly or approximately, zero).

CONCEPTUAL QUESTION 12.72 **Correspondence between current and far-field distributions.** For wire dipole antennas longer than one wavelength, the more zeros in the sinusoidal current distribution along the wire

(A) the more zeros

(B) the fewer zeros

(C) no correspondence

in the far-field distribution (radiation pattern) in an E-plane of the antenna (plane $\phi = \text{const}$ if the antenna is z-directed).

CONCEPTUAL QUESTION 12.73 **Common zero in radiation patterns of all wire dipoles.** Wire dipole antennas of all possible lengths have at least one common zero (direction with no radiation) in the radiation pattern.

(A) True.

(B) False.

CONCEPTUAL QUESTION 12.74 **Electrical lengths of a wire dipole at three different frequencies.** A symmetrical wire dipole antenna of length $l = 3$ m is fed by a current of rms intensity $I_0 = 1$ A, and radiates in free space. We consider the operation of this antenna at three different frequencies. If the operating frequencies are $f_1 = 50$ MHz, $f_2 = 150$ MHz, and $f_3 = 250$ MHz, the corresponding electrical lengths of the wire dipole, l/λ, are

(A) $l/\lambda_1 = 0.5$, $l/\lambda_2 = 1.5$, and $l/\lambda_3 = 2.5$.

(B) $l/\lambda_1 = 0.5$, $l/\lambda_2 = 2.5$, and $l/\lambda_3 = 1.5$.

(C) $l/\lambda_1 = 1.5$, $l/\lambda_2 = 0.5$, and $l/\lambda_3 = 2.5$.

(D) $l/\lambda_1 = 1.5$, $l/\lambda_2 = 2.5$, and $l/\lambda_3 = 0.5$.

(E) $l/\lambda_1 = 2.5$, $l/\lambda_2 = 0.5$, and $l/\lambda_3 = 1.5$.

(F) $l/\lambda_1 = 2.5$, $l/\lambda_2 = 1.5$, and $l/\lambda_3 = 0.5$.

CONCEPTUAL QUESTION 12.75 **Current distributions of a wire dipole at three frequencies.** The current distributions along the wire dipole antenna with $l = 3$ m and $I_0 = 1$ A considered at three different frequencies – for the electrical lengths of the antenna amounting to $l/\lambda_1 = 0.5$, $l/\lambda_2 = 1.5$, and $l/\lambda_3 = 2.5$ – are shown in

(A) Figure 12.11(a), Figure 12.11(b), and Figure 12.11(c),

(B) Figure 12.11(a), Figure 12.11(c), and Figure 12.11(b),

(C) Figure 12.11(b), Figure 12.11(a), and Figure 12.11(c),

(D) Figure 12.11(b), Figure 12.11(c), and Figure 12.11(a),

(E) Figure 12.11(c), Figure 12.11(a), and Figure 12.11(b),

(F) Figure 12.11(c), Figure 12.11(b), and Figure 12.11(a),

respectively.

CONCEPTUAL QUESTION 12.76 **Radiation patterns of a wire dipole at three frequencies.** The normalized E-plane radiation field pattern of the wire dipole antenna at three different frequencies – for the electrical lengths of the dipole equal to $l/\lambda_1 = 0.5$, $l/\lambda_2 = 1.5$, and $l/\lambda_3 = 2.5$ – are given in

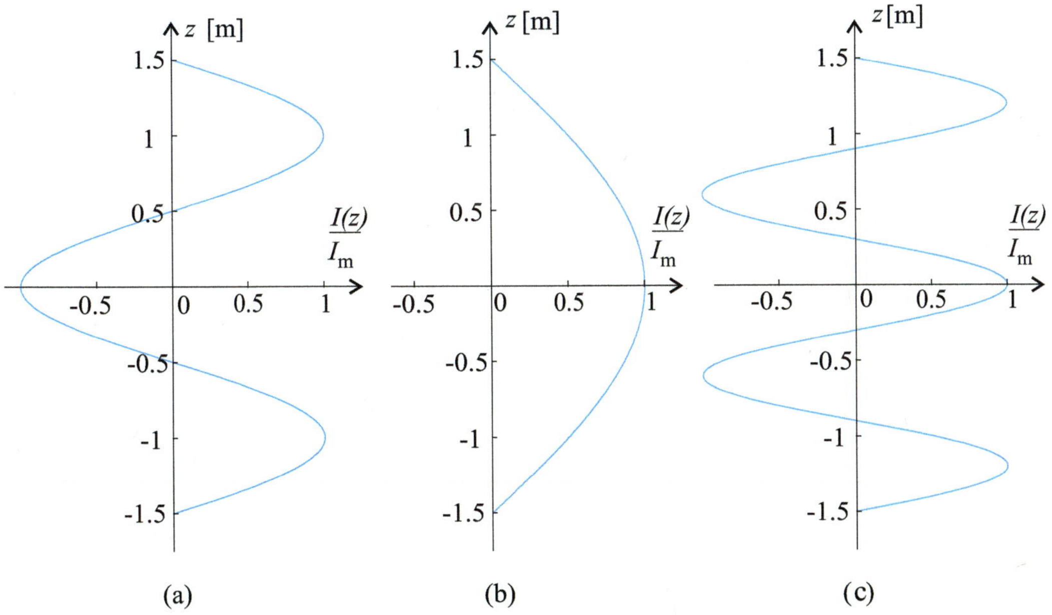

Figure 12.11 Normalized current intensity $\underline{I}(z)/\underline{I}_m$ along a wire dipole antenna at three different frequencies, i.e., for three different electrical lengths (l/λ) of the antenna; for Conceptual Question 12.75.

(A) Figure 12.12(a), Figure 12.12(b), and Figure 12.12(c),

(B) Figure 12.12(a), Figure 12.12(c), and Figure 12.12(b),

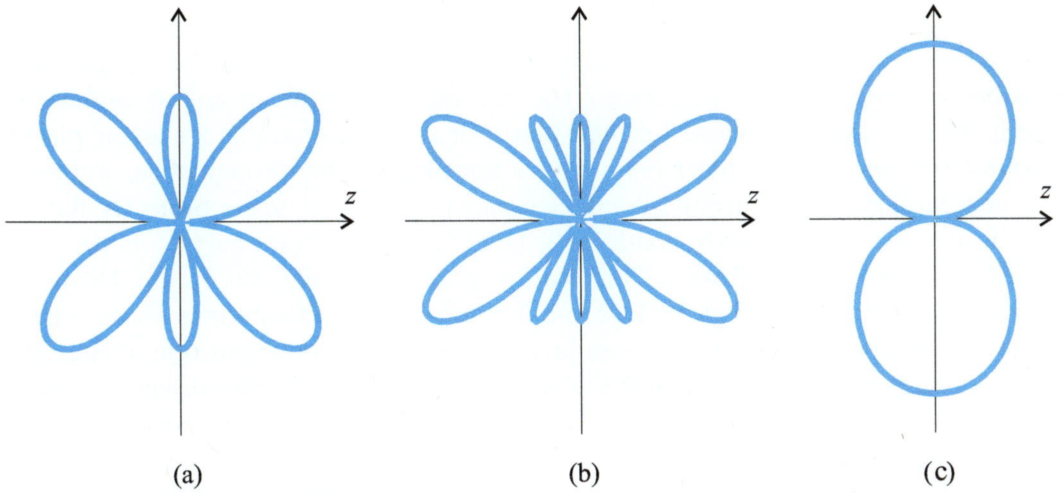

Figure 12.12 Normalized field patterns in an E-plane (plane ϕ = const) of a wire dipole antenna at three different frequencies (for three different electrical lengths, l/λ, of the antenna); for Conceptual Question 12.76.

(C) Figure 12.12(b), Figure 12.12(a), and Figure 12.12(c),

(D) Figure 12.12(b), Figure 12.12(c), and Figure 12.12(a),

(E) Figure 12.12(c), Figure 12.12(a), and Figure 12.12(b),

(F) Figure 12.12(c), Figure 12.12(b), and Figure 12.12(a),

respectively.

CONCEPTUAL QUESTION 12.77 **Copper half-wave dipole treated as a PEC one.** A half-wave dipole antenna cut out of a very thin copper wire

(A) can

(B) cannot

be treated as if it were made of a perfect electric conductor (PEC).

12.7 Image Theory for Antennas above a Perfectly Conducting Ground Plane

Often, we need to analyze antennas in the presence of conducting ground planes. Such a plane can be an approximation of the earth's surface, of a metallic plate (that may be isolated or a part of a larger structure, like an aircraft or an automobile) to which the antenna is attached (ground plate) or placed in parallel (reflector plate), or of a large conducting object (e.g., device housing) in the vicinity of the antenna. In many situations, the ground conductor can be assumed to be perfectly flat, of infinite extent, and perfectly conducting (PEC), so that the analysis model with a PEC ground plane suffices. This section presents image theory, analogous to techniques illustrated in Figures 1.32 and 5.8, for analysis of antennas above a PEC plane. Since an arbitrary antenna can be represented as a superposition of Hertzian dipoles, image theory for a single Hertzian dipole with an arbitrary orientation with respect to the plane, as in Figure 12.13(a), can be readily generalized to radiators with arbitrary current distributions.

Applying the image theory for charges in Figure 1.32 to dipole charges $\underline{Q}$ and $-\underline{Q}$ in Figure 12.13(a), we obtain an equivalent system composed of charges $-\underline{Q}$ and $\underline{Q}$, respectively, shown in Figure 12.13(b). The current $\underline{I}$ and charge $\underline{Q}$ (and $-\underline{Q}$) of the original antenna (dipole) in Figure 12.13(a) are related by the continuity

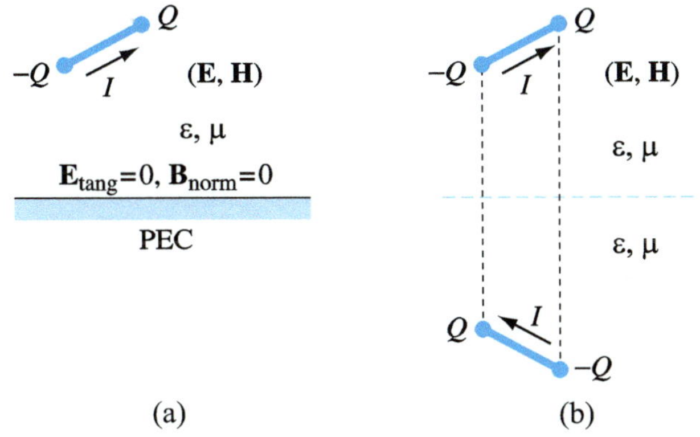

Figure 12.13 Image theory for antennas (or high-frequency current and charge) above a perfectly conducting ground plane: (a) an oblique Hertzian dipole as a constituting element of an arbitrary antenna above the ground and (b) equivalent system with the PEC plane replaced by a negative image of the primary current and charge configuration.

equation, and the same relationship must hold for the current and charge of the image antenna. This means that the image antenna must be a Hertzian dipole as well, which fully and uniquely determines the magnitude and orientation of the image current in Figure 12.13(b). Geometrically, the virtual dipole is the mirror image of the original in the (once PEC) symmetry plane. Electrically, it is a negative image of the charge and current of the actual antenna.

Perhaps the most important application of the image theory in Figures 12.13 is to analyze so-called monopole antennas, i.e., antennas attached to a PEC ground plane and fed against it, as illustrated in Figure 12.14(a). By image theory, the monopole is transformed to the equivalent dipole, in Figure 12.14(b), with the current on the

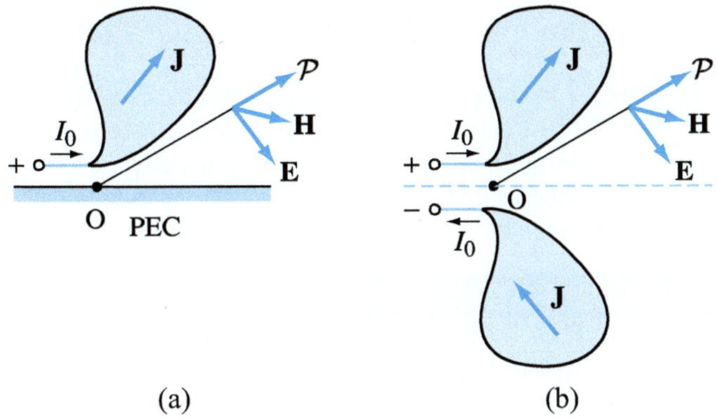

Figure 12.14 Field equivalency in the upper half-space of an arbitrary monopole antenna fed against a PEC plane (a) and the symmetrical dipole in an unbounded medium obtained by image theory (b).

lower half of the dipole being the negative image of the current on its upper half, or of the current on the monopole, and the same electromagnetic field, including near and far field, in the upper half-space (e.g., air) in the two systems. Of course, the analysis of the dipole antenna is normally a much simpler task than the analysis of the antenna composed of the monopole and the ground plane.

CONCEPTUAL QUESTION 12.78 **Image theory for antennas above a PEC ground plane.** In the analysis of an antenna above a PEC ground plane, the plane can be replaced by

(A) a positive image

(B) a negative image

(C) sometimes positive and sometimes negative image

of the antenna in the (once PEC) symmetry plane.

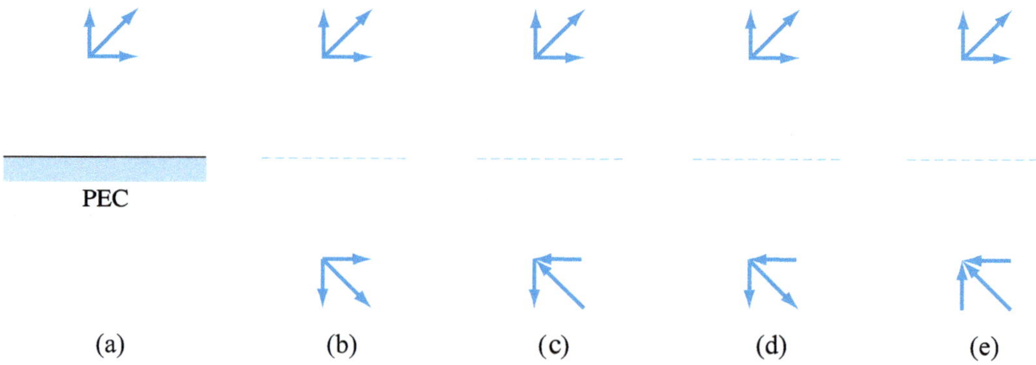

Figure 12.15 Image theory for three characteristic current elements above a PEC ground: original system with the PEC interface (a) and equivalent system with actual and image elements radiating in an unbounded, homogeneous medium (e.g., free space) – four offered sets of images (b)–(e); for Conceptual Question 12.79.

CONCEPTUAL QUESTION 12.79 **Images of three characteristic current elements above a PEC plane.** Consider three characteristic (vertical, horizontal, and oblique) current elements above a PEC plane shown in Figure 12.15(a). Their images in the plane – such that the electric and magnetic fields in the upper half-space are the same in the original system and in an equivalent system obtained using image theory – are as shown in

(A) Figure 12.15(b).

(B) Figure 12.15(c).

(C) Figure 12.15(d).

(D) Figure 12.15(e).

CONCEPTUAL QUESTION 12.80 **Direct wave, reflected wave, and wave due to the image antenna.** Figure 12.16 shows a half-wave wire dipole antenna in air, above a PEC ground plane. Let $\underline{\mathbf{E}}_{\text{direct}}$, $\underline{\mathbf{E}}_{\text{reflected}}$, and $\underline{\mathbf{E}}_{\text{image}}$ stand for the electric field vectors at the receive point P due to the direct wave radiated by the original antenna and propagating parallel to the ground, reflected wave (emanating from the original antenna and bouncing off the ground), and wave launched by the image antenna (introduced by image theory), respectively. The total electric field, $\underline{\mathbf{E}}$, at the point P can be obtained as

(A) $\underline{\mathbf{E}} = \underline{\mathbf{E}}_{\text{direct}}$.

(B) $\underline{\mathbf{E}} = \underline{\mathbf{E}}_{\text{reflected}}$.

(C) $\underline{\mathbf{E}} = \underline{\mathbf{E}}_{\text{image}}$.

(D) $\underline{\mathbf{E}} = \underline{\mathbf{E}}_{\text{direct}} + \underline{\mathbf{E}}_{\text{reflected}} + \underline{\mathbf{E}}_{\text{image}}$.

(E) $\underline{\mathbf{E}} = \underline{\mathbf{E}}_{\text{direct}} + \underline{\mathbf{E}}_{\text{reflected}} = \underline{\mathbf{E}}_{\text{direct}} + \underline{\mathbf{E}}_{\text{image}}$.

(F) $\underline{\mathbf{E}} = \underline{\mathbf{E}}_{\text{reflected}} + \underline{\mathbf{E}}_{\text{image}}$.

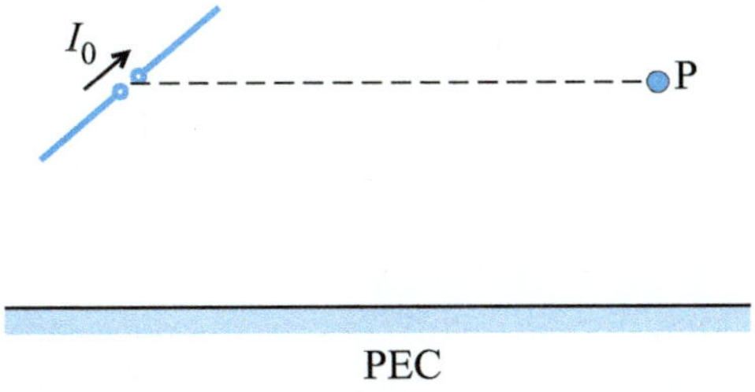

Figure 12.16 Half-wave dipole antenna in air, above a horizontal PEC plane, and a far-field receive point (P) in the same vertical plane with the antenna; for Conceptual Question 12.80.

CONCEPTUAL QUESTION 12.81 **Horizontal half-wave wire dipole above a ground plane.** A horizontal half-wave wire dipole antenna is placed at a height h above a PEC ground plane, in free space. In the analysis, the PEC plane is removed applying the image theory, which results in an equivalent system with two antennas radiating in free space. In this system, the total electric field intensity vector, $\underline{\mathbf{E}}_{\text{tot}} = \underline{\mathbf{E}}_{\text{original}} + \underline{\mathbf{E}}_{\text{image}}$, at a point P in the symmetry plane between the two dipoles is as shown in

(A) Figure 12.17(a).

(B) Figure 12.17(b).

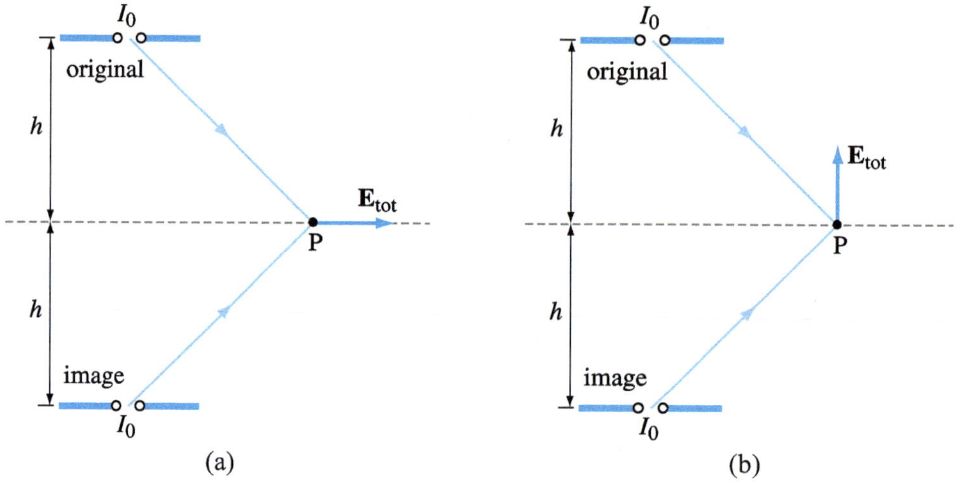

Figure 12.17 Equivalent system with two antennas radiating in free space obtained applying the image theory to a system with a horizontal half-wave dipole antenna above a PEC plane: two offered polarizations (orientations) of the total electric field intensity vector, $\underline{\mathbf{E}}_{\text{tot}}$, at a point P in the symmetry plane between the antennas; for Conceptual Question 12.81.

CONCEPTUAL QUESTION 12.82 **Radiation function of an arbitrary monopole antenna.** Consider an arbitrary monopole antenna fed against a horizontal PEC plane (antenna is above the plane) and the equivalent symmetrical dipole in an unbounded medium obtained by image theory. Let the characteristic radiation functions of the two antennas be $\underline{\mathbf{F}}_{\text{monopole}}$ and $\underline{\mathbf{F}}_{\text{dipole}}$, respectively. At an arbitrary point in the upper half-space, we have that

(A) $\underline{\mathbf{F}}_{\text{monopole}} = \underline{\mathbf{F}}_{\text{dipole}}$.

(B) $\underline{\mathbf{F}}_{\text{monopole}} = 2\underline{\mathbf{F}}_{\text{dipole}}$.

(C) $\underline{\mathbf{F}}_{\text{monopole}} = \underline{\mathbf{F}}_{\text{dipole}}/2$.

(D) $\underline{\mathbf{F}}_{\text{monopole}} = -\underline{\mathbf{F}}_{\text{dipole}}$.

(E) $\underline{\mathbf{F}}_{\text{monopole}} = 0$.

(F) None of the above.

CONCEPTUAL QUESTION 12.83 **Quarter-wave monopole and equivalent half-wave dipole.** Consider a vertical quarter-wave monopole antenna above a horizontal perfectly conducting plane and an equivalent half-wave dipole antenna, as shown in Figure 12.18. The two antennas produce the same field in the upper half-space. If the directivity of the dipole is D, the directivity of the monopole is

(A) $4D$.

(B) $2D$.

(C) D.

(D) $D/2$.

(E) $D/4$.

(F) None of the above.

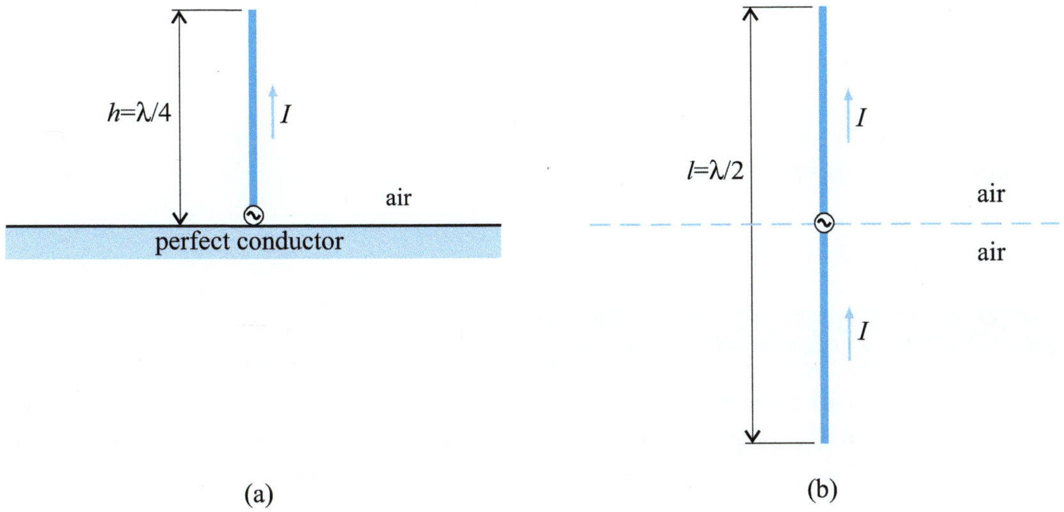

Figure 12.18 (a) Quarter-wave monopole antenna fed against a PEC plane and (b) the symmetrical half-wave dipole antenna in air obtained by image theory; for Conceptual Question 12.83.

CONCEPTUAL QUESTION 12.84 **Decibel directivity of a quarter-wave monopole.** If the directivity in decibels of the $\lambda/2$ dipole in Figure 12.18 is D_{dB}, the dB directivity of the $\lambda/4$ monopole amounts to

(A) $D_{\text{dB}} - 6$ dB.

(B) $D_{\text{dB}} - 3$ dB.

(C) D_{dB}.

(D) $D_{\text{dB}} + 3$ dB.

(E) $D_{\text{dB}} + 6$ dB.

(F) none of the above.

CONCEPTUAL QUESTION 12.85 **Radiation resistance of a quarter-wave monopole.** If the radiation resistance of the dipole antenna in Figure 12.18 is R_{rad}, the radiation resistance of the monopole antenna comes out to be

(A) $4R_{\text{rad}}$.

(B) $2R_{\text{rad}}$.

(C) R_{rad}.

(D) $R_{\text{rad}}/2$.

(E) $R_{\text{rad}}/4$.

(F) none of the above.

CONCEPTUAL QUESTION 12.86 **Oblique monopole and equivalent dipole.** Considering an oblique monopole antenna, namely, a monopole that makes an oblique angle (for instance, 45°) with a PEC plane, as shown in Figure 12.19(a), and the corresponding equivalent dipole antenna (obtained by image theory), in Figure 12.19(b), the directivity of which is D, the directivity of the monopole is

(A) $4D$.

(B) $2D$.

(C) D.

(D) $D/2$.

(E) $D/4$.

(F) none of the above.

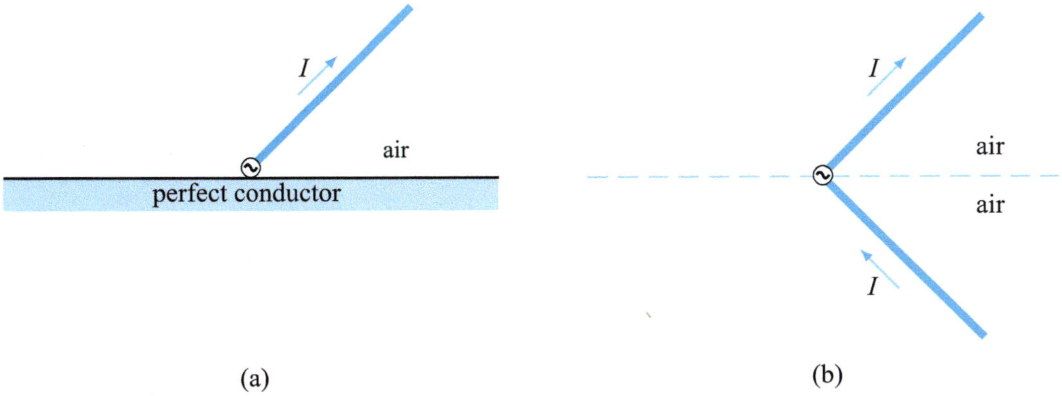

Figure 12.19 (a) Oblique monopole antenna above a PEC plane and (b) the equivalent dipole antenna in air obtained by image theory; for Conceptual Question 12.86.

12.8 Theory of Receiving Antennas. Wireless Links with Nonaligned Wire Antennas

Consider an arbitrary receiving antenna illuminated by a uniform plane time-harmonic wave, of wavelength λ. With reference to Figure 12.20(a), the propagation unit vector of the wave, $\hat{\mathbf{n}}$, is directed toward the global coordinate origin, O, which is usually adopted at the antenna terminals. This wave is, most frequently, originated by another (transmitting) antenna in a wireless link, which is far away from the receiving antenna, Eq. (12.5), so that the uniform-plane-wave approximation of the actual nonuniform spherical wave radiated by the other antenna applies. In addition, let the receiving antenna be terminated in a load of complex impedance $\underline{Z}_{\mathrm{L}}$. With respect to its output terminals, and to the load, the antenna can be replaced by the Thévenin equivalent generator, shown in Figure 12.20(b). The complex internal impedance of the generator, $\underline{Z}_{\mathrm{T}}$, equals the complex input impedance of the antenna in Figure 12.20(a) (with the incident plane wave "turned off"), so it equals the impedance $\underline{Z}_{\mathrm{A}}$ of the antenna [see Eqs. (12.10) and (12.11)] – when transmitting. The emf of the generator, $\underline{\mathcal{E}}_{\mathrm{T}}$, can be found as the voltage $\underline{V}_{\mathrm{oc}}$ across the open terminals (with the load removed) of the antenna (excited by the incoming

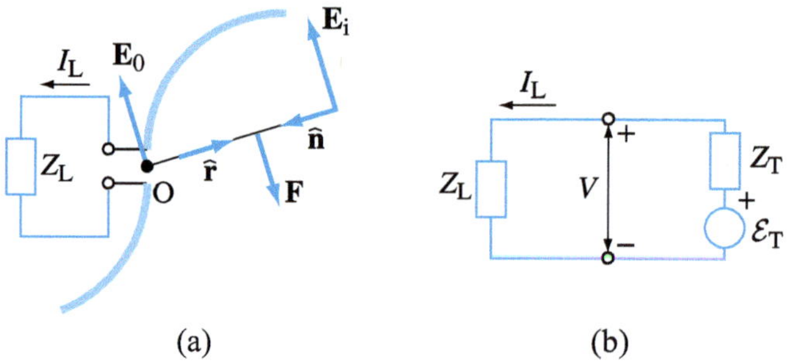

Figure 12.20 (a) Receiving antenna in the field of a uniform plane time-harmonic wave and (b) its Thévenin equivalent representation.

wave) in Figure 12.20(a), and the result is

$$\underline{V}_{\rm oc} = \underline{\mathcal{E}}_{\rm T} = \frac{\lambda}{\pi} \underline{\mathbf{E}}_0 \cdot \underline{\mathbf{F}} \quad \text{(open-circuit voltage of a receiving antenna)}, \tag{12.19}$$

where $\underline{\mathbf{E}}_0$ is the electric field vector of the incident plane wave at the coordinate origin and $\underline{\mathbf{F}}$ is the characteristic radiation function [Eq. (12.12)] that the antenna would have if transmitting in the direction of the wave incidence, i.e., direction defined by $\hat{\mathbf{r}} = -\hat{\mathbf{n}}$, as indicated in Figure 12.20(a). So, the field pattern of a receiving antenna, showing how well it captures the incident signal in different directions in 3-D space, is identical to the radiation pattern of the antenna when in the transmitting mode, Figure 12.7, that is, the transmit and receive patterns of an arbitrary antenna are identical. Finally, the current of the antenna load, $\underline{I}_{\rm L}$, in Figure 12.20(a) and the power delivered to the load are determined from the equivalent circuit in Figure 12.20(b).

In a communication link, the dot product in Eq. (12.19) determines the polarization match (or mismatch) between the two antennas (transmitting and receiving) at the two ends of the link. For example, in the wireless link with two nonaligned wire dipole antennas shown in Figure 12.21, the open-circuit voltage of the receiving dipole, $\underline{V}_{\rm oc}$, is given by Eq. (12.19) with $\underline{\mathbf{E}}_0 = \underline{\mathbf{E}}_{\rm t}$ being the far electric field intensity vector of the transmitting antenna computed at point O′ in Figure 12.21, and $\underline{\mathbf{F}} = \underline{\mathbf{F}}_{\rm r}$

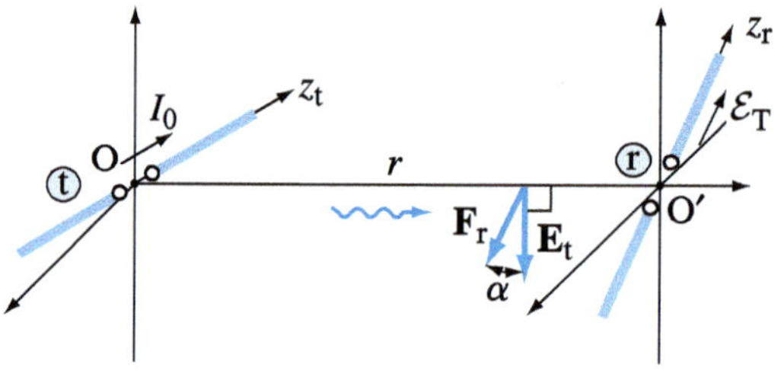

Figure 12.21 Wireless link consisting of two nonaligned wire dipole antennas in free space, with the receiving (r) antenna being in the far field of the transmitting (t) antenna.

the characteristic radiation function of the receiving antenna that it would have if transmitting in the direction toward point O. The voltage is $\underline{V}_{\text{oc}} = (\lambda/\pi)\,\underline{E}_{\text{t}}\underline{F}_{\text{r}}\cos\alpha$, where α is the angle between vectors $\underline{\mathbf{E}}_{\text{t}}$ and $\underline{\mathbf{F}}_{\text{r}}$.

CONCEPTUAL QUESTION 12.87 Transmitting/receiving antennas as passive and active circuits. With respect to their input/output terminals, an arbitrary transmitting antenna and an arbitrary receiving antenna can be replaced by

(A) a passive load and an active circuit, respectively.

(B) an active circuit and a passive load, respectively.

(C) passive loads.

(D) active circuits.

CONCEPTUAL QUESTION 12.88 Received current through short-circuited antenna terminals. Consider a wire dipole antenna whose impedance when transmitting in free space amounts to $\underline{Z}_{\text{A}} = (50 + \text{j}50)\,\Omega$ at a frequency of $f = 1$ GHz. When illuminated by an incident uniform plane wave at $f = 1$ GHz in free space, and with its terminals open-circuited, this antenna receives an rms voltage of $V_{\text{oc}} = 1$ V across the terminals. When illuminated by the same wave, in exactly the same way, but with the terminals short-circuited, the rms current through the terminals of the antenna, I_{sc}, equals

(A) $I_{\text{sc}} = 0$.

(B) $I_{\text{sc}} = (1/377)$ A.

(C) $I_{\text{sc}} = [1/(50\sqrt{2})]$ A.

(D) $I_{\text{sc}} \to \infty$.

(E) The current I_{sc} and voltage V_{oc} are not related to each other.

CONCEPTUAL QUESTION 12.89 Far field of the transmitting antenna in a wireless link. In the wireless link with two nonaligned half-wave wire dipole antennas shown in Figure 12.22, the far electric field intensity vector of the transmitting antenna $\underline{\mathbf{E}}_{\text{t}}$ is computed using (namely, $\underline{\mathbf{E}}_{\text{t}}$ is proportional to) the characteristic radiation function of a half-wave dipole antenna given by $\underline{\mathbf{F}}_{\text{t}} = F(\theta_{\text{t}})(-\hat{\mathbf{x}}') = \cos[(\pi/2)\cos\theta_{\text{t}}]\,(-\hat{\mathbf{x}}')/\sin\theta_{\text{t}}$, with

(A) $\theta_t = \gamma_1$.

(B) $\theta_t = 90° - \gamma_1$.

(C) $\theta_t = \gamma_2$.

(D) $\theta_t = 90° - \gamma_2$.

(E) $\theta_t = 0$.

(F) $\theta_t = 90°$.

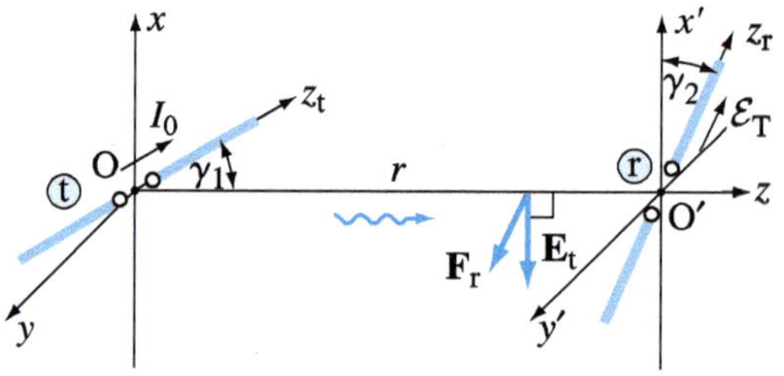

Figure 12.22 Wireless link with two nonaligned half-wave dipole antennas at a distance $r \gg \lambda$: the transmitting antenna is placed in the xz-plane at an angle γ_1 with respect to the z-axis, while the receiving antenna lies in the $x'y'$-plane, where it makes an angle γ_2 with the x'-axis; for Conceptual Question 12.89.

CONCEPTUAL QUESTION 12.90 Local θ angle for the radiation function of a receiving antenna. In the wireless link in Figure 12.22, the magnitude of the characteristic radiation function of the receiving antenna (for the direction toward the transmitting antenna), $\underline{\mathbf{F}}_r$, equals the function $F(\theta_r)$ of a half-wave dipole, $F(\theta_r) = \cos[(\pi/2)\cos\theta_r]/\sin\theta_r$, with

(A) $\theta_r = \gamma_1$.

(B) $\theta_r = 90° - \gamma_1$.

(C) $\theta_r = \gamma_2$.

(D) $\theta_r = 90° - \gamma_2$.

(E) $\theta_r = 0$.

(F) $\theta_r = 90°$.

CONCEPTUAL QUESTION 12.91 Antenna polarization mismatch angle in a wireless link. In the wireless link in Figure 12.22, the open-circuit voltage of the receiving dipole amounts to $\underline{V}_{\text{oc}} = (\lambda/\pi)\, \underline{E}_{\text{t}} \underline{F}_{\text{r}} \cos\alpha$, where the angle α between vectors $\underline{\mathbf{E}}_{\text{t}}$ and $\underline{\mathbf{F}}_{\text{r}}$ equals

(A) $\alpha = \gamma_1$.

(B) $\alpha = 90° - \gamma_1$.

(C) $\alpha = \gamma_2$.

(D) $\alpha = 90° - \gamma_2$.

(E) $\alpha = 0$.

(F) $\alpha = 90°$.

CONCEPTUAL QUESTION 12.92 Maximization of the received voltage in a wireless system. The received voltage in the wireless system in Figure 12.22 can be maximized, for the two antennas used and the given length of the link, input power, and frequency, by orienting the dipoles such that

(A) $\gamma_1 = \gamma_2 = 0$.

(B) $\gamma_1 = \gamma_2 = 45°$.

(C) $\gamma_1 = 0°$ and $\gamma_2 = 90°$.

(D) $\gamma_1 = 90°$ and $\gamma_2 = 0$.

(E) $\gamma_1 = \gamma_2 = 90°$.

CONCEPTUAL QUESTION 12.93 Switching places of transmitting and receiving antennas. Assume that in the wireless link with two nonaligned half-wave dipole antennas shown in Figure 12.22, the transmitting and receiving antennas switch places. Namely, the dipole on the right-hand side (lying in the $x'y'$-plane) is now transmitting, while the terminals of the dipole on the left-hand side (in the xz-plane) are left open, as depicted in Figure 12.23. Under these circumstances, consider the open-circuit voltage $\underline{V}_{\text{oc}}^{\text{new}}$ received by the latter antenna and determine: what is the angle α^{new} between new vectors $\underline{\mathbf{E}}_{\text{t}}^{\text{new}}$ and $\underline{\mathbf{F}}_{\text{r}}^{\text{new}}$, which can be used to obtain the new received voltage as $\underline{V}_{\text{oc}}^{\text{new}} = (\lambda/\pi)\, \underline{\mathbf{E}}_{\text{t}}^{\text{new}} \cdot \underline{\mathbf{F}}_{\text{r}}^{\text{new}} = (\lambda/\pi)\, \underline{E}_{\text{t}}^{\text{new}} \underline{F}_{\text{r}}^{\text{new}} \cos\alpha^{\text{new}}$?

(A) $\alpha^{\text{new}} = \gamma_1$.

(B) $\alpha^{\text{new}} = 90° - \gamma_1$.

(C) $\alpha^{\text{new}} = \gamma_2$.

(D) $\alpha^{\text{new}} = 90° - \gamma_2$.

(E) $\alpha^{\text{new}} = 0$.

(F) $\alpha^{\text{new}} = 90°$.

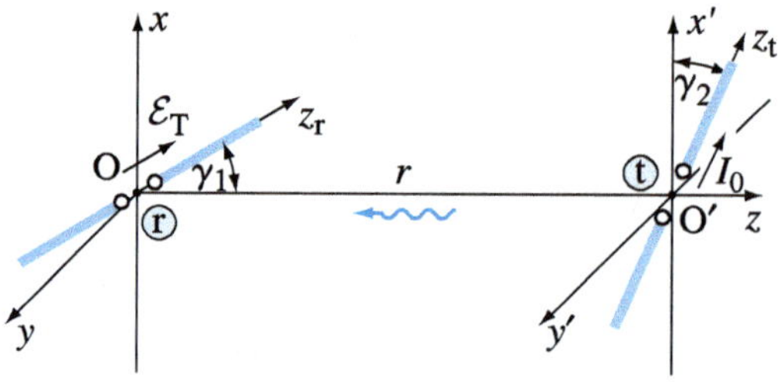

Figure 12.23 Wireless link with the transmitting and receiving antenna places switched; for Conceptual Question 12.93.

CONCEPTUAL QUESTION 12.94 **Link with two nonaligned wire dipoles, received voltage.** In a wireless link in free space, both antennas are half-wave wire dipoles. They are positioned in two parallel planes such that the line between their centers is perpendicular to the planes. The angle between the dipoles, i.e., the angle between the first dipole and the projection of the second dipole on the plane of the first one, is γ, as shown in Figure 12.24. For a given input power of the transmitting antenna, the magnitude of the open-circuit voltage of the receiving antenna is proportional to

(A) $\sin\gamma$.

(B) $\cos\gamma$.

(C) $\sin\gamma\cos\gamma$.

(D) $\sin^2\gamma$.

(E) none of the above.

(F) Need more information.

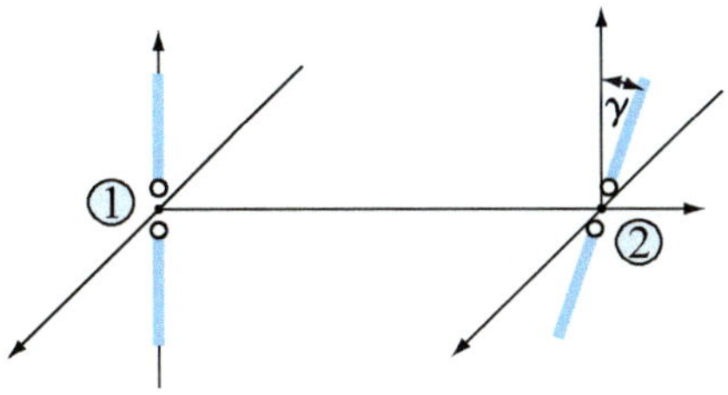

Figure 12.24 Wireless link with two half-wave wire dipole antennas positioned in two parallel planes but nonaligned, with an angle, $\gamma \neq 0$, between the first dipole and the projection of the second dipole on the plane of the first one; for Conceptual Question 12.94.

CONCEPTUAL QUESTION 12.95 **Magnitude of the radiation function of a receiving antenna.** In computing the received voltage in the wireless system in Figure 12.24, the magnitude of the characteristic radiation function of the receiving antenna is taken to be

(A) unity.

(B) a function of γ.

(C) a constant (not a function of γ), but not unity.

CONCEPTUAL QUESTION 12.96 **Reception of a circularly polarized wave by a wire dipole.** At the transmit end of a wireless link, two half-wave wire dipole antennas positioned along the x- and y-axis, as shown in Figure 12.25(a), are fed with currents of complex rms intensities $\underline{I}_{01} = I_0$ and $\underline{I}_{02} = \mathrm{j}I_0$, respectively, and the same frequency, f. At the receive end, $r = 100\lambda$ away from the crossed dipoles, another half-wave dipole lies in the plane $x'y'$ and makes an angle γ with the x'-axis [Figure 12.25(b)]. The total complex rms voltage across the open terminals of the receiving dipole is given by

(A) $\underline{V}_{\mathrm{oc}} = \lambda \underline{E}_{\mathrm{t1}} \cos\gamma / \pi$,

(B) $\underline{V}_{\mathrm{oc}} = \lambda \underline{E}_{\mathrm{t1}} \sin\gamma / \pi$,

(C) $\underline{V}_{\mathrm{oc}} = \lambda \underline{E}_{\mathrm{t1}} \cos\gamma\,(1+\mathrm{j})/\pi$,

(D) $\underline{V}_{\mathrm{oc}} = \lambda \underline{E}_{\mathrm{t1}} \sin\gamma\,(1+\mathrm{j})/\pi$,

(E) $\underline{V}_{\mathrm{oc}} = \lambda \underline{E}_{\mathrm{t1}} (\cos\gamma + \sin\gamma)/\pi$,

(F) $\underline{V}_{oc} = \lambda \underline{E}_{t1}(\cos\gamma - j\sin\gamma)/\pi$,

where $\underline{E}_{t1}$ is the complex rms electric field intensity radiated by dipole 1 at the receive location (point O′) in Figure 12.25.

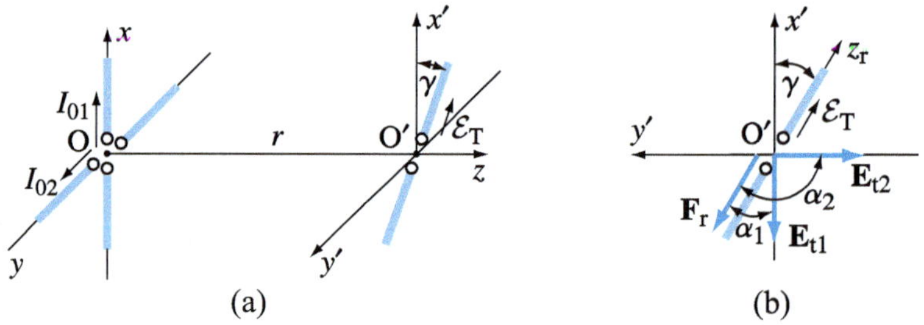

Figure 12.25 Wireless communication system with three half-wave dipole antennas: (a) two crossed dipoles radiating a circularly polarized electromagnetic wave ($\underline{I}_{02} = j\underline{I}_{01}$) and a receiving dipole in a transverse plane of the wave, and (b) detail at the receive end, showing the two incident electric field vectors and characteristic radiation function of the receiving antenna; for Conceptual Question 12.96.

CONCEPTUAL QUESTION 12.97 **Magnitude of the received voltage – due to a CP antenna system.** The magnitude of the total received voltage in Figure 12.25, $|\underline{V}_{oc}|$, is

(A) the same for all angles γ.

(B) a function of γ.

CONCEPTUAL QUESTION 12.98 **Two collinear or parallel half-wave dipoles above ground.** Both the transmitting and receiving antennas in a free-space wireless link at a frequency of $f = 500$ GHz are half-wave dipoles parallel to a PEC ground plane, at the same height, $h = 20$ m, with respect to it. The distance between dipole centers is $r = 40$ m, and the input power of the transmitting antenna (dipole 1) is $P_{in} = 1$ W. Let the open-circuit rms voltage (magnitude) of the receiving antenna (dipole 2) due to the direct and reflected waves be denoted by V_{direct} and $V_{reflected}$, respectively. How do these voltages compare to each other if the two dipoles are oriented to be (a) collinear (coaxial) with respect to each other, as in Figure 12.26(a), or (b) parallel to each other, as in Figure 12.26(b)?

(A) $V_{direct} > V_{reflected}$ in both cases (a) and (b).

(B) $V_{\text{direct}} > V_{\text{reflected}}$ in case (a) and $V_{\text{direct}} < V_{\text{reflected}}$ in case (b).

(C) $V_{\text{direct}} < V_{\text{reflected}}$ in case (a) and $V_{\text{direct}} > V_{\text{reflected}}$ in case (b).

(D) $V_{\text{direct}} < V_{\text{reflected}}$ in both cases (a) and (b).

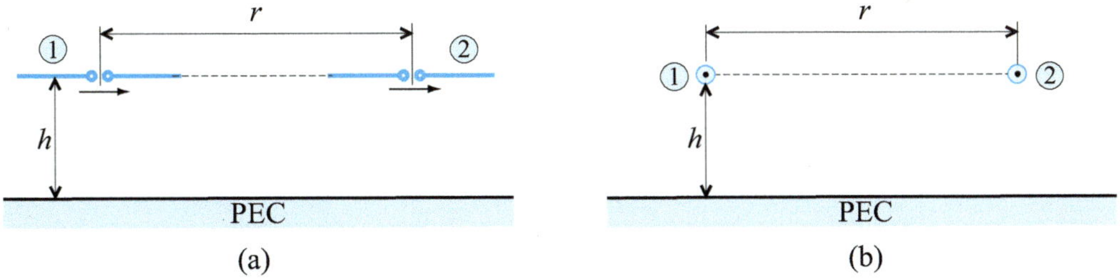

Figure 12.26 Communication between two half-wave dipole antennas above a ground plane: (a) collinear and (b) parallel dipoles; for Conceptual Question 12.98.

CONCEPTUAL QUESTION 12.99 Communication between two monopole antennas via ionosphere. A quarter-wave ($h = \lambda/4$) vertical wire monopole transmitting antenna (antenna 1) is fed at its base against the earth's surface by a time-harmonic current of frequency $f = 8$ MHz. The input power of the antenna is $P_{\text{in}} = 1$ kW. Both the earth and ionosphere can be considered as PEC planes, and the perpendicular distance between them, the so-called virtual height of the ionosphere (symbolized by h_{v}), is $h_{\text{v}} = 100$ km, as shown in Figure 12.27. The receiving antenna, at a receive point (P) on the ground at a distance $r = 500$ km from the transmitting antenna, is another quarter-wave monopole (antenna 2). There is no direct line of sight between the antennas. How should the receiving monopole be oriented (what should the angle α in Figure 12.27 be) to maximize the reception in this system?

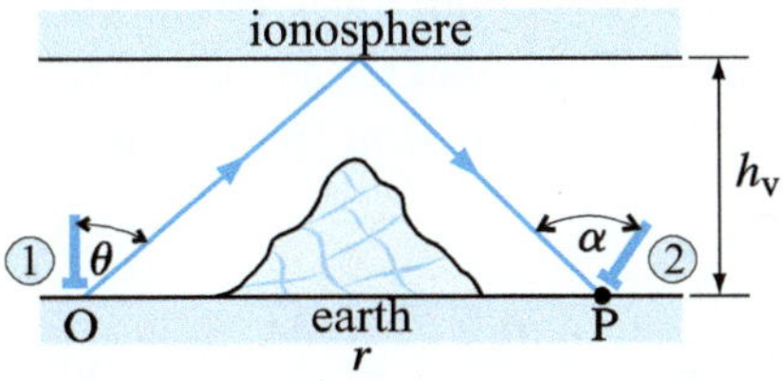

Figure 12.27 Communication system with two quarter-wave monopole antennas at the earth's surface in the presence of ionosphere and no direct line of sight: approximation of both the earth and ionosphere by PEC planes, separated by a distance equal to the virtual height of the ionosphere (h_{v}), and ray path of the ionospheric wave, bouncing off the ionosphere; for Conceptual Question 12.99.

(A) $\alpha = 0$.

(B) $\alpha = 90°$.

(C) $\alpha = 180°$.

(D) $\alpha = \theta$ (angle θ is shown in Figure 12.27).

(E) $\alpha = 90° - \theta$.

(F) any value (the received voltage does not depend on α).

12.9 Antenna Effective Aperture

One of the basic parameters of receiving antennas is the so-called effective aperture of an antenna in a given direction. With reference to Figure 12.20(a), it is defined as the ratio of the power received by the load at the antenna output terminals and the surface power density of the incoming electromagnetic wave. In other words, it represents a portion of the incident wavefront from which the antenna, effectively, extracts power and delivers it to the load, and can thus be thought of as the effective (equivalent) collecting area of the antenna. For given antenna parameters and incident wave, the time-average power delivered to the load is maximum under the conjugate matching condition, namely, when the load impedance is $\underline{Z}_{\mathrm{L}} = \underline{Z}_{\mathrm{T}}^* = \underline{Z}_{\mathrm{A}}^*$ in Figure 12.20(b), and this maximum received power amounts to $P_{\mathrm{r}} = P_{\mathrm{L}} = |\underline{\mathcal{E}}_{\mathrm{T}}|^2/(4R_{\mathrm{A}}) = |\underline{V}_{\mathrm{oc}}|^2/(4R_{\mathrm{A}})$, where $R_{\mathrm{A}} = R_{\mathrm{rad}} + R_{\mathrm{ohmic}}$ is the real part of the antenna input impedance, $\underline{Z}_{\mathrm{A}}$. To maximize the power further, we also assume that the antenna is polarization-matched to the wave, so that $|\underline{V}_{\mathrm{oc}}| = \lambda|\underline{\mathbf{E}}_0||\underline{\mathbf{F}}|/\pi$ [Eq. (12.19) with $\alpha = 0$ or $180°$], and $P_{\mathrm{r}} = \lambda^2|\underline{\mathbf{E}}_0|^2|\underline{\mathbf{F}}|^2/(4\pi^2 R_{\mathrm{A}})$. Using Eqs. (6.27), (6.26), and (7.12), the time-average Poynting vector of the incident wave is given by $(\mathcal{P}_{\mathrm{i}})_{\mathrm{ave}} = |\underline{\mathbf{E}}_0|^2/\eta$, with which P_{r} can be written as a product of $(\mathcal{P}_{\mathrm{i}})_{\mathrm{ave}}$ and a surface area (measured in m^2). This exactly is the antenna effective aperture, A_{eff}, which, having as well in mind the expression for the antenna gain, G, in Eq. (12.15), is computed as

$$P_{\mathrm{r}} = (\mathcal{P}_{\mathrm{i}})_{\mathrm{ave}} A_{\mathrm{eff}} \quad \longrightarrow \quad A_{\mathrm{eff}} = \frac{\lambda^2 \eta |\underline{\mathbf{F}}|^2}{4\pi^2 R_{\mathrm{A}}} = \frac{\lambda^2}{4\pi} G \quad \text{(antenna effective aperture)} . \tag{12.20}$$

Figure 12.28 illustrates the concept of effective aperture for three characteristic types of antennas. For antennas that have obvious physical apertures (openings), such as horn antennas and parabolic reflector antennas, A_{eff} can be written as $A_{\mathrm{eff}} = \eta_{\mathrm{aperture}} A_{\mathrm{physical}}$, where A_{physical} is the physical aperture area of the antenna and

η_{aperture} is a dimensionless coefficient termed the aperture efficiency ($0 \leq \eta_{\text{aperture}} \leq 1$). Combining this relationship with Eq. (12.20), we realize that an antenna has to be physically large, and thus have a large effective aperture, to attain a high gain. However, large A_{eff} is irrelevant if considered alone, isolated from the operating frequency of the wave, since $G \propto A_{\text{eff}}/\lambda^2$, and it is actually the electrical effective aperture, defined as the ratio of A_{eff} to the wavelength squared, which determines the gain, and vice versa.

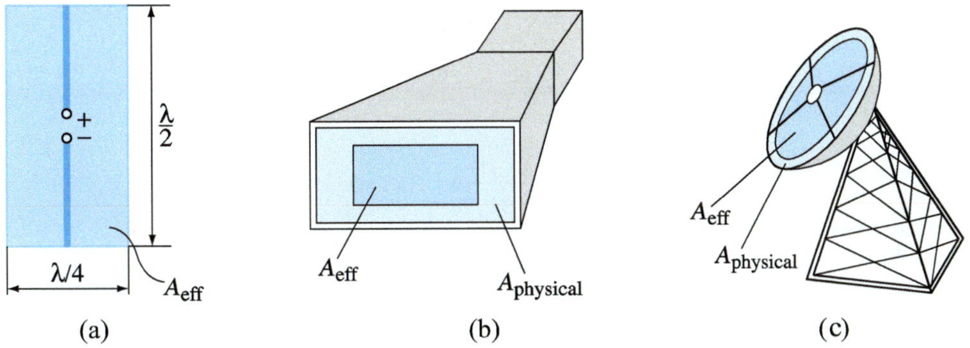

Figure 12.28 Illustration of the concept of antenna effective aperture (A_{eff}), Eq. (12.20), as an equivalent area from which an antenna extracts incident power and delivers it to a load (the presented shapes of A_{eff} are aimed to just illustrate the typical relative sizes of A_{eff} as compared to physical dimensions of antennas) – for a half-wave wire dipole (a), pyramidal horn antenna (b), and circular parabolic reflector antenna (c). For cases (b) and (c), the antenna physical aperture area (A_{physical}) is also shown.

CONCEPTUAL QUESTION 12.100 Load impedance for the maximum power transfer. Consider an arbitrary receiving antenna illuminated in air by a uniform plane time-harmonic wave and terminated in a load of complex impedance $\underline{Z}_{\text{L}}$. Let $\underline{Z}_{\text{A}} = R_{\text{A}} + \text{j}X_{\text{A}}$ be the complex input impedance of the antenna from its transmitting mode of operation. The time-average power delivered to the load, P_{L}, is maximum when the load impedance is given by

(A) $\underline{Z}_{\text{L}} = R_{\text{A}} + \text{j}0$.

(B) $\underline{Z}_{\text{L}} = R_{\text{A}} + \text{j}X_{\text{A}}$.

(C) $\underline{Z}_{\text{L}} = R_{\text{A}} - \text{j}X_{\text{A}}$.

(D) $\underline{Z}_{\text{L}} = |\underline{Z}_{\text{A}}| + \text{j}0$.

(E) $\underline{Z}_{\text{L}} = (377 + \text{j}0)\ \Omega$.

(F) $\underline{Z}_{\text{L}} = (50 + \text{j}0)\ \Omega$.

CONCEPTUAL QUESTION 12.101 **Effective aperture of a receiving antenna.** In a wireless link, the distance between a transmitting antenna and a receiving antenna is $r = 100$ m and the radiated wavelength is $\lambda = 50$ cm. The effective aperture of the receiving antenna is $A_{\text{eff}} = 1\ \text{m}^2$ and the time-average intensity of the Poynting vector at the receiving location is $\mathcal{P}_{\text{ave}} = 1\ \text{mW/m}^2$. The total time-average power received is

(A) $P_{\text{received}} = 10$ W.

(B) $P_{\text{received}} = 125.66$ W.

(C) $P_{\text{received}} = 1$ mW.

(D) $P_{\text{received}} = 250\ \mu$W.

(E) Need more information.

CONCEPTUAL QUESTION 12.102 **Effective aperture vs. dimensions of a horn antenna.** An increase in the dimensions of the aperture (opening) of a pyramidal horn antenna results in an increase in the effective aperture of the antenna.

(A) True.

(B) False.

CONCEPTUAL QUESTION 12.103 **Effective aperture of a wire dipole antenna.** The effective aperture of a half-wave wire dipole antenna cannot be defined since it does not have any physical aperture (opening) and is not a surface antenna.

(A) True.

(B) False.

CONCEPTUAL QUESTION 12.104 **Relationship between the effective aperture and gain.** Two antennas operate at different frequencies and have different effective apertures. For these two antennas, we can say that

(A) the one with the larger effective aperture also has a higher gain.

(B) the one with the larger effective aperture has a lower gain.

(C) Need more information.

CONCEPTUAL QUESTION 12.105 **Gain vs. electrical size of an antenna.** In general, antennas with high gains (and narrow beams in radiation patterns) are electrically

(A) large.

(B) small.

(C) medium sized.

12.10 Friis Transmission Formula for a Wireless Link

In practice, we normally aim at ensuring antenna matching and orientation conditions that would maximize the power transfer in a wireless system, and in this section we restrict our attention to such an ideal case. Let us consider a general wireless communication link consisting of two antennas at a far distance r, Eq. (12.5), in free space, as shown in Figure 12.29. For an ideal case, we then assume an impedance match of the load to the receiving antenna, polarization match between the antennas, and that both antennas are aligned and pointed toward each other for maximum

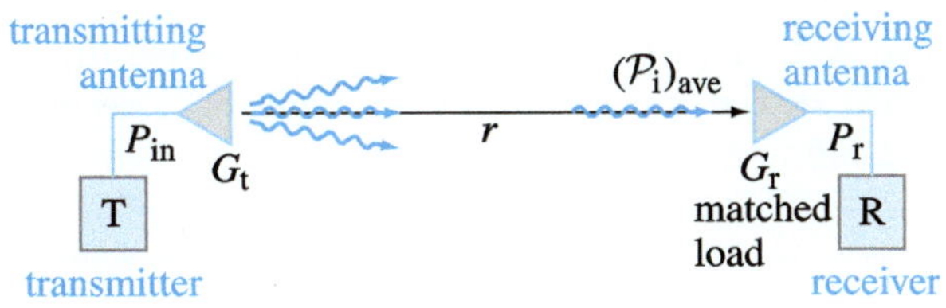

Figure 12.29 Wireless communication link with ideal antenna matching and orientation conditions (load impedance match to the receiving antenna, polarization match of antennas, and orientation of both antennas for maximum gains) – for deriving the Friis transmission formula, Eq. (12.21).

gains, which equal G_t and G_r for the transmitting and receiving antennas, respectively. To completely determine the power budget in this link, let us find the ratio of the time-average power received by the load, P_r, to the time-average input power that the transmitting antenna accepts at its terminals, P_{in}. Using Eqs. (12.20) and (12.15), both powers can be expressed in terms of the time-average Poynting vector magnitude $(\mathcal{P}_i)_{ave}$ of the wave radiated by the transmitting antenna computed at the receiving antenna terminals, $(\mathcal{P}_i)_{ave}$ thus being the connecting point between the two antennas, which is eliminated in the power ratio,

$$P_r = \frac{\lambda^2 G_r}{4\pi}(\mathcal{P}_i)_{ave} \quad \text{and} \quad P_{in} = \frac{4\pi r^2}{G_t}(\mathcal{P}_i)_{ave} \quad \longrightarrow \quad \frac{P_r}{P_{in}} = \left(\frac{\lambda}{4\pi r}\right)^2 G_t G_r$$

(Friis transmission formula) , (12.21)

where λ is the operating wavelength in the system. This expression for P_r/P_{in} is known as the Friis transmission formula, and it is extremely useful in evaluations of signal power levels in antenna communication systems, as part of the analysis or design of such systems.

CONCEPTUAL QUESTION 12.106 **Orientation of two antennas for maximum gains.** Although nonaligned, the two wire dipole antennas in the wireless system shown in Figure 12.30 are pointed toward each other for their maximum gains.

(A) True.

(B) False.

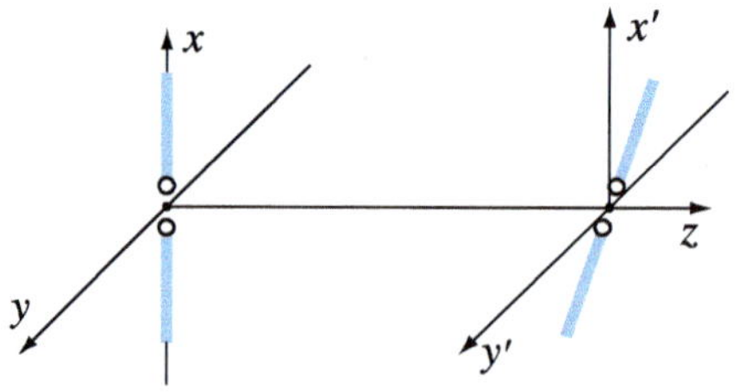

Figure 12.30 Wireless system with two nonaligned half-wave dipole antennas (the second antenna is positioned in the $x'y'$-plane, at an angle with respect to the x'-axis); for Conceptual Question 12.106.

CONCEPTUAL QUESTION 12.107 **Power budget of a radio link in terms of antenna gains.** In a radio link, the maximum gain of the transmitting antenna is G_t and that of the receiving antenna is G_r. The two antennas are pointed toward each other for maximum gains. The time-average power received

(A) is proportional to G_t/G_r.

(B) is proportional to G_tG_r.

(C) is proportional to G_t and does not depend on G_r.

(D) is proportional to G_r and does not depend on G_t.

(E) does not depend on either G_t or G_r.

CONCEPTUAL QUESTION 12.108 **Power budget of a satellite system using antenna effective apertures.** In a direct television satellite system, the transmitting and receiving antennas are pointed toward each other for maximum gains, and their effective apertures are $(A_{\text{eff}})_t$ and $(A_{\text{eff}})_r$, respectively. The ratio of the received and input powers in the system, P_r/P_{in},

(A) is proportional to $(A_{\text{eff}})_r / (A_{\text{eff}})_t$.

(B) is proportional to $(A_{\text{eff}})_t / (A_{\text{eff}})_r$.

(C) is proportional to $(A_{\text{eff}})_t (A_{\text{eff}})_r$.

(D) is proportional to $(A_{\text{eff}})_r$ and does not depend on $(A_{\text{eff}})_t$.

(E) is proportional to $1/ (A_{\text{eff}})_t$ and does not depend on $(A_{\text{eff}})_r$.

(F) does not depend on either $(A_{\text{eff}})_t$ or $(A_{\text{eff}})_r$.

CONCEPTUAL QUESTION 12.109 **Attenuation of a wireless link in decibels.** In decibels, the power budget in a free-space wireless link can be expressed as a positive attenuation, A_{dB}, between the transmit and receive ends in the link, namely, as $A_{\text{dB}} = 10\log(P_{\text{in}}/P_r)$, where P_{in} and P_r are the input and received powers, respectively, in the link. The dependence of A_{dB} on the electrical separation between the antennas, r/λ, can be described as

(A) $A_{\text{dB}} = \text{const} - 20\log(r/\lambda)$.

(B) $A_{\text{dB}} = \text{const} - 10\log(r/\lambda)$.

(C) $A_{\text{dB}} = \text{const} + 10\log(r/\lambda)$.

(D) $A_{\text{dB}} = \text{const} + 20\log(r/\lambda)$.

(E) another dependence on r/λ.

(F) A_{dB} does not depend on r/λ.

CONCEPTUAL QUESTION 12.110 **Attenuation in free space.** Consider the expression for the dB attenuation of a free-space wireless link, $A_{\text{dB}} = 10\log(P_{\text{in}}/P_{\text{r}})$, with P_{in} and P_{r} standing for the input and received powers, respectively, in the link. A part of this expression representing A_{dB} for $G_{\text{t}} = G_{\text{r}} = 1$ (both antennas isotropic radiators, with uniform radiation/reception in all directions) is referred to as the attenuation in free space and is denoted as $(A_{\text{dB}})_{\text{free space}}$, as it is completely independent of the particular antennas in the system. Having in mind that $20\log(4\pi) = 22$, $(A_{\text{dB}})_{\text{free space}}$ is always (for any separation between the antennas and any frequency, provided that the antennas are in the far field of each other)

(A) larger than

(B) equal to

(C) smaller than

22 dB.

CONCEPTUAL QUESTION 12.111 **Link attenuation in decibels in terms of dB gains of antennas.** The dB attenuation of a free-space wireless link, A_{dB}, can be expressed in terms of the attenuation in free space, $(A_{\text{dB}})_{\text{free space}}$, that is, A_{dB} for isotropic radiators, independent of the particular antennas in the system, and dB gains of the transmitting and receiving antennas, $(G_{\text{t}})_{\text{dB}}$ and $(G_{\text{r}})_{\text{dB}}$, respectively, as follows:

(A) $A_{\text{dB}} = (A_{\text{dB}})_{\text{free space}} + (G_{\text{t}})_{\text{dB}} + (G_{\text{r}})_{\text{dB}}$.

(B) $A_{\text{dB}} = (A_{\text{dB}})_{\text{free space}} + (G_{\text{t}})_{\text{dB}} - (G_{\text{r}})_{\text{dB}}$.

(C) $A_{\text{dB}} = (A_{\text{dB}})_{\text{free space}} - (G_{\text{t}})_{\text{dB}} + (G_{\text{r}})_{\text{dB}}$.

(D) $A_{\text{dB}} = (A_{\text{dB}})_{\text{free space}} - (G_{\text{t}})_{\text{dB}} - (G_{\text{r}})_{\text{dB}}$.

(E) None of the above.

CONCEPTUAL QUESTION 12.112 **Distance doubled and frequency halved in a radio link.** Assume that the distance between the transmitting and receiving antennas in a radio link is doubled and that the operating frequency of the link is halved, as well as that the gains of both antennas do not change upon the reduction in frequency. As a result, the dB attenuation, A_{dB}, between the transmit and receive ends in the link

(A) increases by 6 dB.

(B) increases by 3 dB.

(C) remains the same.

(D) decreases by 3 dB.

(E) decreases by 6 dB.

(F) decreases by 12 dB.

CONCEPTUAL QUESTION 12.113 **Both distance and frequency halved.** If the separation between the transmitting and receiving antennas in a wireless link is halved, along with halving the frequency of the link (antenna gains are unchanged), the attenuation A_{dB} in the link

(A) increases by 6 dB.

(B) increases by 3 dB.

(C) remains the same.

(D) decreases by 3 dB.

(E) decreases by 6 dB.

(F) decreases by 12 dB.

12.11 Antenna Arrays

Antenna arrays are spatial arrangements of identical antennas (array elements), equally oriented in space (e.g., wire array elements are parallel to each other or collinear), and excited independently, with feed currents of generally different magnitudes and phases, but of the same frequency. Most frequently, centers of array elements lie along a straight line – linear arrays. We consider a linear array of N point sources (having uniform radiation in all directions), with complex rms current intensities $\underline{I}_k = I_k\,\mathrm{e}^{\mathrm{j}\alpha_k}$ ($I_k = |\underline{I}_k|$), $k = 1, 2, \ldots, N$, placed along the z-axis and radiating in an ambient medium for which the phase coefficient is β, as shown in Figure 12.31. The locations of sources in the array are defined by their coordinates z_k ($k = 1, 2, \ldots, N$), which can be arbitrary. This array is simply an antenna with discrete spatial current distribution, and we can evaluate its radiation (far-zone) magnetic vector potential, $\underline{\mathbf{A}}$, using the discrete form of the radiation integral in Eq. (12.8), with the integral along the length l of the antenna in Figure 12.3 now becoming a sum over array constituents,

$$\underline{F}_{\mathrm{a}} = \mathrm{AF} = \sum_{k=1}^{N} \underline{I}_k\,\mathrm{e}^{\mathrm{j}\beta z_k \cos\theta} = \sum_{k=1}^{N} I_k\,\mathrm{e}^{\mathrm{j}(\beta z_k \cos\theta + \alpha_k)} \quad \text{(array factor)}\,. \tag{12.22}$$

Based on Eq. (12.8), the total $\underline{\mathbf{A}}$ due to an antenna array can be represented as $\underline{\mathbf{A}} = \underline{\mathbf{A}}_{\mathrm{element}}\,\underline{F}_{\mathrm{a}}$, namely, as a product of the vector potential $\underline{\mathbf{A}}_{\mathrm{element}}$ that a single array element antenna (point source in this case) with current $\underline{I}_0 = 1\,\mathrm{e}^{\mathrm{j}0}$ A would radiate if placed at the coordinate origin (reference point) and a complex scalar function $\underline{F}_{\mathrm{a}}$, in Eq. (12.22). This function provides a complete far-field characterization of the array itself (regardless of the characteristics of its elements), and is called accordingly the array factor (AF). Most importantly, $\underline{\mathbf{A}}_{\mathrm{element}}$ can be replaced by the far-zone potential due to an arbitrary antenna as the reference antenna (array element).

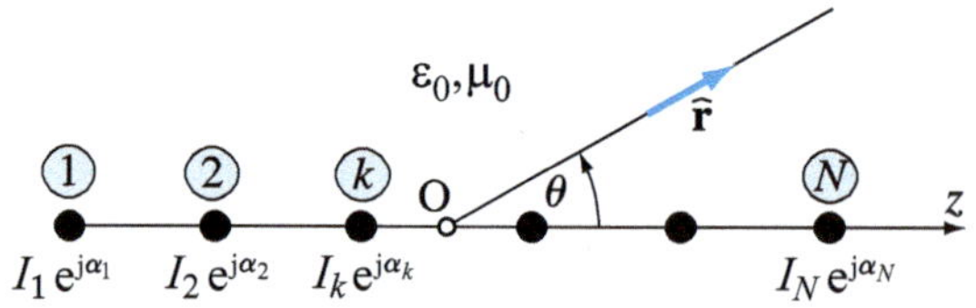

Figure 12.31 Linear antenna array with N point source elements along the z-axis radiating in free space.

Alternatively, the pattern of an antenna array can be predicted also by tracking the individual electromagnetic waves (rays) launched by the array elements, as indicated in Figure 12.32 in an example of a two-element array of point sources with feed currents of equal magnitudes and initial phases ($I_1 = I_2$ and $\alpha_1 = \alpha_2 = 0$), and a half-wave interelement spacing, $d = \lambda/2$ [Figure 12.32(a)]. Namely, looking into the positive z direction, wave 1 (emanating from source 1) travels a path equal to $d = \lambda/2$, and thus acquires an extra phase factor of $\mathrm{e}^{-\mathrm{j}\beta d} = \mathrm{e}^{-\mathrm{j}\pi} = -1$, before it reaches source 2 and joins wave 2. Consequently, the two waves propagating from

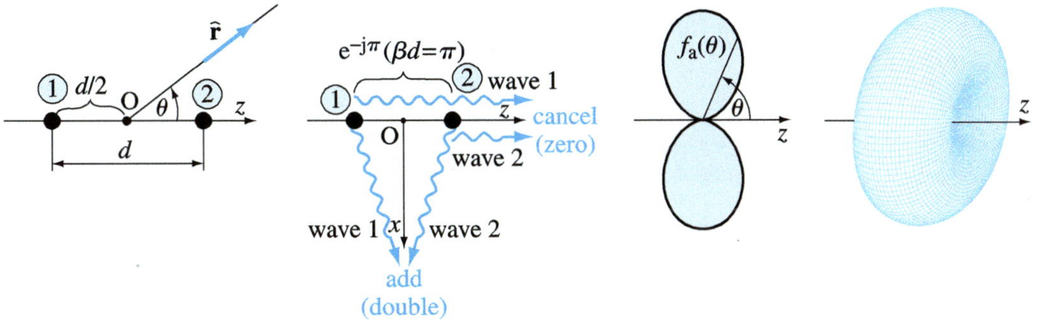

(a) (b) (c) (d)

Figure 12.32 Radiation analysis of a broadside two-element antenna array (point sources with equal input powers, in phase, and half-wave apart): (a) array geometry, (b) determination of nulls and maxima of the array radiation in directions of perfect cancelations and additions, respectively, of individual waves launched by array elements, (c) 2-D polar plot of the normalized array factor, and (d) 3-D radiation pattern.

there on together to the right in Figure 12.32(b) arrive at the far zone in counterphase (due to the half-wave interelement spacing) and with equal magnitudes (coming from the equal magnitudes of feed currents), causing a perfect cancelation of their fields and pattern null for $\theta = 0$. The same occurs in the negative z direction ($\theta = 180°$). On the other hand, for any direction normal to the array axis (so-called broadside direction) in Figure 12.32(b), waves 1 and 2 travel exactly the same paths from the respective sources to a far-field point. In addition, given the equal initial phases of the feed currents, the rays arrive at the far field in phase, so that their fields add up together constructively (the total field is double that of one source). In other words, we have a perfect addition of far fields and pattern maximum (main lobe) for $\theta = 90°$, and hence this array is termed a broadside antenna array. Finally, since the phase difference between waves 1 and 2 varies smoothly from 0 to 180° as the observation point moves from a broadside direction to an axial direction (along the array axis) over a far-zone sphere centered at the coordinate origin, there is a smooth pattern variation between the maximum and null points, making up the pattern curve and 3-D plot in Figures 12.32(c) and (d), respectively. Of course, the same patterns are obtained using the array factor in Eq. (12.22).

The far electric field vector, $\underline{\mathbf{E}}$, due to the antenna array is obtained from the potential $\underline{\mathbf{A}}$, as for any transmitting antenna, using Eqs. (12.9). Since $\underline{F}_{\mathrm{a}}$ in Eq. (12.22) is a scalar, taking the transverse components of $\underline{\mathbf{A}}$ in Eqs. (12.9) applies only to $\underline{\mathbf{A}}_{\mathrm{element}}$, and hence the total field of the array is the field of the reference element, $\underline{\mathbf{E}}_{\mathrm{element}}$, times the array factor. Having in mind Eq. (12.12), the same can be written for the characteristic radiation function of the antenna array,

$$\underline{\mathbf{F}} = \underline{\mathbf{F}}_{\mathrm{element}}\, \underline{F}_{\mathrm{a}} \quad \text{(pattern multiplication for antenna arrays)}\,, \tag{12.23}$$

and this is known as the pattern multiplication theorem for antenna arrays. Simply, multiplying the element radiation function (pattern) and array factor we obtain the overall array pattern.

CONCEPTUAL QUESTION 12.114 **Array with a half-wave separation between in-phase sources.** Consider a two-element array of point sources (isotropic radiators) in free space, fed with time-harmonic currents of equal complex intensities, so equal magnitudes and phases, and spaced a half-wavelength apart ($d = \lambda/2$), as shown in Figure 12.33. In a spherical coordinate system adopted as in Figure 12.33, the array factor of this array is given by

(A) $\underline{F}_a(\theta) = e^{-j\pi\cos\theta} + e^{j\pi\cos\theta}$.

(B) $\underline{F}_a(\theta) = -e^{-j\pi\cos\theta} + e^{j\pi\cos\theta}$.

(C) $\underline{F}_a(\theta) = e^{-j(\pi/2)\cos\theta} + e^{j(\pi/2)\cos\theta}$.

(D) $\underline{F}_a(\theta) = -e^{-j(\pi/2)\cos\theta} + e^{j(\pi/2)\cos\theta}$.

(E) $\underline{F}_a(\theta) = e^{-j[(\pi/4)\cos\theta-\pi/4]} + e^{j[(\pi/4)\cos\theta-\pi/4]}$.

(F) $\underline{F}_a(\theta) = e^{-j(\pi/4)\cos\theta} + e^{j(\pi/4)\cos\theta}$.

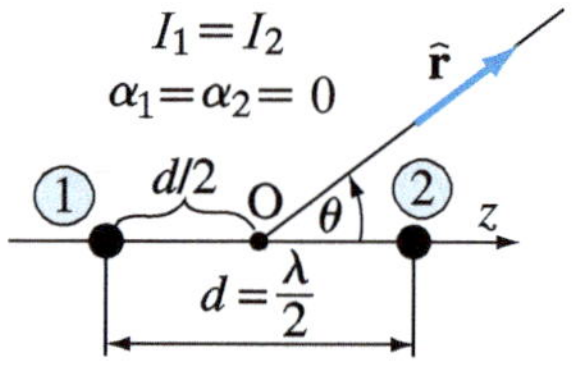

Figure 12.33 Two-element array of point sources with half-wave interelement spacing and in-phase excitation; for Conceptual Question 12.114.

CONCEPTUAL QUESTION 12.115 **Half-wave interelement spacing and counter-phase excitation.** The array factor of a two-element array of point sources in counter-phase, namely, with the phase shift between feed currents of array elements amounting to $\delta = \alpha_2 - \alpha_1 = \pm 180°$, whereas the elements are spaced a half-wavelength apart and fed with same current magnitudes, Figure 12.34, comes out to be

(A) $\underline{F}_a(\theta) = e^{-j\pi\cos\theta} + e^{j\pi\cos\theta}$.

(B) $\underline{F}_a(\theta) = -e^{-j\pi\cos\theta} + e^{j\pi\cos\theta}$.

(C) $\underline{F}_a(\theta) = e^{-j(\pi/2)\cos\theta} + e^{j(\pi/2)\cos\theta}$.

(D) $\underline{F}_{\rm a}(\theta) = -\,{\rm e}^{-{\rm j}(\pi/2)\cos\theta} + {\rm e}^{{\rm j}(\pi/2)\cos\theta}$.

(E) $\underline{F}_{\rm a}(\theta) = {\rm e}^{-{\rm j}[(\pi/4)\cos\theta-\pi/4]} + {\rm e}^{{\rm j}[(\pi/4)\cos\theta-\pi/4]}$.

(F) $\underline{F}_{\rm a}(\theta) = {\rm e}^{-{\rm j}(\pi/4)\cos\theta} + {\rm e}^{{\rm j}(\pi/4)\cos\theta}$.

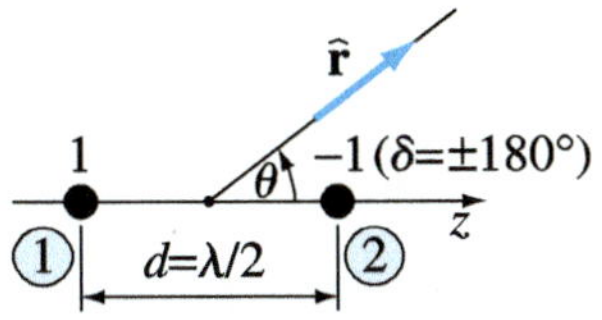

Figure 12.34 Array with a half-wave separation between point sources in counter-phase; for Conceptual Question 12.115.

CONCEPTUAL QUESTION 12.116 Full-wave separation between in-phase array elements. Consider a two-element array of point sources with a full-wave interelement spacing ($d = \lambda$); elements are fed with same current magnitudes and in phase, as depicted in Figure 12.35. The array factor is

(A) $\underline{F}_{\rm a}(\theta) = {\rm e}^{-{\rm j}\pi\cos\theta} + {\rm e}^{{\rm j}\pi\cos\theta}$.

(B) $\underline{F}_{\rm a}(\theta) = -\,{\rm e}^{-{\rm j}\pi\cos\theta} + {\rm e}^{{\rm j}\pi\cos\theta}$.

(C) $\underline{F}_{\rm a}(\theta) = {\rm e}^{-{\rm j}(\pi/2)\cos\theta} + {\rm e}^{{\rm j}(\pi/2)\cos\theta}$.

(D) $\underline{F}_{\rm a}(\theta) = -\,{\rm e}^{-{\rm j}(\pi/2)\cos\theta} + {\rm e}^{{\rm j}(\pi/2)\cos\theta}$.

(E) $\underline{F}_{\rm a}(\theta) = {\rm e}^{-{\rm j}[(\pi/4)\cos\theta-\pi/4]} + {\rm e}^{{\rm j}[(\pi/4)\cos\theta-\pi/4]}$.

(F) $\underline{F}_{\rm a}(\theta) = {\rm e}^{-{\rm j}(\pi/4)\cos\theta} + {\rm e}^{{\rm j}(\pi/4)\cos\theta}$.

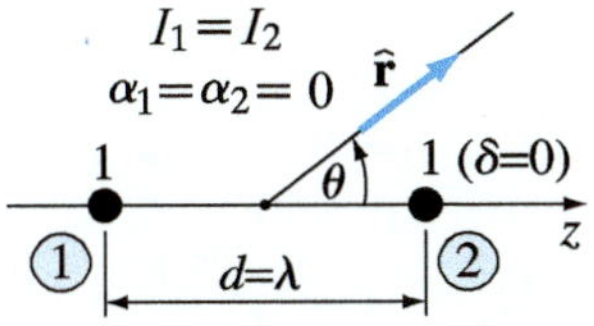

Figure 12.35 Array of point sources with full-wave separation and in-phase excitation; for Conceptual Question 12.116.

CONCEPTUAL QUESTION 12.117 **Point sources full-wave apart and fed in counter-phase.** The array factor of an array (along the z-axis) of two point sources with a full-wave separation in counter-phase ($d = \lambda$, $\delta = \pm 180°$, currents of equal magnitudes) can be expressed – in a spherical coordinate system adopted as in Figure 12.35 – as

(A) $\underline{F}_{\rm a}(\theta) = {\rm e}^{-{\rm j}\pi\cos\theta} + {\rm e}^{{\rm j}\pi\cos\theta}$.

(B) $\underline{F}_{\rm a}(\theta) = -\,{\rm e}^{-{\rm j}\pi\cos\theta} + {\rm e}^{{\rm j}\pi\cos\theta}$.

(C) $\underline{F}_{\rm a}(\theta) = {\rm e}^{-{\rm j}(\pi/2)\cos\theta} + {\rm e}^{{\rm j}(\pi/2)\cos\theta}$.

(D) $\underline{F}_{\rm a}(\theta) = -\,{\rm e}^{-{\rm j}(\pi/2)\cos\theta} + {\rm e}^{{\rm j}(\pi/2)\cos\theta}$.

(E) $\underline{F}_{\rm a}(\theta) = {\rm e}^{-{\rm j}[(\pi/4)\cos\theta-\pi/4]} + {\rm e}^{{\rm j}[(\pi/4)\cos\theta-\pi/4]}$.

(F) $\underline{F}_{\rm a}(\theta) = {\rm e}^{-{\rm j}(\pi/4)\cos\theta} + {\rm e}^{{\rm j}(\pi/4)\cos\theta}$.

CONCEPTUAL QUESTION 12.118 **Array elements quarter-wave apart and in time-phase quadrature.** For a two-element array of point sources with a quarter-wave interelement spacing ($d = \lambda/4$) and in time-phase quadrature, where element 2 lags by 90° in phase with respect to element 1, $\delta = \alpha_2 - \alpha_1 = -90°$ (elements are fed with same current magnitudes), Figure 12.36, we have that

(A) $\underline{F}_{\rm a}(\theta) = {\rm e}^{-{\rm j}\pi\cos\theta} + {\rm e}^{{\rm j}\pi\cos\theta}$.

(B) $\underline{F}_{\rm a}(\theta) = -\,{\rm e}^{-{\rm j}\pi\cos\theta} + {\rm e}^{{\rm j}\pi\cos\theta}$.

(C) $\underline{F}_{\rm a}(\theta) = {\rm e}^{-{\rm j}(\pi/2)\cos\theta} + {\rm e}^{{\rm j}(\pi/2)\cos\theta}$.

(D) $\underline{F}_{\rm a}(\theta) = -\,{\rm e}^{-{\rm j}(\pi/2)\cos\theta} + {\rm e}^{{\rm j}(\pi/2)\cos\theta}$.

(E) $\underline{F}_{\rm a}(\theta) = {\rm e}^{-{\rm j}[(\pi/4)\cos\theta-\pi/4]} + {\rm e}^{{\rm j}[(\pi/4)\cos\theta-\pi/4]}$.

(F) $\underline{F}_{\rm a}(\theta) = {\rm e}^{-{\rm j}(\pi/4)\cos\theta} + {\rm e}^{{\rm j}(\pi/4)\cos\theta}$.

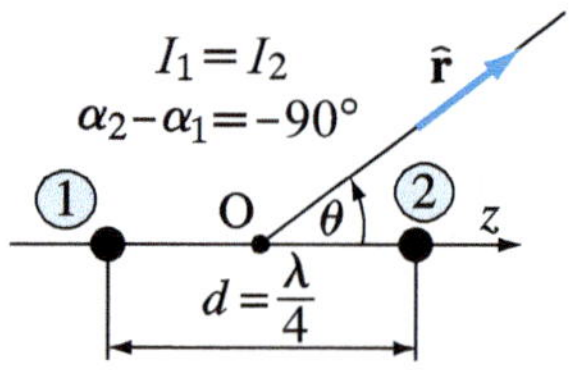

Figure 12.36 Two-element array of point sources with quarter-wave separation and in time-phase quadrature; for Conceptual Question 12.118.

CONCEPTUAL QUESTION 12.119 **Normalized array factor of a two-element antenna array.** The function $f_{\rm a}(\theta) = \cos[(\pi/2)\cos\theta]$ is the normalized array factor of an array whose factor $\underline{F}_{\rm a}(\theta)$ is given as one of the following six expressions. Which one?

(A) $\underline{F}_{\rm a}(\theta) = {\rm e}^{-{\rm j}\pi\cos\theta} + {\rm e}^{{\rm j}\pi\cos\theta}$.

(B) $\underline{F}_{\rm a}(\theta) = -\,{\rm e}^{-{\rm j}\pi\cos\theta} + {\rm e}^{{\rm j}\pi\cos\theta}$.

(C) $\underline{F}_{\rm a}(\theta) = {\rm e}^{-{\rm j}(\pi/2)\cos\theta} + {\rm e}^{{\rm j}(\pi/2)\cos\theta}$.

(D) $\underline{F}_{\rm a}(\theta) = -\,{\rm e}^{-{\rm j}(\pi/2)\cos\theta} + {\rm e}^{{\rm j}(\pi/2)\cos\theta}$.

(E) $\underline{F}_{\rm a}(\theta) = {\rm e}^{-{\rm j}[(\pi/4)\cos\theta-\pi/4]} + {\rm e}^{{\rm j}[(\pi/4)\cos\theta-\pi/4]}$.

(F) $\underline{F}_{\rm a}(\theta) = {\rm e}^{-{\rm j}(\pi/4)\cos\theta} + {\rm e}^{{\rm j}(\pi/4)\cos\theta}$.

CONCEPTUAL QUESTION 12.120 **Normalized array factor of another two-element array.** Consider the normalized array factor $f_{\rm a}(\theta) = |\sin[(\pi/2)\cos\theta]|$. Which one is the corresponding function $\underline{F}_{\rm a}(\theta)$?

(A) $\underline{F}_{\rm a}(\theta) = {\rm e}^{-{\rm j}\pi\cos\theta} + {\rm e}^{{\rm j}\pi\cos\theta}$.

(B) $\underline{F}_{\rm a}(\theta) = -\,{\rm e}^{-{\rm j}\pi\cos\theta} + {\rm e}^{{\rm j}\pi\cos\theta}$.

(C) $\underline{F}_{\rm a}(\theta) = {\rm e}^{-{\rm j}(\pi/2)\cos\theta} + {\rm e}^{{\rm j}(\pi/2)\cos\theta}$.

(D) $\underline{F}_{\rm a}(\theta) = -\,{\rm e}^{-{\rm j}(\pi/2)\cos\theta} + {\rm e}^{{\rm j}(\pi/2)\cos\theta}$.

(E) $\underline{F}_{\rm a}(\theta) = {\rm e}^{-{\rm j}[(\pi/4)\cos\theta-\pi/4]} + {\rm e}^{{\rm j}[(\pi/4)\cos\theta-\pi/4]}$.

(F) $\underline{F}_{\rm a}(\theta) = {\rm e}^{-{\rm j}(\pi/4)\cos\theta} + {\rm e}^{{\rm j}(\pi/4)\cos\theta}$.

CONCEPTUAL QUESTION 12.121 **Plotting the array factor, half-wave spacing, in-phase currents.** The normalized array factor $f_{\rm a}(\theta) = \cos[(\pi/2)\cos\theta]$ is plotted in

(A) Figure 12.37(a).

(B) Figure 12.37(b).

(C) Figure 12.37(c).

(D) Figure 12.37(d).

(E) Figure 12.37(e).

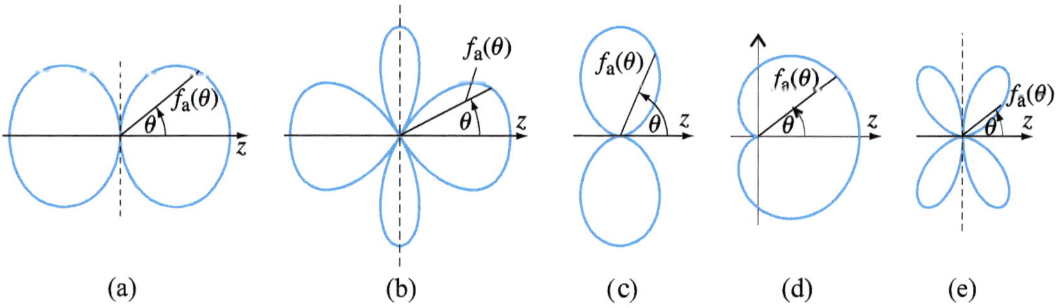

Figure 12.37 2-D polar plots of the normalized array factor, $f_a(\theta)$, for five different two-element arrays; for Conceptual Question 12.121.

CONCEPTUAL QUESTION 12.122 **Array factor plot, half-wave spacing, counter-phase currents.** The plot of the normalized array factor $f_a(\theta) = |\sin[(\pi/2)\cos\theta]|$ is as shown in

(A) Figure 12.37(a).

(B) Figure 12.37(b).

(C) Figure 12.37(c).

(D) Figure 12.37(d).

(E) Figure 12.37(e).

CONCEPTUAL QUESTION 12.123 **Array factor plot, full-wave spacing, in-phase currents.** Which one of the five offered 2-D polar plots in Figure 12.37 corresponds to the normalized array factor given by $f_a(\theta) = |\cos(\pi\cos\theta)|$?

(A) The one in Figure 12.37(a).

(B) The one in Figure 12.37(b).

(C) The one in Figure 12.37(c).

(D) The one in Figure 12.37(d).

(E) The one in Figure 12.37(e).

CONCEPTUAL QUESTION 12.124 **Array factor plot, full-wave spacing, counter-phase currents.** For the normalized array factor $f_a(\theta) = |\sin(\pi \cos\theta)|$, the plot is as shown in

(A) Figure 12.37(a).

(B) Figure 12.37(b).

(C) Figure 12.37(c).

(D) Figure 12.37(d).

(E) Figure 12.37(e).

CONCEPTUAL QUESTION 12.125 **Array factor plot, quarter-wave spacing, currents in quadrature.** The plot of $f_a(\theta) = \cos[(\pi/4)(\cos\theta - 1)]$ is given in

(A) Figure 12.37(a).

(B) Figure 12.37(b).

(C) Figure 12.37(c).

(D) Figure 12.37(d).

(E) Figure 12.37(e).

CONCEPTUAL QUESTION 12.126 **2-D vs. 3-D radiation pattern plots for an array.** Which one of the five 2-D polar plots of normalized array factors in Figure 12.37 represents a cut of the 3-D pattern plot shown in Figures 12.38?

(A) Figure 12.37(a).

(B) Figure 12.37(b).

(C) Figure 12.37(c).

(D) Figure 12.37(d).

(E) Figure 12.37(e).

(F) None of the plots.

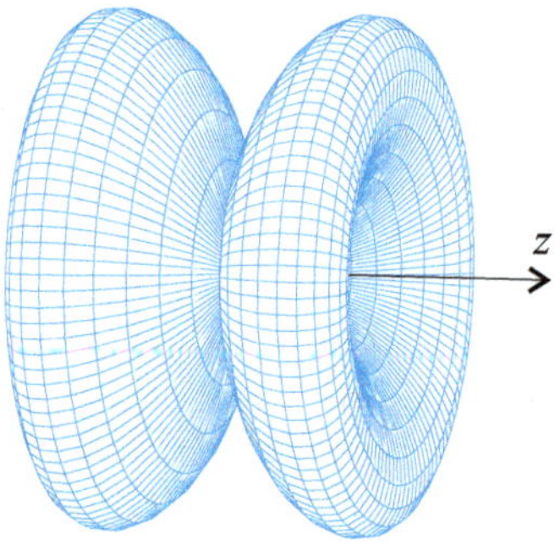

Figure 12.38 3-D polar plot of the normalized array factor of a two-element array; for Conceptual Question 12.126.

CONCEPTUAL QUESTION 12.127 **Broadside antenna array.** Consider a two-element array of point sources fed with equal powers and spaced a half-wavelength apart. What phase shift between feed currents of the array elements, $\delta = \alpha_2 - \alpha_1$, makes this a broadside array, with peak radiation in directions normal to the array axis?

(A) $\delta = -90°$.

(B) $\delta = 0$.

(C) $\delta = 45°$.

(D) $\delta = 90°$.

(E) $\delta = 180°$.

(F) None of the above.

CONCEPTUAL QUESTION 12.128 **Endfire antenna array.** Determine the phase shift δ that results in main lobe maxima along the array axis, that is, in an endfire array.

(A) $\delta = -90°$.

(B) $\delta = 0$.

(C) $\delta = 45°$.

(D) $\delta = 90°$.

(E) $\delta = 180°$.

(F) None of the above.

CONCEPTUAL QUESTION 12.129 **Tracking waves launched by point sources, endfire radiation.** Figure 12.39 illustrates the technique of tracking individual waves launched by array elements to identify directions of their perfect cancelations and additions (nulls and maxima of the array radiation pattern) applied to a two-element array of point sources with feed currents of equal magnitudes in free space. Based on the figure, what are the interelement spacing and phase shift, d and $\delta = \alpha_2 - \alpha_1$, respectively, for this array?

(A) $d = \lambda$ and $\delta = 0$.

(B) $d = \lambda$ and $\delta = 180°$.

(C) $d = \lambda/2$ and $\delta = 0$.

(D) $d = \lambda/2$ and $\delta = 180°$.

(E) $d = \lambda/4$ and $\delta = 0$.

(F) $d = \lambda/4$ and $\delta = -90°$.

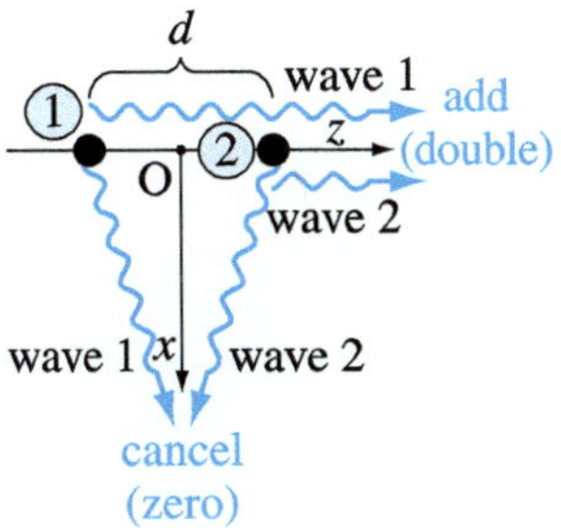

Figure 12.39 Radiation analysis of a two-element antenna array by tracking individual waves emanating from array elements (point sources); for Conceptual Question 12.129.

CONCEPTUAL QUESTION 12.130 **Tracking waves launched by point sources, cardioid pattern.** Find d and δ of the array whose wave tracking diagram is shown in Figure 12.40. Note that the resulting 2-D radiation pattern curve in a plane containing the array axis (z-axis) is the so-called cardioid pattern (it resembles the heart shape).

(A) $d = \lambda$ and $\delta = 0$.

(B) $d = \lambda$ and $\delta = 180°$.

(C) $d = \lambda/2$ and $\delta = 0$.

(D) $d = \lambda/2$ and $\delta = 180°$.

(E) $d = \lambda/4$ and $\delta = 0$.

(F) $d = \lambda/4$ and $\delta = -90°$.

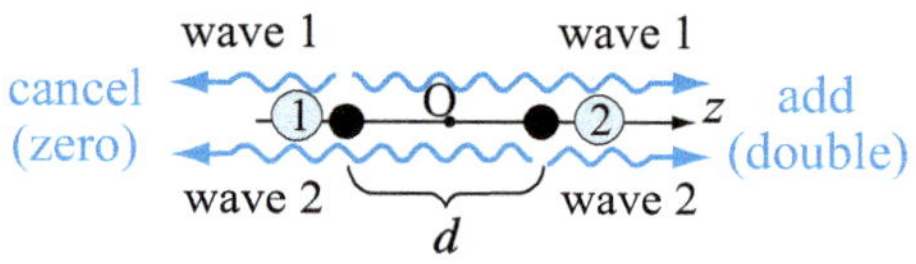

Figure 12.40 Radiation analysis of an array by tracking waves launched by point sources; for Conceptual Question 12.130.

CONCEPTUAL QUESTION 12.131 **Tracking waves launched by point sources, grating lobes.** Which d and δ are for the array with the wave tracking diagram shown in Figure 12.41? Note that, assuming that this array is principally meant as a broadside array, with a broadside 3-D main lobe, its endfire radiation lobes are considered as side lobes. However, they are of the same intensity at their peaks as the main lobe. Such additional "main" lobes are called grating lobes, and in the majority of array applications they are undesirable.

(A) $d = \lambda$ and $\delta = 0$.

(B) $d = \lambda$ and $\delta = 180°$.

(C) $d = \lambda/2$ and $\delta = 0$.

(D) $d = \lambda/2$ and $\delta = 180°$.

(E) $d = \lambda/4$ and $\delta = 0$.

(F) $d = \lambda/4$ and $\delta = -90°$.

CONCEPTUAL QUESTION 12.132 **Finding a radiation pattern null at an oblique angle.** What is the angle θ_0 in Figure 12.41, defining the direction of a radiation pattern null for the array?

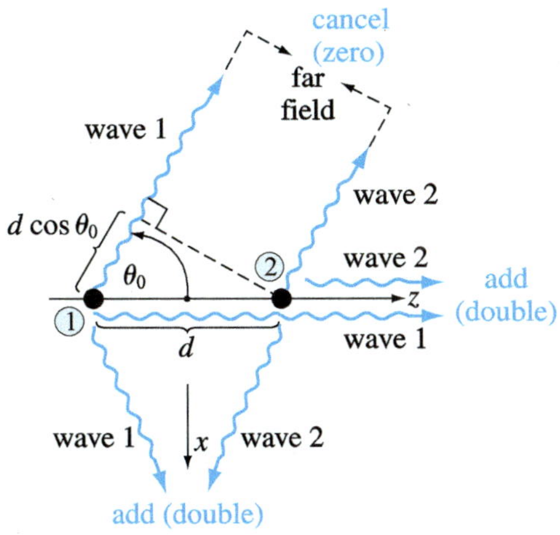

Figure 12.41 Array radiation analysis by tracking waves emanating from point sources; for Conceptual Question 12.131.

(A) $\theta_0 = 15°$.

(B) $\theta_0 = 30°$.

(C) $\theta_0 = 45°$.

(D) $\theta_0 = 60°$.

(E) $\theta_0 = 75°$.

(F) None of the above.

CONCEPTUAL QUESTION 12.133 **Tracking waves launched by point sources, multiple radiation lobes.** What are the interelement spacing and phase shift, d and δ, of the array for which Figure 12.42 illustrates the technique of tracking individual waves launched by array elements?

(A) $d = \lambda$ and $\delta = 0$.

(B) $d = \lambda$ and $\delta = 180°$.

(C) $d = \lambda/2$ and $\delta = 0$.

(D) $d = \lambda/2$ and $\delta = 180°$.

(E) $d = \lambda/4$ and $\delta = 0$.

(F) $d = \lambda/4$ and $\delta = -90°$.

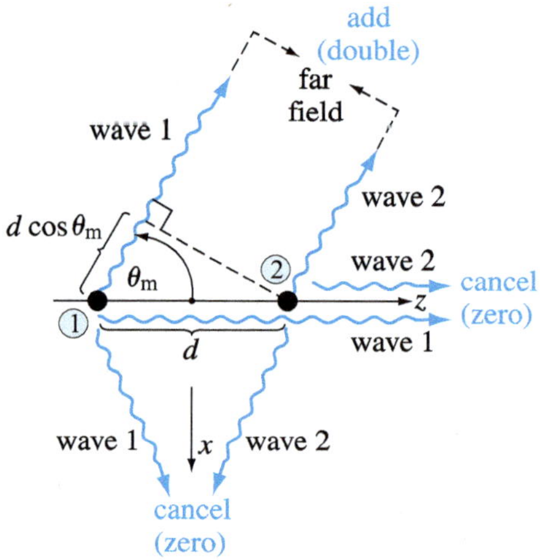

Figure 12.42 Tracking waves radiated by individual elements of an array; for Conceptual Question 12.133.

CONCEPTUAL QUESTION 12.134 **Number of radiation lobes with equal intensities at peaks.** How many radiation lobes of the same intensity at their peaks (pattern maxima) exist in the resulting 2-D radiation pattern (in a plane containing the z-axis) of the array in Figure 12.42?

(A) One.

(B) Two.

(C) Three.

(D) Four.

(E) Six.

(F) Eight.

CONCEPTUAL QUESTION 12.135 **Array of two parallel Hertzian dipoles – pattern multiplication.** An array of two parallel Hertzian dipole antennas (the dipole axes are perpendicular to the array axis) whose centers are spaced a half-wavelength apart radiates in free space, as depicted in Figure 12.43(a). The dipoles are fed with time-harmonic currents of equal complex intensities. Figures 12.43(b) and (c) show the plots of the element (Hertzian dipole) pattern and array factor, respectively, in the plane in which dipoles lie (xz-plane). The total radiation pattern of the array (of dipole antennas) plotted in the xz-plane is as shown in

(A) Figure 12.43(d).

(B) Figure 12.43(e).

(C) Figure 12.43(f).

(D) Figure 12.43(g).

(E) Figure 12.43(h).

(F) None of the above.

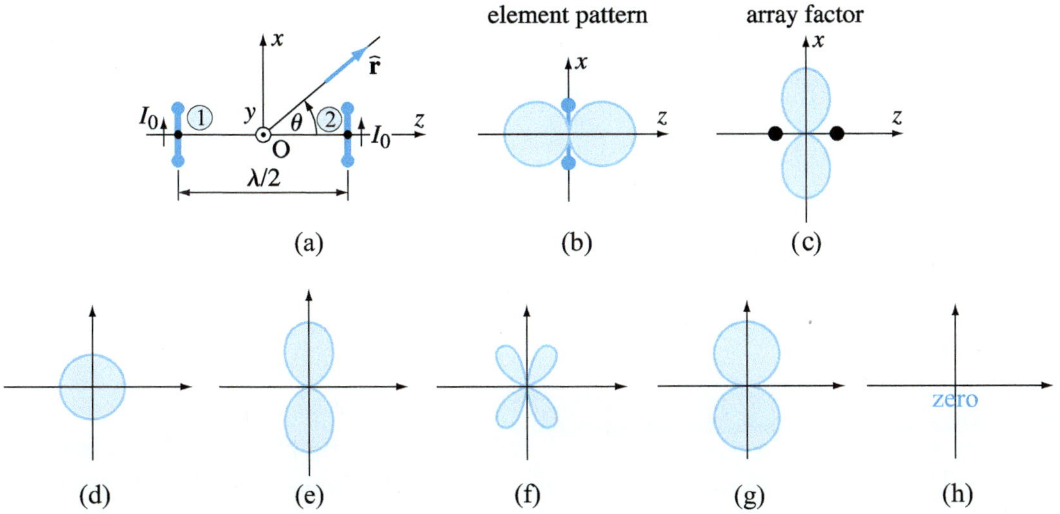

Figure 12.43 Radiation by an array of two parallel Hertzian dipole antennas: (a) array configuration and adopted coordinate system, (b) element pattern in the xz-plane, (c) array factor in the xz-plane, and (d)–(h) five offered plots for the total radiation pattern of the array of dipoles; for Conceptual Question 12.135.

CONCEPTUAL QUESTION 12.136 Array of parallel dipoles – element pattern in different cuts. For the array of parallel Hertzian dipole antennas in Figure 12.43(a), the element pattern plots (radiation patterns of a Hertzian dipole) in the plane perpendicular to dipoles (yz-plane) and the plane perpendicular to the array axis (xy-plane) are as shown in

(A) Figure 12.43(d) (plots are the same).

(B) Figure 12.43(g).

(C) Figure 12.43(d) and Figure 12.43(g), respectively.

(D) Figure 12.43(g) and Figure 12.43(d), respectively.

(E) Figure 12.43(d) and Figure 12.43(f), respectively.

(F) Figure 12.43(f) and Figure 12.43(g), respectively.

CONCEPTUAL QUESTION 12.137 **Array of parallel dipoles – array factor in different cuts.** For the array of parallel in-phase dipoles in Figure 12.43(a), the array factor plots in planes yz and xy are as shown in

(A) Figure 12.43(d) (plots are the same).

(B) Figure 12.43(e).

(C) Figure 12.43(d) and Figure 12.43(e), respectively.

(D) Figure 12.43(e) and Figure 12.43(d), respectively.

(E) Figure 12.43(d) and Figure 12.43(f), respectively.

(F) Figure 12.43(h) and Figure 12.43(d), respectively.

CONCEPTUAL QUESTION 12.138 **Array of parallel dipoles – total pattern in different cuts.** The total radiation pattern of the array of parallel in-phase Hertzian dipoles in Figure 12.43(a) plotted in planes yz and xy is as shown in

(A) Figure 12.43(d) and Figure 12.43(e),

(B) Figure 12.43(e) and Figure 12.43(d),

(C) Figure 12.43(h) and Figure 12.43(g),

(D) Figure 12.43(g) and Figure 12.43(f),

(E) Figure 12.43(e) and Figure 12.43(g),

(F) Figure 12.43(g) and Figure 12.43(e),

respectively.

CONCEPTUAL QUESTION 12.139 **Array of two parallel dipoles with counter-phase excitation.** Assume that the two parallel Hertzian dipole antennas in Figure 12.43(a) are fed in counter-phase. For this new situation, the total radiation pattern of the array in the xy-plane is as shown in

(A) Figure 12.43(d).

(B) Figure 12.43(e).

(C) Figure 12.43(f).

(D) Figure 12.43(g).

(E) Figure 12.43(h).

CONCEPTUAL QUESTION 12.140 **Array of three parallel dipoles – pattern multiplication, first cut.** Shown in Figure 12.44(a) is an array of three parallel half-wave wire dipole antennas. The dipoles all lie in one plane, with a quarter-wave separation between adjacent element points. The feed currents of adjacent dipoles are in time-phase quadrature (90° out of phase with respect to each other). The current magnitudes are in the ratio 1 : 2 : 1 along the array, so this is a nonuniform array – more precisely, a nonuniformly excited, equally spaced array. Figure 12.44(b) gives the plot of the element (half-wave dipole) pattern and Figure 12.44(c) depicts the array factor in the plane of dipoles (xz-plane). The cut of the overall array pattern (of the actual array of dipoles) in the xz-plane is as sketched in

(A) Figure 12.44(d).

(B) Figure 12.44(e).

(C) Figure 12.44(f).

(D) Figure 12.44(g).

(E) Figure 12.44(h).

CONCEPTUAL QUESTION 12.141 **Array of three parallel dipoles – pattern multiplication, second cut.** For the array of three parallel half-wave wire dipole antennas with $d = \lambda/4$, $\delta = 90°$, and $1 : 2 : 1$ current magnitude ratio in Figure 12.44(a), which one of the five offered pattern plots in Figure 12.44(d)–(h) represents the cut of the overall array pattern in the plane perpendicular to the dipoles (xy-plane)?

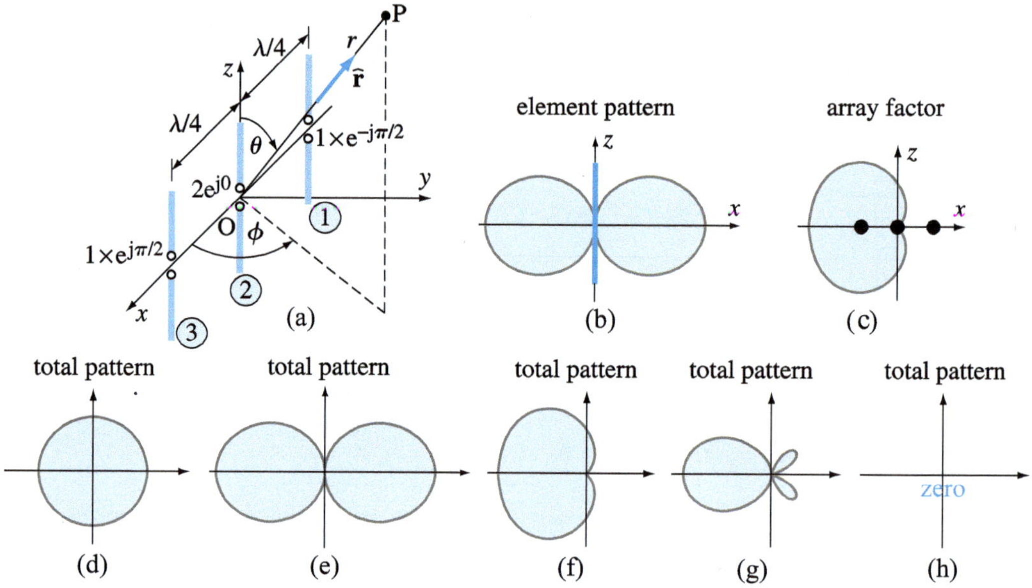

Figure 12.44 Analysis of a nonuniform array of three parallel half-wave wire dipole antennas: (a) quarter-wave spacing and excitation in time-phase quadrature of adjacent dipoles, (b) element pattern in the xz-plane, (c) array factor in the xz-plane, and (d)–(h) five offered plots for the total radiation pattern of the antenna array; for Conceptual Question 12.140.

(A) The one in Figure 12.44(d).

(B) The one in Figure 12.44(e).

(C) The one in Figure 12.44(f).

(D) The one in Figure 12.44(g).

(E) The one in Figure 12.44(h).

CONCEPTUAL QUESTION 12.142 **Array of three parallel dipoles – pattern multiplication, third cut.** Considering the five offered pattern plots in Figure 12.44(d)–(h), the total array pattern cut in the plane perpendicular to the array axis (yz-plane) is as in

(A) Figure 12.44(d).

(B) Figure 12.44(e).

(C) Figure 12.44(f).

(D) Figure 12.44(g).

(E) Figure 12.44(h).

APPENDIX A: QUANTITIES, SYMBOLS, UNITS, CONSTANTS

Symbol	Quantity or Parameter	Alternative Notation*	SI Unit (Value)
A	Scalar function (variable)		Appropriate unit
A	Scalar constant (single-valued), e.g., j, e		Appropriate unit
$\mathbf{A}$	Vector	$\vec{A}$	Appropriate unit
$\hat{\mathbf{A}}$	Unit vector of $\mathbf{A}$	$\mathbf{a}_A$, $\mathbf{i}_A$, $\mathbf{u}_A$	Appropriate unit
$\underline{A}$	Complex (phasor) quantity	$\tilde{A}$, A_s	Appropriate unit
$\underline{\mathbf{A}}$	Complex vector	$\tilde{\mathbf{A}}$, $\mathbf{A}_s$	Appropriate unit
$\mathbf{A}$	Magnetic vector potential		$\mathrm{T \cdot m}$ (tesla-m)
A_{dB}	Decibel attenuation (attenuation in dB)		dB (decibel)
$\mathbf{B}$	Magnetic flux density vector		T (tesla)
C	Capacitance		F (farad)
C'	Capacitance per unit length (p.u.l.)	C	F/m (F/meter)
c	Velocity of electromagnetic (EM) waves		m/s (s – second)
c_0	Velocity of EM waves in free space	c	299,792,458 m/s
$\mathbf{D}$	Electric flux density (displacement) vect.		$\mathrm{C/m^2}$ (see Q)
D	Antenna directivity		Dimensionless
e	Charge of electron, magnitude		1.602×10^{-19} C
$\mathbf{E}$	Electric field intensity vector		V/m (volt/m)
E_{cr}	Dielectric strength of a material		V/m
E_{cr0}	Dielectric strength of air		3 MV/m
$\mathbf{E}_{\mathrm{ind}}$	Induced electric field intensity vector		V/m
$\mathcal{E}$, e	Electromotive force (emf) of a generator		V (volt)
e_{ind}	Induced emf	V_{emf}	V
$\mathbf{F}_{\mathrm{e}}$	Electric (Coulomb) force		N (newton)
$\mathbf{F}_{\mathrm{m}}$	Magnetic force		N
f	Frequency	ν	Hz (hertz)
f_{p}	Plasma frequency		Hz
f_{res}	Resonant frequency of an EM resonator		Hz
f_{c}	Cutoff frequency of a waveguide mode		Hz
$\underline{\mathbf{F}}$	Antenna characteristic radiation function		Dimensionless
f	Antenna normalized field pattern		Dimensionless
$\underline{F}_{\mathrm{a}}$	Antenna array factor (AF)		Dimensionless
G	Conductance		S
G'	Conductance per unit length	G	S/m
G	Antenna gain		Dimensionless
$\mathbf{H}$	Magnetic field intensity vector		A/m (amp/m)
I, i	Current intensity (or current)		A (ampere, amp)
j or i	Imaginary unit ($\mathrm{j} = \sqrt{-1}$)		Dimensionless
$\mathbf{J}$	Current density vect. (conduction curr.)		$\mathrm{A/m^2}$
$\mathbf{J}_{\mathrm{d}}$	Displacement current density vector		$\mathrm{A/m^2}$
$\mathbf{J}_{\mathrm{s}}$	Surface current density vector	$\mathbf{K}$	A/m

**Alternative notations used by one or more other electromagnetics texts.*

Symbol	Quantity or Parameter	Alternative Notation	SI Unit (Value)
l	Length		m (meter)
L	Self-inductance		H (henry)
L_{21}	Mutual inductance	M	H
L'	Inductance per unit length	L	H/m
$\mathbf{m}$	Magnetic dipole moment		$\mathrm{A \cdot m^2}$
m	Mass		kg (kilogram)
n	Index of refraction		Dimensionless
$\mathbf{p}$	Electric dipole moment		$\mathrm{C \cdot m}$
P	Power (instantaneous)		W (watt)
P_{J}	Power of Joule's losses or ohmic losses		W
P_{ave}	Time-average power		W
$\boldsymbol{\mathcal{P}}$	Poynting vector	$\mathbf{S}$	$\mathrm{W/m^2}$
$\underline{\boldsymbol{\mathcal{P}}}$	Complex Poynting vector	$\underline{\mathbf{S}}$, $\tilde{\mathbf{S}}$	$\mathrm{W/m^2}$
$\boldsymbol{\mathcal{P}}_{\mathrm{ave}}$	Time-average Poynting vector	$\mathbf{S}_{\mathrm{ave}}$	$\mathrm{W/m^2}$
Q, q	Charge		C (coulomb)
Q'	Line charge density (charge p.u.l.)	ρ_{l}	C/m
Q	Quality factor of a resonator		Dimensionless
R	Source-to-field distance	R'	m
R	Resistance		Ω (ohm)
R_{rad}	Radiation resistance of an antenna		Ω
R_{ohmic}	Ohmic resistance of an antenna		Ω
R_{s}	Surface resistance of a good conductor		Ω/square (or Ω)
R'	Resistance per unit length	R	Ω/m
r	Normalized transmission-line resistance		Dimensionless
r	Radial distance cylindrical	ρ	m
r	Radial distance spherical	R	m
S	Surface area	s, A, a	$\mathrm{m^2}$
s	Standing wave ratio (SWR)		Dimensionless
t	Time		s (second)
T	Period of time-harmonic oscillation		s
T	One-way time delay period of a tr. line		s
T	Temperature		K (kelvin) or °C
v	Volume	V, τ	$\mathrm{m^3}$
V	Electric scalar potential (at a point)	Φ	V
V, v	Voltage (between two points)		V
V_{cr}	Breakdown voltage		V
$\underline{V}$	Complex rms (root-mean-square) voltage		V
$\underline{V}_{\mathrm{oc}}$	Open-circuit voltage of a receiv. antenna		V
$\mathbf{v}$	Velocity (vector)	$\mathbf{u}$	m/s
v_{p}	Phase velocity		m/s
v_{g}	Group velocity		m/s
W	Work or energy	A (work)	J (joule)
W_{e}	Electric energy		J
W_{m}	Magnetic energy		J
W_{em}	Electromagnetic energy		J
w_{e}	Electric energy density		$\mathrm{J/m^3}$
w_{m}	Magnetic energy density		$\mathrm{J/m^3}$

Symbol	Quantity or Parameter	Alternative Notation	SI Unit (Value)
X	Reactance		Ω
x	Normalized transmission-line reactance		Dimensionless
$\underline{Y}$	Complex admittance		S
Y_0	Characteristic admittance of a tr. line	Y_c	S
$\underline{Z}$	Complex impedance		Ω
Z_0	Characteristic impedance of a tr. line	Z_c	Ω
$Z_{\rm TEM}$	TEM wave impedance		Ω
$Z_{\rm TE}$	TE wave impedance		Ω
$Z_{\rm TM}$	TM wave impedance		Ω
$\underline{z}_{\rm n}$	Normalized transmission-line impedance		Dimensionless
α	Attenuation coefficient		Np/m (neper/m)
$\alpha_{\rm c}$	Attenuat. coeff. due to conductor losses		Np/m
$\alpha_{\rm d}$	Attenuation coeff. due to dielectric losses		Np/m
β	Phase coefficient or wavenumber	k	rad/m
$\underline{\gamma}$	Complex propagation coefficient		m^{-1}
$\underline{\Gamma}$	Reflection coefficient	R, ρ	Dimensionless
δ	Skin depth		m
$\tan\delta_{\rm d}$	Loss tangent		Dimensionless
ε	Permittivity of a dielectric material		F/m
ε_0	Permittivity of a vacuum (free space)		8.8542 pF/m
$\varepsilon_{\rm r}$	Relative permittivity		Dimensionless
$\varepsilon_{\rm reff}$	Effective relative permittivity of a tr. line		Dimensionless
η	Intrinsic impedance of a medium		Ω
η_0	Intrinsic impedance of free space		$\approx 120\pi\ \Omega \approx 377\ \Omega$
$\underline{\eta}$	Complex intrinsic impedance		Ω
$\underline{\eta}_{\rm w}$	Wave impedance		Ω
$\eta_{\rm rad}$	Radiation efficiency of an antenna		Dimensionless
λ	Wavelength		m
λ_0	Free-space wavelength		m
λ_z	Wavelength along a tr. line or waveguide		m
μ	Permeability of a magnetic material		H/m
μ_0	Permeability of a vacuum (free space)		$4\pi \cdot 10^{-7}$ H/m
$\mu_{\rm r}$	Relative permeability		Dimensionless
ρ	Volume charge density	$\rho_{\rm v}$	$\mathrm{C/m^3}$
$\rho_{\rm s}$	Surface charge density	σ	$\mathrm{C/m^2}$
ρ	Resistivity of a medium		$\Omega \cdot \mathrm{m}$
σ	Conductivity of a medium		S/m
τ	Relaxation time or time constant		s
$\underline{\tau}$	Transmission coefficient	$\underline{T}$	Dimensionless
Φ	Magnetic flux	Ψ, λ	Wb (weber)
ψ	Phase angle of a reflection coefficient		rad (radian)
ω	Angular or radian frequency		rad/s

Powers of Ten as Multipliers of Fundamental Units

Multiple	Prefix	Symbol	Multiple	Prefix	Symbol
10^{18}	Exa	E	10^{-2}	centi	c
10^{15}	Peta	P	10^{-3}	milli	m
10^{12}	Tera	T	10^{-6}	micro	μ
10^{9}	Giga	G	10^{-9}	nano	n
10^{6}	Mega	M	10^{-12}	pico	p
10^{3}	kilo	k	10^{-15}	femto	f
10^{2}	hecto	h	10^{-18}	atto	a

APPENDIX B: MATHEMATICAL FACTS AND IDENTITIES

B.1 Trigonometric Identities

$$\sin(\alpha \pm \beta) = \sin\alpha\cos\beta \pm \cos\alpha\sin\beta\,, \quad \cos(\alpha \pm \beta) = \cos\alpha\cos\beta \mp \sin\alpha\sin\beta$$

$$2\sin\alpha\sin\beta = \cos(\alpha-\beta) - \cos(\alpha+\beta)\,, \quad 2\sin\alpha\cos\beta = \sin(\alpha+\beta) + \sin(\alpha-\beta)\,,$$

$$2\cos\alpha\cos\beta = \cos(\alpha+\beta) + \cos(\alpha-\beta)\,; \quad \sin\alpha \pm \sin\beta = 2\sin\frac{\alpha\pm\beta}{2}\cos\frac{\alpha\mp\beta}{2}\,,$$

$$\cos\alpha + \cos\beta = 2\cos\frac{\alpha+\beta}{2}\cos\frac{\alpha-\beta}{2}\,, \quad \cos\alpha - \cos\beta = -2\sin\frac{\alpha+\beta}{2}\sin\frac{\alpha-\beta}{2}$$

$$\sin^2\alpha = \frac{1-\cos 2\alpha}{2}\,, \quad \cos^2\alpha = \frac{1+\cos 2\alpha}{2}\,, \quad \sin^2\alpha + \cos^2\alpha = 1\,,$$

$$\sin 2\alpha = 2\sin\alpha\cos\alpha\,, \quad \cos 2\alpha = \cos^2\alpha - \sin^2\alpha\,, \quad \sin(-\alpha) = -\sin\alpha\,,$$

$$\cos(-\alpha) = \cos\alpha\,, \quad \sin(\alpha \pm 90^\circ) = \pm\cos\alpha\,; \quad \cos(\alpha \pm 90^\circ) = \mp\sin\alpha\,;$$

$$\tan\alpha = \frac{\sin\alpha}{\cos\alpha}\,, \quad \cot\alpha = \frac{1}{\tan\alpha}$$

$$\sin 0 = 0\,, \quad \sin 30^\circ = \frac{1}{2}\,, \quad \sin 45^\circ = \frac{\sqrt{2}}{2}\,, \quad \sin 60^\circ = \frac{\sqrt{3}}{2}\,, \quad \sin 90^\circ = 1\,,$$

$$\sin 180^\circ = 0\,, \quad \cos 0 = 1\,, \quad \cos 30^\circ = \frac{\sqrt{3}}{2}\,, \quad \cos 45^\circ = \frac{\sqrt{2}}{2}\,, \quad \cos 60^\circ = \frac{1}{2}\,,$$

$$\cos 90^\circ = 0\,, \quad \cos 180^\circ = -1$$

$$c^2 = a^2 + b^2 - 2ab\cos\gamma \quad \text{(cosine formula, arbitrary triangle; } \gamma \text{ is opposite to } c\text{)}\,,$$

$$c^2 = a^2 + b^2 \quad \text{(Pythagorean theorem, right triangle; } c \text{ is hypotenuse)}$$

B.2 Exponential, Logarithmic, and Hyperbolic Identities

$$\mathrm{e}^x\,\mathrm{e}^y = \mathrm{e}^{x+y}\,, \quad (\mathrm{e}^x)^a = \mathrm{e}^{ax}\,, \quad \mathrm{e}^{\mathrm{j}x} = \cos x + \mathrm{j}\sin x \quad \text{(Euler's identity)}\,,$$

$$\mathrm{e}^{\mathrm{j}x} + \mathrm{e}^{-\mathrm{j}x} = 2\cos x\,, \quad \mathrm{e}^{\mathrm{j}x} - \mathrm{e}^{-\mathrm{j}x} = 2\mathrm{j}\sin x\,, \quad \mathrm{e} = 2.71828\,,$$

$$\mathrm{j} = \sqrt{-1} \quad \text{(imaginary unit)} \quad \log x = \log_{10} x \quad \text{(common logarithm)}\,,$$

$$\ln x = \log_{\mathrm{e}} x \quad \text{(natural logarithm)}\,, \quad \ln \mathrm{e}^x = x \quad \log(xy) = \log x + \log y\,,$$

$$\log\frac{x}{y} = \log x - \log y\,, \quad \log x^a = a\log x \quad \text{(logarithm of any base)}$$

$$\sinh x = \frac{e^x - e^{-x}}{2}\,, \quad \cosh x = \frac{e^x + e^{-x}}{2} \quad \text{(hyperbolic sine and cosine)}$$

$$\sinh jx = j\sin x\,, \quad \cosh jx = \cos x$$

$$\tanh x = \frac{\sinh x}{\cosh x}\,, \quad \coth x = \frac{1}{\tanh x} \quad \text{(hyperbolic tangent and cotangent)}$$

B.3 Solution of Quadratic Equation

$$ax^2 + bx + c = 0 \quad \longrightarrow \quad x = \frac{-b \pm \sqrt{b^2 - 4ac}}{2a}$$

B.4 Approximations for Small Quantities

$$\text{For } |x| \ll 1\,, \quad (1+x)^a \approx 1 + ax\,, \quad \sin x \approx x\,, \quad \cos x \approx 1 - \frac{x^2}{2}\,, \quad e^x \approx 1 + x$$

B.5 Derivatives

$$\frac{d}{dx}x^c = cx^{c-1}\,, \quad \frac{dc}{dx} = 0 \quad (c = \text{const})\,, \quad \frac{d}{dx}e^x = e^x\,, \quad \frac{d}{dx}\ln x = \frac{1}{x}\,,$$

$$\frac{d}{dx}\sin x = \cos x\,, \quad \frac{d}{dx}\cos x = -\sin x\,, \quad \frac{d}{dx}\tan x = \frac{1}{\cos^2 x}$$

$$\text{For } f = f(x) \text{ and } g = g(x)\,, \quad \frac{d}{dx}(cf) = c\frac{df}{dx}\,, \quad \frac{d}{dx}(f+g) = \frac{df}{dx} + \frac{dg}{dx}\,,$$

$$\frac{d}{dx}(fg) = \frac{df}{dx}g + f\frac{dg}{dx}\,, \quad \frac{d}{dx}\left(\frac{f}{g}\right) = \frac{\frac{df}{dx}g - f\frac{dg}{dx}}{g^2}\,,$$

$$\frac{d}{dx}f[g(x)] = \frac{df}{dg}\frac{dg}{dx} \quad \text{(chain rule for taking derivatives)}$$

B.6 Integrals

$$\int x^c\,dx = \frac{x^{c+1}}{c+1} + C \quad (c \neq -1)\,, \quad \int\frac{dx}{x} = \ln x + C\,, \quad \int e^x\,dx = e^x + C\,,$$

$$\int \sin x\,dx = -\cos x + C\,, \quad \int \cos x\,dx = \sin x + C\,; \quad \int cf\,dx = c\int f\,dx\,,$$

$$\int (f+g)\,dx = \int f\,dx + \int g\,dx\,, \quad \int f\,dg = fg - \int g\,df \quad \text{(integration by parts)}$$

B.7 Vector Algebraic Identities

For vectors $\mathbf{a}$ and $\mathbf{b}$, and angle α between them, $\mathbf{a}\cdot\mathbf{b} = |\mathbf{a}||\mathbf{b}|\cos\alpha$ (dot product of vectors), $\mathbf{a}\times\mathbf{b} = |\mathbf{a}||\mathbf{b}|\sin\alpha\,\hat{\mathbf{n}}$ (cross product of vectors; $\hat{\mathbf{n}}$ is the unit vector normal to the plane of $\mathbf{a}$ and $\mathbf{b}$, directed according to the right-hand rule when $\mathbf{a}$ is rotated by the shortest route toward $\mathbf{b}$), $\mathbf{a}\cdot\mathbf{a} = |\mathbf{a}|^2 = a^2$, $\hat{\mathbf{a}} = \dfrac{\mathbf{a}}{a}$ (unit vector of $\mathbf{a}$; $|\hat{\mathbf{a}}| = 1$), $\mathbf{a}\cdot\mathbf{b} = \mathbf{b}\cdot\mathbf{a}$, $\mathbf{a}\times\mathbf{b} = -\mathbf{b}\times\mathbf{a}$,

$$(\mathbf{a}\times\mathbf{b})\cdot\mathbf{c} = (\mathbf{b}\times\mathbf{c})\cdot\mathbf{a} = (\mathbf{c}\times\mathbf{a})\cdot\mathbf{b} \quad \text{(scalar triple product)}$$

$$\mathbf{a}\times(\mathbf{b}\times\mathbf{c}) = \mathbf{b}(\mathbf{a}\cdot\mathbf{c}) - \mathbf{c}(\mathbf{a}\cdot\mathbf{b}) \quad \text{(vector triple product)}$$

For vectors in the Cartesian coordinate system, $\mathbf{a} = a_x\,\hat{\mathbf{x}} + a_y\,\hat{\mathbf{y}} + a_z\,\hat{\mathbf{z}}$ ($\hat{\mathbf{x}}$, $\hat{\mathbf{y}}$, and $\hat{\mathbf{z}}$ are coordinate unit vectors), $a = |\mathbf{a}| = \sqrt{a_x^2 + a_y^2 + a_z^2}$,

$$\mathbf{a}+\mathbf{b} = (a_x + b_x)\,\hat{\mathbf{x}} + (a_y + b_y)\,\hat{\mathbf{y}} + (a_z + b_z)\,\hat{\mathbf{z}}\,, \quad \mathbf{a}\cdot\mathbf{b} = a_x b_x + a_y b_y + a_z b_z\,,$$

$$\mathbf{a}\times\mathbf{b} = (a_y b_z - a_z b_y)\,\hat{\mathbf{x}} + (a_z b_x - a_x b_z)\,\hat{\mathbf{y}} + (a_x b_y - a_y b_x)\,\hat{\mathbf{z}}$$

B.8 Vector Calculus Identities

For a scalar function f (and g) and a vector function $\mathbf{a}$ (and $\mathbf{b}$), $\nabla f \equiv \operatorname{grad} f$ (gradient of f), $\nabla\cdot\mathbf{a} \equiv \operatorname{div}\mathbf{a}$ (divergence of $\mathbf{a}$), $\nabla\times\mathbf{a} \equiv \operatorname{curl}\mathbf{a}$ (curl of $\mathbf{a}$), $\nabla\cdot(\nabla f) \equiv \nabla^2 f$ (Laplacian of f), $\nabla\times(\nabla f) = 0$, $\nabla\cdot(\nabla\times\mathbf{a}) = 0$, $\nabla(fg) = (\nabla f)\,g + f\nabla g$, $\nabla f(g) = \dfrac{\mathrm{d}f}{\mathrm{d}g}\nabla g$ (chain rule), $\nabla\cdot(f\mathbf{a}) = (\nabla f)\cdot\mathbf{a} + f\nabla\cdot\mathbf{a}$, $\nabla\times(f\mathbf{a}) = (\nabla f)\times\mathbf{a} + f\nabla\times\mathbf{a}$, $\nabla\cdot(\mathbf{a}\times\mathbf{b}) = \mathbf{b}\cdot(\nabla\times\mathbf{a}) - \mathbf{a}\cdot(\nabla\times\mathbf{b})$, $\nabla\times(\nabla\times\mathbf{a}) = \nabla(\nabla\cdot\mathbf{a}) - \nabla^2\mathbf{a}$ ($\nabla^2\mathbf{a}$ – Laplacian of $\mathbf{a}$)

$$\int_v \nabla\cdot\mathbf{a}\,\mathrm{d}v = \oint_S \mathbf{a}\cdot\mathrm{d}\mathbf{S} \quad \text{(divergence theorem; } S \text{ is the boundary surface of } v\text{)}\,,$$

$$\int_S (\nabla\times\mathbf{a})\cdot\mathrm{d}\mathbf{S} = \oint_C \mathbf{a}\cdot\mathrm{d}\mathbf{l} \quad \text{(Stokes' theorem; } C \text{ is the boundary contour of } S\text{)}$$

B.9 Gradient, Divergence, Curl, Laplacian in Orthogonal Coordinate Systems

Cartesian coordinate system

$$[f(x,y,z)\,,\quad \mathbf{a}=a_x(x,y,z)\,\hat{\mathbf{x}}+a_y(x,y,z)\,\hat{\mathbf{y}}+a_z(x,y,z)\,\hat{\mathbf{z}}]\,,$$

$$\nabla=\frac{\partial}{\partial x}\,\hat{\mathbf{x}}+\frac{\partial}{\partial y}\,\hat{\mathbf{y}}+\frac{\partial}{\partial z}\,\hat{\mathbf{z}}\ \ \text{(del operator)}\,,\quad \nabla f=\frac{\partial f}{\partial x}\,\hat{\mathbf{x}}+\frac{\partial f}{\partial y}\,\hat{\mathbf{y}}+\frac{\partial f}{\partial z}\,\hat{\mathbf{z}}\,,$$

$$\nabla\cdot\mathbf{a}=\frac{\partial a_x}{\partial x}+\frac{\partial a_y}{\partial y}+\frac{\partial a_z}{\partial z}\,,\quad \nabla\times\mathbf{a}=\left(\frac{\partial a_z}{\partial y}-\frac{\partial a_y}{\partial z}\right)\hat{\mathbf{x}}+\left(\frac{\partial a_x}{\partial z}-\frac{\partial a_z}{\partial x}\right)\hat{\mathbf{y}}$$

$$+\left(\frac{\partial a_y}{\partial x}-\frac{\partial a_x}{\partial y}\right)\hat{\mathbf{z}}\,,\quad \nabla^2 f=\frac{\partial^2 f}{\partial x^2}+\frac{\partial^2 f}{\partial y^2}+\frac{\partial^2 f}{\partial z^2}\,,\quad \nabla^2\mathbf{a}=\nabla^2 a_x\,\hat{\mathbf{x}}+\nabla^2 a_y\,\hat{\mathbf{y}}+\nabla^2 a_z\,\hat{\mathbf{z}}$$

Cylindrical coordinate system

$$[f(r,\phi,z)\,,\quad \mathbf{a}=a_r(r,\phi,z)\,\hat{\mathbf{r}}+a_\phi(r,\phi,z)\,\hat{\boldsymbol{\phi}}+a_z(r,\phi,z)\,\hat{\mathbf{z}}]\,,$$

$$\nabla f=\frac{\partial f}{\partial r}\,\hat{\mathbf{r}}+\frac{1}{r}\frac{\partial f}{\partial \phi}\,\hat{\boldsymbol{\phi}}+\frac{\partial f}{\partial z}\,\hat{\mathbf{z}}\,,\quad \nabla\cdot\mathbf{a}=\frac{1}{r}\frac{\partial}{\partial r}(ra_r)+\frac{1}{r}\frac{\partial a_\phi}{\partial \phi}+\frac{\partial a_z}{\partial z}\,,$$

$$\nabla\times\mathbf{a}=\left(\frac{1}{r}\frac{\partial a_z}{\partial \phi}-\frac{\partial a_\phi}{\partial z}\right)\hat{\mathbf{r}}+\left(\frac{\partial a_r}{\partial z}-\frac{\partial a_z}{\partial r}\right)\hat{\boldsymbol{\phi}}+\frac{1}{r}\left[\frac{\partial}{\partial r}(ra_\phi)-\frac{\partial a_r}{\partial \phi}\right]\hat{\mathbf{z}}\,,$$

$$\nabla^2 f=\frac{1}{r}\frac{\partial}{\partial r}\left(r\frac{\partial f}{\partial r}\right)+\frac{1}{r^2}\frac{\partial^2 f}{\partial \phi^2}+\frac{\partial^2 f}{\partial z^2}\,,\quad \nabla^2\mathbf{a}=\nabla(\nabla\cdot\mathbf{a})-\nabla\times(\nabla\times\mathbf{a})$$

Spherical coordinate system

$$[f(r,\theta,\phi)\,,\quad \mathbf{a}=a_r(r,\theta,\phi)\,\hat{\mathbf{r}}+a_\theta(r,\theta,\phi)\,\hat{\boldsymbol{\theta}}+a_\phi(r,\theta,\phi)\,\hat{\boldsymbol{\phi}}]\,,\quad \nabla f=\frac{\partial f}{\partial r}\,\hat{\mathbf{r}}+\frac{1}{r}\frac{\partial f}{\partial \theta}\,\hat{\boldsymbol{\theta}}$$

$$+\frac{1}{r\sin\theta}\frac{\partial f}{\partial \phi}\,\hat{\boldsymbol{\phi}}\,,\quad \nabla\cdot\mathbf{a}=\frac{1}{r^2}\frac{\partial}{\partial r}\left(r^2 a_r\right)+\frac{1}{r\sin\theta}\frac{\partial}{\partial \theta}(\sin\theta\, a_\theta)+\frac{1}{r\sin\theta}\frac{\partial a_\phi}{\partial \phi}\,,$$

$$\nabla\times\mathbf{a}=\frac{1}{r\sin\theta}\left[\frac{\partial}{\partial \theta}(\sin\theta\, a_\phi)-\frac{\partial a_\theta}{\partial \phi}\right]\hat{\mathbf{r}}+\frac{1}{r}\left[\frac{1}{\sin\theta}\frac{\partial a_r}{\partial \phi}-\frac{\partial}{\partial r}(ra_\phi)\right]\hat{\boldsymbol{\theta}}$$

$$+\frac{1}{r}\left[\frac{\partial}{\partial r}(ra_\theta)-\frac{\partial a_r}{\partial \theta}\right]\hat{\boldsymbol{\phi}}\,,\quad \nabla^2 f=\frac{1}{r^2}\frac{\partial}{\partial r}\left(r^2\frac{\partial f}{\partial r}\right)+\frac{1}{r^2\sin\theta}\frac{\partial}{\partial \theta}\left(\sin\theta\frac{\partial f}{\partial \theta}\right)$$

$$+\frac{1}{r^2\sin^2\theta}\frac{\partial^2 f}{\partial \phi^2}\,,\quad \text{for } \nabla^2\mathbf{a}\text{, see above}$$

B.10 Vector Algebra and Calculus Index

REFERENCES

[1] Notaroš, B. M., *Electromagnetics*, Prentice Hall, Upper Saddle River, NJ, 2010.

[2] Notaroš, B. M., *MATLAB®-Based Electromagnetics*, Prentice Hall, Upper Saddle River, NJ, 2013.

[3] Ulaby, F. T., E. Michielssen, and U. Ravaioli, *Fundamentals of Applied Electromagnetics*, Prentice Hall, Upper Saddle River, NJ, 2010, 6th edition.

[4] Sadiku, M. N. O., *Elements of Electromagnetics*, Oxford University Press, New York, 2009, 5th edition.

[5] Hayt, W. H. Jr., and J. A. Buck, *Engineering Electromagnetics*, McGraw-Hill, New York, 2011, 8th edition.

[6] Griffiths, D. J., *Introduction to Electrodynamics*, Addison Wesley, Upper Saddle River, NJ, 2012, 4th edition.

[7] Cheng, D. K., *Field and Wave Electromagnetics*, Addison-Wesley, Reading, MA, 1989, 2nd edition.

[8] Popović, Z., and B. D. Popović, *Introductory Electromagnetics*, Prentice Hall, Upper Saddle River, NJ, 2000.

[9] Rao, N. N., *Elements of Engineering Electromagnetics*, Prentice Hall, Upper Saddle River, NJ, 2004, 6th edition.

[10] Inan U. S., and A. S. Inan, *Engineering Electromagnetics*, Addison Wesley Longman, Menlo Park, CA, 1999.

[11] Wentworth, S. M., *Fundamentals of Electromagnetics with Engineering Applications*, John Wiley & Sons, New York, 2005.

[12] Paul, C. R., *Electromagnetics for Engineers with Applications*, John Wiley & Sons, New York, 2004.

[13] Demarest, K. R., *Engineering Electromagnetics*, Prentice Hall, Upper Saddle River, NJ, 1998.

[14] Kraus, J. D., and D. A. Fleisch, *Electromagnetics with Applications*, McGraw-Hill, New York, 1999, 5th edition.

[15] Iskander, M. F., *Electromagnetic Fields and Waves*, Waveland Press, Prospect Hills, IL, 2000.

[16] Lonngren, K. E., S. V. Savov, and R. J. Jost, *Fundamentals of Electromagnetics with MATLAB®*, SciTech Publishing, Raleigh, NC, 2007, 2nd edition.

[17] Johnk, C. T. A., *Engineering Electromagnetic Fields and Waves*, John Wiley & Sons, New York, 1988, 2nd edition.

[18] Ida, N., *Engineering Electromagnetics*, Springer, New York, 2004, 2nd edition.

[19] Stutzman, W. L., and G. A. Thiele, *Antenna Theory and Design*, John Wiley & Sons, New York, 1998, 2nd edition.

[20] Balanis, C. A., *Antenna Theory: Analysis and Design*, John Wiley & Sons, New York, 1997, 2nd edition.

[21] Kraus, J. D., and R. J. Marhefka, *Antennas for All Applications*, McGraw-Hill, New York, 2002, 3rd edition.

[22] Pozar, D. M., *Microwave Engineering*, John Wiley & Sons, New York, 2005, 3rd edition.

[23] Inan U. S., and A. S. Inan, *Electromagnetic Waves*, Prentice Hall, Upper Saddle River, NJ, 2000.

[24] Collin, R. E., *Field Theory of Guided Waves*, Wiley-IEEE Press, New York, 1990, 2nd edition.

[25] Ramo, S., J. R. Whinnery, and T. Van Duzer, *Fields and Waves in Communication Electronics*, John Wiley & Sons, New York, 1994, 3rd edition.

[26] Haus, H. A., and J. R. Melcher, *Electromagnetic Fields and Energy*, Prentice Hall, Upper Saddle River, NJ, 1989.

[27] Claycomb, J. R., *Applied Electromagnetics Using QuickField and MATLAB®*, Jones and Bartlett Publishers, Sudbury, MA, 2009.

[28] Balanis, C. A., *Advanced Engineering Electromagnetics*, John Wiley & Sons, New York, 2012, 2nd edition.

[29] Harrington, R. F., *Time-Harmonic Electromagnetic Fields*, Wiley-IEEE Press, New York, 2001 (classic reissue).

[30] Stratton, J. A., *Electromagnetic Theory*, Wiley-IEEE Press, New York, 2007 (classic reissue).

[31] Jin, J., *Theory and Computation of Electromagnetic Fields*, Wiley-IEEE Press, New York, 2010.

[32] Sadiku, M. N. O., *Numerical Techniques in Electromagnetics with MATLAB®*, CRC Press, Boca Raton, FL, 2009, 3rd edition.

[33] Warnick, K. F., *Numerical Methods for Engineering: An Introduction Using MATLAB and Computational Electromagnetics Examples*, SciTech Publishing, Raleigh, NC, 2011.

[34] Peterson, A. F., S. L. Ray, and R. Mittra, *Computational Methods for Electromagnetics*, Wiley-IEEE Press, New York, 1997.

INDEX

Note: Page numbers in **bold** type refer to main discussions. Un-bolded page numbers refer to conceptual questions or problems.

A

D

E

F

N

O

P

S

T